Theory, Processes and Applications of Biotechnology

Theory, Processes and Applications of Biotechnology

Editor: Suzy Hill

Callisto Reference
www.callistoreference.com

Callisto Reference,
118-35 Queens Blvd., Suite 400,
Forest Hills, NY 11375, USA

Visit us on the World Wide Web at:
www.callistoreference.com

ISBN: 978-1-63239-922-9 (Hardback)

Printed in China.

Cataloging-in-Publication Data

Theory, processes and applications of biotechnology / edited by Suzy Hill.
p. cm.
Includes bibliographical references and index.
ISBN 978-1-63239-922-9
1. Biotechnology. 2. Genetic engineering. I. Hill, Suzy.
TP248.2 .T44 2018
660.6--dc23

Table of Contents

Preface

Biotechnology plays a crucial role in diverse areas such as agriculture, medicine and food production. This book elucidates the theories, processes and applications of biotechnology. It uses living systems to develop products or applications such as recombinant gene technique, development of pharmaceutical therapies, etc. At times, it overlaps with fields such as bioengineering, biomanufacturing and molecular engineering. The book presents researches and studies performed by experts across the globe. Coherent flow of topics, student-friendly language and extensive use of examples make it an invaluable source of knowledge. Through this book, we attempt to further enlighten the readers about the new concepts of this field.

The information shared in this book is based on empirical researches made by veterans in this field of study. The elaborative information provided in this book will help the readers further their scope of knowledge leading to advancements in this field.

Finally, I would like to thank my fellow researchers who gave constructive feedback and my family members who supported me at every step of my research.

Editor

1

Innate Defense Regulator Peptide 1018 in Wound Healing and Wound Infection

Lars Steinstraesser[1*⁋], Tobias Hirsch[1⁋], Matthias Schulte[1], Maximilian Kueckelhaus[1], Frank Jacobsen[1], Evgenija A. Mersch[1], Ingo Stricker[2], Nicole Afacan[3], Havard Jenssen[4], Robert E. W. Hancock[3], Jason Kindrachuk[3]

1 Department of Plastic Surgery, BG University Hospital Bergmannsheil, Ruhr University Bochum, Bochum, Germany, **2** Institut for Pathology, BG University Hospital Bergmannsheil, Ruhr University Bochum, Bochum, Germany, **3** Centre for Microbial Diseases and Immunity Research, University of British Columbia, Vancouver, British Columbia, Canada, **4** Department of Science, Systems, and Models, Roskilde University, Roskilde, Denmark

Abstract

Innate defense regulators (IDRs) are synthetic immunomodulatory versions of natural host defense peptides (HDP). IDRs mediate protection against bacterial challenge in the absence of direct antimicrobial activity, representing a novel approach to anti-infective and anti-inflammatory therapy. Previously, we reported that IDR-1018 selectively induced chemokine responses and suppressed pro-inflammatory responses. As there has been an increasing appreciation for the ability of HDPs to modulate complex immune processes, including wound healing, we characterized the wound healing activities of IDR-1018 *in vitro*. Further, we investigated the efficacy of IDR-1018 in diabetic and non-diabetic wound healing models. In all experiments, IDR-1018 was compared to the human HDP LL-37 and HDP-derived wound healing peptide HB-107. IDR-1018 was significantly less cytotoxic *in vitro* as compared to either LL-37 or HB-107. Furthermore, administration of IDR-1018 resulted in a dose-dependent increase in fibroblast cellular respiration. *In vivo*, IDR-1018 demonstrated significantly accelerated wound healing in *S. aureus* infected porcine and non-diabetic but not in diabetic murine wounds. However, no significant differences in bacterial colonization were observed. Our investigation demonstrates that in addition to previously reported immunomodulatory activities IDR-1018 promotes wound healing independent of direct antibacterial activity. Interestingly, these effects were not observed in diabetic wounds. It is anticipated that the wound healing activities of IDR-1018 can be attributed to modulation of host immune pathways that are suppressed in diabetic wounds and provide further evidence of the multiple immunomodulatory activities of IDR-1018.

Editor: Johanna M. Brandner, University Hospital Hamburg-Eppendorf, Germany

Funding: Financial Support to REWH from the Foundation for the National Institutes of Health and Canadian Institutes for Health Research through the Grand Challenges in Global Health Initiative is gratefully acknowledged. REWH is the recipient of a Canada Research Chair. JK was supported by a fellowship from the Canadian Cystic Fibrosis Foundation. The work was additionally supported by the Deutsche Forschungsgemeinschaft (Ste1099 to LS and HI1279/4–1 to TH). The funders had no role in study design, data collection and analysis, decision to publish, or preparation of the manuscript.

Competing Interests: Peptide IDR-1018 has been filed for patent protection by Hancock and assigned to his employer, the University of British Columbia. It has been licenced to Elanco for development as a therapy in animal health applications, and has been approved by Cystic Fibrosis Canada for pre-clinical development as an anti-inflammatory treatment for CF lung disease.

* E-mail: lars.steinstraesser@ruhr-uni-bochum.de

⁋ These authors contributed equally to this work.

Introduction

Cutaneous wound repair is a specialized, multifactorial process that involves a large number of factors that are involved in the regulation of this process [1]. Although initially characterized for their direct antimicrobial activities [2,3,4], there is an increasing appreciation for the immunomodulatory activities of cationic host defense peptides (HDPs; also termed antimicrobial peptides). HDPs are ubiquitous in nature and form central components of the innate immune system of eukaryotes. The immunomodulatory activities of such natural peptides, especially the human cathelicidin LL-37 and human β-defensin (hBD) -3, are extremely diverse and include, but are not limited to, the stimulation of epithelial cell migration, promotion of angiogenesis, and suppression of pro-inflammatory responses [3,5,6,7,8]. For example, the human cathelicidin peptide LL-37 attracts neutrophils, monocytes, mast cells, and T lymphocytes [9,10], and also induces the production of neutrophil and monocyte chemoattractants by many cell types [3,11,12,13,14]. Recently, HDPs have also been implicated as regulators of cutaneous wound repair by regulating inflammation, angiogenesis, and extracellular tissue deposition and remodeling [3,15,16,17,18].

The highly diverse sequences and versatile antimicrobial and immunomodulatory activities of natural HDPs provides a wide range of templates for the design and development of synthetic peptides with activities tailored for clinical applications [3]. A number of peptides and peptidomimetic compounds are currently undergoing clinical trials; however, most HDP clinical trials are aimed at primarily at topical applications of peptides and are based on their direct microbicidal properties (reviewed in [3,19,20]).

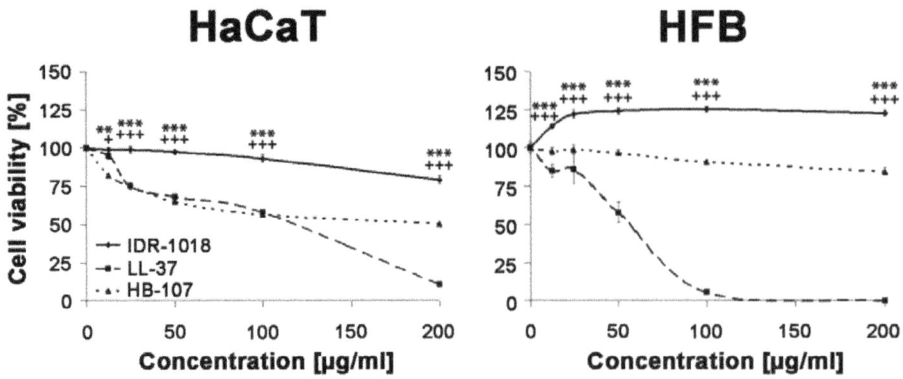

Figure 1. Lack of IDR-1018 toxicity. The effects of IDR-1018, LL-37 and HB-107 on viability/metabolic activity were determined using an MTT-Assay with HaCaT cells (left) and primary human fibroblasts (HFB; right) at the indicated concentrations. Cells were incubated with peptides for 24 hours. Standard deviation was calculated out of three independent experiments and assessed as SEM (Statistical significance IDR-1018 vs. LL-37: * = $p<0.05$; ** = $p<0.01$; *** = $p<0.005$; IDR-1018 vs. HB-107^{+} = $p<0.05$).

It has been shown that the influence of HDPs on wound repair is not dependent on antimicrobial function and provides a potential novel clinical application for HDPs [21]. For example, HB-107, a synthetic HDP originally derived from the antimicrobial cecropin B, was shown to promote wound healing [21]. A new group of synthetic variants of HDPs, termed innate defense regulators (IDRs), which provide broad-spectrum protection against systemic infections with multidrug-resistant bacteria, have recently been described [22,23,24,25]. IDR-1 and IDR 1002 confer protection against microbial challenges by enhancing innate immune defenses of the host while suppressing potentially harmful excessive inflammatory responses [22,23,24,25]. Such selective enhancement of innate immunity represents a novel approach to anti-infective therapy with many advantages over directly microbicidal compounds [26,27,28]. Although many HDPs exert their direct antimicrobial activity through disruption of microbial membranes, IDRs appear to exert their immunomodulatory activities through non membrane-lytic mechanisms. Thus issues of hemolytic activity and cytotoxicity toward mammalian cells are minimized, as supported by the minimal toxicity of peptide IDR-1 compared with natural cathelicidins such as LL-37 [25,28].

Despite the increasing interest in utilizing the immunomodulatory activities of HDPs and IDRs for clinical applications, investigations into optimizing and enhancing peptide immunomodulatory properties have thus far been limited. However, it is apparent that the ability to modulate chemotactic activity and chemokine induction are important aspects of the immunomodulatory activities shared by many natural HDPs. Recently, we screened a series of more than 100 distantly related peptide variants of bovine HDP Bac2A (unrelated by sequence to IDR-1, [22]) for improved immunomodulatory activity as assessed by chemokine production in human peripheral blood monocytic cells. The most active peptide observed was IDR-1018 (VRLIVAVRIWRR-NH_2) which was shown to form an α-helix in neutral membranes [29].

In the current work, we have expanded on the characterization of the immunomodulatory activities of IDR-1018 and demonstrated that it also promotes wound healing activities. IDR-1018 promoted wound healing in normal and infected wounds, and its effects were favourably compared to those of naturally occurring LL-37 and synthetic HB-107 as prospective positive controls.

Results

Effects of IDR-1018 on cell viability and migration

To investigate of effects of IDR-1018 on cell oxidative metabolism, an MTT-Assay was performed on HaCaT (immortalized human keratinocyte cell line) and primary human fibroblasts (HFB). In this assay, toxicity is assessed as residual metabolic activity through the production of MTT formazan crystals by mitochondrial dehydrogenases in living cells. The HDPs LL-37 and HB-107 served as peptide controls for naturally occurring and synthetic HDPs with wound healing activity [21,30]. IDR-1018 treated HaCaT cells maintained cell viability better than those treated with LL-37 or HB-107 at all concentrations analyzed in this study (Fig. 1). In all samples, cell viability was affected in a concentration dependent manner. LL-37 showed a significant decrease in cell viability at higher concentrations (10% compared to untreated cells at 200 µg/ml ($p<0.0001$), consistent with literature data demonstrating that this peptide is cytotoxic at higher concentrations. Moreover, HB-107 exhibited a reduction in HaCaT cell viability of up to 50% compared to untreated cells ($p<0.0001$). In contrast, IDR-1018 administration had very little effect on the metabolic activity of HaCaT cells as compared to LL-37 and HB-107 (78% residual metabolic activity at 200 µg/ml; $p<0.0001$ compared to LL-37 and or $p<0.001$ compared to HB-107 or $p=0.0013$ compared to untreated cells).

In HFB, incubation with LL-37 or HB-107 led to significant decreases in cell viability. At a concentration of 200 µg/ml, LL-37 led to a complete inhibition of cell metabolism ($p<0.0001$). Administration of HB-107 slightly decreased cell viability as compared to non-treated cells ($p<0.0001$). In contrast, IDR-1018 administration was completely non-cytotoxic and instead appeared to increase cell viability. Indeed, even low doses of IDR-1018 increased cell metabolic activity compared to untreated HFBs (25 µg/ml: 121% compared to untreated cells; $p=0.0185$), or HFBs treated with other peptides, ($p=0.0004$).

Dose-dependent effect of IDR-1018 on wound healing

To analyze the effects of IDR-1018 in the complex process of wound healing we performed a dose response study using a splint-model in non-diabetic mice (n = 5 for each group). The application of IDR-1018 led to significant increase in wound closure in a dose-dependent manner as compared to vehicle control (Fig. 2). Treatment with the highest dose of IDR-1018 (200 µg/ml) resulted in the greatest acceleration of wound closure. This dosage of IDR-1018 also resulted in the most significant difference in wound closure as compared to vehicle control. Wound closure in IDR-1018 treated animals led to 14 28% smaller wound areas as compared to controls from days 2–8 ($p<0.0001$ to $p=0.0087$; IDR-1018 treated wounds vs. untreated wounds). Moreover, although the rate of wound closure was slower in animals treated with lower doses, even small amounts of IDR-1018 (2 μg/ml) improved wound closure on days 8–10 (Fig 2).

IDR-1018 induced faster wound closure than previously described wound healing peptides

To further characterize the wound healing activities of IDR-1018, we compared it to two peptides with previously characterized wound healing activities, the human cathelicidin LL-37 and the synthetic peptide HB-107. At the 200 μg/ml (3 μg/dose) level, treatment of wounds with IDR-1018 led to significantly higher rates of re-epithelialization in non-diabetic mice as compared to LL-37 and HB-107. By day 2, a significantly faster wound closure was measured in IDR-1018 treated wounds compared to untreated wounds (14%, $p<0.0001$). Further, IDR-1018 treated wounds were 28% smaller than control wounds on day 8 (d8, $p=0.0018$). In contrast, neither HB-107 (which led to a slight but non-significant decrease in the rate of wound healing; $p=0.0577$–0.4415 from day 2–14; Fig. 3) nor LL-37 ($p=0.1586$–0.9243; day 2 14) improved re-epithelialization compared to carrier control treated wounds in diabetic mice. Although there was a tendency towards better re-epithelialization with LL-37 on day 4 ($p=0.1586$ respectively) this tendency was lost over time (Fig. 3).

IDR-1018 showed no effect in diabetic mouse wound healing

Interestingly, the effects of IDR-1018 on experimental diabetic (type-II) wounds contrasted with the positive impact on wound healing in non-diabetic mice. IDR-1018 treated diabetic mice ($n=5$) demonstrated a time-course of wound-closure comparable to PBS-treated wounds (±4% compared to PBS-treated wounds, $p=0.1122$ to 0.7627, days 0–14). This was also demonstrated for LL-37 as diabetic wounds treated with LL-37 ($n=5$) were not significantly different from PBS-treated wounds (±4.6% compared to PBS-treated wounds, $p=0.1586$–0.9243; days 0–14). In contrast, HB-107 ($n=5$) exhibited a larger, but not significant, wound

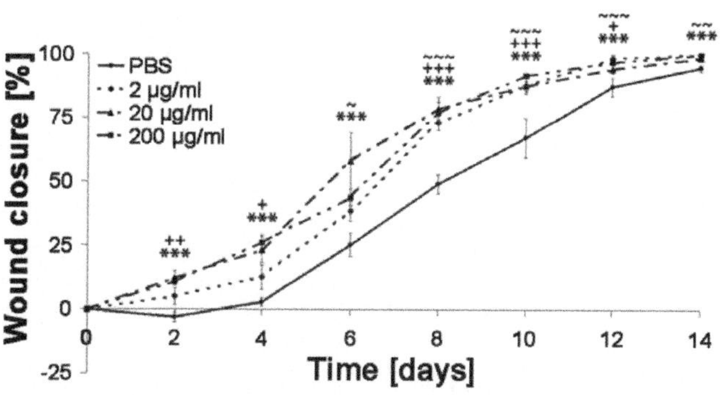

Figure 2. Efficacy in wound healing of IDR-1018 peptide in a murine dose-response model. Peptides were administered every 48 hours with the indicated concentrations of IDR-1018 or PBS (negative control). Re-epithelialization/wound closure was measured every second day. Standard deviation was assessed as SEM (Superscript symbols *, +, and ~ represent 15 μl doses of 3, 0.3 and 0.03 μg respectively; *,+,~ $=p<0.05$; **,++,~~ $=p<0.01$; ***,+++,~~~ $=p<0.005$ compared to vehicle control).

area compared to control wounds throughout the experiment (day 10: 30% larger wound area in HB-107 treated group, $p=0.057$; Fig. 3).

IDR-1018 in *S. aureus* infected wounds in pigs

We next sought to examine the wound healing activities of IDR-1018 in a different animal species. Using a porcine wound healing model we assessed both anti-infective activity and wound repair function. Full-thickness skin wounds ($n=12$ per animal) were made on the backs of two pigs and were subsequently inoculated with *S. aureus*. Wounds were treated with peptide every 48 h and evaluated for wound diameter and bacterial burden over time. We also compared LL-37 and vehicle alone control treatment.

On day 1 post-infection, all wounds showed clinical signs of inflammation including swelling and redness (Fig 4a). By day 4 the wound fluid was cloudy and a fibrinous- purulent wound clot formed on top of the wounds. In all groups, representative cross-sectional biopsies were taken on days 6 and 10 after surgery and demonstrated that the high dose of IDR-1018 resulted in almost complete healing by day 10.

IDR-1018 treated wounds demonstrated significantly enhanced wound healing/re-epithelialization as compared to PBS treated wounds (day 4: 28% improvement with a 20 μg/ml IDR-1018; $p=0.0069$ and 17% improvement with the 20 μg/ml IDR-1018; $p=0.0289$) (Fig 4b). LL-37 (200 μg/ml) treated wounds also showed a higher rate of re-epithelialization as compared to the PBS treated wounds (12–21% faster re-epithelialization on days 2–10, $p=0.0733$–0.1982). Epidermal healing in IDR-1018 treated wounds occurred in a dose dependent manner (60–104% greater re-epithelialization of wounds treated with 200 μg/ml IDR-18 compared to 20 μg/ml; $p=0.0319$–0.3410). In addition, a significantly higher rate of re-epithelialization was detected in wounds treated with IDR-1018 compared to the same dose of LL-37 (200 μg/ml; days 2–6; $p=0.007$–0.03). The highest dose of IDR-1018 led to an enhancement in re-epithelialization of up to 30% compared to PBS treated control wounds. A lower dose of IDR-1018 (20 μg/ml) resulted in a similar pattern of re-epithelialization as compared to the highest dose of LL-37 (200 μg/ml) providing further evidence for the potent wound healing activities of IDR-1018.

Hematoxylin-eosin staining demonstrated that both IDR-1018 and LL-37 evenly formed new epithelium evenly including connective tissue papillae. In contrast, wounds treated with PBS showed decreased organization of newly formed epithelium (Fig 5). IDR-1018 treated wounds exhibited advanced, concentration-dependent, re-epithelialization with keratinocytes migrating from the wound edges towards the centre of the wound forming epithelial tongues. LL-37 treatment resulted in reduced re-epithelialization compared to an equal dosage of IDR-1018.

Interestingly, assessment of bacterial colonization in the wound-tissue showed no significant differences between the treatment groups. On day ten post-infection, bacterial quantification revealed 8×10^5 cfu (colony forming units)/g tissue for the IDR-1018 20 μg/ml treatment group ($n=3$; $p=0.4047$ cf. PBS-control) and 7×10^5 cfu/g tissue for the 200 μg/ml group ($n=3$; $p=0.2998$ cf. PBS-control). In contrast LL-37 application resulted in a colonization of 1.8×10^6 cfu/g tissue ($n=3$; $p=0.5968$ cf. PBS-control) while PBS-control ($n=3$) itself had 1.3×10^6 cfu/g tissue; Fig 6).

Discussion

In addition to their ability to modulate innate and adaptive immunity, HDPs have also been implicated as positive effectors of

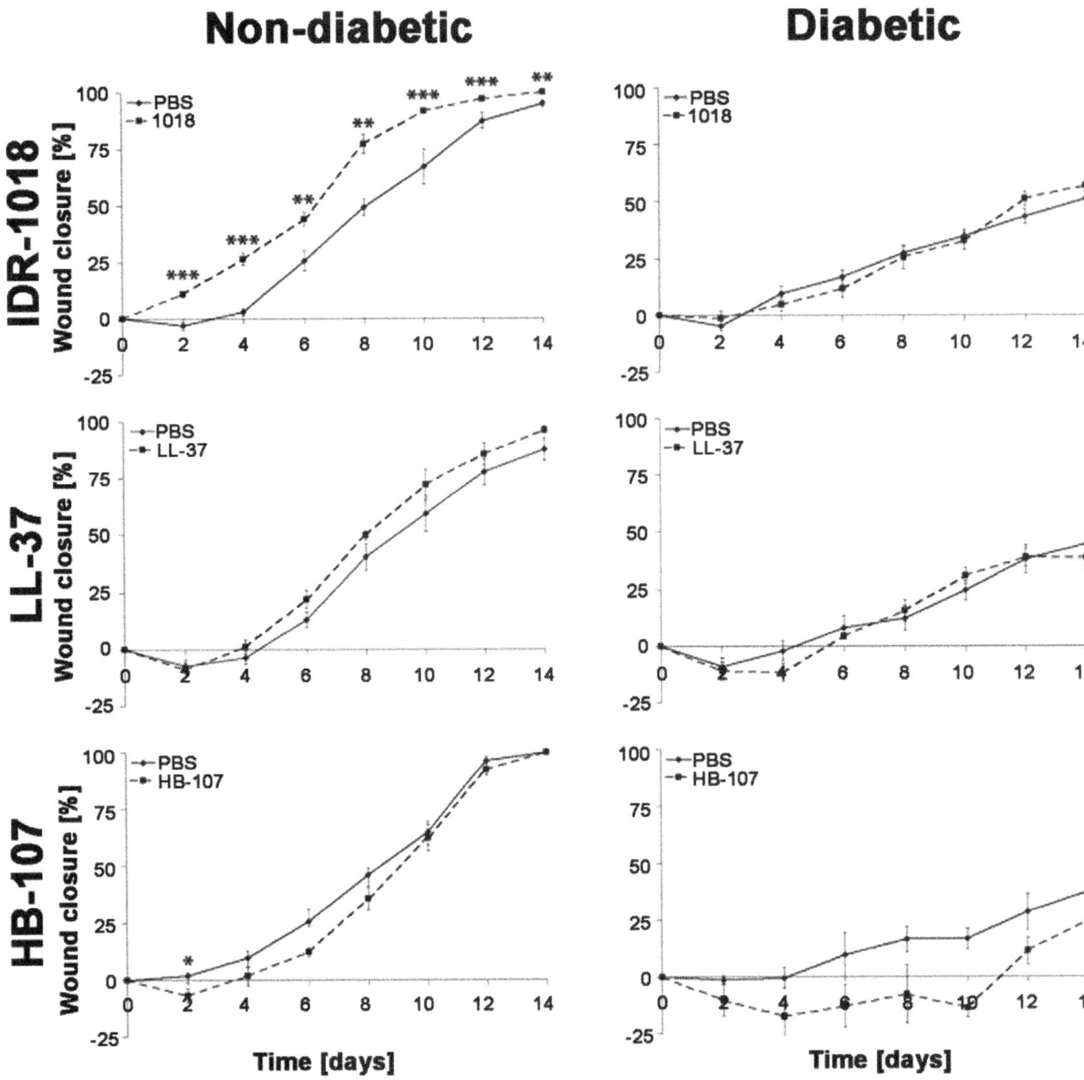

Figure 3. Efficacy in murine wound healing of IDR-1018 compared to LL-37 and HB-107. On day 0, mice were wounded on two dorsolateral areas. Treatments were administered every 48 hours for 14 days with either 15 µl of vehicle control (PBS; wound one) or 200 µg/ml (3 µg/dose) of the indicated peptides (wound two). Re-epithelialization was measured every second day. The bars represent standard error of the mean wound closure (* = $p<0.05$; ** = $p<0.01$; *** = $p<0.005$ compared to vehicle control).

wound repair [18,31,32,33]. Recent structure guided investigations of HDPs have focused on identifying the particular sequence motifs responsible for these immunomodulatory activities in an effort to create optimized peptide sequences of minimal sequence length and complexity, the IDRs. Indeed, recently developed optimized IDRs have been demonstrated to elicit strong immunomodulatory activities *in vitro* and *in vivo* and confer protection in murine models of bacterial infection in the absence of significant direct antimicrobial activity [24,25]. Their mechanism of action largely involves enhanced chemokine production accompanied by suppression of the production of pro-inammatory cytokines, such as TNFα, and is mediated through modulation of specific intracellular signalling pathways.

Peptide based immunotherapies are still in their infancy and there has been a paucity of clinical investigations from which to derive knowledge. This study investigated the new designer peptide IDR-1018 for its ability to influence wound repair. IDR-1018 was selected from a library of scrambled bovine HDP bactenecin derivatives due to its potency in inducing chemokines *in vitro* [29]. Normal wound repair involves the precise orchestration of inflammation, epithelialization, tissue granulation and remodeling. It has been demonstrated that HDPs influence all of these mechanisms lending further credence to our investigations [17,19,24]. For example, the cathelicidin PR-39 possesses anti-inflammatory qualities by inhibiting neutrophil oxidase activity, and induces syndecans, heparin sulfate proteoglycans important in wound repair [17,18]. Another member of the cathelicidin family, the human HDP LL-37, has been demonstrated to suppress pro-inflammatory responses [17] and positively influences the re-epithelialization of human skin wounds [7,33]. Furthermore, human β-defensin 3 (hBD-3), promotes wound healing in infected porcine diabetic wounds [5]. As these peptides also possess intrinsic antimicrobial activities it has been difficult to discern whether antimicrobial activity is integral to their positive influence on wound repair. Further, therapeutic applications for LL-37 and hBD-3 are limited due to their large size, and thus high associated costs of production, as well as their sequence complexity (hBD-3 disulfide bridges) and high cytotoxicities. In contrast, HB-107,

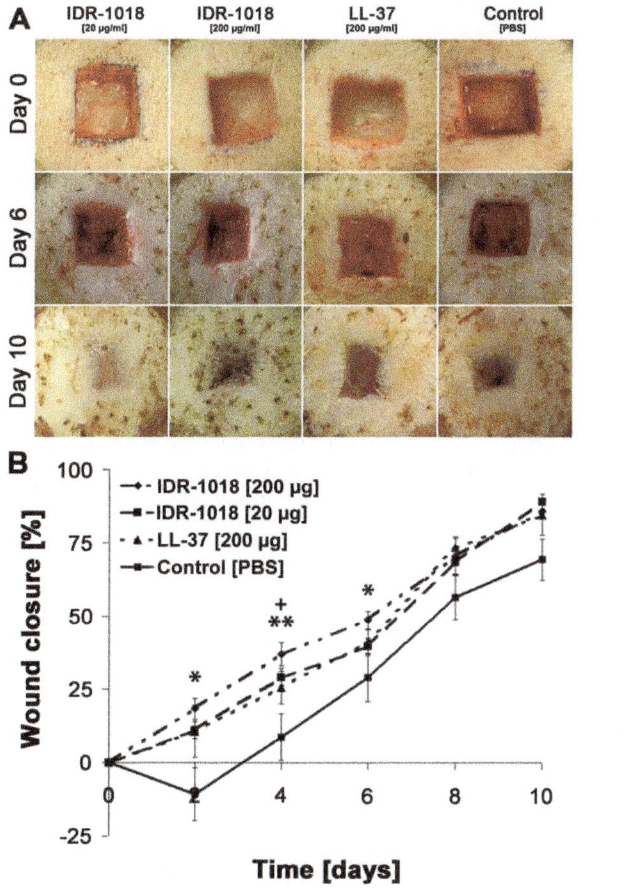

Figure 4. Efficacy in porcine wound healing of IDR-1018 compared to LL-37. (A) Efficacy in an infected porcine model of wound healing by IDR-1018. Digital photographic overview of porcine wounds treated with IDR-1018 (20 or 200 µg/ml every 48 hours) compared to LL-37 (200 µg/ml, positive control) and PBS (carrier control). (B) Quantification of re-epithelialization in wounds treated with IDR-1018. Re- epithelialization was assessed with IDR-1018 (20 and 200 µg/ml), LL-37 (200 µg/ml) or PBS. For day 0 to day 6 each value was calculated as mean out of six wounds while values of day 8 and 10 were calculated out of three different wounds. The bars represent standard error of the mean wound closure (*,+ $=p<0.05$; **,++ $=p<0.01$ (*IDR-1018 200 µg/ml; +IDR-1018 20 µg/ml) compared to vehicle control).

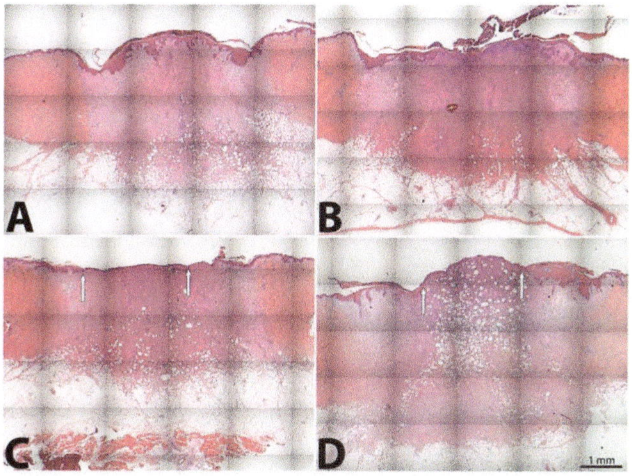

Figure 5. IDR-1018 promoted early keratinocyte proliferation in porcine wounds. Representative examples of paraffin embedded, hematoxylin-eosin-stained sections of porcine wounds at day 10 of the study. Sections show enhanced re-epithelialization in the IDR-1018–treated wounds (A: 200 µg/ml and B: 20 µg/ml) compared to LL-37 (200 µg/ml) (C) and PBS (D) treated wounds at wound edge (white arrow).

a synthetic HDP derived from the moth antimicrobial cecropin B [34], was found to aid in wound repair by stimulating keratinocyte proliferation and enhancing chemokine-induced leukocyte infiltration despite its lack of antimicrobial activity [21].

In this study we have demonstrated that in addition to its previously reported *in vitro* immunomodulatory activities, IDR-1018 also possesses potent wound healing activities *in vivo*. IDR-1018 elicited stronger wound healing activities in terms of wound closure than either HB-107 and LL-37 in a murine wound healing model. Furthermore, IDR-1018 treatment demonstrated a significantly higher rate of re-epithelialization compared to LL-37 and HB-107 and IDR-1018 showed lower cytotoxicity towards skin cells *in vitro* being significantly less cytotoxic than LL-37 and HB-107, consistent with previous studies with PBMC (peripheral blood mononuclear cells) [29].

A dose response study in a murine splint model demonstrated a dose-dependent effect of IDR-1018 on re-epithelialization. Wound treatment with as little as 30 ng/dose of IDR-1018 was associated with a comparable effect to 100-fold higher doses of LL-37 or HB-107. Higher concentrations of IDR-1018 (0.3 or 3 µg/doses) led to a significantly higher re-epithelialization compared to the same dose of LL-37 or HB-107.

To study the wound healing properties of IDR-1018 in another species and a more complex wound environment, IDR-1018 was tested in a *S. aureus* infected porcine wound model. The data obtained for IDR-1018 indicated it also a dose-dependent effect in re-epithelialization of infected wounds. Histological analysis of IDR-1018 treated wounds revealed a significantly increased re-epithelialization in IDR-1018 treated wounds compared to controls, and indicated that IDR-1018 may influence keratinocyte proliferation or migration.

Quantification of bacterial colonization did not exhibit any significant differences between treatment groups. Together with data from enhanced re-epithelialization in IDR-1018 treated

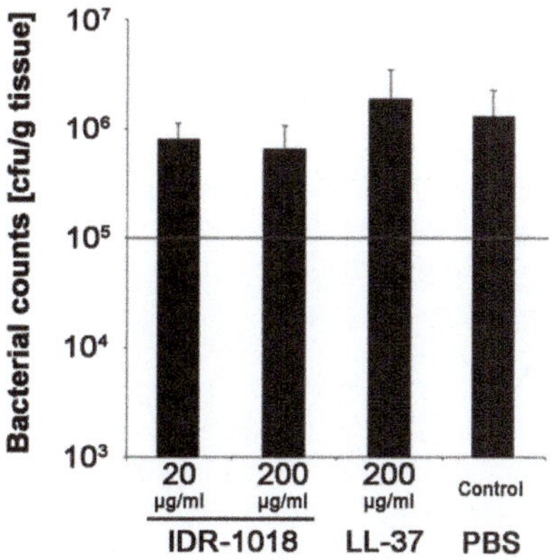

Figure 6. Assessment of bacterial colonization in *S. aureus* infected, porcine skin on day 10 after infection. Wounds were treated with the indicated concentrations of IDR-1018 or LL-37 (positive control). PBS served as negative control. Standard deviation was assessed as SEM.

groups of non-diabetic mice and pigs, this finding confirms data about lack of direct anti-microbial and enhanced immunomodulatory activity of IDRs described previously [23,25].

Diabetic wounds are associated with a high rate of complications in wound healing independent of infection [35]. Analysis of the wound healing activities of IDR-1018 in diabetic wounds was performed in db/db mice. These db/db mice are leptin receptor deficient and represent a model of type II diabetes mellitus characterized by hyperglycemia, obesity, hyperinsulinemia, and impaired wound healing [36]. Interestingly, data from this study did not demonstrate any effect of IDR-1018 on diabetic wound healing, suggesting the wound healing promoting mechanism is absent or suppressed in diabetic wounds. Previous studies have indicated that diabetes negatively influences immune responses, while IDR peptides work by modulating these responses; thus it is likely that the compromised immune system in diabetic animals blocks one or more signaling pathways by which IDR peptides exert their effects.

In conclusion, we have demonstrated here that in addition to the anti-infective and anti-inflammatory properties demonstrated by IDR peptides, the novel designer peptide IDR-1018 promotes wound healing in two different wound healing models without any significant antibacterial activity in this context.

Materials and Methods

Cell culture

Human skin was obtained in the operating room of the BG University Hospital Bergmannsheil. All patients gave their written, informed consent, and the study was approved by the ethics committee of the BG University Hospital Bergmannsheil.

Primary human fibroblasts (HFB) were obtained by incubating small pieces of dermis from human volunteers in 1% collagenase type II (Cell Systems, St. Katharinen, Germany) in PBS for 2 hours with shaking at 37°C. The solution was then filtered through a 100 µm cell strainer, centrifuged and resuspended in DMEM containing 10% fetal bovine serum (FBS) and 1% penicillin/streptomycin (PAA, Coelbe, Germany). Medium was changed every second day. All cell types, including the HaCaT keratinocytes cell line (kindly supplied by Prof. Fusenig, University of Heidelberg) [37], were cultured in DMEM containing 10% FBS and 1% penicillin/streptomycin at 37°C in a humidified atmosphere of 5% CO_2.

Peptides

IDR-1018, LL-37 and HB-107 were synthesized by GenScript (Piscataway, NJ, USA) or The Nucleic Acid/Protein Synthesis Unit at UBC using solid phase Fmoc chemistry and purified to a purity >95% using reversed phase HPLC. Peptide mass and lack of contaminating peptides were confirmed by mass spectrometry.

MTT-Assay

HFB and HaCaT cells were seeded in a 96-well microtiterplate (Omnilab, Bremen, Germany). After 24 hours, peptides were added at the indicated concentrations. Cells were cultured for an additional 24 hours as described above, and 0.5 mg/ml 3-[4,5-dimethylthiazol-2-yl\-2,5-diphenyl tetrazolium bromide (MTT) (Sigma, Taufkirchen, Germany) was added. After four hours 0.01N HCl containing 10% sodium dodecyl sulfate (SDS) was added to stop the reaction, cells were lysed with dimethylsulfoxide (DMSO) and adsorbance at 562nm measured using an ELISA plate reader (BioTek Instruments, Bad Friedrichshall, Germany).

Murine wound healing model

The protocol utilized complied with all regulations relating to animal use and other federal statutes. It was conducted in compliance with the principles in the 'Guide for the Care and Use of Laboratory Animals' from the German Animal Welfare Act. Six week old non-diabetic (C57BLKS) and diabetic (C57BLKS-LepR; leptin receptor deficient; "db/db") mice (Charles River, Sulzfeld, Germany) were used. Mice were anesthetized by isoflurane inhalation. The dorsal surface was shaved, sterilized with betadine and draped. Two full-thickness wounds, one on each side of midline, were made using a 4 mm punch biopsy created. A donut-shaped splint with a diameter of 1 cm was fashioned from a 0.5 mm-thick silicone sheet (Grace Bio-Laboratories, Bend, USA). An immediate-bonding adhesive was used to fix the splint to the skin followed by interrupted nylon sutures (Ethicon, Inc., Somerville, NJ). Tegaderm (transparent polyurethane bandage) was used as an occlusive wound dressing (Tegaderm, 3M, Neuss, Germany).

Wounds were treated with 15 µl of peptide solution containing 2, 20 or 200 µg/ml (i.e. 30, 300 or 3000 ng) of peptide or PBS as the carrier control every 48 hours ($n = 5$ animals per group; one treatment and control wound per animal). On the 14th post operative day, mice were euthanized, wounds excised, bisected and stored for histological analysis.

Porcine wound healing model

Two female Goettinger Minipigs (Ellegaard, Dalmose, Denmark) weighing 20–25 kg at arrival were allowed to acclimatize for 4 weeks prior to initiation of the experiment. Anesthesia was induced with Ketamine/Xylazine (Xyla-Ject, Phoenix, St Josephs, MO, USA) via intramuscular injection. Animals were then transferred to general anaesthesia with 2–3% isoflurane in oxygen (Novaplus, Hospira Inc.). Prior to surgery, the porcine dorsum was shaved and 12 squares measuring 1.5×1.5 cm were outlined using a template and a tattoo gun (Tattoo Age, Dietenheim, Germany). The paraspinal area was disinfected using 10% povidone iodine paint. Twelve full thickness wounds per animal (1.5×1.5×0.5 cm) were created. An adhesive polyurethane chamber (Coloplast, Hamburg, Germany) was placed over each wound.

For bacterial inoculation, a methicillin-sensitive strain of *S. aureus* (ATCC29523) was used. A single colony was inoculated into Laura Bertani culture media (Becton & Dickinson, Heidelberg, Germany), and incubated for 18 hours at 37°C. The culture was centrifuged and resuspended in sterile PBS and adjusted to a final amount of 6.7×10^9 colony-forming units (cfu)/ml by use of the equation (cfu/ml $= OD_{600nm} \times 2.5 \times 10^8$). To immerse the enclosed wound surface, 75 µl of the bacterial suspension was injected into each chamber.

Wounds were treated with 85 µl of peptide solution ($n = 3$ wounds per group per animal) every 48 hours. On day 6 and 10 post operation, pigs were euthanized, wounds excised, bisected and stored for further analysis.

Histology

Samples from each experiment were fixed in 4% buffered paraformaldehyde. Tissue samples were stained with haematoxylin/eosin, analyzed by three independent investigators. Images were taken with an AxioCam camera on an Axioplan microscope (Carl Zeiss GmbH, Oberkochen, Germany).

Wound analysis

Digital photographs were taken every second day. Wound closure was calculated by measuring the wound area from the computerized image with Photoshop CS4 Pro software (Adobe, San Jose, USA) using the formula (Wound area)/(Inner area of polyurethane chambers or silicon splints). Wound area was calculated as the percent area of the original wound.

Assessment of wound infection

Tissue samples were homogenized in PBS (PT3100 Polytron, Kinematica, Littau-Luzern, Switzerland). Serially diluted aliquots of homogenate were quantified on Luria–Bertani agar (IDC, Bury, Lancashire, UK) after incubation for 18 h at 37°C. cfu/g tissue was calculated as described previously [38]. A stable wound in fection was defined as 10^5 colony forming units (cfu)/g tissue.

Statistical analysis

Data were analyzed using analysis of variance and independent sample t-test with the software package software package SPSS 14.0 (SPSS, Chicago, IL, USA). A p-value of less than 0.05 was considered significant.

Acknowledgments

We sincerely thank Andrea Rittig for excellent technical support and Sammy Al-Benna for carefully proofreading this manuscript.

Author Contributions

Conceived and designed the experiments: LS JK TH FJ REWH. Performed the experiments: TH MS MK FJ EM. Analyzed the data: MS TH MK FJ. Contributed reagents/materials/analysis tools: HJ NA IS. Wrote the paper: TH LS MS FJ JK REWH.

References

1. Singer AJ, Clark RA (1999) Cutaneous wound healing. N Engl J Med 341: 738–746.
2. Hirsch T, Jacobsen F, Steinau HU, Steinstraesser L (2008) Host defense peptides and the new line of defence against multiresistant infections. Protein Pept Lett 15: 238–243.
3. Steinstraesser L, Kraneburg UM, Hirsch T, Kesting M, Steinau HU, et al. (2009) Host defense peptides as effector molecules of the innate immune response: a sledgehammer for drug resistance? Int J Mol Sci 10: 3951–3970.
4. Steinstraesser L, Oezdogan Y, Wang SC, Steinau HU (2004) Host defense peptides in burns. Burns 30: 619–627.
5. Hirsch T, Spielmann M, Zuhaili B, Fossum M, Metzig M, et al. (2009) Human beta-defensin-3 promotes wound healing in infected diabetic wounds. J Gene Med 11: 220–228.
6. Otte JM, Zdebik AE, Brand S, Chromik AM, Strauss S, et al. (2009) Effects of the cathelicidin LL-37 on intestinal epithelial barrier integrity. Regul Pept 156: 104–117.
7. Steinstraesser L, Koehler T, Jacobsen F, Daigeler A, Goertz O, et al. (2008) Host defense peptides in wound healing. Mol Med 14: 528–537.
8. Steinstraesser L, Ring A, Bals R, Steinau HU, Langer S (2006) The human host defense peptide LL37/hCAP accelerates angiogenesis in PEGT/PBT biopolymers. Ann Plast Surg 56: 93–98.
9. De Y, Chen Q, Schmidt AP, Anderson GM, Wang JM, et al. (2000) LL-37, the neutrophil granule- and epithelial cell-derived cathelicidin, utilizes formyl peptide receptor-like 1 (FPRL1) as a receptor to chemoattract human peripheral blood neutrophils, monocytes, and T cells. J Exp Med 192: 1069–1074.
10. Niyonsaba F, Iwabuchi K, Someya A, Hirata M, Matsuda H, et al. (2002) A cathelicidin family of human antibacterial peptide LL-37 induces mast cell chemotaxis. Immunology 106: 20–26.
11. Braff MH, Hawkins MA, Di Nardo A, Lopez-Garcia B, Howell MD, et al. (2005) Structure-function relationships among human cathelicidin peptides: dissociation of antimicrobial properties from host immunostimulatory activities. J Immunol 174: 4271–4278.
12. Elssner A, Duncan M, Gavrilin M, Wewers MD (2004) A novel P2X7 receptor activator, the human cathelicidin-derived peptide LL37, induces IL-1 beta processing and release. J Immunol 172: 4987–4994.
13. Scott MG, Davidson DJ, Gold MR, Bowdish D, Hancock REW (2002) The human antimicrobial peptide LL-37 is a multifunctional modulator of innate immune responses. J Immunol 169: 3883–3891.
14. Tjabringa GS, Aarbiou J, Ninaber DK, Drijfhout JW, Sorensen OE, et al. (2003) The antimicrobial peptide LL-37 activates innate immunity at the airway epithelial surface by transactivation of the epidermal growth factor receptor. J Immunol 171: 6690–6696.
15. Li J, Post M, Volk R, Gao Y, Li M, et al. (2000) PR39, a peptide regulator of angiogenesis. Nat Med 6: 49–55.
16. Oono T, Shirafuji Y, Huh WK, Akiyama H, Iwatsuki K (2002) Effects of human neutrophil peptide-1 on the expression of interstitial collagenase and type I collagen in human dermal fibroblasts. Arch Dermatol Res 294: 185–189.
17. Shi J, Ross CR, Leto TL, Blecha F (1996) PR-39, a proline-rich antibacterial peptide that inhibits phagocyte NADPH oxidase activity by binding to Src homology 3 domains of p47 phox. Proc Natl Acad Sci U S A 93: 6014–6018.
18. Gallo RL, Ono M, Povsic T, Page C, Eriksson E, et al. (1994) Syndecans, cell surface heparan sulfate proteoglycans, are induced by a proline-rich antimicrobial peptide from wounds. Proc Natl Acad Sci U S A 91: 11035–11039.
19. Hancock REW, Sahl HG (2006) Antimicrobial and host-defense peptides as new anti-infective therapeutic strategies. Nat Biotechnol 24: 1551–1557.
20. Zhang L, Falla TJ (2009) Host defense peptides for use as potential therapeutics. Curr Opin Investig Drugs 10: 164–171.
21. Lee PH, Rudisill JA, Lin KH, Zhang L, Harris SM, et al. (2004) HB-107, a nonbacteriostatic fragment of the antimicrobial peptide cecropin B, accelerates murine wound repair. Wound Repair Regen 12: 351–358.
22. Easton DM, Nijnik A, Mayer ML, Hancock RE (2009) Potential of immunomodulatory host defense peptides as novel anti-infectives. Trends Biotechnol 27: 582–590.
23. Lee HY, Bae YS (2008) The anti-infective peptide, innate defense-regulator peptide, stimulates neutrophil chemotaxis via a formyl peptide receptor. Biochem Biophys Res Commun 369: 573–578.
24. Nijnik A, Madera L, Ma S, Waldbrook M, Elliott MR, et al. (2010) Synthetic cationic peptide IDR-1002 provides protection against bacterial infections through chemokine induction and enhanced leukocyte recruitment. J Immunol 184: 2539–2550.
25. Scott MG, Dullaghan E, Mookherjee N, Glavas N, Waldbrook M, et al. (2007) An anti-infective peptide that selectively modulates the innate immune response. Nat Biotechnol 25: 465–472.
26. Bowdish DM, Davidson DJ, Lau YE, Lee K, Scott MG, et al. (2005) Impact of LL-37 on anti-infective immunity. J Leukoc Biol 77: 451–459.
27. Goldman MJ, Anderson GM, Stolzenberg ED, Kari UP, Zasloff M, et al. (1997) Human beta-defensin-1 is a salt-sensitive antibiotic in lung that is inactivated in cystic fibrosis. Cell 88: 553–560.
28. Johansson J, Gudmundsson GH, Rottenberg ME, Berndt KD, Agerberth B (1998) Conformation-dependent antibacterial activity of the naturally occurring human peptide LL-37. J Biol Chem 273: 3718–3724.
29. Wieczorek M, Jenssen H, Kindrachuk J, Scott WR, Elliott M, et al. (2010) Structural studies of a peptide with immune modulating and direct antimicrobial activity. Chem Biol 17: 970–980.
30. Carretero M, Escamez MJ, Garcia M, Duarte B, Holguin A, et al. (2008) In vitro and in vivo wound healing-promoting activities of human cathelicidin LL-37. J Invest Dermatol 128: 223–236.
31. Dorschner RA, Pestonjamasp VK, Tamakuwala S, Ohtake T, Rudisill J, et al. (2001) Cutaneous injury induces the release of cathelicidin anti-microbial peptides active against group A Streptococcus. J Invest Dermatol 117: 91–97.
32. Gallo RL, Murakami M, Ohtake T, Zaiou M (2002) Biology and clinical relevance of naturally occurring antimicrobial peptides. J Allergy Clin Immunol 110: 823–831.
33. Heilborn JD, Nilsson MF, Kratz G, Weber G, Sorensen O, et al. (2003) The cathelicidin anti-microbial peptide LL-37 is involved in re-epithelialization of human skin wounds and is lacking in chronic ulcer epithelium. J Invest Dermatol 120: 379–389.
34. Steiner H, Hultmark D, Engstrom A, Bennich H, Boman HG (1981) Sequence and specificity of two antibacterial proteins involved in insect immunity. Nature 292: 246–248.
35. Brem H, Tomic-Canic M (2007) Cellular and molecular basis of wound healing in diabetes. J Clin Invest 117: 1219–1222.
36. Coleman DL (1982) Diabetes-obesity syndromes in mice. Diabetes 31: 1–6.
37. Fusenig NE, Boukamp P (1998) Multiple stages and genetic alterations in immortalization, malignant transformation, and tumor progression of human skin keratinocytes. Mol Carcinog 23: 144–158.
38. Jacobsen F, Baraniskin A, Mertens J, Mittler D, Mohammadi-Tabrisi A, et al. (2005) Activity of histone H1.2 in infected burn wounds. J Antimicrob Chemother 55: 735–741.

2

Detection of Phospho-Sites Generated by Protein Kinase CK2 in CFTR: Mechanistic Aspects of Thr1471 Phosphorylation

Andrea Venerando[1], Cinzia Franchin[1,2], Natasha Cant[3], Giorgio Cozza[1], Mario A. Pagano[4], Kendra Tosoni[5], Ateeq Al-Zahrani[3¤], Giorgio Arrigoni[1,2], Robert C. Ford[3], Anil Mehta[5], Lorenzo A. Pinna[1]*

1 Department of Biomedical Sciences, University of Padova, and CNR Institute of Neurosciences, Padova, Italy, **2** Proteomic Center of Padova University, VIMM, and Padova University Hospital, Padova, Italy, **3** Faculty of Life Sciences, University of Manchester, Manchester, United Kingdom, **4** Department of Molecular Medicine, University of Padova, Padova, Italy, **5** Division of Cardiovascular and Diabetes Medicine, Ninewells Hospital and Medical School, University of Dundee, Dundee, United Kingdom

Abstract

By mass spectrometry analysis of mouse Cystic Fibrosis Transmembrane-conductance Regulator (mCFTR) expressed in yeast we have detected 21 phosphopeptides accounting for 22 potential phospho-residues, 12 of which could be unambiguously assigned. Most are conserved in human CFTR (hCFTR) and the majority cluster in the Regulatory Domain, lying within consensus sequences for PKA, as identified in previous mammalian studies. This validates our yeast expression model. A number of phospho-residues were novel and human conserved, notably mouse Ser670, Ser723, Ser737, and Thr1467, that all lie in acidic sequences, compatible with their phosphorylation by protein kinase CK2. Thr1467 is localized in the C-terminal tail, embedded in a functionally important and very acidic sequence (EETEEE) which displays an optimal consensus for protein kinase CK2. Herein, we show that Thr1467, homologous to human Thr1471 is readily phosphorylated by CK2. Indeed a 42 amino acid peptide encompassing the C-terminal segment of human CFTR is readily phosphorylated at Thr1471 with favorable kinetics (Km 1.7 μM) by CK2 holoenzyme, but neither by its isolated catalytic subunit nor by other acidophilic Ser/Thr kinases (CK1, PLK2/3, GCK/FAM20C). Our finding that by treating CFTR expressing BHK cells with the very specific CK2 inhibitor CX4945, newly synthesized wild type CFTR (and even more its Phe508del mutant) accumulates more abundantly than in the absence of CK2 inhibitor, supports the conclusion that phosphorylation of CFTR by CK2 correlates with decreased stability of the protein.

Editor: Alexander V. Ljubimov, Cedars-Sinai Medical Center, United States of America

Funding: LAP was supported by the Fondazione per la Ricerca sulla Fibrosi Cistica (grant #3/2011 adopted by Delegazione FFC della Valdadige and Associazione Trentina FC Fiaba "Il Villaggio di Natale" in ricordo di Massimiliano e Sebastiano) and Associazione Italiana per la Ricerca sul Cancro. This work was funded also by Wellcome Trust (grant numbers 088929 to LAP with AM and 069150 to AM). RCF acknowledges the Cystic Fibrosis Foundation for funding under the auspices of its CFTR 3D Structure Consortium. NC was funded by a BBSRC PhD studentship. The authors are grateful to the Cassa di Risparmio di Padova e Rovigo (Cariparo) Holding for funding the acquisition of the LTQ-Orbitrap XL mass spectrometer. The funders had no role in study design, data collection and analysis, decision to publish, or preparation of the manuscript.

Competing Interests: The authors have declared that no competing interests exist.

* E-mail: lorenzo.pinna@unipd.it

¤ Current address: Al-Qunfudhah University College, Umm Al-Qura University, Makkah Al-Mukarramah, Saudi Arabia

Introduction

Cystic fibrosis (CF) is caused by mutations affecting a polytopic integral membrane protein, termed Cystic Fibrosis Transmembrane-conductance Regulator (CFTR) functioning as a phosphorylation stimulated, ATP-dependent anion channel. Pathogenic mutations of CFTR either impair its anion channel functions or make the protein unstable determining its premature degradation. This latter event is promoted by the commonest mutation of CFTR, a deletion of the amino acid residue phenylalanine 508 (Phe508delCFTR), which is found in 70–90% of CF patients. Phe508delCFTR is only partially functional when it is induced to traffic to the cell membrane [1,2], but >99% of the protein undergoes premature degradation being targeted to the proteasome by the ER quality control machinery before it reaches the membrane [3]. Even the minor aliquot of Phe508delCFTR that reaches the membrane, moreover, displays a half-life much shorter than that of the wild type CFTR [4], further highlighting the intrinsic instability of this mutant.

In an attempt to disclose molecular events that may underlie deregulation of CFTR functionality and/or stability, attention has been focused on post-translational events, with special reference to phosphorylation. These studies clearly demonstrated that multiple phosphorylation of the regulatory domain of CFTR by PKA and probably also by other basophilic protein kinases such as PKC is essential to confer the full channel functionality to CFTR. The same residues phosphorylated in vitro by PKA [5] (and in some cases by PKC) were also found to undergo phosphorylation in living cells [6,7], thus providing the clear-cut demonstration of the physiological occurrence of these events. It was also found in mutation studies that no single PKA site was dominant towards CFTR function and even a mutant with all PKA sites deleted retained some CFTR function.

Intriguingly however CFTR also includes several potential phospho-acceptor sites that display the consensus for the most pleiotropic acidophilic protein kinase, CK2 (an acronym derived from the old misnomer "casein kinase 2" [8]), but none of these has been ever reported to be phosphorylated in vivo, although the phosphorylation of a couple of these could be demonstrated in vitro [9]. While the PKA sites, whose phosphorylation has been validated in vivo are concentrated in the R domain (where they play a prominent role in the activation of the channel function) the potential CK2 sites are also localized outside the R domain and, by analogy with other CK2 targets whose phosphorylation commits them to degradation (see e.g. [10–12]), they may be implicated in the premature proteolysis of CFTR, also consistent with the outcome of recent mutational studies [13,14]. It should be noted in fact that, although wild type CFTR is much more stable than its Phe508del counterpart, it nevertheless undergoes a very stringent quality control as well, resulting in perhaps 50% of the protein being discarded and proteolytically degraded [15]. Such a process is exceedingly complicated involving a network of proteins which includes chaperones [16,17], glycan processing enzymes as well as the basal trafficking machinery [18,19]. In this context CK2 may play a critical role in promoting accelerated degradation of CFTR, thus accounting for the failure to detect the phospho-form of CK2 phosphoacceptor sites in full length CFTR from living cells.

We reasoned that to circumvent this problem, advantage could be taken of CFTR expressed in yeast, where CK2 (as well as PKA and many other Ser/Thr protein kinases) is abundantly expressed and where a methodology for the expression of CFTR by a protocol that overcomes proteolysis problems has been recently set up [20]. By this method the degradation of CFTR is minimized providing good yields of pure CFTR exploitable for structural studies, but also useful for evaluation of its phosphorylation by endogenous Ser/Thr kinases minimizing complications arising from premature degradation.

By MS analysis of CFTR expressed in yeast we have now been able to show that besides all the "canonical" PKA sites concentrated in the R domain (whose detection demonstrates the reliability of the "yeast" approach) we additionally observe that some novel phospho-residues previously never found in CFTR from animal cells, are present in mouse CFTR expressed in yeast. One of these is the best predicted site for CK2, Thr1467 that is conserved in human CFTR (Thr1471): this residue is also readily phosphorylated in vitro with favourable kinetics by CK2 holoenzyme once included in a recombinant fragment of human CFTR encompassing its C-terminal segment. This outcome is described in the present report together with pharmacological evidence supporting the view that CFTR phosphorylation by CK2 may correlate with accelerated degradation of the protein.

Materials and Methods

Materials

All chemicals, cell culture media, anti-β tubulin antibody, and phosphatase inhibitor cocktails 2 and 3 were from Sigma-Aldrich. [γ-^{33}P]ATP (3000 Ci/mmol) as well as HRP (horseradish peroxidase)-conjugated antibodies were purchased from PerkinElmer. Mouse monoclonal anti-CFTR antibody (10B6.2) that recognizes a specific epitope on NBD1 domain was from CFTR folding consortium (www. cftrfolding.org). Complete protease inhibitor cocktail tablets were from Roche Diagnostic. Okadaic acid was purchased from Alexis Biochemicals. Methotrexate was from Santa Cruz Biotechnology. CX4945 was from SYN|thesis med chem.

Cloning, Expression and Purification of CFTR

Human CK2, both the α-catalytic subunit and the $\alpha_2\beta_2$-holoenzyme were expressed in *Escherichia coli*. CK2$\alpha_2\beta_2$ holoenzyme was reconstituted by mixing pellets expressing CK2α and CK2β independently. Cells were lysed by French press at 4°C in Tris/HCl 25 mM, pH 8, NaCl 0.4 M, DTT 1 mM, proteases inhibitors cocktail (Roche). The soluble fraction obtained after 30 min centrifugation at 15000 g was applied on Heparin-Sepharose and the proteins were separated using a gradient elution (0.4–1 M NaCl). Fractions containing CK2 were further purified by size-exclusion chromatography using a buffer composed by Tris/HCl 25 mM, pH 8, NaCl 0.5 M, DTT 1 mM.

CK1 isoforms and PLK 2 and 3 were obtained as described in [21], and [22], respectively.

Recombinant mouse full length CFTR (mCFTR) was cloned, expressed and purified as described in [20]. A 42-mer peptide reproducing the C-terminal tail of human CFTR (residues 1439–1480) was expressed in *Escherichia coli* with an N-terminal His tag and purified using affinity chromatography (Talon) followed by cleavage of the His tag by thrombin. Further details are given in the Material S1.

Synthetic peptide encompassing the last 20 residues of human CFTR C-terminal domain (KPQIAALKEETEEEVQDTRL, residues 1461–1480) was obtained by solid phase peptide synthesis method using a multiple peptide synthesizer (SyroII, MultiSynTech GmbH) on p-Benzyloxybenzyl alcohol resin (100–200 mesh) HMP resin (Novabiochem). The fluoren-9-ylmethoxycarbonyl (Fmoc) strategy was used throughout the peptide chain assembly, utilizing O-(7-Azabenzotriazol-1-yl)-N,N,N′,N′-tetramethyluronium hexafluorophosphate (HATU) as coupling reagent [23]. The side-chain protected amino acid building blocks used were: N-α-Fmoc-Nω-(2,2,4,6,7-pentamethyldihydrobenzofuran-5-sulfonyl)-L-arginine, N-α-Fmoc-γ-tert-butyl-L-glutamic acid, N-α-Fmoc-β-tert-butyl-L-aspartic acid, N-α-Fmoc-O-tert-butyl-L-threonine, N-α-Fmoc-Nε-(tert-butyloxycarbonyl)-L-lysine, N-α-Fmoc-N-γ-trityl-L-glutamine. Cleavage of the peptide was performed by reacting the peptidyl-resin with a mixture containing TFA/ethanedithiol/phenol 5% for 2.5 h. Crude peptide were purified by a preparative reverse phase HPLC. Molecular mass of the peptide were confirmed by mass spectroscopy on a MALDI TOF-TOF mass spectrometer (model 4800– AB Sciex). The purity (>90%) was evaluated by analytical reverse phase HPLC.

MS Analysis

Trypsin digestion. Murine CFTR and the C-terminal peptide from human CFTR were digested in-gel with trypsin. Briefly, gel band was cut in small pieces of about 1 square mm and washed with several changes of 50% acetonitrile (ACN)/25 mM NH_4HCO_3. Gel pieces were then dried under vacuum and rehydrated with 50–100 µL of freshly prepared 10 mM DTT in 25 mM NH_4HCO_3. Reduction was carried on for 1 h at 56°C. DTT solution was discarded and a freshly prepared solution of 55 mM iodoacetamide in 25 mM NH_4HCO_3 was added to the gel pieces to alkylate cysteine residues for 45 min at room temperature in the dark. Gel pieces were then washed 3 times (10 min every time) alternating 25 mM NH_4HCO_3 and 50% ACN. The last washing step was done with 100% ACN and gel pieces were dried under vacuum. 20 µL of a solution 12.5 ng/µL in 25 mM NH_4HCO_3 of sequencing grade trypsin (Promega) was added to the gel sample and rehydration was carried on for 45 min at 4°C. Digestion was then performed overnight at 37°C. Peptides were extracted with 3 changes of 50% ACN/0.1% formic acid (30 min between changes). Sample was dried under vacuum kept at −20°C until MS analyses were performed.

Phosphopeptides enrichment. Phosphopeptides derived from the trypsin digestion of CFTR were enriched with micro-columns of TiO_2 prepared in house as described in [24]. Sample was dissolved in 20 µL of loading buffer consisting of 80% ACN/ 6% trifluoroacetic acid (TFA). The micro-column was conditioned twice (20 µL each time) with ACN and twice with loading buffer. Sample was loaded and slowly pushed through the column, which was then washed twice (20 µL each) with loading buffer and twice with a washing buffer (0.1% TFA). Phosphopeptides bound to TiO2 were eluted with 20 µL of freshly prepared 5% NH_4OH and subsequently with 20 µL of 50% ACN/0.1% formic acid. Sample was immediately acidified by adding 2 µL of 100% formic acid and dried under vacuum.

LC-MS/MS and data analysis. Mass spectrometry analysis of phosphopeptides was performed with a LTQ-Orbitrap XL hybrid mass spectrometer (Thermo Fisher Scientific) coupled online with a nano-HPLC Ultimate 3000 (Dionex-Thermo Fisher Scientific). Sample was dissolved in 20 µL of 0.1% formic acid (FA) and for every analysis 5 µL of sample were loaded at a flow rate of 8 µl/min into a trap column (300 µm I.D., 300 Å, C18, 3 µm; SGE Analytical Science). Sample was then injected with a flow rate of 250 nL/min into a 10 cm pico-frit capillary column (75 µm I.D., 10 µm tip; New Objectives) home-packed with C18 resin (Reprosil 300Å, 3 µm). Peptides were separated using a linear gradient of ACN/0.1% FA from 3% to 40% in 20 min.

MS analyses were performed with 3 different acquisition methods, using the same chromatographic conditions. A MS2 data dependent acquisition (1 MS scan on the Orbitrap with a resolution of 60000, followed by MS/MS spectra acquired in the linear ion trap for the 10 most abundant peptides); a MS3 neutral loss-triggered dependent acquisition (1 MS scan on the Orbitrap with a resolution of 60000, followed by MS/MS scans on the 3 most intense ions and by MS3 upon detection of neutral loss of phosphoric acid in MS2 spectra); a Multi Stage Acquisition (MSA) (1 MS scan at a resolution of 60000 followed by MS/MS scans on the 3 most abundant ions with the activation of neutral loss product without an additional isolation cycle).

Raw data files were analyzed with the Proteome Discoverer software version 1.3 (Thermo Fisher Scientific) connected to a Mascot Server version 2.2.4 (Matrix Science, UK) and a Sequest search engine version 28.0 (Thermo Fisher Scientific) against the Uniprot/SwissProt Database (version 20110526). Trypsin was set as digesting enzyme with up to 1 missed-cleavage. Cabamido-methyl cysteine was set as fixed modification, while phosphorylation of Ser/Thr/Tyr and methionine oxidation were set as variable modifications. Peptide and fragment tolerance were 5 ppm and 0.6 Da respectively. False Discovery Rate (FDR) based on the search against the corresponding randomized database and Percolator were used to discriminate between correct and decoy spectrum identifications. The software phosphoRS [25] was used to help in the correct assignment of phosphorylation sites.

Files were analyzed using both Mascot and Sequest and searched for MS2 and MS3 spectra; all results were combined directly by Proteome Discoverer into a single multi-report file. Spectra of phosphorylated peptides were manually inspected for further confirmation of the results.

Cell Culture

Baby hamster kidney cells (BHK) stably expressing human CFTR (hCFTR) either wild type (WT), or Phe508del, or WT T1471A were growth to reach 70–80% confluency [13]. After 2 h-long treatment with 100 µg/mL cyclohexamide, cells were washed with PBS and incubated with 2 µM CX4945 or its vehicle for the indicated time. At each time-point cells were washed with ice-cold PBS, scraped from the plates, pelleted by centrifugation (1200 g for 5 min at room tempertaure) and lysed in a buffer containing 1% (v/v) Nonidet P40, 50 mM Tris/HCl (pH 7.5) and 150 mM NaCl with fresh protease inhibitor tablets (Roche), 1 µM Okadaic acid (Alexis) and phosphatases inhibitors cocktails 2 and 3 (Sigma). Cells were lysed by incubation on ice for 30 min and then the lysates were cleared by centrifugation at 17000 g for 30 min at 4°C. The supernatant was collected and subjected to protein concentration quantification.

Western blot analysis was performed as described in [14] using a modified Laemmli buffer containing 105 mM dithiothreitol (final concentration) to enhance the solubilization of CFTR. 20–30 µg of total proteins were separated on 4–12% precast NuPage® Bis-Tris gradient gels (Invitrogen) according to the manufacturer's instructions and transferred on to Optitran BA-S 83 nitrocellulose membranes (Whatman) with a semi-dry blotter (Biometra) for 1 h at 150 mA. Membranes were saturated for 1 h with 5% skimmed milk in Tris-buffered saline (TBS) containing 0.2% Tween 20. Immunoblots were developed by enhanced chemiluminescence, detected on Kodak Image Station 440 cf and analyzed by the Kodak 1D image software. Tubulin was used as loading control.

Phosphorylation Assay

Peptides derived from human CFTR C-terminal region were phosphorylated by incubation in a 20-µl volume containing 50 mM Tris/HCl, pH 7.5, 10 mM $MgCl_2$, 100 mM NaCl, and 50 µM [γ-^{33}P]ATP (specific radioactivity 1,500–2,000 cpm/ pmol). The reaction was started by the addition of the indicated protein kinases. The reaction mixtures were incubated for 10 minutes at 37°C and stopped by the addition of 2× concentrated Tricine sample buffer (450 mM Tris HCl, pH 8.45, 12% (v/v) glycerol, 4% (w/v) SDS, 0.0025% (w/v) Coomassie Blue G, 0.0025% (w/v) Phenol Red and 105 mM DTT). Samples were separated on Tricine-SDS-PAGE following the Schägger's protocol [26] in a running buffer composed of 100 mM Tris/HCl, pH 8.3, 100 mM Tricine, 0.1% (w/v) SDS. ^{33}P incorporation was evaluated by gels autoradiography on PerkinElmer's Cyclone Plus Storage Phosphor System. Initial rate data were analyzed with the program Prism (GraphPad Software, La Jolla, CA).

In Silico Analysis

Structural model of the last 42 aminoacids of CFTR was built using ab initio Rosetta tool [27] and generating 50000 models of the target sequence. These models were clusterized using the cluster application in order to identify the most frequently sampled conformations. The final model was submitted to a 100 ns Molecular Dynamics (MD).

Protein-protein docking analysis was performed using two FFT-based docking software PIPER [28] and Zdock [29]. The procedures were performed using the C-terminal model of CFTR as the probe and CK2 tetrameric structure (PDB code: 4DGL) as the target protein. 1000 complexes were obtained from both docking algorithms and clusterized using the pairwise RMSD (Root Mean Square Deviation) into 5 largest clusters. The final complex was chosen according to the energy scoring function.

Molecular dynamics (MD) simulations of the final complex (parameterized with AMBER99) were performed with NAMD 2.8 [30] in order to verify their stability over time; in particular a 100 ns of NPT (1 atm, 300 K) MD simulation were performed after an equilibration phase of 1 ns (positional restraints were applied on carbon atoms to equilibrate the solvent around the protein).

Results

Full length mouse CFTR (mCFTR) expressed in *Saccharomyces cerevisiae* was subjected to tryptic digestion followed by TiO_2 phosphopeptide enrichment prior to MS analysis as detailed in the Experimental section. From the MS spectra (Material S1, Figures S1 and S2) it was possible to identify 21 phosphopeptides derived from CFTR (see Table S1) leading to the identification of 22 potential phospho-sites, 12 of which could be unambiguously assigned (see Table 1 and Figure 1). In a few cases it was not possible to assign the phosphate to individual Ser/Thr residue(s) within the phosphopeptides, thus hindering the precise localization of the phospho-residue(s). Almost all the phospho-sites identified in mouse CFTR are conserved in human CFTR (hCFTR), as also indicated in Table 1. The majority belong to the Regulatory (R) domain and display the consensus for PKA. Many of these residues were found to be phosphorylated in CFTR expressed in *Xenopus* oocytes [6] and/or in HEK293S cells [7]. The same residues were also phosphorylated in vitro by PKA [5,6].

Several phospho-residues identified by our analysis however were not previously reported to be phosphorylated in vivo: one or more are located in the N-terminal segment (Ser45 and/or Ser50) and at the end of the R domain (Ser808 and Ser811), most of which are conserved in hCFTR. Also conserved is Ser670 that lies at the end of the R domain, whose in vitro phosphorylation by both PKA [6] and CK2 [9] was reported but which hitherto has never been observed in vivo.

Phospho-Thr1467, close to the C-terminus and sitting in a region which is 100% conserved in hCFTR (Thr1471), attracted our attention for three reasons: i) it conforms to the optimal sequence recognized by CK2, ii) it is adjacent to the PDZ interacting motif which is essential for interaction with AMP-activated protein kinase [31] and with the PDZ domain of the scaffold protein NHERF-1 [32,33] and iii) it is embedded in an acidic cluster that plays a role in regulating CFTR gating and appears to increase the amount of intracellular protein [34].

To corroborate the implication of CK2 as a phosphorylating agent of CFTR Thr1471 a 42 amino acid-long recombinant fragment encompassing the C-terminal segment of hCFTR was generated and tested for its phosphorylation by CK2. As shown in Figure 2A the peptide is readily phosphorylated by CK2 holoenzyme, composed of two catalytic subunits assembled with a dimer of the regulatory β-subunits, but not by the isolated catalytic α-subunit alone, a behavior typical of class III substrates of CK2 [8] that require the structural β-subunit for recognition by the holoenzyme (see below). Phosphorylation occurs at Thr1471 as revealed by MS analysis of the phosphorylated 42 aa fragment (Material S1, Figure S3). Consistently a synthetic peptide encompassing the last 20 residues of hCFTR, in which the only phosphorylatable residue is Thr1471 can also be phosphorylated by CK2 holoenzyme, although much less efficiently (Figure 3). To note in this respect the quite low Km of the 42 amino acids fragment (1.4 μM) a value denoting high affinity and typically displayed by physiological targets rather than by short peptide substrates. These biochemical data are in very good agreement with an in silico study whose outcome is summarized in Figure 4. This predicts that the binding of the hCFTR C-terminal tail to CK2 holoenzyme is critically mediated by interactions between an acidic cluster present in the N-terminal region of the CK2 β-subunit (residues Glu57, Asp59, Glu60, Glu63, Asp64) and a number of basic side chains clustered in CFTR up-stream from Thr1471 (residues Arg1446, Lys1448, His1452, Arg1453, Lys1457). Of note, the latter are present in the recombinant fragment of 42 residues, but not in the shorter 20 amino acids peptide. It is hence quite expected that optimal phosphorylation of Thr1471 will require both the integrity of the CFTR C-terminal segment and the presence of the β-subunit, as highlighted by the experiments in Figure 2A.

We also determined whether other acidophilic Ser/Thr protein kinases, besides CK2, are able to phosphorylate Thr1471. While protein kinases CK1 and the genuine Golgi apparatus casein kinase (GCK), recently identified with FAM20C protein [35,36] proved unable to phosphorylate the 42 amino acid peptide (not shown), a weak phosphorylation was observed using PLK2 and PLK3 recently shown to recognize residues specified by acidic determinants [37,38]. Such a phosphorylation however is negligible as compared to that catalyzed by CK2 under comparable conditions (Figure 2B).

Given the location of Thr1471 close to the PDZ-interacting motif that favors retention of CFTR within the cell by interacting with Golgi associated CAL (Cheng J et al JBC 2002) [39] and its competitor MAST205 [40], we argued that its phosphorylation

MQKSPLEKAS FISKLFFSWT TPILRKGYRH HLELSDIYQA PSADSADHLS EKLEREWDRE QASKKNPQLI HALRRCFFWR FLFYGILLYL GEVTKAVQPV LLGRIIASYD PENKVERSIA IYLGIGLCLL FIVRTLLLHP 140

AIFGLHRIGM QMRTAMFSLI YKKTLKLSSR VLDKISIGQL VSLLSNNLNK FDEGLALAHF IWIAPLQVTL LMGLLWDLLQ FSAFCGLGLL IILVIFQAIL GKMMVKYRDQ RAAKINERLV ITSEIIDNIY SVKAYCWESA 280

MEKMIENLRE VELKMTRKAA YMRFFTSSAF FFSGFFVVFL SVLPYTVING IVLRKIFTTI SFCIVLRMSV TRQFPTAVQI WYDSFGMIRK IQDFLQKQEY KVLEYNLMTT GIIMENVTAF WEEGFGELLE KVQQSNGDRK 420

HSSDENNVSF SHLCLVGNPV LKNINLNIEK GEMLAITGST GSGKTSLLML ILGELEASEG IIKHSGRVSF CSQFSWIMPG TIKENIIFGV SYDEYRYKSV VKACQLQQDI TKFAEQDNTV LGEGGVTLSG GQRARISLAR 560

AVYKDADLYL LDSPFGYLDV FTEEQVFESC VCKLMANKTR ILVTSKMEHL RKADKILILH QGSSYFYGTF SELQSLRPDF SSKLMGYDTF DQFTEERRSS ILTETLRRFS VDDSSAPWSK PKQSFRQTGE VGEKRKNSIL 700

NSFSSVRKIS IVQKTPLCID GESDDLQEKR LSLVPDSEQG EAALPRSNMI ATGPTFPGRR RQSVLDLMTF TPNSGSSNLQ RTRTSIRKIS LVPQISLNEV DVYSRRLSQD STLNITEEIN EEDLKECFLD DVIKIPPVTT 840

WNTYLRYFTL HKGLLLVLIW CVLVFLVEVA ASLFVLWLLK NNPVNSGNNG TKISNSSYVV IITSTSFYYI FYIYVGVADT LLALSLFRGL PLVHTLITAS KILHRKMLHS ILHAPMSTIS KLKAGGILNR FSKDIAILDD 980

FLPLTIFDFI QLVFIVIGAI IVVSALQPYI FLATVPGLVV FILLRAYFLH TAQQLKQLES EGRSPIFTHL VTSLKGLWTL RAFRRQTYFE TLFHKALNLH TANWFMYLAT LRWFQMRIDM IFVLFFIVVT FISILTTGEG 1120

EGTAGIILTL AMNIMSTLQW AVNSSIDTDS LMRSVSRVFK FIDIQTEESM YTQIIKELPR EGSSDVLVIK NEHVKKSDIW PSGGEMVVKD LTVKYMDDGN AVLENISFSI SPGQRVGLLG RTGSGKSTLL SAFLRMLNIK 1260

GDIEIDGVSW NSVTLQEWRK AFGVITQKVF IFSGTFRQNL DPNGKWKDEE IWKVADEVGL KSVIEQFPGQ LNFTLVDGGY VLSHGHKQLM CLARSVLSKA KIILLDEPSA HLDPITYQVI RRVLKQAFAG CTVILCEHRI 1400

EAMLDCQRFL VIEESNVWQY DSLQALLSEK SIFQQAISSS EKMRFFQGRH SSKHKPRTQI TALKEETEEE VQETRL 1476

Figure 1. The sequence of mouse CFTR (accession number P26361) with all the unambiguously assigned phospho-sites identified.

Table 1. Phospho-sites identified in mouse CFTR expressed in yeast.

phospho-residue number						
Mouse CFTR (P26361)	**Human CFTR (P13569)**	**Domain**	**Sequence (mouse)**	**Phosphorylated in vivo**	**Phosphorylated in vitro by PKA**	**Phosphorylated in vitro by CK2**
S670	S670	R-domain	TETLRRF**S**VDDSSAP		[6]	[9]
S698	S700	R-domain	VGEKRKN**S**ILNSFSS	[6,7]	[5,6]	
S710	S712	R-domain	FSSVRKI**S**IVQKTPL	[6,7]	[5,6]	
S723	S728	R-domain	PLCIDGE**S**DDLQEKR			
S732	S737	R-domain	DLQEKRL**S**LVPDSEQ	[6,7]	[5,6]	
S737	S742	R-domain	RLSLVPD**S**EQGEAAL			
S763	S768	R-domain	FPGRRRQ**S**VLDLMTF	[6]	[5,6]	
S790	S795	R-domain	RTSIRKI**S**LVPQISL	[6,7]	[5,6]	
S808	S813	R-domain	DVYSRRL**S**QDSTLNI		[5,6]	
S811	T816	R-domain	SRRLSQD**S**TLNITEE			
S1183	-	-	KELPREG**S**SDVLVIKN			
T1467	T1471	C-term	ITALKEE**T**EEEVQET			This paper

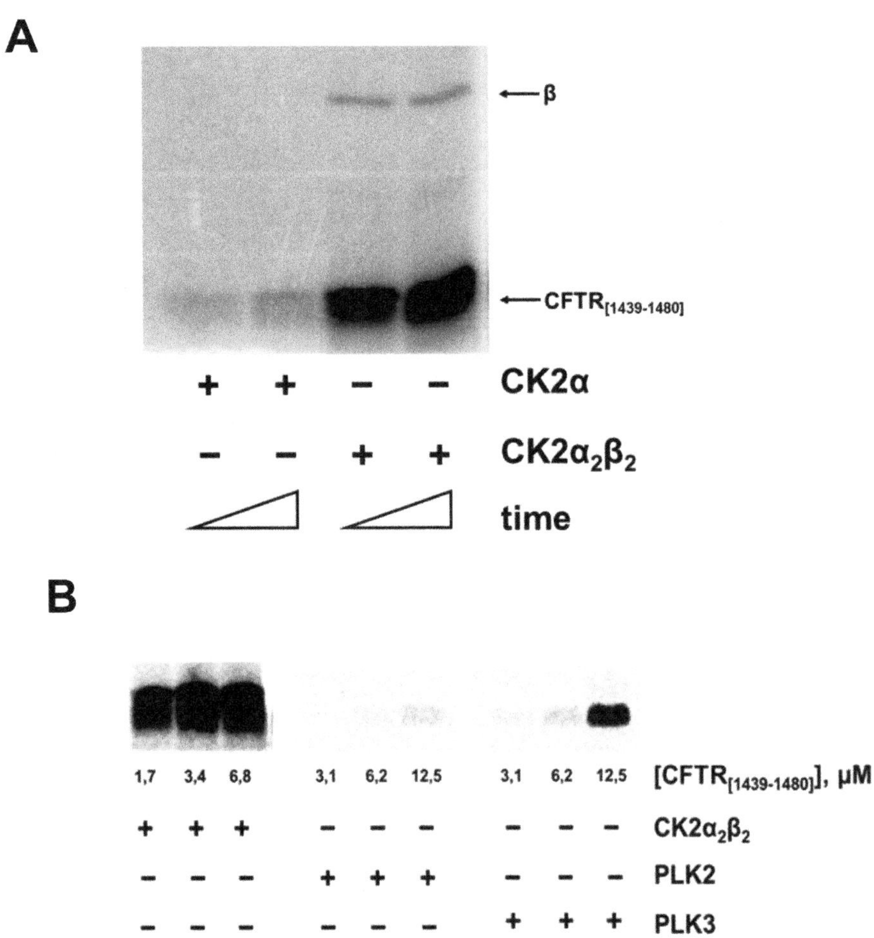

Figure 2. C-terminal domain of human CFTR (hCFTR) is readily phosphorylated by CK2 holoenzyme. (A) 0.4 μg of recombinant C-terminal tail of CFTR encompassing the last 42 residues of the human sequence was phosphorylated as described in Materials and Methods section in the presence of either CK2 α-catalytic subunit or CK2 holoenzyme (CK2 $\alpha_2\beta_2$). The arrows indicate 42-mer CFTR peptide ($CFTR_{[1439-1480]}$) phosphorylation and β-subunit autophosphorylation, respectively. (B) Phosphorylation of the 42-mer peptide in the same conditions as in (A) by PLK2 and PLK3, is very low as compared to that obtained with CK2 holoenzyme.

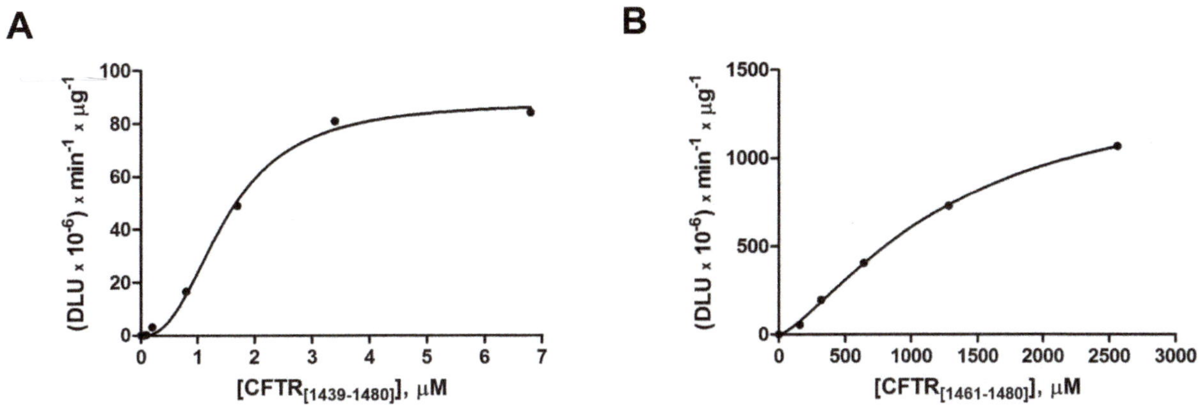

C

Peptide	Kprime (μM)	Khalf (μM)	Vmax ($DLUx10^{-6}$)
$CFTR_{[1439-1480]}$	3.13	1.49	90.5
$CFTR_{[1461-1480]}$	1430	1290	1477

Figure 3. Kinetics of the recombinant 42-mer peptide (A) and of a synthetic peptide reproducing the last 20 residues of the hCFTR C-term ($CFTR_{[1461-1480]}$) (B). Data obtained did not follow the Michaelis-Menten equation but fitted to an allosteric (sigmoidal) model. Kinetic parameters displayed in (C) were calculated by GraphPad Prism program. Kprime (also described as $Khalf^h$, where h is the Hill slope) is related to Km but it is not equal to it unless h = 1.

could affect the stability of such interactions and consequently the turnover of CFTR itself. Accordingly a hCFTR mutant no longer susceptible to such a phosphorylation (T1471A) was overexpressed in BHK cells as compared to hCFTR wild type. This mutant was previously shown to fragment in a manner similar to the wild type protein [14].

To provide evidence about the implication of CK2 in processes affecting the stability of CFTR and/or of its Phe508del mutant

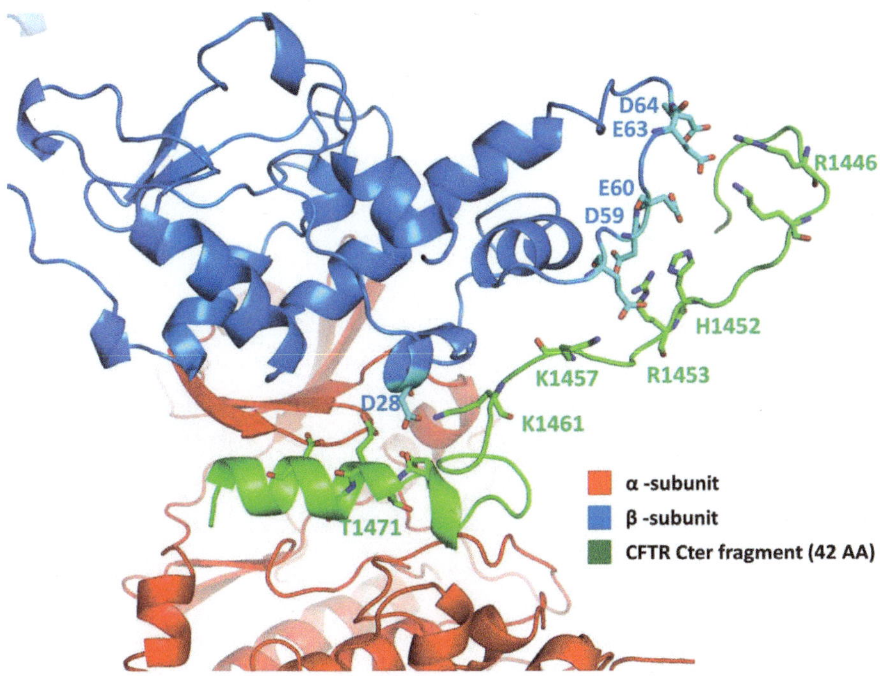

Figure 4. Protein-protein docking complex between the tetrameric structure of CK2 (PDB code: 4DGL) and the hCFTR peptide containing the last 42 amino acids.

which is the commonest cause of Cystic Fibrosis, advantage has been taken of the most selective cell permeable CK2 inhibitor available, CX4945. In order to evaluate the efficacy of CX4945 on the level of de novo synthesized CFTR, protein synthesis was inhibited by two hours cell treatment with cycloheximide (CHX). Blockade was then removed by washing off CHX and newly synthesized CFTR was quantified after 2, 4, and 8 hrs incubation either in the absence or presence of CX4945. As shown in Figure 5 treatment with CHX causes a fast and almost complete disappearance of band B, corresponding to immature core-glycosylated CFTR, whereas the decrease of band C, corresponding to mature fully glycosylated CFTR, where present (Panels A and B), is much slower, its degradation being partially compensated by the conversion of band B into C, a process which is not prevented by CHX. This point is further highlighted by comparing CHX and Brefeldin A, a compound that blocks the escape of core-glycosylated CFTR from Endoplasmic Reticulum to the Golgi, displaying opposite effects on the fate of band B as shown in Figure S4. To note that in the case of Phe508delCFTR (Figure 5, Panel C) band C is undetectable and therefore the disappearance of band B is more dramatic being not accompanied by any generation of band C. Both these behaviors were expectable, considering the inability of Phe508delCFTR to undergo maturation and its special proneness to premature degradation. Once blockade is removed de novo synthesis of CFTR wild type is significantly stimulated if endogenous CK2 is inhibited by cell treatment with CX4945 (Panel A), a similar effect of CX4945 being also detectable if wild type CFTR is replaced by its T1471A mutant, which is no longer susceptible to phosphorylation at Thr1471 (Panel B). By contrast the efficacy of inhibiting CK2 is enhanced if the de novo synthesis of Phe508delCFTR is monitored (Panel C). This point has been corroborated by performing the experiment in triplicate and drawing from it quantification of Phe508delCFTR band B, as reported in Panel D.

It can be concluded from the data of Figure 5 that inhibition of CK2 correlates with increased stability of CFTR, which is particularly evident in the case of Phe508delCFTR. Phosphorylation of Thr1471 seems not to be instrumental to such an effect, at least in a CFTR wild type background.

Discussion

Expression of CFTR in yeast was primarily aimed at producing sufficient amount of the protein for structural studies [20]. We reasoned however that an added value of this method could be that of making possible the detection of phosphorylated residues which may be lost if the expression is performed in animal cells. We were impressed in particular by the observation that, although several potential phosphoacceptor sites for CK2 are present in CFTR and some of these are readily phosphorylated by CK2 in vitro [9] none of these have been ever reported to become phosphorylated in vivo, in sharp contrast with potential PKA phosphoacceptor sites for many of which a good correlation has been shown to exist between in vitro and in vivo phosphorylation [5–7]. Considering on the one hand the stringent quality control experienced by CFTR, leading to the destruction of almost half of the newly synthesized protein, and, on the other, the role frequently played by CK2 mediated phosphorylation to commit proteins to degradation (e.g. [10–12]) a possibility would be that in its physiological environment CFTR phosphorylation by CK2 cannot be detected because it is immediately followed by CFTR fragmentation.

Based on this supposition we wanted to see if CFTR expressed in a foreign environment like yeast, where CK2 (as well as PKA and many other important Ser/Thr protein kinases) are abundant,

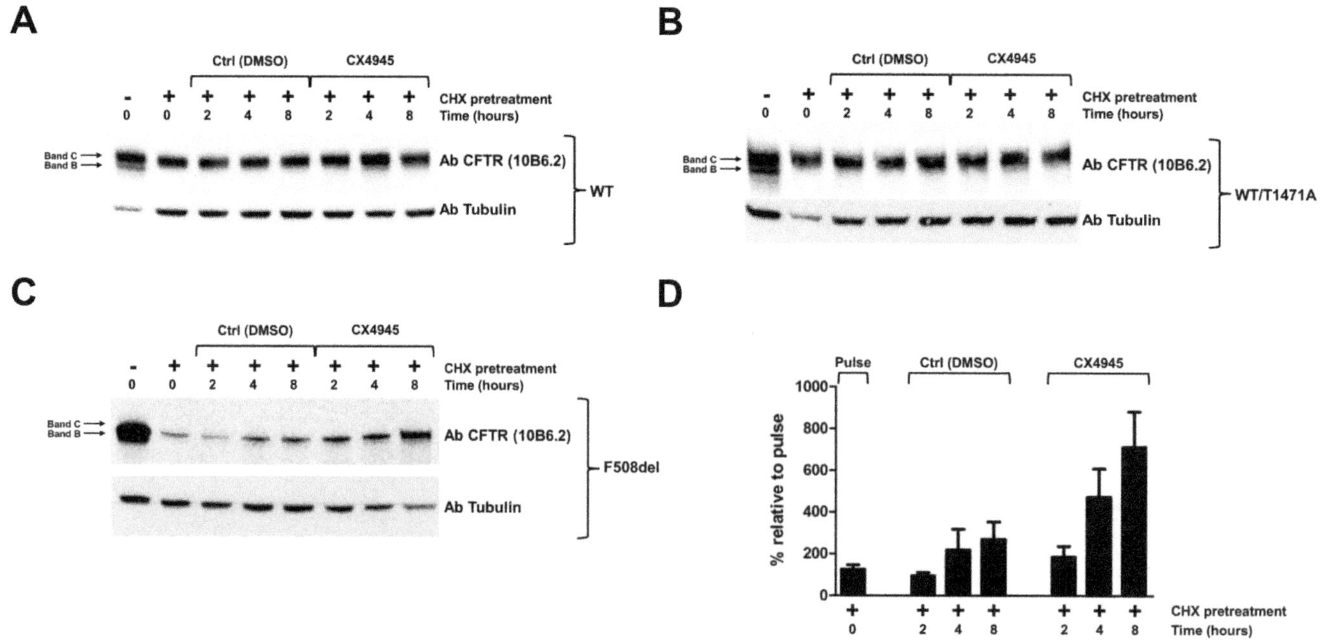

Figure 5. Effect of CK2 inhibition on Phe508delCFTR synthesis. BHK cells expressing human CFTR wild type (A), T1471A (B) and Phe508del (C) were treated for 2 h with CHX to stop protein synthesis (compare first and second lanes). Cells were washed with PBS and incubated for the indicated time with either CK2 specific inhibitor CX4945 or its vehicle. Proteins from each sample were separated by gel electrophoresis and blotted on nitrocellulose membranes (see Materials and Methods). Blots were developed for the indicated antibodies. In (D) quantitation of Phe508delCFTR band B expression after the indicated treatment is shown. The figure is representative of experiments performed in triplicate. Due to the low expression level, Phe508delCFTR signal was obtained by overexposed conditions.

but where the homeostatic mechanism leading to CFTR degradation is lacking and proteolysis has been prevented by a number of devices, CFTR is phosphorylated, and at which residues, by CK2.

The study presented here led to the identification of 22 potential phospho-residues (12 of which were unambiguously assigned) in mouse CFTR expressed in yeast, all of which, with just a few exceptions, are conserved in human CFTR. Seven of these coincide with phospho-residues previously identified in human CFTR expressed in mammalian cells and/or *Xenopus* oocytes [6,7], thus corroborating the reliability of the yeast expression approach. These phospho-residues, with only one exception are located within the Regulatory (R) domain of CFTR and they display the consensus for being phosphorylated by PKA. Accordingly these residues are also included in the list of those which are phosphorylated in vitro by PKA [5,6].

To the best of our knowledge the other phospho-residues detected by us in CFTR were not reported before: one however (Ser670) was previously reported to undergo in vitro phosphorylation by both PKA [6] and CK2 [9] but was never reported as an in vivo site. It actually displays the consensus for both kinases (RR-x-S, and S-x-x-E/D): whether PKA or CK2 or maybe both are committed with its phosphorylation in vivo remains a matter of conjecture.

At variance with the previously identified CFTR phospho-sites, which are almost exclusively located in the R domain and display the PKA consensus, several of the newly identified ones described here are located close to acidic residues which may suggest their targeting by CK2 and some are located outside the R domain: in particular one or two are located in the N-terminal segment, but could not be assigned with good confidence to individual positions (see Material S1), though it is possible that serines 45 and 50 are both phosphorylated. None of these conforms to either the PKA or CK2 consensuses, while another phospho-site outside the R domain, around phospho-Thr1467 (100% conserved in human CFTR as Thr1471) displays an optimal sequence for being phosphorylated by CK2. Two other phospho-residues which, though not fulfilling the canonical CK2 consensus are close to multiple acidic side chains which could nevertheless determine their phosphorylation by CK2 [41] are Ser723, homologous to hCFTR Ser728 which is located inside a PEST region [42] where it could affect degradation, and Ser737 whose human equivalent (Ser742) is surrounded by an identical sequence and was never reported to be phosphorylated before.

Our attention was mainly focused on phospho-Thr1471 (human numbering), since besides being highly conserved in human CFTR and displaying an optimal sequence for CK2 phosphorylation, it is adjacent to the PDZ interacting motif which is essential for interactions with the scaffold protein NHERF-1, anchoring CFTR to the cytoskeleton [32,33] and with the Golgi associated CAL favoring CFTR retention into the cell [39]. To validate the implication of CK2 in the phosphorylation of CFTR Thr1471 we have firstly shown that a recombinant fragment of human CFTR encompassing it is readily phosphorylated, with favorable kinetics, at Thr1471 by CK2 holoenzyme but not by a number of other acidophilic protein kinases. Secondly we demonstrated that a shorter synthetic peptide encompassing Thr1471 is also phosphorylated by CK2 albeit with lower efficiency, as expected from in silico analysis showing that optimal binding of the CFTR C-terminal tail to CK2 holoenzyme requires the integrity of a segment which is present in the recombinant fragment but not in the synthetic peptide. These interactions take place outside the catalytic subunit of CK2, involving elements of the β-subunit: this includes CFTR in the list of class III CK2 substrates [8] whose phosphorylation is readily catalyzed by the holoenzyme (composed by two catalytic and two β-subunits) but not by the isolated catalytic α-subunits.

To get clues about the possible consequences of CFTR phosphorylation by CK2, advantage has been taken of the most selective CK2 inhibitor available to date, CX-4945 [43] and of a CFTR mutant in which Thr1471 has been replaced by alanine (T1471A) [13].

By treating CFTR expressing BHK cells with cycloheximide, in order to block protein synthesis and then restoring protein synthesis by washing away the inhibitor we observed that the re-appearance of CFTR reached higher levels if cells were incubated with the CK2 inhibitor CX-4945, as compared to control. This indicates that inhibition of CK2 results in larger accumulation of newly synthesized CFTR, consistent with diminished degradation of the protein. Interestingly such an effect is definitely more pronounced in the case of Phe508delCFTR with a two to three-fold increment of band B upon treatment with the CK2 inhibitor (see Figure 5D), suggesting that CK2 specifically cooperates with the mechanism which leads to premature degradation of this mutant. On the other hand the effect of CK2 on CFTR stability seems not to be mediated by Thr1471, at least in a wild type background, since a similar effect of CX4945 was observed with CFTR wild type and its T1471A mutant (Figure 5, compare Panels A and B). It cannot be ruled out however that in a Phe508del background phosphorylation of Thr1471 may play a role in committing CFTR to premature degradation. Other possibilities worty of consideration are that the detrimental effect of CK2 on CFTR stability as outlined by the experiments shown in Figure 5 is promoted through the phosphorylation of residues other than Thr1471, with special reference to Ser728 (see Table 1), located near to a PEST region and/or indirectly, by altering the functionality of proteins committed to quality control, folding and degradation of CFTR [44], several of which have been shown to undergo phosphorylation by CK2 [45 and unpublished data].

Although additional work is needed in order to understand the significance of CFTR phosphorylation by CK2, with special reference to the enigmatic role of Thr1471, our data disclose the importance that a number of previously undetected phosphoacceptor sites may have in determining the stability and the functionality of CFTR. Our experiments also highlight the role of the master kinase CK2 in these processes, providing, from an empirical stand point, a new strategy to increase the yield of functional CFTR in CF cells, possibly by combining CK2 manipulation with usage of correctors and/or potentiators.

Supporting Information

Figure S1 The sequence of mouse CFTR with all phosphopeptides identified is reported together with peptide identification confidence and phosphorylation site probability that are indicated with a color code. Table at the end of the file summarizes the data, by listing all identified phosphopeptides with their corresponding phosphorylation sites.

Figure S2 Annotated fragmentation spectra of the phosphopeptides identified in this study.

Figure S3 MS/MS spectrum of phosphopeptide EE-TEEEVQDTR from Human CFTR.

Figure S4 BHK cells expressing WT, WT/T1471A or Phe508delCFTR were exposed to either cycloheximide (CHX, 100 μg/ml) or Brefeldin A (BFA, 200 ng/ml). CFTR synthesis was blocked by two hours treatment with CHX as it is shown by the attenuation/disappearance of the ER-resident band B of CFTR. On the contrary the fully mature band C is not affected by CHX treatment due to its slower turnover. Conversely, when the escape from ER is prevented by using Brefeldin A, band B increases [a]. [a] Glozman R, Okiyoneda T, Mulvihill CM, Rini JM, Barriere H, Lukacs GL (2009) N-glycans are direct determinants of CFTR folding and stability in secretory and endocytic membrane traffic. J. Cell Biol. 184, 847–862.

Table S1 Phosphopeptides identified from mouse CFTR. For each peptide the following details are reported: aminoacid sequence, modifications, phosphorylation site/s, pRS score, pRS probability and pRS site probability, q-value, PEP score and number of missed-cleavages.

Acknowledgments

We thank Carlos M. Farinha (Lisboa, Portugal) for the BHK cell mutants.

Author Contributions

Conceived and designed the experiments: LAP AV AM. Performed the experiments: AV CF GC MAP KT. Analyzed the data: LAP AV GA AM RCF. Contributed reagents/materials/analysis tools: NC KT AA RCF. Wrote the paper: LAP AM AV.

References

1. Pissarra LS, Farinha CM, Xu Z, Schmidt A, Thibodeau PH, et al. (2008) Solubilizing mutations used to crystallize one CFTR domain attenuate the trafficking and channel defects caused by the major cystic fibrosis mutation. Chem Biol 15: 62–69.
2. Schultz BD, Singh AK, Devor DC, Bridges RJ (1999) Pharmacology of CFTR chloride channel activity. Physiol Rev 79: S109–144.
3. Jensen TJ, Loo MA, Pind S, Williams DB, Goldberg AL, et al. (1995) Multiple proteolytic systems, including the proteasome, contribute to CFTR processing. Cell 83: 129–135.
4. Heda GD, Tanwani M, Marino CR (2001) The Delta F508 mutation shortens the biochemical half-life of plasma membrane CFTR in polarized epithelial cells. Am J Physiol, Cell Physiol 280: C166–174.
5. Neville DC, Rozanas CR, Price EM, Gruis DB, Verkman AS, et al. (1997) Evidence for phosphorylation of serine 753 in CFTR using a novel metal-ion affinity resin and matrix-assisted laser desorption mass spectrometry. Protein Sci 6: 2436–2445.
6. Csanády L, Seto-Young D, Chan KW, Cenciarelli C, Angel BB, et al. (2005) Preferential phosphorylation of R-domain Serine 768 dampens activation of CFTR channels by PKA. J Gen Physiol 125: 171–186.
7. McClure M, DeLucas LJ, Wilson L, Ray M, Rowe SM, et al. (2012) Purification of CFTR for mass spectrometry analysis: identification of palmitoylation and other post-translational modifications. Protein Eng Des Sel 25: 7–14.
8. Pinna LA (2002) Protein kinase CK2: a challenge to canons. J Cell Sci 115: 3873–3878.
9. Pagano MA, Arrigoni G, Marin O, Sarno S, Meggio F, et al. (2008) Modulation of protein kinase CK2 activity by fragments of CFTR encompassing F508 may reflect functional links with cystic fibrosis pathogenesis. Biochemistry 47: 7925–7936.
10. Torres J, Pulido R (2001) The tumor suppressor PTEN is phosphorylated by the protein kinase CK2 at its C terminus. Implications for PTEN stability to proteasome-mediated degradation. J Biol Chem 276: 993–998.
11. Kato T Jr, Delhase M, Hoffmann A, Karin M (2003) CK2 Is a C-Terminal IkappaB Kinase Responsible for NF-kappaB Activation during the UV Response. Mol Cell 12: 829–839.
12. Scaglioni PP, Yung TM, Cai LF, Erdjument-Bromage H, Kaufman AJ, et al. (2006) A CK2-dependent mechanism for degradation of the PML tumor suppressor. Cell 126: 269–283.
13. Luz S, Kongsuphol P, Mendes AI, Romeiras F, Sousa M, et al. (2011) Contribution of casein kinase 2 and spleen tyrosine kinase to CFTR trafficking and protein kinase A-induced activity. Mol Cell Biol 31: 4392–4404.
14. Tosoni K, Stobbart M, Cassidy DM, Venerando A, Pagano MA, et al. (2013) CFTR mutations altering CFTR fragmentation. Biochem J 449: 295–305.
15. Kopito RR (1999) Biosynthesis and Degradation of CFTR. Physiol Rev 79: S167–S173.
16. Meacham GC, Lu Z, King S, Sorscher E, Tousson A, et al. (1999) The Hdj-2/Hsc70 chaperone pair facilitates early steps in CFTR biogenesis. EMBO J 18: 1492–1505.
17. Farinha CM, Amaral MD (2005) Most F508del-CFTR is targeted to degradation at an early folding checkpoint and independently of calnexin. Mol Cell Biol 25: 5242–5252.
18. Gentzsch M, Chang X-B, Cui L, Wu Y, Ozols VV, et al. (2004) Endocytic trafficking routes of wild type and DeltaF508 cystic fibrosis transmembrane conductance regulator. Mol Biol Cell 15: 2684–2696.
19. Peters KW, Qi J, Johnson JP, Watkins SC, Frizzell RA (2001) Role of snare proteins in CFTR and ENaC trafficking. Pflugers Arch 443 Suppl 1: S65–69.
20. O'Ryan L, Rimington T, Cant N, Ford RC (2012) Expression and purification of the cystic fibrosis transmembrane conductance regulator protein in Saccharomyces cerevisiae. J Vis Exp. 61: e3860.
21. Venerando A, Marin O, Cozza G, Bustos VH, Sarno S, et al. (2010) Isoform specific phosphorylation of p53 by protein kinase CK1. Cell Mol Life Sci 67: 1105–1118.
22. Salvi M, Trashi E, Marin O, Negro A, Sarno S, et al. (2012) Superiority of PLK-2 as α-synuclein phosphorylating agent relies on unique specificity determinants. Biochem Biophys Res Commun 418: 156–160.
23. Carpino LA, Henklein P, Foxman BM, Abdelmoty I, Costisella B, et al. (2001) The solid state and solution structure of HAPyU. J Org Chem 66: 5245–5247.
24. Thingholm TE, Jørgensen TJD, Jensen ON, Larsen MR (2006) Highly selective enrichment of phosphorylated peptides using titanium dioxide. Nat Protoc 1: 1929–1935.
25. Taus T, Köcher T, Pichler P, Paschke C, Schmidt A, et al. (2011) Universal and Confident Phosphorylation Site Localization Using phosphoRS. J Proteome Res 10: 5354–5362.
26. Schägger H (2006) Tricine-SDS-PAGE. Nat Protoc 1: 16–22.
27. Rohl CA, Strauss CEM, Misura KMS, Baker D (2004) Protein structure prediction using Rosetta. Meth Enzymol 383: 66–93.
28. Kozakov D, Brenke R, Comeau SR, Vajda S (2006) PIPER: an FFT-based protein docking program with pairwise potentials. Proteins 65: 392–406.
29. Chen R, Li L, Weng Z (2003) ZDOCK: An initial-stage protein-docking algorithm. Proteins: Structure, Function, and Bioinformatics 52: 80–87.
30. Phillips JC, Braun R, Wang W, Gumbart J, Tajkhorshid E, et al. (2005) Scalable molecular dynamics with NAMD. J Comput Chem 26: 1781–1802.
31. Hallows KR, Raghuram V, Kemp BE, Witters LA, Foskett JK (2000) Inhibition of cystic fibrosis transmembrane conductance regulator by novel interaction with the metabolic sensor AMP-activated protein kinase. J Clin Invest 105: 1711–1721.
32. Short DB, Trotter KW, Reczek D, Kreda SM, Bretscher A, et al. (1998) An apical PDZ protein anchors the cystic fibrosis transmembrane conductance regulator to the cytoskeleton. J Biol Chem 273: 19797–19801.
33. Wang S, Raab RW, Schatz PJ, Guggino WB, Li M (1998) Peptide binding consensus of the NHE-RF-PDZ1 domain matches the C-terminal sequence of cystic fibrosis transmembrane conductance regulator (CFTR). FEBS Lett 427: 103–108.
34. Ostedgaard LS, Randak C, Rokhlina T, Karp P, Vermeer D, et al. (2003) Effects of C-terminal deletions on cystic fibrosis transmembrane conductance regulator function in cystic fibrosis airway epithelia. Proc Natl Acad Sci USA 100: 1937–1942.
35. Tagliabracci VS, Engel JL, Wen J, Wiley SE, Worby CA, et al. (2012) Secreted kinase phosphorylates extracellular proteins that regulate biomineralization. Science 336: 1150–1153.
36. Lolli G, Cozza G, Mazzorana M, Tibaldi E, Cesaro L, et al. (2012) Inhibition of Protein Kinase CK2 by Flavonoids and Tyrphostins. A Structural Insight. Biochemistry 51: 6097–6107.
37. Salvi M, Trashi E, Cozza G, Franchin C, Arrigoni G, et al. (2012) Investigation on PLK2 and PLK3 substrate recognition. Biochim Biophys Acta 1824: 1366–1373.
38. Johnson EF, Stewart KD, Woods KW, Giranda VL, Luo Y (2007) Pharmacological and Functional Comparison of the Polo-like Kinase Family: Insight into Inhibitor and Substrate Specificity. Biochemistry 46: 9551–9563.
39. Cheng J, Moyer BD, Milewski M, Loffing J, Ikeda M, et al. (2002) A Golgi-associated PDZ domain protein modulates cystic fibrosis transmembrane regulator plasma membrane expression. J Biol Chem 277: 3520–3529.
40. Ren A, Zhang W, Yarlagadda S, Sinha C, Arora K, et al. (2013) MAST205 Competes with Cystic Fibrosis Transmembrane Conductance Regulator (CFTR)-associated Ligand for Binding to CFTR to Regulate CFTR-mediated Fluid Transport. J Biol Chem 288: 12325–12334.
41. Meggio F, Pinna LA (2003) One-thousand-and-one substrates of protein kinase CK2? FASEB J 17: 349–368.

42. Chen EY, Clarke DM (2002) The PEST sequence does not contribute to the stability of the cystic fibrosis transmembrane conductance regulator. BMC Biochem 3: 29.
43. Battistutta R, Cozza G, Pierre F, Papinutto E, Lolli G, et al. (2011) Unprecedented selectivity and structural determinants of a new class of protein kinase CK2 inhibitors in clinical trials for the treatment of cancer. Biochemistry 50: 8478–8488.
44. Wang X, Venable J, LaPointe P, Hutt DM, Koulov AV, et al. (2006) Hsp90 cochaperone Aha1 downregulation rescues misfolding of CFTR in cystic fibrosis. Cell 127: 803–815.
45. Venerando A, Pagano MA, Cozza G, Arrigoni G, Mehta A, Pinna LA (2011) Subverted signalling by protein kinase CK2 in ΔF508 CFTR expressing cells. Poster presentation. FEBS Journal, 278: 387.

3

A Synthetic Peptide with the Putative Iron Binding Motif of Amyloid Precursor Protein (APP) Does Not Catalytically Oxidize Iron

Kourosh Honarmand Ebrahimi, Peter-Leon Hagedoorn, Wilfred R. Hagen*

Department of Biotechnology, Delft University of Technology, Delft, The Netherlands

Abstract

The β-amyloid precursor protein (APP), which is a key player in Alzheimer's disease, was recently reported to possess an Fe(II) binding site within its E2 domain which exhibits ferroxidase activity [Duce et al. 2010, Cell 142: 857]. The putative ligands of this site were compared to those in the ferroxidase site of ferritin. The activity was indirectly measured using transferrin, which scavenges the Fe(III) product of the reaction. A 22-residue synthetic peptide, named FD1, with the putative ferroxidase site of APP, and the E2 domain of APP were each reported to exhibit 40% of the ferroxidase activity of APP and of ceruloplasmin. It was also claimed that the ferroxidase activity of APP is inhibited by Zn(II) just as in ferritin. We measured the ferroxidase activity indirectly (i) by the incorporation of the Fe(III) product of the ferroxidase reaction into transferrin and directly (ii) by monitoring consumption of the substrate molecular oxygen. The results with the FD1 peptide were compared to the established ferroxidase activities of human H-chain ferritin and of ceruloplasmin. For FD1 we observed no activity above the background of non-enzymatic Fe(II) oxidation by molecular oxygen. Zn(II) binds to transferrin and diminishes its Fe(III) incorporation capacity and rate but it does not specifically bind to a putative ferroxidase site of FD1. Based on these results, and on comparison of the putative ligands of the ferroxidase site of APP with those of ferritin, we conclude that the previously reported results for ferroxidase activity of FD1 and – by implication – of APP should be re-evaluated.

Editor: Maria A. Deli, Biological Research Centre of the Hungarian Academy of Sciences, Hungary

Funding: This work was supported by a research grant from the Dutch National Research School Combination - Catalysis Controlled by Chemical Design (NRSC-Catalysis). The funder had no role in study design, data collection and analysis, decision to publish, or preparation of the manuscript.

Competing Interests: The authors have declared that no competing interests exist.

* E-mail: w.r.hagen@tudelft.nl

Introduction

Human β-amyloid precursor protein (APP) is generally thought to play a key role in Alzheimer's disease as the source of plaque-forming β-amyloid peptides (Aβ) [1,2]. APP is a transmembrane protein made up of a large, multidomain extracellular extension, a small, single-pass transmembrane part, and a small intracellular extension (Figure 1A). Alternative exon splicing of the APP gene affords eight different mRNAs that translate into eight APP isoforms whose length range from 365 to 770 amino acid residues [3], with APP695 as the dominant form in the brain. Sequential processing of APP by the proteolytic enzymes β-secretase and γ-secretase liberates Aβ: a sequence of typically 40 or 42 residues originally located partially in the membrane and for the remainder extracellularly. It is not known whether Aβ in Alzheimer's disease is a causative agent or a resulting product. The physiological role(s) of APP in healthy cells has also not been firmly established.

Metal ions, notably iron, copper, and zinc have been implicated in the normal physiological functioning of APP [4,5–6], in the regulation of APP expression [7,8], in the processing of APP affording Aβ [8,9], and in Aβ-plaque related pathogeny [5,10]. The complete APP has not been crystallized, but the 3D structure of two extracellular domains, E1 and E2, have been determined. Two subdomains of the N-terminal E1 domain, the 'growth factor like domain' GFD and the 'copper binding domain' CuBD, have separately been crystallized [11,12,13]. In none of these four crystal structures any metal has been found. When the CuBD crystal is soaked in 100 mM $CuCl_2$ a single site is found with Cu^{2+} square-pyramidal coordination by His147, His151, Tyr168 and two waters [12]. A physiological role, if any, of this copper-binding site has yet to be established. Other putative copper (and zinc) binding sites have been proposed in E1 [5], and physiological studies have implicated copper in APP trafficking [4].

An interaction of iron and the Aβ domain of APP has been reported [14] but a direct interaction with the E2 domain of APP has not been explicitly considered until the recent proposal that APP exhibits an Fe(II) binding site within the E2 domain (Fig. 1B–C). It was proposed that this site has homology with the ferroxidase center in the iron-storing ferritins (Fig. 1D), and it was claimed that the site possesses catalytic iron oxidation activity (ferroxidase activity) [15]. Steady-state kinetics of ferroxidase activity were reported for APP695, for the isolated E2 domain, for a synthetic 22 amino acid peptide named FD1 (duplicating a stretch in the E2 domain with the putative ferroxidase site), and for the human copper protein ceruloplasmin (as a positive control) [15]. Ferroxidase activity of APP was proposed to be functionally relevant in interaction with the iron exporter ferroportin to prevent intracellular iron accumulation and associated oxidative stress. Like in ferritins [16–20], the APP ferroxidase activity was reported to be inhibited by binding of zinc to the APP, its E2 domain, or the FD1 peptide [15]. The present study, which elaborates on the proposed ferroxidase activity of the FD1 peptide [15], was incited by five distinct aspects of the previous work: (i) the ferroxidase activity was measured with what appears to be an unvalidated procedure involving transfer of Fe(III) to the iron transport protein transferrin, and reported activities for ceruloplasmin, APP, the E2 domain of APP, and the FD1 peptide were claimed to be orders of magnitude higher than ferritin ferroxidase; (ii) reported Michaelis-Menten plots for ferroxidase activity of each of the four studied proteins had shapes that appeared to be incompatible with the hyperbolic shape of Michelis-Menten kinetics, and V_{max} values that were read as saturation values from the data in figures differed by circa 40% from the values obtained from fitting the Michaelis-Menten equation to the non-hyperbolic data sets; (iii) the k_{cat} value reported in the figures for the FD1 peptide was identical to the values reported for APP and ceruloplasmin, and it was not 40% of the k_{cat} values reported for APP and ceruloplasmin as claimed in the text; (iv) reported ferroxidase activities of the E2 domain for two experiments performed under the same conditions differed by circa 50% while error bars of <1% were claimed(v) the proposed homology between APP and human H-chain ferritin appeared to be based on a mistake of assignment of iron coordinating amino acids [15]. We have therefore re-evaluated the transferrin-based ferroxidase assay by comparison with direct measurements of O_2 consumption, and we have compared ferroxidase activity of the putative ferroxidase peptide FD1 from APP with those of ceruloplasmin and human H-chain ferritin.

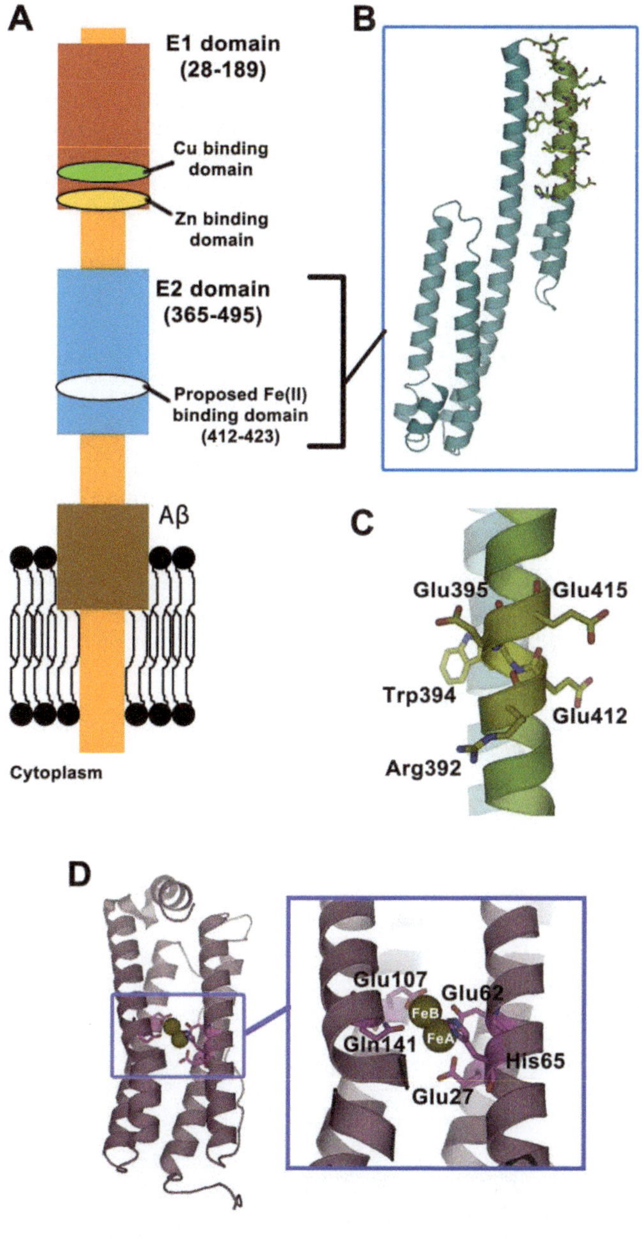

Figure 1. Topology of β amyloid precursor protein (APP) and comparison of its putative ferroxidase site with the ferroxidase site of human H-chain ferritin. (A) A schematic representation of APP and its metal binding domains (APP770 isoform). (B) Structure of the E2 domain of APP751 (PDB 1RW6) [13] and (C) the putative ligands of the ferroxidase site (APP770 numbering). The residues in green show the section of the E2 domain that contains the putative ferroxidase site of APP and the residues that were used to synthesize the FD1 peptide. (D) The diiron catalytic center, the ferroxidase center, of human H-chain ferritin (PDB 1FHA).

Results

Significance of non-enzymatic oxidation of Fe(II) in the presence of transferrin

Osaki [21,22] used the Fe(III) binding protein transferrin to study the ferroxidase activity, i.e. enzymatic oxidation of Fe(II) by molecular oxygen, of ceruloplasmin. In the absence of ceruloplasmin and for an Fe(II) concentration of 100 μM, the overall initial rate of non-enzymatic oxidation of Fe(II) by molecular oxygen and subsequent incorporation of the Fe(III) product into transferrin at pH 7.3 was significant, namely, 24 μM Fe(III) formed per minute [22]. More recently, Duce et al used transferrin in attempts to test the ferroxidase activity of human β-amyloid precursor protein (APP), of the E2 domain of APP, and of a 22-residue sub-domain of E2 in the form of the synthetic peptide, FD1 [15]. In that study the initial rate of non-enzymatic oxidation of Fe(II) in the presence of transferrin was reported to be zero for an Fe(II) concentration of 0–200 μM and at pH 7.2. This observation contradicts the original results reported by Osaki. Therefore, we checked the significance of non-enzymatic oxidation of Fe(II) and incorporation of the Fe(III) product into transferrin at pH 7.0 in the absence of any other proteins and at various concentrations of Fe(II) (Fig. 2). Kinetics with apparent positive cooperativity was observed and the Hill equation was used to fit the data with a Hill constant of 1.9 and a V_{max} of 21 μM/min. The observed V_{max} is comparable to all previous reported values by others [21–23], and is within experimental error comparable to the values reported by Duce et al [15] for the putative ferroxidase activity of the APP protein, the E2 domain, and the FD1 peptide. The rate of non-enzymatic oxidation of Fe(II) and formation of diferric-transferrin dramatically increased with increasing the pH from 6.2 to 8.2. The initial rate of Fe(III) formation (μM per minute) was 0.08±0.01 at pH 6.2, 2.6±0.4 at pH 7.2, and 80.2±8.5 at pH 8.2 for an initial Fe(II) concentration of 40 μM. Thus, the initial rate of non-enzymatic oxidation of Fe(II) in the presence of transferrin depends on the concentration of Fe(II) and on the pH.

The use of transferrin to study the ferroxidase activity of H-chain ferritin and ceruloplasmin

Before using the transferrin assay for measurement of the ferroxidase activity of the FD1 peptide with the putative ferroxidase site of APP, we tested this assay by measuring the established ferroxidase activities of human H-chain ferritin

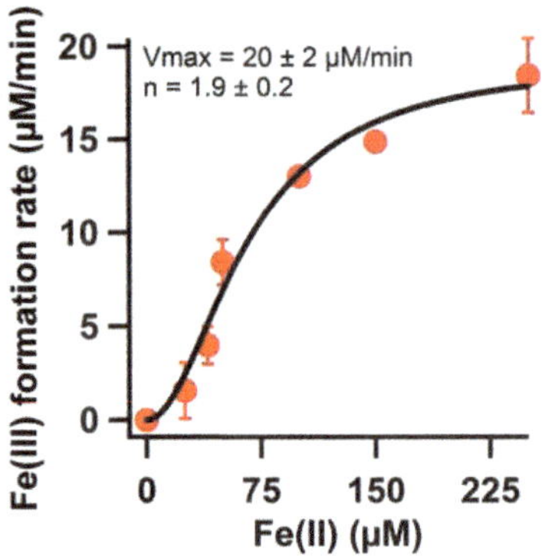

Figure 2. Effect of concentration of Fe(II) on non-enzymatic oxidation of Fe(II) and incorporation of the Fe(III) product into transferrin. Progress curves were recorded at 460 nm for formation of diferric-transferrin and the initial rates were plotted versus concentration of Fe(II). The error bars represent standard deviation of three independent measurements. Concentration of apo-transferrin was 100 µM. Buffer was 100 mM Hepes, 100 mM NaCl, pH 7.2. Measurements were performed at 37°C.

(HuHF) and ceruloplasmin. Bakker and Boyer measured the iron oxidation activity of horse spleen ferritin (HoSF) at pH 6.2 in the presence of human transferrin [23]. They reported the activity as formation of diferric-transferrin (i.e. sequestering of Fe(III) by transferrin from ferritin) under the – then unproven – assumption that the ferroxidase activity of ferritin was the rate-limiting step in the overall assay, and under the assumption that all the 24 subunits of HoSF were catalytically active. Subsequent studies have shown that HoSF consists of two distinct subunits, 80% L 'light' and 20% H 'heavy' subunits [24]; and the catalytic center is only present in the H subunit [25,26]. Measurements of kinetics of iron oxidation by HuHF have shown that oxidation of Fe(II) occurs within seconds [27]. In this process two Fe(II) bind simultaneously to the catalytic site of ferritin, which is commonly known as the ferroxidase site (FC), and they are oxidized by molecular oxygen at a high rate. This reaction has typically been measured by following the UV-visible absorbance spectrum of ferritin-bound Fe(III) between 300 nm and 500 nm [28]. Using this procedure a $k_{cat} = 3.3\ sec^{-1}$ has been reported for HuHF [26] which is comparable to the values reported by Duce et al for ceruloplasmin and not orders of magnitude lower as claimed by them [15]. We determined whether the oxidation rate of Fe(II) by the FC of HuHF in one turnover, i.e. oxidation of 2 Fe(II) per FC, is comparable to the rate of diferric-transferrin complex formation. Absorbance at 650 nm was recorded for the formation and decay of a blue intermediate, the proposed µ-peroxodiferric complex, which is an indicator of the oxidation of Fe(II) in the ferroxidase site of HuHF [29,30], and at 460 nm for the formation of diferric-transferrin complex. An amount of 2 Fe(II) per FC was added to apo-HuHF in the presence of transferrin and, using a stopped-flow spectrometer, the spectra were recorded (Fig. 3A). Within five seconds the blue intermediate with maximum absorbance at 650 nm forms and decays. This rate of formation and decay for the blue intermediate is comparable to the previously reported values for HuHF [29]. The rise in absorbance at 460 nm in the first five seconds is due to the broad absorbance spectrum of ferritin-bound Fe(III) from the UV region to circa 500 nm [28]. Subsequently, the absorbance at 460 nm slowly increases because transferrin scavenges ferritin-bound Fe(III). Thus the rate limiting step is the transfer of the Fe(III) from ferritin to transferrin, and – contrary to the proposal of Bakker and Boyer [23] – the rate of diferric-transferrin complex formation under these conditions is not equal to the rate of the ferroxidase activity. Therefore, the transferrin assay is not a suitable method to quantitatively measure ferroxidase kinetics of the iron-storage protein ferritin.

Then, we measured the ferroxidase activity of ceruloplasmin as a function of Fe(II) concentrations using the transferrin assay (Fig. 3B). The specific activity of ceruloplasmin was calculated from the initial rate of diferric-transferrin formation which was obtained from the progress curves at 460 nm using an extinction coefficient of $4.56\ mM^{-1}\ cm^{-1}$ for the diferric-transferrin complex. The activity was measured for Fe(II) concentrations between 0 and 160 µM. Addition of ceruloplasmin significantly increased the rate, and at an Fe(II) concentration of 40 µM a maximum rate of 240 µM Fe(III) formed per minute per µM of ceruloplasmin was observed (Fig. 3B). The data were fitted to the Michaelis-Menten equation affording $K_m = 10 \pm 3$ µM and $V_{max} = 3.1 \pm 0.3$ U/mg

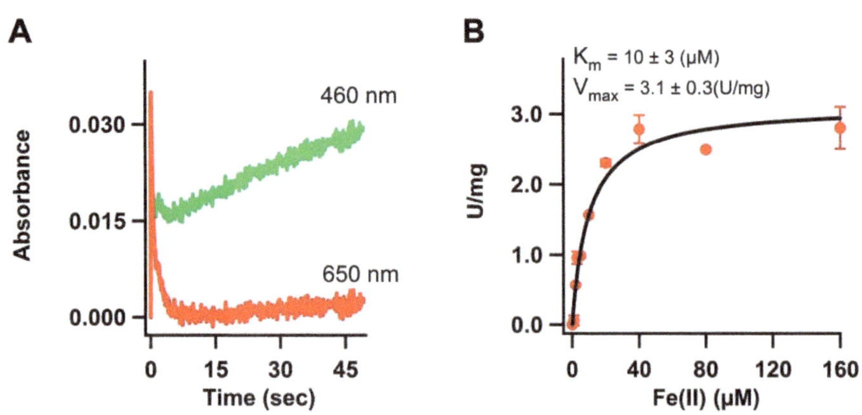

Figure 3. Use of transferrin to measure the ferroxidase activity of ferritin and ceruloplasmin. (A) Effect of transferrin on Fe(II) oxidation by human H-chain ferritin (HuHF) measured at 650 nm for formation and decay of the blue intermediate, and at 460 nm for formation of diferric transferrin complex. Concentration of apo-HuHF was 1.7 µM (24-mer) and that of transferrin was 70 µM. Two Fe(II) per ferroxidase center, i.e. 48 Fe(II) per 24-mer which is equal to a final concentration of 81.6 µM Fe(II), were added and measurements were done at 34°C in 400 mM Mops buffer, 100 mM NaCl, pH 7.0. (B) Ferroxidase activity of ceruloplasmin was measured using the transferrin assay at 460 nm. Progress curves were recorded and initial rate of Fe(III) formation was plotted versus concentration of Fe(II). Concentration of ceruloplasmin was 0.1 µM. Measurements were performed at 37°C in triplicate. Buffer was 100 mM Mops, 100 mM NaCl, pH 7.1.

(μmol of Fe(II) oxidized per minute per mg of ceruloplasmin). It can be concluded that unlike HuHF, ceruloplasmin does not store iron and the resulting Fe(III) product of the ferroxidase reaction is rapidly scavenged by transferrin and therefore the transferrin assay can be used to measure the activity of ceruloplasmin.

The FD1 peptide does not have ferroxidase activity in the transferrin-coupled assay

The reported ferroxidase activity of APP was tested using the 22-residue synthetic peptide FD1, which carries the putative ferroxidase active site of APP. Duce et al have reported that this peptide possesses 40% of the activity of APP or ceruloplasmin [15]. The initial rate of diferric-transferrin complex formation due to oxidation of Fe(II) by FD1 was measured and was compared (i) to that of the non-enzymatic oxidation in the presence of transferrin, (ii) to that of BSA which is known not to oxidize Fe(II), (iii) to that of HuHF which should decrease the rate because it stores the Fe(III), and finally (iv) to that of ceruloplasmin (Fig. 4A). At an Fe(II) concentration of circa 40 μM the rate of the non-enzymatic reaction is circa 3 μM Fe(III) formed per minute (Fig. 4A and see Fig. 2). Ceruloplasmin shows a rate that is 8-fold above that of the non-enzymatic reaction. This rate of 24 μM Fe(III) per minute (i.e. a specific activity of 3.1 U/mg) is comparable to the originally reported value for this protein [22]. With the FD1 peptide, however, the rate of diferric-transferrin formation is within experimental error equal to that of non-enzymatic oxidation of Fe(II) by molecular oxygen either in the presence or in the absence of BSA. To test if the pH can have an effect on the rate of Fe(II) oxidation by FD1, the ferroxidase activity of FD1 was examined at three different pH values, 6.2, 7.2, and 8.2 (Fig. 4B). The results were within experimental error identical to those of the non-enzymatic oxidation of Fe(II). The presence of FD1 did not change the rate of iron oxidation and diferric-transferrin complex formation at any of the tested pH values.

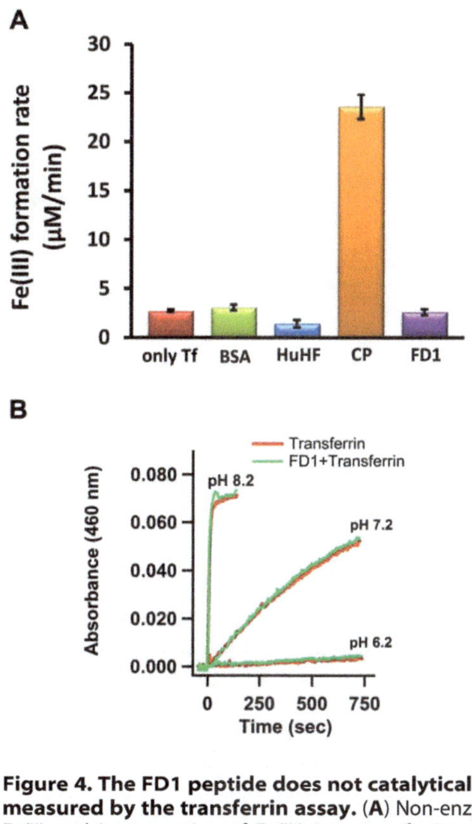

Figure 4. The FD1 peptide does not catalytically oxidize iron as measured by the transferrin assay. (**A**) Non-enzymatic oxidation of Fe(II) and incorporation of Fe(III) into transferrin was compared with that of BSA, HuHF, ceruloplasmin (CP) and FD1. 52 μM of apo-transferrin was mixed with 2 μM BSA or 0.18 μM HuHF (24-mer) or 2 μM FD1 peptide, or 0.1 μM ceruloplasmin, and then an aliquot of 5 μl anaerobic solution of ferrous sulphate was added. Measurements were done in 100 mM Hepes, 100 mM NaCl, pH 7.2. Final concentration of Fe(II) was 40 μM. (B) Effect of pH on non-enzymatic oxidation of Fe(II) in the presence of transferrin was checked at three different pH values in the presence or absence of FD1 peptide (2 μM): pH 6.2, 100 mM Mes, 100 mM NaCl; pH 7.2, 100 mM Hepes, 100 mM NaCl, and pH 8.2, 100 mM Tris, 100 mM NaCl. Concentration of apo-transferrin was 52 μM. Final concentration of Fe(II) was 40 μM. Measurements were performed at 37°C in triplicate with two different batches of transferrin.

The FD1 peptide does not have oxygen-consumption activity in the presence of Fe(II)

A second procedure that has been used to follow the ferroxidase activity of proteins such as ceruloplasmin and ferritin, and one that provides information about the stoichiometry of the reaction, is to amperometrically measure the consumption of the substrate molecular oxygen [31,32,33]. Therefore oxygen consumption by FD1 was measured and was compared with that of ceruloplasmin and HuHF as positive controls. BSA does not consume oxygen and was used as a negative control. It has been shown that when two Fe(II) per subunit are added to HuHF, one molecular oxygen is consumed at a high rate for the oxidation of two Fe(II) in the ferroxidase center [26,34]. When the amount of Fe(II) added per subunit of ferritin is increased, the Fe(II)/O_2 ratio increases, and at Fe(II) per subunit ratios greater than 15 circa 4 Fe(II) are oxidized per molecular oxygen [35]. When we added 10 Fe(II) per subunit to HuHF (Fig. 5A) a stoichiometry of 2.4 Fe(II) oxidized per one O_2 consumed was found, which is in agreement with previous studies [26,36]. At Fe(II) concentration of 200 μM the rate of oxygen consumption of ceruloplasmin was comparable to that of HuHF (Fig. 5B). As previously reported [22] a stoichiometry of one O_2 consumed per 4 Fe(II) oxidized was obtained for ceruloplasmin, which suggests that molecular oxygen is reduced

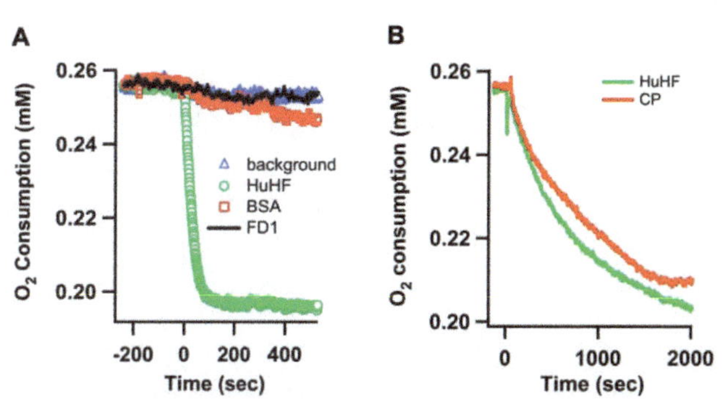

Figure 5. Oxygen consumption as a monitor of Fe(II) oxidation. (A) Consumption of oxygen as a result of oxidation of Fe(II) was measured for HuHF, FD1, and BSA and was compared with that of the non-enzymatic oxidation of Fe(II). Concentrations of HuHF (subunit), FD1, and BSA were 14 μM. Buffer was 100 mM Hepes, 100 mM NaCl, pH 7.2. Final concentration of Fe(II) was 140 μM. Each measurement was performed in triplicate. (B) Comparison of oxygen consumption in the ferroxidase reaction of HuHF and of ceruloplasmin. Final concentration of HuHF (subunit) and ceruloplasmin (CP) was 1.1 μM. Concentration of Fe(II) was 200 μM. Measurements were performed in 100 mM Mops buffer, 100 mM NaCl, pH 7.1. For all measurements the temperature was 22°C.

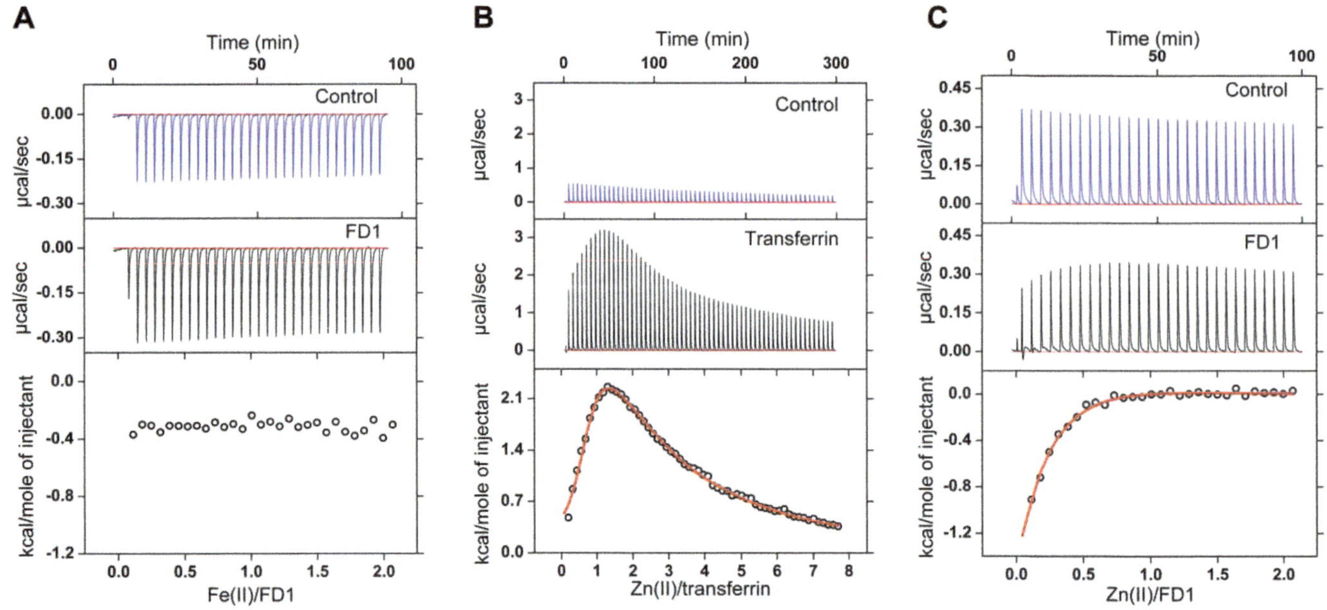

Figure 6. Fe(II) and Zn(II) binding to transferrin and to FD1 measured by isothermal titration calorimetry. (A) Anaerobic Fe(II) titration of the FD1 peptide. Concentration of FD1 in the cell was 70 µM and that of Fe(II) in the syringe was 2.23 mM. The Fe(II) solution was prepared in 100 mM Mops, 100 mM NaCl pH 5.8, and FD1 was in 100 mM Mops, 100 mM NaCl, pH 7.0. The data for the integrated heat of binding were corrected for the heat of dilution due to titration of Fe(II) to buffer (control). Measurements were performed at 25°C. (B) Zn(II) binding to transferrin and (C) to the FD1 peptide. For experiments with transferrin concentration of protein in the cell was 126 µM and the concentration of Zn(II) in the syringe was 12 mM. For experiments with the FD1 peptide, concentration of peptide in the cell was 70 µM and that of Zn(II) in the syringe was 3.35 mM. Transferrin, FD1 peptide, and Zn(II) were prepared in 100 mM Mops, 100 mM NaCl, pH 7.0. Measurements were performed at 25°C. The data for the integrated heat of binding were corrected for the heat of dilution due to titration of Zn(II) to buffer (control).

to water by ceruloplasmin. Comparison of oxygen consumption of FD1 with that in the non-enzymatic reaction (no peptide or protein present) and that of HuHF, ceruloplasmin and BSA (Fig. 5) provides further independent evidence that FD1 does not have ferroxidase activity. The rate of oxygen consumption in the presence of FD1 is not different from that in the presence of BSA and is equal to non-enzymatic oxidation of Fe(II). From the combined results of iron oxidation and oxygen consumption measurements we conclude that FD1 does not have ferroxidase activity.

Fe(II) and Zn(II) binding to transferrin and to the FD1 peptide

Because FD1 does not show any iron oxidation activity, we checked if it binds Fe(II) at all. We measured anaerobic binding of Fe(II) to the FD1 peptide using isothermal titration calorimetry (ITC) (Fig. 6A). No Fe(II) binding to the FD1 peptide was observed within the detection limit of ITC.

Then, using ITC binding of Zn(II) to transferrin and to the FD1 peptide was measured (Fig. 6B–C). Statistical analysis of the integrated heat of binding data was performed to obtain the thermodynamic parameters. For transferrin a model of two independent binding sites was required to obtain a fit to the data of integrated heat (Fig. 6B). A binding with $K_{a1} = (6.0 \pm 1.8) \cdot 10^4$ (M^{-1}), (Table 1) and with a stoichiometry of circa one Zn(II) per transferrin was found. A second site with a stoichiometry of circa one and with a lower affinity $K_{a2} = (1.3 \pm 0.6) \cdot 10^3$ (M^{-1}) (Table 1) was also observed. Zn(II) has been shown to bind to human serum transferrin as well [37]. Previous measurements of Zn(II) binding to transferrin using difference UV monitoring in the presence of competing ligands have suggested two strong binding sites for Zn(II) [38]. Unlike the previous study we measured Zn(II) binding to transferrin directly in the absence of any competing ligands. For analysis of Zn(II) binding to FD1 peptide (Fig. 6C) a model of a single binding site was sufficient. A binding with $K_a = (4.1 \pm 1.0) \cdot 10^4$ M^{-1} and stoichiometry of circa 0.5 Zn(II) per FD1 peptide was obtained. This stoichiometry suggests that two FD1 molecules create one coordinating site for a single Zn(II) (Table 1). The observation that Zn(II) apparently does not bind to the putative ferroxidase site of a single peptide is consistent with the results of a recent X-ray crystal structure determination of monomeric E2 domain of APP in the presence of Zn(II) or Cd(II)

Table 1. Thermodynamic parameters for Zn(II) binding to transferrin and FD1 peptide.

	Transferrin	FD1
N_1	0.8±0.06	0.5±0.2
K_{a1} (M^{-1})	$(6.0 \pm 1.8) \cdot 10^4$	$(5.1 \pm 1.9) \cdot 10^4$
ΔH_1 (kJ/mol)	−1.2±0.2	−2.5±0.6
ΔG_1 (kJ/mol)	−27.7±0.5	−26.6±1.0
N_2	1±0.5	----
K_{a2} (M^{-1})	$(1.3 \pm 0.6) \cdot 10^3$	----
ΔH_2 (kJ/mol)	45±16	----
ΔG_2 (kJ/mol)	−17.4±1.2	----

Binding of Zn(II) to transferrin and FD1 peptide was measured by isothermal titration calorimetry. N is the stoichiometry of Zn(II) binding per protein. Measurements were performed at 25°C. A two independent binding site model or a single binding site model was used to fit the data points for integrated heat of binding. Standard deviations were obtained from triplicate experiments.

in which no metal binding to the ligands of the putative ferroxidase site has been observed [39].

Zn(II) bound to transferrin inhibits binding of the Fe(III) product of Fe(II) oxidation

Because FD1 does not bind Fe(II) and does not catalytically oxidize Fe(II), we surmise that the previously observed inhibitory effect of Zn(II) on the Fe(II) oxidation rate in the presence of transferrin [15] could be due to binding of Zn(II) to transferrin and thus to a decrease in the rate of diferric-transferrin formation. Therefore, we measured the effect of Zn(II) concentration on the rate of non-enzymatic oxidation of Fe(II) (200 μM) and incorporation of the Fe(III) product into transferrin (Fig. 7A). The overall rate decreased considerably in the presence of Zn(II) presumably because Zn(II) binds to the Fe(III) binding sites of transferrin. The addition of one Zn(II) per transferrin was enough to decrease the rate by approximately 80%. The reason for observation of this residual rate in the presence of Zn(II) is possibly the fact that Fe(III) binds transferrin orders of magnitude stronger than Zn(II). Thus Fe(III) can displace Zn(II) from its binding sites. If Zn(II) inhibition of putative ferroxidase activity of APP would be specific, i.e. because of binding of Zn(II) to the APP and not to transferrin, then Zn(II) should not inhibit ferroxidase activity of ceruloplasmin

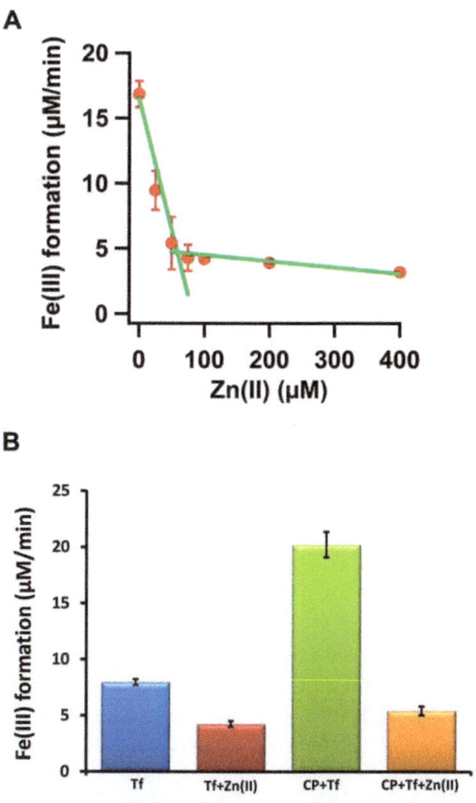

Figure 7. Inhibitory effect of Zn(II) on diferric-transferrin complex formation. (A) Initial rates of diferric-transferrin formation at 460 nm are plotted versus concentration of Zn(II). Concentration of apo-transferrin was 80 μM and that of Fe(II) was 200 μM. (B) Comparison of the inhibitory effect of Zn(II) on the diferric-transferrin complex formation for the non-enzymatic oxidation of Fe(II) by molecular oxygen in the presence of transferrin (Tf 70 μM) and for the ferroxidase activity of ceruloplasmin (CP 0.08 μM) measured with the transferrin assay. Final concentration of Fe(II) was 60 μM and that of Zn(II) was 70 μM. Measurements were done in 100 mM Hepes buffer, 100 mM NaCl, pH 7.2 in triplicate at 37°C.

as indeed claimed by Duce et al. [15]. However, we find with the transferrin assay (Fig. 7B) that in the presence of Zn(II) the ferroxidase activity of ceruloplasmin is diminished and is comparable to the residual rate of non-enzymatic oxidation of Fe(II) in the presence of at least one Zn(II) per transferrin. This decrease in rate is apparently due to binding of Zn(II) to the transferrin and inhibition of Fe(III)-transferrin complex formation.

Discussion

In human H-chain ferritin the ferroxidase reaction occurs in the ferroxidase center made up of five residues [40] that can bind two Fe(II) simultaneously through eight coordination bonds to afford their oxidation by molecular oxygen (Fig. 1). On the contrary, the crystal structure of the E2 domain of APP [13] shows (Fig. 1) that the RExxE putative ferroxidase site of APP [15] consists of only two putative metal-binding residues, namely E412 and E415 in APP770. The side chain of these two residues point to the surface of the protein, i.e. there do not appear to be additional residues with which to form a metal binding site. In the previous work of Duce et al. the two residues E412 and E415, have been arbitrarily aligned with two metal-coordinating residues of the ferroxidase centre of human H-chain ferritin, namely E62 and H65 [15]. However, the two glutamates in the RExxE motif in ferritin are *not* ligands to the metals in the ferroxidase center. In other words, there is no significant overall similarity between the structure of the ferroxidase centre of ferritin and the RExxE putative ferroxidase activity carrying motif of APP.

We argue that the study reported by Duce et al. [15] on ferroxidase activity of purified APP695, its E2 domain, the 22-residue synthetic peptide FD1, and tissues containing APP, and the inhibition of these activities by Zn(II) has several important deficiencies: (i) The authors find that non-enzymatic oxidation of up to 200 μM Fe(II) in the presence of transferrin is negligible at pH 7.2 [15]. This is clearly in disagreement with several previous reports [21–23] and with our current results. The rate of non-enzymatic oxidation of Fe(II) by molecular oxygen and incorporation of the resulting Fe(III) product into transferrin is significant at pH 7.2. This rate is not dependent on the presence of the FD1 peptide, indicating that FD1 does not have any ferroxidase activity. Increasing the pH from 6.2 to 8.2 drastically increases the rate of non-enzymatic oxidation of Fe(II) and formation of diferric-transferrin. The effect of pH is crucial since results based on small differences or changes in the pH could easily be mistaken for enzymatic activity. We find that the overall non-enzymatic rate of oxidation of Fe(II) and incorporation of the Fe(III) product into transferrin at pH 7.0 is within experimental uncertainty close to the previously claimed specific activity values for APP, its E2 domain, and FD1 [15]. (ii) In the previous study the fact that Zn(II) can bind to transferrin and inhibit the formation of the diferric-transferrin complex has not been considered [15]. It has been suggested that Zn(II) binds to the putative ferroxidase site of APP but not to ceruloplasmin and that it inhibits ferroxidase activity of APP only. However, in agreement with other previous studies [38,41] we have shown that Zn(II) binds to transferrin. Binding of Zn(II) to transferrin inhibits formation of the diferric-transferrin complex and therefore decreases the ferroxidase activity of ceruloplasmin in the transferrin assay. (iii) Duce et al. have not measured oxygen consumption by APP as an independent method for ferroxidase activity assay and for determination of the stoichiometry of the ferroxidase reaction of APP, however, they have proposed a model in which APP can catalyse the oxidation of four iron by one molecular oxygen [15]. The oxygen consumption measurements reported here show that FD1 does not consume oxygen in the presence of iron. (iv) Duce et al. have not measured Fe(II) binding to APP, its E2 domain or FD1 peptide. Using isothermal titration calorimetry we observed that the putative catalytic site of the FD1 peptide does not bind Fe(II).

We therefore conclude that the FD1 peptide with the RExxE ferroxidase consensus motif of APP does not carry ferroxidase activity. In view of these results and of the observation that Zn(II) binding to transferrin directly interferes with the measurement of ferroxidase activity using this protein, we suggest that the results of Duce et al [15] for the ferroxidase activity of the FD1 peptide and – by implication – Of the E2 domain and the full-lenght APP should be re-evaluated.

Materials and Methods

Chemicals

All chemicals were reagent grade and were purchased from Sigma Aldrich. Human ceruloplasmin (40 U/mg) was also obtained from Sigma Aldrich.

Expression and purification of human H chain ferritin

The *Escherichia coli* Top10 strain (Invitrogen) was transformed with the plasmid containing the coding sequence for expression of human H chain ferritin (HuHF) pET12a. After two hours of growth to an $OD_{600\ nm}$ of 0.6, production of HuHF was induced by the addition of 0.1 mM IPTG. After eight hours cells were collected and were disrupted using a cell disruptor system (Constant systems). The supernatant was collected using centrifugation and was subjected to a heat step at 85°C for 10 min. Denatured proteins were separated using a centrifugation step. HuHF was made apo as described previously [32] and the final preparation contained less than one iron per 24-meric protein as determined with the ferene assay. The concentration of protein was measured using the bicinchoninic acid (BCA) assay using bovine serum albumin as standard.

Preparation of apo-transferrin

Apo-bovine transferrin (>98% pure) was purchased from Sigma Aldrich. The lyophilized powder was dissolved in 10 mM Mops, 100 mM NaCl, pH 7.0 and was dialyzed for at least 3 days against the same buffer using dialysis tube (Spectrumlabs) with a cut-off of 10 kDa. Subsequently consecutive dilution and concentration steps using an ultrafiltration membrane with a cut-off of 30 kDa (Millipore) were used to clean transferrin from possible metal complexing molecules which may have remained in the transferrin lyophilized powder from the manufacturing process and to concentrate the protein. The concentration of transferrin was measured using the bicinchoninic acid (BCA) assay and the purified apo-transferrin was directly used in experiments.

Preparation of the synthetic peptide (FD1)

The peptide FD1 (90% pure, HPLC) was purchased from GenScript. 4 mg of peptide was dissolved in 1 ml of 10 mM Hepes, 100 mM NaCl, pH 7.2, and dilutions were used directly in experiments. The sequence HRERMSQVMREWEEAER-QAKNL was confirmed by low energy collision induced dissociation tandem mass spectrometry using a hybrid quadrupole time-of-flight mass spectrometer (Waters QTof Premier).

Preparation of ferrous sulphate solution

The Fe(II) solution was prepared as explained previously [17,32].The pH of Milli.Q water was set to 2.5 using HCl. The solution was made anaerobic by flushing with high purity argon gas (99.999%), and then it was added to ferrous sulphate salt in an anaerobic glove box (Coy Laboratory Products).

Steady state kinetics of Fe(II) oxidation in the presence of transferrin

Kinetics of Fe(II) oxidation and Fe(III) uptake by transferrin were measured at 460 nm using a molar extinction coefficient for diferric transferrin of $\varepsilon_{460\ nm} = 4.56\ mM^{-1}\ cm^{-1}$ [23]. Measurements were carried out on a fiber-optics spectrophotometer (Avantes) using 1 ml glass cuvettes with a path length of 1 cm. Measurements were done at 37°C in 100 mM Hepes, 100 mM NaCl, pH 7.2 unless otherwise stated. Progress curves were recorded for circa 5 minutes and initial rates were obtained from the slope of a line that was fitted to the data points for the first 50–100 seconds. Each experiment was repeated at least three times with two different batches of transferrin. For kinetic measurements to each cuvette the following additions were made in order (total volume of 1 ml): Hepes buffer, aliquot of transferrin between 50 to 100 µl to reach the desired final concentration, 3.5 µl of 1.4 mM FD1 peptide (as control experiment no peptide was added), and aliquots of $FeSO_4$ solution, between 0.5 µl and 20 µl. To determine the inhibitory effect of zinc, aliquots of 0.5 µl to 20 µl of $ZnSO_4$ solution were added 5 minutes before addition of $FeSO_4$.

Oxygen consumption measurements

To measure consumption of oxygen a Clark electrode was used as described before [32]. The final reaction volume in the cell was 2 ml. To the cell the following was added in order: buffer, protein or peptide. Subsequently aliquots of 5 µl anaerobic solution of ferrous sulfate were added using a gastight syringe. The buffer was 100 mM Hepes, 100 mM NaCl, pH 7.2. Measurements were performed at 22°C.

Fast kinetics of Fe(II) oxidation

Fe(II) oxidation by human H chain ferritin was followed using a Scientific PQ/SF-53 preparative quench/stopped-flow instrument with an EG&G Princeton Applied Research 1,024-element photodiode-array detector. The instrument was equipped with a nitrogen gas flow on top of the sample cells to prevent oxidation of the Fe(II) solution before mixing with protein. The instrument was set to zero with rapid mixing of protein with anaerobic water, pH 2.4, before each measurement. The buffer was 400 mM Mops, 100 mM NaCl, pH 7.0.

Isothermal titration calorimetry

Zn(II) binding to transferrin was measured using isothermal titration calorimetry with a VP-ITC microcalorimeter (GE healthcare). A control experiment in the absence of protein or peptide was performed to measure the heat generated due to Fe(II) or Zn(II) dilution in the buffer. The conditions for the control experiments were identical to the conditions for titration of Fe(II) or Zn(II) to transferrin or to FD1 peptide. For anaerobic Fe(II) binding to FD1, the ITC machine was placed in a polyethylene bag (Atmosbag, Sigma Aldrich). The bag was made anaerobic by high purity argon gas (99.999%). All solutions were made anaerobic using the same argon gas. The Fe(II) solution was prepared in 200 mM Mops, 100 mM NaCl pH 5.8 and the protein was in the same buffer with pH 7.0. Measurements were performed at 25°C in duplicate. The parameter settings of the ITC machine were number of injections 30 or 60, volume of each injection 3 µl, stirring rate 307 rpm, and time spacing 200 sec. The volume of the first injection was 2 µl and the resulting data point was excluded from the fit of integrated heat of binding. The results of ITC were analyzed using Origin 7.0 software under a two independent binding sites model or a single binding site model.

Acknowledgments

We are grateful to Dr. Martijn W.H. Pinkse (Delft University of Technology, The Netherlands) for confirmation of the sequence of the synthetic peptide FD1 by mass spectrometry. We thank Dr. Paolo Arosio (Università degli Studi di Brescia, Italy) for providing the construct for production of recombinant HuHF.

Author Contributions

Conceived and designed the experiments: KHE WRH. Performed the experiments: KHE. Analyzed the data: KHE PLH WRH. Contributed reagents/materials/analysis tools: KHE. Wrote the paper: KHE PLH WRH.

References

1. O'Brien RJ, Wong PC (2011) Amyloid precursor protein processing and Alzheimer's Disease. Annu Rev Neurosci 34: 185–204.
2. Zheng H, Koo EH (2011) Biology and pathophysiology of the amyloid precursor protein. Mol Neurodegener 6. Doi:10.1186/1750-1326-6-27
3. Venkataramani V, Rossner C, Iffland L, Schweyer S, Tamboli IY, et al. (2010) Histone deacetylase inhibitor valproic acid inhibits cancer cell proliferation via down-regulation of the Alzheimer amyloid precursor protein. J Biol Chem 285: 10678–10689.
4. Acevedo KM, Hung YH, Dalziel AH, Li Q-X, Laughton K, et al. (2011) Copper promotes the trafficking of the Amyloid Precursor Protein. J Biol Chem 286: 8252–8262.
5. Duce JA, Bush AI (2010) Biological metals and Alzheimer's disease: Implications for therapeutics and diagnostics. Prog Neurobiol 92: 1–18.
6. Wan L, Nie G, Zhang J, Zhao B (2012) Overexpression of wild-type Amyloid-β protein precursor decreases the iron content and increases the oxidative stress of neuroblastoma SH-SY5Y cells. J Alz Dis 30: 523–530
7. Cho H-H, Cahill CM, Vanderburg CR, Scherzer CR, Wang B, et al. (2010) Selective translational control of the Alzheimer amyloid precursor protein transcript by iron regulatory protein-1. J Biol Chem 285: 31217–31232.
8. Hung Y, Bush A, Cherny R (2010) Copper in the brain and Alzheimer's disease. J Biol Inorg Chem 15: 61–76.
9. Watt NT, Whitehouse IJ, Hooper NM (2011) The role of zinc in Alzheimer's Disease. Int J Alz Dis 2011. doi:10.4061/2011/971021
10. Tabner BJ, Mayes J, Allsop D (2011) Hypothesis: soluble AB oligomers in association with redox-active metal ions are the optimal generators of reactive oxygen species in Alzheimer's Disease. Int J Alz Dis 2011. doi:10.4061/2011/546380
11. Rossjohn J, Cappai R, Feil SC, Henry A, McKinstry WJ, et al. (1999) Crysttal structure of the N-terminal, growth factor-like domain of Alzheimer amyloid precursor protein. Nat Struct Biol 6: 327–331.
12. Kong GKW, Adams JJ, Harris HH, Boas JF, Curtain CC, et al. (2007) Structural studies of the Alzheimer's amyloid precursor protein copper-binding domain reveal how it binds copper ions. J Mol Biol 367: 148–161.
13. Wang Y, Ha Y (2004) The X-Ray structure of an antiparallel dimer of the human amyloid precursor protein E2 domain. Molecular Cell 15: 343–353.
14. Atwood CS, Moir RD, Huang X, Scarpa RC, Bacarra NME, et al. (1998) Dramatic aggregation of Alzheimer Aβ by Cu(II) is induced by conditions representing physiological acidosis. J Biol Chem 273: 12817–12826
15. Duce JA, Tsatsanis A, Cater MA, James SA, Robb E, et al. (2010) Iron-Export ferroxidase activity of [beta]-amyloid precursor protein is inhibited by zinc in Alzheimer's Disease. Cell 142: 857–867.
16. Sun S, Chasteen ND (1992) Ferroxidase kinetics of horse spleen apoferritin. J Biol Chem 267: 25160–25166.
17. Honarmand Ebrahimi K, Hagedoorn P-L, Hagen W (2010) Inhibition and stimulation of formation of the ferroxidase center and the iron core in *Pyrococcus furiosus* ferritin. J Biol Inorg Chem 15: 1243–1253.
18. Treffry A, Bauminger ER, Hechel D, Hodson NW, Nowik I, et al. (1993) Defining the roles of the threefold channels in iron uptake, iron oxidation and iron-core formation in ferritin: a study aided by site-directed mutagenesis. Biochem J 296: 721–728.
19. Treffry A, Zhao Z, Quail MA, Guest JR, Harrison PM (1998) The use of zinc(II) to probe iron binding and oxidation by the ferritin (EcFtnA) of *Escherichia coli*. J Biol Inorg Chem 3: 682–688.
20. Bou-Abdallah F, Arosio P, Levi S, Janus-Chandler C, Chasteen ND (2003) Defining metal ion inhibitor interactions with recombinant human H- and L-chain ferritins and site-directed variants: an isothermal titration calorimetry study. J Biol Inorg Chem 8: 489–497.
21. Osaki S (1966) Kinetic studies of ferrous Ion oxidation with crystalline human ferroxidase (Ceruloplasmin). J Biol Chem 241: 5053–5059.
22. Osaki S, Johnson DA, Frieden E (1966) The possible significance of the ferrous oxidase activity of ceruloplasmin in normal human serum. J Biol Chem 241: 2746–2751.
23. Bakker GR, Boyer RF (1986) Iron incorporation into apoferritin. The role of apoferritin as a ferroxidase. J Biol Chem 261: 13182–13185.
24. Santambrogio P, Levi S, Cozzi A, Rovida E, Albertini A, et al. (1993) Production and characterization of recombinant heteropolymers of human ferritin H and L chains. J Biol Chem 268: 12744–12748.
25. Levi S, Luzzago A, Cesareni G, Cozzi A, Franceschinelli F, et al. (1988) Mechanism of ferritin iron uptake: activity of the H-chain and deletion mapping of the ferro-oxidase site. A study of iron uptake and ferro-oxidase activity of human liver, recombinant H-chain ferritins, and of two H-chain deletion mutants. J Biol Chem 263: 18086–18092.
26. Sun S, Arosio P, Levi S, Chasteen ND (1993) Ferroxidase kinetics of human liver apoferritin, recombinant H-chain apoferritin, and site-directed mutants. Biochemistry 32: 9362–9369.
27. Bou-Abdallah F, Zhao G, Biasiotto G, Poli M, Arosio P, et al. (2008) Facilitated diffusion of iron(II) and dioxygen substrates into human H-Chain ferritin. A fluorescence and absorbance study employing the ferroxidase center substitution Y34W. J Am Chem Soc 130: 17801–17811.
28. Snow CL, Martineau LN, Hilton RJ, Brown S, Farrer J, et al. (2011) Ferritin iron mineralization proceeds by different mechanisms in MOPS and imidazole buffers. J Inorg Biochem 105: 972–977.
29. Bou-Abdallah F, Papaefthymiou GC, Scheswohl DM, Stanga D, Arosio P, et al. (2002) mu-1,2-peroxobridged diiron (III) dimer formation in human H-chain ferritin. Biochem J 364: 57–63.
30. Treffry A, Zhao Z, Quail MA, Guest JR, Harrison PM (1995) Iron(II) Oxidation by H Chain Ferritin: Evidence from Site-Directed Mutagenesis That a Transient Blue Species Is Formed at the Dinuclear Iron Center. Biochemistry 34: 15204–15213.
31. Taylor AB, Stoj CS, Ziegler L, Kosman DJ, Hart PJ (2005) The copper-iron connection in biology: Structure of the metallo-oxidase Fet3p. Proc Natl Acad Sci U S A 102: 15459–15464.
32. Honarmand Ebrahimi K, Hagedoorn P-L, Jongejan J, Hagen W (2009) Catalysis of iron core formation in *Pyrococcus furiosus* ferritin. J Biol Inorg Chem 14: 1265–1274.
33. Pouvreau LAM, Strampraad MJF, Berloo SV, Kattenberg JH, de Vries S (2008) NO, N2O, and O2 Reaction Kinetics: Scope and Limitations of the Clark Electrode. In: Robert KP, editor. Methods in Enzymology: Academic Press. pp. 97–112.
34. Yang X, Chen-Barrett Y, Arosio P, Chasteen ND (1998) Reaction paths of iron oxidation and hydrolysis in horse spleen and recombinant human ferritins. Biochemistry 37: 9743–9750.
35. Zhao G, Bou-Abdallah F, Arosio P, Levi S, Janus-Chandler C, et al. (2003) Multiple Pathways for Mineral Core Formation in Mammalian Apoferritin. The Role of Hydrogen Peroxide. Biochemistry 42: 3142–3150.
36. Bou-Abdallah F, Biasiotto G, Arosio P, Chasteen ND (2004) The Putative "Nucleation site" in suman H-chain ferritin is not required for mineralization of the iron core. Biochemistry 43: 4332–4337.
37. Harris WR, Stenback JZ (1988) The bicarbonate-dependence of zinc(II)-transerrin binding. J Inorg Biochem 33: 211–223.
38. Harris WR (1983) Thermodynamic binding constants of the zinc-human serum transferrin complex. Biochemistry 22: 3920–3926.
39. Dahms SO, Könning I, Roeser D, Gührs KH, Mayer MC, et al. (2012) Metal binding dictates conformational and function of the amyloid precursor protein (APP) E2 domain. J Mol Biol 416: 438–452
40. Bou-Abdallah F (2010) The iron redox and hydrolysis chemistry of the ferritins. Biochimica et Biophysica Acta (BBA) - General Subjects 1800: 719–731.
41. Tinoco AD, Valentine AM (2005) Ti(IV) binds to human serum transferrin more tightly than does Fe(III). J Ame Chem Soc 127: 11218–11219.

4

Anti-Aquaporin-1 Autoantibodies in Patients with Neuromyelitis Optica Spectrum Disorders

John S. Tzartos[1,2]*, Christos Stergiou[1], Konstantinos Kilidireas[3], Paraskevi Zisimopoulou[1], Thomas Thomaidis[2], Socrates J. Tzartos[1,4]*

1 Department of Biochemistry, Hellenic Pasteur Institute, Athens, Greece, **2** Department of Neurology, General Hospital "Red Cross", Athens, Greece, **3** Department of Neurology, Aeginition Hospital, School of Medicine, National and Kapodistrian University, Athens, Greece, **4** Department of Pharmacy, University of Patras, Patras, Greece

Abstract

Autoantibodies against aquaporin-4 (AQP4), a water channel in CNS astrocytes, are detected in ~50–80% of patients with neuromyelitis optica spectrum disorders (NMOsd), characterized by longitudinally extensive transverse myelitis (LETM) and/or optic neuritis. Although these autoantibodies present an invaluable biomarker for NMOsd and for the differential diagnosis of multiple sclerosis (MS), diagnosis of anti-AQP4-seronegative NMOsd remains challenging. We hypothesized that seronegative NMOsd patients might have autoantibodies against aquaporin-1 (AQP1), another water channel in CNS astrocytes. We initially developed a radioimmunoprecipitation assay to search for anti-AQP1 antibodies in sera from 632 individuals. Anti-AQP1 or anti-AQP4 autoantibodies were detected in 16.7% and 12%, respectively, of 348 patients with suspected NMOsd. Anti-AQP1 specificity was confirmed by competition, protein immunoblotting and ELISA assays, whereas epitope localization was studied by immunoadsorption on intact cells expressing AQP1 and peptide mapping experiments. Most anti-AQP1 autoantibodies were of the complement-activating IgG1 subclass and the majority bound to the extracellular domain of AQP1, suggesting a possible pathogenic role. Five out of 42 MS patients had anti-AQP1 antibodies, but 2 of them also had spinal cord lesions, while the anti-AQP1 antibodies in the other 3 bound to the cytoplasmic domain of AQP1. Anti-AQP1 antibodies were not detected in 100 healthy individuals or 142 patients with non-demyelinating neuroimmune diseases. Analysis of 17 anti-AQP1+/anti-AQP4- patients with suspected NMOsd showed that 5 had NMO and 11 had LETM. 12/17 of these sera bound predominantly to the extracellular AQP1 loop-A. Overall, we found that anti-AQP1 autoantibodies are present in a subgroup of patients with chronic demyelination in the CNS and similarities with anti-AQP4-seronegative NMOsd, offering a novel potential biomarker for CNS demyelination disorders.

Editor: Jun-ichi Kira, Kyushu University, Japan

Funding: This work was supported by overheads of the Department of Biochemistry of the Hellenic Pasteur Institute and co-financed by the EU (European Social Fund) and Greek national funds, through the Program "Education and Lifelong Learning" of the NSRF - Research Funding Programs: "Thales" and "Aristeia". The funders had no role in study design, data collection and analysis, decision to publish, or preparation of the manuscript.

Competing Interests: The authors have read the journal's policy and have the following conflicts: J. Tzartos, Ch. Stergiou, K. Kilidireas, and S. Tzartos are the inventors in a patent application that relates to the detection of antibodies to aquaporin-1 in demyelinating diseases of the CNS. S. Tzartos is currently in the process of establishing a private diagnostics laboratory. P. Zisimopoulou and Th. Thomaidis report no disclosures.

* E-mail: jtzartos@gmail.com (JST); tzartos@pasteur.gr (SJT)

Introduction

Patients suffering from chronic inflammatory demyelinating diseases of the central nervous system (CNS) present a variable disease course, pathology, and response to treatment [1]. As a consequence, specific biomarkers are needed to classify these patients and adopt the most appropriate treatment. Currently, autoantibodies to aquaporin-4 (AQP4), a water channel abundantly expressed in CNS astrocytes, are a widely accepted biomarker in patients suffering from neuromyelitis optica spectrum disorders (NMOsd). The NMOsd group of demyelinating disorders includes the classic neuromyelitis optica (NMO), with major characteristics longitudinally extensive transverse myelitis (LETM) and optic neuritis [2], only LETM, only recurrent optic neuritis and related disorders [3]. In a high proportion of NMOsd patients, brain lesions are also present [4,5], thus complicating diagnosis, and the detection of anti-AQP4 antibodies is therefore vital for the early differentiation of NMO from MS [6] and for appropriate treatment [7]. Unfortunately, about 20–50% of NMOsd patients are seronegative for anti-AQP4 autoantibodies [8–10].

AQP4 belongs to the large AQP family with 13 members (AQP0-12), which form tetramers in the cell membrane, each monomer acting as a water channel [11]. AQPs have been identified in several mammalian tissues [12]. Several reports have shown that, in addition to AQP4, human CNS astrocytes also abundantly express AQP1 on their surface [13–17]. Specifically, it has been shown that AQP1 is highly expressed in areas prone to develop NMO lesions (spinal cord, optic nerves and brain white matter) [17]. Moreover, both AQP1 and AQP4 are overexpressed in the brain in some neurological diseases, such as MS, compared to "normal" brains [13], possibly for maintenance of water homeostasis. We therefore wondered whether patients with NMOsd, but seronegative for anti-AQP4 antibodies, had antibodies to AQP1. NMO-related autoantibodies (named NMO-IgG) were initially identified using an immunofluorescence assay in which serum from NMO patients was applied to mouse brain sections, resulting in a characteristic pattern [18]. Subsequently, it was found that NMO-IgG were anti-AQP4 antibodies [19]. Interestingly, anti-AQP1 antibodies would not be detected in the usual immunofluorescence assay for NMO-IgG using rodent brain tissue because AQP1, although expressed in rodent brain, it is not detected in their astrocytes [16,20–22] and pre-treatment of serum with liver powder (known to contain AQP1 [23]), used as a blocking agent in this assay, selectively removes anti-AQP1 antibodies, as shown in this manuscript.

Here, we screened sera from 348 patients with suspected NMOsd and found that a higher percentage contained anti-AQP1 antibodies (~17%) than anti-AQP4 antibodies (12%). Anti-AQP1 antibodies were not detected in 242 control sera. The clinical and radiological characteristics for 17 anti-AQP1-seropositive patients with suspected NMOsd suggest remarkable similarities between anti-AQP1-seropositive patients and anti-AQP4-seronegative NMOsd patients.

Materials and Methods

Serum samples

Serum samples were obtained from 348 patients whose doctors had requested testing for anti-AQP4 autoantibodies. Additionally, sera from 42 patients diagnosed with MS (fulfilling the McDonald 2010 MS criteria [24]), 142 sera from patients with other neuroimmune diseases, and 100 sera from healthy individuals were used as controls (total 632 serum samples). The 142 patients with other neuroimmune diseases consisted of 101 patients with anti-acetylcholine receptor antibody-positive myasthenia gravis (MG), 30 patients with anti-MuSK antibody-positive MG, 5 patients with antibodies against Ca^{2+} channels (responsible for Lambert-Eaton myasthenic syndrome), and 6 patients with other autoantibodies involved in neuroimmunological diseases (antibodies against K^{+}-channels, gangliosides GM2, GQ1b or GD1A, or MAG and YO antigens).

Ethics Statement

Sera were used after the approval of the Bioethics Committee of the Hellenic Pasteur Institute (last approval ref. no 1288/2012) and the participants provided written informed consent to participate in this study in a form approved by the above Bioethics Committee.

Plasmid constructs and AQP proteins

pEGFP-N1 was purchased from Clontech Laboratories, CA, USA. The pCMV6-AQP1-tGFP (AQP1-GFP) and pCMV6-AQP4-M23-tGFP (AQP4-GFP) were both purchased from OriGene Technologies Inc. (Rockville, MD, USA) as ready to use plasmids. The cDNAs encoding human AQP1 and human AQP4 were also subcloned into pPICZα and pPIC9, respectively (Life Technologies, Invitrogen), in order to be expressed in yeast (with 6xHis and 8xHis tags for AQP1 and AQP4, respectively) and were purified by Ni-NTA chromatography and gel filtration. In addition, a commercially available fusion protein of human recombinant AQP1 fused with a glutathione-S-transferase tag (AQP1-GST) was expressed in the wheat germ cell-free protein expression system and affinity purified on immobilized glutathione (Novus Biologicals, Cambridge, UK). The yeast-expressed AQP1 was generated at a later stage and was used on selected sera (including all sera in Table 1) to confirm the findings.

Maintenance and transfection of human embryonic kidney cells (HEK 293)

The HEK293 cell line was grown in Dulbecco's modified eagle's medium (DMEM; Life Technologies, Gibco) supplemented with 10% FCS and 100 units/ml each of penicillin G and streptomycin (Life Technologies, Invitrogen) at 37°C in an atmosphere of 5% CO_2. HEK-293 cells were seeded and allowed to grow for at least 24 hours on 10 cm plates when used for competition and cell-immunoadsorption assays. Cells were transiently transfected with Lipofectamine 2000 (Life Technologies, Invitrogen) at 40–50% confluency.

Detection of antibody binding to AQPs by RIPA and two-step RIPA

RIPA. Human AQPs were indirectly labeled with ^{125}I by incubating 0.25 nmol of AQP with 5 nmol of NHS-PEO_4-Biotin on ice for 2 h, then incubating approximately 0.1 μg of biotinylated AQP for 1 h on ice with 50,000 cpm of ^{125}I-streptavidin radioiodinated by the chloramine-T method to a specific activity of 1600–1870 Ci/mmol. For the RIPA, 5 μl or less of serum was incubated for 2 h at 25°C with 50 μl of phosphate-buffered saline (PBS) containing 0.5% Triton X-100, pH 7.4 (assay buffer) and 0.1 μg of the indirectly labeled AQP, then 50 μl of goat anti-human immunoglobulin (Ig) was added and the mixture incubated for 2 h at 25°C and centrifuged, and the pellet washed twice with 1 ml of assay buffer and the radioactivity in the pellet measured in a gamma-counter.

The average value for sera from 30 healthy human individuals plus 3 standard deviations, SD (~250+3×35 = ~355 cpm) was arbitrarily multiplied by 1.4. This value (~500 cpm) was considered to be the cut-off for a positive result, and the average value plus 3 SD arbitrary multiplied by 1.1 (~390 cpm) was taken as the cut-off for an "ambiguous" result; the cut-offs correspond to antibody titers of 2.5 and 1.4 nM, respectively.

Two-step RIPA. In addition to the above RIPA, a more sensitive two-step RIPA used to detect anti-AQP4 autoantibodies, and in selected cases of low anti-AQP1 titer sera, was also used to confirm the anti-AQP1 positivity of the sera. This assay was based on a method involving semi-purification of the specific autoantibodies from a large volume of serum (~150 μl) by non-stringent affinity chromatography and their use in the standard RIPA (described in Trakas et al. 2011 [25]).

Detection of antibody binding to AQPs by ELISA

Antibody binding to yeast-expressed AQP1, was also studied by an ELISA assay. 96-well plates were coated with 0.5 μg peptide, or BSA as control, (in 50 μL of 0.01 M carbonate buffer, pH 9.6) per well. After overnight incubation at 4°C, plates were washed 3 times with PBS/0.05% Tween-20 and once with PBS and the remaining sites were blocked by incubation with Superblock blocking buffer (Thermo Scientific, Rockford, USA) for 1 h at room temperature. After 3 washes of the plates with PBS/0.05% Tween 20, and once with PBS, sera were added at a 1/100 dilution in PBS/2% BSA and the plates incubated overnight at 4°C. After washing the plates as above, horseradish peroxidase-conjugated rabbit anti-human IgG antibodies diluted 1/5000 in 2% BSA/PBS were added and the plates incubated for 1 h at room temperature, washed as above, and incubated for 15 min at room temperature with 3,3′,5,5′-tetramethylbenzidine (TMB). Color development was quantified at 450 nm after stopping the reaction with 0.2 M sulfuric acid. Values were considered positive when they were 50% or more higher than the mean OD_{450} for normal control sera, a value which, using serum samples from 44

Table 1. Laboratory and clinical data for anti-AQP1 seropositive patients.

Patient no.	Sex	Age at onset (years)	Disease duration (years)	Anti-AQP4 Abs[a]	Anti-AQP1 titer (nM)[b]	Abs bound to intact AQP1-expressing cells (%+SD)[c]	Titer (nM)[d]	Predominant epitope[e]	MRI and clinical data
1	F	34	8	no	1.9	NT	NT	N-term	LETM and optic neuritis (NMO)
2	F	18	3	no	2.7	5±2	0.1	-	LETM and optic neuritis (NMO)
3	F	47	2	no	2.5	96±6	2.4	Loop-C	LETM and optic neuritis (NMO)
4	M	23	20	no	3.7	88±10	3.3	Loop-A	LETM and optic neuritis (NMO)
5	F	39	28	no	3.9	79±3	3.1	Loop-A	LETM and optic neuritis (NMO)
6	F	27	1	no	1.6	NT	NT	N-term	LETM
7	F	38	2	no	2.7	13±3	0.4	Loop-B	LETM
8	F	34	2	no	1.8	30±3	0.5	Loop-E	LETM
9	F	28	6	no	2.6	99±2	2.6	Loop-A	LETM
10	F	39	12	no	2.0	99±6	2.0	Loop-A	LETM
11	M	34	3	no	5.8	88±1	5.1	Loop-A	LETM
12	M	33	9	no	4.2	16±5	0.7	Loop-A	LETM
13	F	40	10	no	10.5	92±6	9.7	Loop-A	LETM
14	F	23	14	no	2.8	82±5	2.3	Loop-A	LETM
15	M	30	18	no	2.6	8±10	0.2	Loop-B	LETM.Hodgkin lymphoma
16	M	43	20	no	2.7	77+10	2.1	Loop-A	LETM and brain lesions (fulfilled Barkhof criteria)
17	M	39	1	no	2.5	93±2	2.3	Loop-A	Transverse myelitis. Kidney neoplasm
18	F	42	5	no	4.0	100±1	4.0	Loop-E	MS with predominant spinal cord lesions
19	F	44	25	no	3.8	92±6	3.5	Loop-A	MS with predominant spinal cord lesions and axonal neuropathy. Mammary neoplasm
20	F	17	26	no	3.9	0±1	0.0	Loop-B	MS
21	F	18	19	no	3.0	6±1	0.2	Loop-B	MS
22	F	35	8	no	2.6	2±4	0.1	Loop-B	MS

F, female; LETM, longitudinally extensive transverse myelitis; M, male; MS, multiple sclerosis; NMO, neuromyelitis optica; NT, not tested.

[a]All 22 patients tested negative for serum anti-AQP-4 autoantibodies in the two-step RIPA and confirmed by both a commercial and an in-house CBA.

[b]The anti-AQP1 positivity of the sera was also confirmed by the more sensitive two-step RIPA for anti-AQP1 antibodies (not shown).

[c]According to the Methods and Figure 8.

[d]Calculated by multiplying anti-AQP1 titers (in 6th column) by the percentages of the extracellularly binding antibodies (in 7th column).

[e]Binding of sera to synthetic peptides corresponding to AQP1 segments (from Figure 9B). Underlined loops are extracellular.

healthy controls, was higher than the mean OD_{450} of the controls plus 3SD. Specifically, the average OD_{450} for these control sera for AQP1 was 0.24 with SD 0.03, and thus the cut-off for positive values was 0.36. The O.D.450 by all tested sera for plated BSA was 0.23±0.06.

Effect of guinea pig liver powder on the AQP1-RIPA

Five microliter samples of test sera were diluted in 100 µl of 0.2% bovine serum albumin (BSA) in PBS and pretreated with 20 mg of guinea pig liver powder (Rockland Inc, Gilbertsville, PA.) for 1 h at room temperature. After centrifugation, the supernatants were tested by AQP1-RIPA and AQP4-RIPA and the results compared to those obtained using untreated sera.

Competition RIPA experiments

To test whether the antibodies bound to the AQP1-GST fusion protein or to ^{125}I-streptavidin, 10 AQP1-antibody positive sera and negative controls were tested for direct binding to the same amount of ^{125}I-streptavidin. To test whether the antibodies bound to the AQP1 moiety of the AQP1-GST fusion protein, 10 µl of sera positive for anti-AQP1 autoantibodies was preincubated for 3 h at 4°C with excess GST (1.6 µg) immobilized on Sepharose-glutathione beads (GE Healthcare), then 5 µl of the treated samples was tested in the RIPA using ^{125}I-streptavidin-labeled AQP1-GST.

Competition tests in which commercial AQP1-GST and AQP1 expressed in-house in HEK293 cells or yeast competed for binding to serum antibodies were performed as follows: Five microliter samples of anti-AQP1 antibody-containing sera were preincubated for 3 h at 4C (i) with extracts of nontransfected HEK293 cells or HEK293 cells transfected with human AQP1-GFP prepared by solubilizing 500,000 cells in 500 µl of homogenization buffer (100 mM NaCl, 100 mM KCl, 5 mM EDTA, 5 mM EGTA, 50 mM Tris pH 7.5, and 1% Triton X-100)] or (ii) with 10 µl of assay buffer containing 0.2 µg of either AQP1 or AQP4 expressed in yeast cells and purified, or with 0.2 µg of BSA as a control. These mixtures were then tested in the regular RIPA using indirectly labeled commercial AQP1-GST.

Detection of antibody binding to AQP1 by Western blotting

Samples (0.45 µg of protein per lane) of yeast-expressed AQP1 or a control protein (MuSK) were electrophoresed on a 4–12% gradient SDS-polyacrylamide gel and transferred onto nitrocellulose membranes. The membranes were then incubated for 1 h at 25°C with 3% BSA in PBS, then for 2 h at 4°C with a commercial rabbit anti-AQP1 polyclonal antibody (AB3272; Millipore, Temecula, CA), anti-AQP1-positive sera, or sera from healthy individuals, all diluted (30-500 times) in 0.2% BSA in PBS. The membranes were then washed and incubated for 90 min at 4°C with a 1:1000 dilution of horseradish peroxidase (HRP)-conjugated anti-rabbit or HRP-conjugated anti-human antibody, washed with PBS, and were with the chromomeric substrate 3,3'-diaminobenzidine (DAB) to visualize antibody binding.

Investigation of cross-reactivity between anti-AQP1 and anti-AQP4 antibodies

Double-positive sera were depleted of one type of antibody as follows: 0.2 µg of either AQP1 or AQP4 (or control BSA) was immobilized in each ELISA well, then 150 µl of 2% BSA in PBS containing a suitable volume of test serum was added and the mixture incubated for 16 h at 4°C. The supernatants were then tested in parallel RIPAs for the presence of anti-AQP1 and anti-AQP4 antibodies to determine the level of the remaining antibody against each antigen.

Determination of anti-AQP1 antibody class and subclass

The Ig subclass of the anti-AQP1 autoantibodies was investigated using the RIPA with the following modification. After the incubation of labeled antigen with the serum sample, 5 µl of sheep anti-human IgG subclass or sheep anti-human IgM antibodies (1 mg/ml; Binding Site, Birmingham, UK) was added and the mixture incubated overnight at 4°C, then 20 µl of rabbit anti-sheep Ig antiserum (Sigma-Aldrich, Munich, Germany) was added and the mixture incubated for 2 h at room temperature, then the pellets were washed and their radioactivity counted.

Measurement of antibodies binding to the extracellular region of AQP1

HEK293 cells were transfected with AQP1-GFP, AQP4-GFP and pEGFP-N1. The culture medium used after transfection was supplemented with 10^{-7} M secretin (Bachem, Bubendorf, Switzerland) which induces surface expression of AQP1 [4] whereas in parallel transfections with the same plasmids no secretin was added. The cells were collected 48 h after transfection, washed twice with PBS, and counted. Five or 10 microliter samples of anti-AQP1 antibody-containing sera were preincubated for 2 h at room temperature with 50,000–500,000 intact transfected HEK293 cells in 50 µl of PBS buffer, then the mixtures were centrifuged and the supernatants tested in the regular RIPA to determine the titre of unbound antibodies.

Peptides and peptide mapping experiments

Six synthetic peptides corresponding to the extracellular and cytoplasmic segments of human AQP1 were synthesized and HPLC-purified (>80% purity) by JPT Peptide Technologies GmbH (Berlin). These corresponded to the extracellular segments 36–49 (including Loop-A), 116–137 (including Loop-C), and 183–188/199–210 (including Loop-E) and the intracellular segments 1–22 (N-terminus), 81–100 (Loop-B), and 248–270 (C-terminus).

Antibody binding to the peptides was studied by ELISA. Ninety-six-well plates were coated by overnight incubation at 4°C with 0.2 µg peptide, or BSA as control, in 50 µL of 0.01 M carbonate buffer, pH 9.6, per well, then were washed 3 times with PBS/0.05% Tween-20 and once with PBS, and the remaining sites blocked by incubation for 1 h at room temperature with 4% BSA in PBS. The following steps were as described above with the ELISA with intact AQP1. Values were considered positive when they were 50% or more higher than the mean OD_{450} for normal control sera, a value which, using serum samples from 12 healthy controls, was always higher than the mean OD_{450} of the controls plus 3SD. In practice, the average OD_{450} for these 12 control sera for any peptide (and for BSA instead of peptide) was between 0.25–0.29 (many more sera from healthy controls subsequently tested gave very similar values), the SDs were between 0.03–0.05, and the cut-off for positive values was ~0.45.

Cell based assays (CBA)

In house CBA for anti-AQP4 antibodies was performed according to Leite et al. [9]. In addition, a commercial AQP4-CBA kit (Euroimmun, Luebeck, Germany) was also used according to the manufacturer's instructions.

Results

Anti-AQP1 autoantibodies are present in sera of patients with suspected NMOsd at a similar frequency to anti-AQP4 autoantibodies

We developed a RIPA for anti-AQP1 autoantibodies using, as labeled antigen, a human AQP1-GST fusion protein biotinylated and indirectly labeled with ^{125}I-streptavidin. This RIPA for anti-AQP1 antibodies was used to examine sera from 348 patients whose doctors had requested screening for anti-AQP4 antibodies. These sera had previously been tested for anti-AQP4 antibodies using the very sensitive two-step RIPA and 42 had been classed as anti-AQP4 antibody-positive. In the present study, 44 of the 306 anti-AQP4 antibody-negative (14.4%) were found to be positive for anti-AQP1 antibodies, as were 14 of the 42 anti-AQP4 antibody-positive sera (33%) (Figure 1). Overall, 16.7% of the tested sera (58/348) were anti-AQP1 antibody-positive. The titer range of anti-AQP1 antibodies in our anti-AQP1-positive sera was similar to that of the anti-AQP4 titers in our anti-AQP4-positive sera; specifically, after excluding the top 10% of titers in each group which caused very high standard deviations, the average values of the remaining 52 anti-AQP1-positive and 40 anti-AQP4-positive sera were 5.3±2.2 nM and 6.3±10.2 nM (not shown), respectively. 142 sera from patients with autoantibodies related to non-demyelinated neuroimmune diseases (the 131 with MG) and 100 sera from healthy controls were negative (Figure 1). However, 5 of 42 sera from patients diagnosed with MS had low anti-AQP1 antibody titers, but, as described below, these patients either had concurrent spinal cord lesions or their antibodies did not bind to intact AQP1-expressing cells but rather bound to cytoplasmic epitopes. The female/male ratio of the 44 patients with exclusively anti-AQP1 antibodies was 1.9:1.

The above RIPA assay and the identified anti-AQP1-positive sera were then used to investigate whether guinea pig liver powder, which is routinely used as a blocking agent in immunofluorescence assays for the detection of NMO-IgG, removed anti-AQP1 antibodies from the anti-AQP1-antibody containing sera, thus preventing their detection in these assays. When anti-AQP1 and anti-AQP4-positive sera were pretreated with guinea pig liver powder and the supernatants assessed by RIPA, the liver powder completely removed the anti-AQP1 antibodies, but had no effect on the anti-AQP4 antibodies (Figure 2), apparently because liver contains AQP1 [23]. Thus, anti-AQP1 autoantibodies would escape detection in the immunofluorescence assay for NMO-IgG.

Several experiments confirmed the specificity of the anti-AQP1 RIPA assay. Initially, we confirmed that the antibodies bound exclusively to the AQP1 moiety of the AQP1-GST-^{125}I-streptavidin complex by showing that "anti-AQP1 antibodies" in 10 positive sera did not bind to either ^{125}I-streptavidin (not shown) nor the GST moiety of the AQP1-GST fusion protein (Figure 3A). We then checked whether antibodies binding to the human AQP1-GST preparation, expressed in a cell-free system, also bind to human AQP1 expressed in alternative systems, in the following experiments: First, binding of antibodies to commercial AQP1-GST labeled with ^{125}I-streptavidin was inhibited by pre-incubation of the sera with extracts from AQP1-GFP expressing HEK293 cells, but not with extracts from nontransfected HEK293 cells (Figure 3B) or with purified yeast-expressed AQP1, but not with purified yeast-expressed AQP4 (Figure 3C). Second, when yeast-expressed AQP1 was used as the labeled antigen in the RIPA, the results were very similar to those obtained using the commercial AQP1 (Figure 3D). These experiments show that antibody binding is independent of the source of AQP1 antigen used.

The RIPA results were then confirmed by Western blotting. All 4 anti-AQP1 antibody-containing sera tested bound to electrophoresed denatured yeast-expressed AQP1 (Figure 4). This approach provides an alternative assay for anti-AQP1 autoanti-

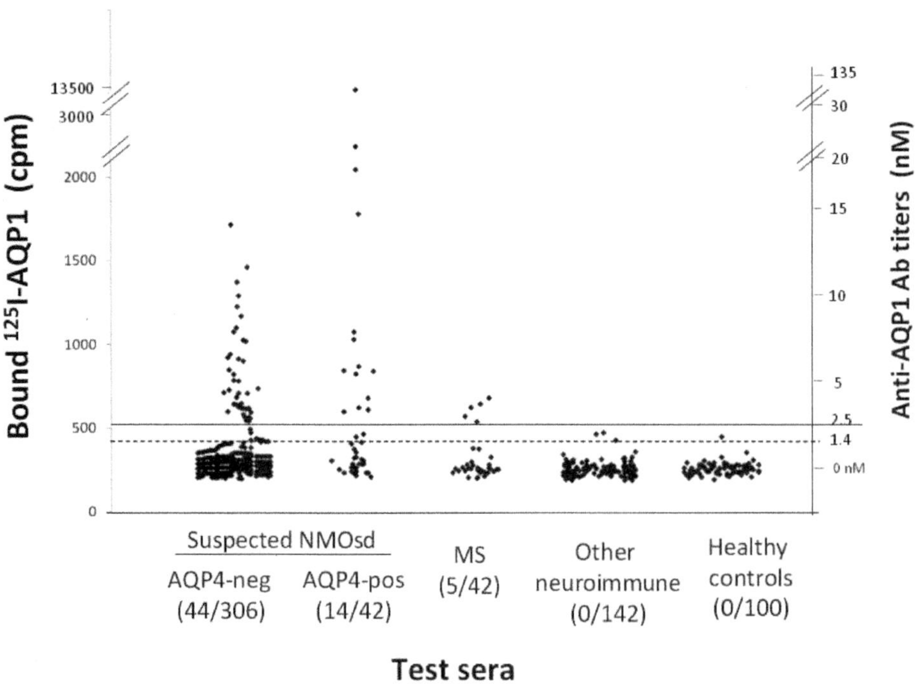

Figure 1. Measurement of anti-AQP1 antibodies in the sera of different patient groups and healthy controls by RIPA. The serum samples are (left to right) from patients with suspected NMOsd without (AQP4-neg) or with (AQP4-pos) anti-AQP4 antibodies, MS, other neuroimmune diseases (131/142 with MG), or healthy controls. The numbers in parenthesis below the group are the number of anti-AQP1 antibody-positive sera and the total number of test sera for each group of patients. The dashed and solid horizontal lines denote the cut-off values for ambiguous and positive titers. The left and right y axis show, respectively, the precipitated cpm and the estimated antibody titer.

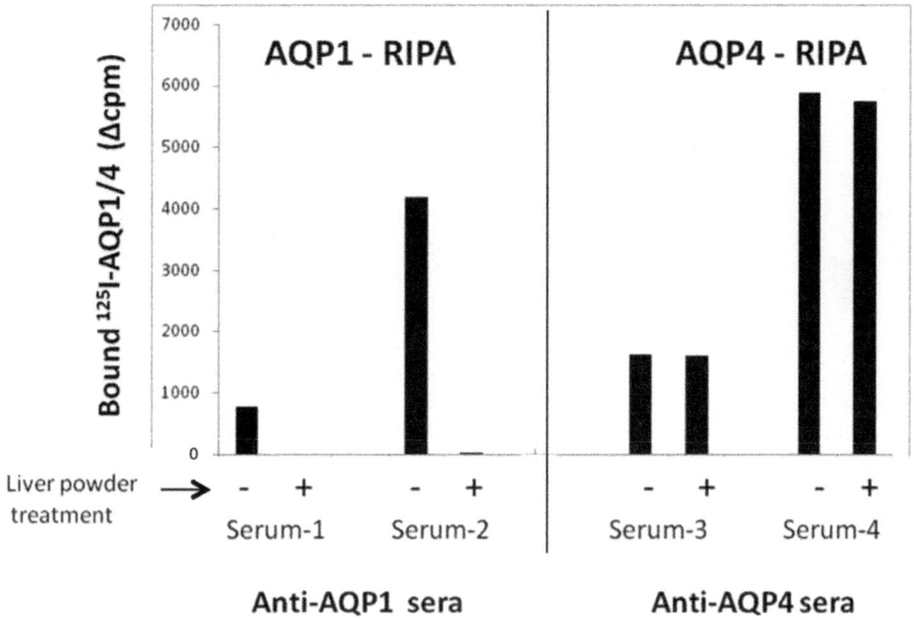

Figure 2. Liver powder removes anti-AQP1 antibodies. Two exclusively anti-AQP1-positive and two exclusively anti-AQP4-positive serum samples were pretreated with guinea pig liver powder, then the supernatants were tested by RIPA using indirectly radiolabeled AQP1 (left panel) or AQP4 (right panel). Key: +, pretreated serum; -, untreated serum.

bodies and shows that serum antibodies also bind to denatured AQP1.

Finally, we developed an ELISA with immobilized purified yeast-expressed AQP1 to further confirm the RIPA results and possibly provide an easier method for future routine diagnosis. We tested three serum dilutions (1/25, 1/100 and 1/500) of which the 1/100 dilution gave the best results and was adopted for the following experiments. Figure 5 shows that there is very good correlation between the results of RIPA and ELISA: all tested 79 sera previously found negative for AQP1 antibodies by the RIPA (44 from healthy controls, 5 from anti-AQP4+/AQP1- sera and 30 from anti-AQP1-negative MS patients) were clearly negative, whereas 29/31 sera previously found anti-AQP1 positive by the RIPA (including one double-positive and two MS sera) were also positive by the ELISA. Only 2/31 sera (i.e. 6.5%) were positive by RIPA and negative by ELISA.

Anti-AQP1 and anti-AQP4 autoantibodies are specific for the corresponding antigen

We then tested whether the double-positivity (anti-AQP1 and anti-AQP4) seen with some sera was due to cross-reactivity by preincubating selected double-positive sera with one of the antigens immobilized on ELISA plates, then examining whether the unbound material was able to bind to either of the antigens. The results showed that immobilized AQP1 removed the anti-AQP1 antibodies, but not the anti-AQP4 ones (Figure 6A), while immobilized AQP4 removed only the anti-AQP4-antibodies (Figure 6B). We subsequently tested whether there was a correlation between the concentrations of the two antibodies in the double-positive sera, but no such correlation was detected (Figure 6C). However, sera with a relatively high titer of antibodies for one AQP were more likely to be double-positive than lower titer sera (not shown). In conclusion, the double-positive sera contain two sets of antibodies, each specific for one AQP.

The vast majority of anti-AQP1 antibodies are IgG1

We then examined the Ig class and subclass of the anti-AQP1 antibodies in sera from 7 patients (4 sera from patients no. 11-14 in Table 1 with available clinical data, and the other 3 randomly selected from the remaining anti-AQP1 antibody-positive sera). All 7 sera showed a marked predominance of the complement-activating IgG1 subclass (Figure 7).

Most anti-AQP1 antibodies of NMO-suspected patients bind to the extracellular region of AQP1

We then examined whether the anti-AQP1 antibodies could bind to intact HEK293 cells expressing AQP1. We used a quantitative approach to determine the fraction of antibodies that bind to the extracellular region of AQP1. This involved a combined immunoadsorption with RIPA experiment in which the test sera were preincubated with intact AQP1-GFP-transfected or EGFP-transfected HEK293 cells followed by RIPA on the depleted sera. To enhance surface expression of AQP1, the cells were pretreated with the hormone secretin [26]. As shown in Figure 8, secretin-treated AQP1-GFP transfected HEK293 cells efficiently removed the anti-AQP1 antibodies from 3 of the 4 test sera (sera no. 1-3), while EGFP-transfected secretin-treated cells or AQP4-transfected cells had no effect (only representative results are shown in Figure 8). Serum no. 4 was practically not depleted of its antibodies by the intact AQP1-transfected cells; however its preincubation with detergent-solubilised AQP1-transfected cells efficiently inhibited antibody binding in the RIPA suggesting that most anti-AQP1 antibodies of this serum bind to the cytoplasmic domain of AQP1. As shown in Figure 8, 500.000 AQP1-transfected secretin-treated cells were necessary and sufficient for efficient immunoadsorption of the anti-AQP1 antibodies. In the absence of secretin, immunoadsorption of anti-AQP1 antibodies by 500.000 cells was insignificant (~0-10%) but it was efficient when we used 4 million cells per reaction (~50-75% immunoadsorption for 3 tested sera; not shown). AQP4-transfected cells (but

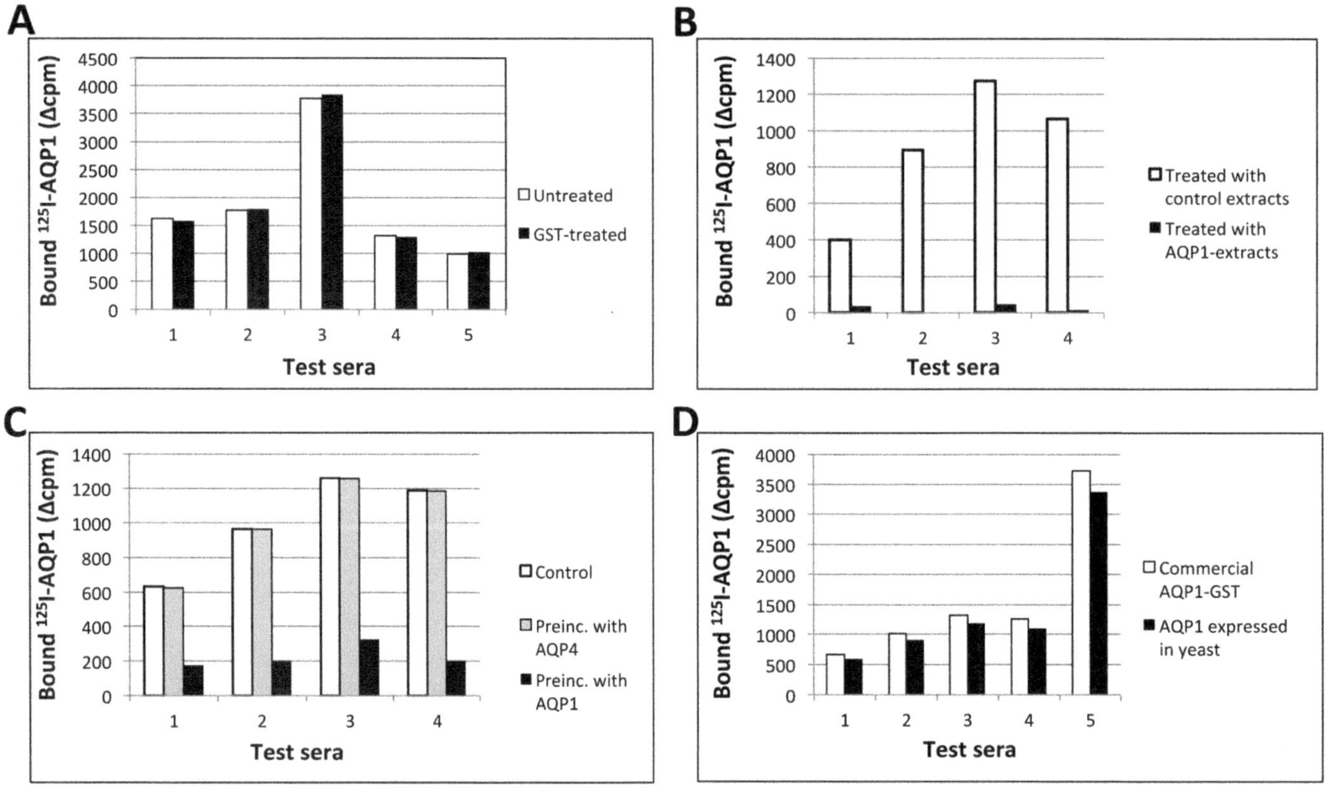

Figure 3. Anti-AQP1 specificity of the identified autoantibodies. (A) Patient's autoantibodies recognize the AQP1 moiety of AQP1-GST. Five sera that had tested positive for binding to the AQP1-GST fusion protein were preincubated with an excess of GST immobilized on Sepharose-glutathione beads, then were tested by RIPA using ^{125}I-streptavidin labeled AQP1-GST. (B) Binding of anti-AQP1 autoantibodies is specifically inhibited by an extract from AQP1-expressing HEK293 cells, but not control HEK293 cells. Four anti-AQP1-positive sera were preincubated with extracts prepared from either EGFP-transfected or AQP1-GFP-transfected HEK293 cells, then were tested by RIPA for binding to the commercial AQP1 preparation. (C) Binding of anti-AQP1 autoantibodies is specifically inhibited by yeast-expressed human AQP1. Four exclusively anti-AQP1-positive sera were preincubated with human AQP1 or AQP4 that had been expressed in yeast and purified or with BSA as control, then were tested by RIPA using ^{125}I-streptavidin-labeled commercial AQP1-GST fusion protein. (D) AQP1 autoantibody binding is independent of the source of AQP1. Both the commercial AQP1-GST fusion protein and the in house AQP1 purified from yeast were biotinylated, indirectly labeled by preincubation with ^{125}I-streptavidin, and used in the RIPA. Five anti-AQP1-positive sera and one serum sample from a healthy control (HC) were tested.

not AQP1-transfected or EGFP-transfected cells) were very efficient in removing anti-AQP4 antibodies in similar assays for the detection of anti-AQP4 antibodies (in fact, they were ~10 times more efficient than the AQP1-transfected secretin-treated cells since only 50,000 AQP4-transfected cells (without secretin-treatment) were sufficient to deplete the relevant sera; data not shown). These results suggest that antibodies to both the extracellular and the cytoplasmic domains of AQP1 may exist in anti-AQP1-positive sera.

Epitope mapping of the anti-AQP1 antibodies

To confirm the above results using a different approach and to locate more precisely the epitopes recognized by the anti-AQP1 antibodies, we used ELISA to test the binding of several anti-AQP1 sera to synthetic peptides corresponding to the 3 extracellular loops (Loops A, C, and E) and the 3 main cytoplasmic segments (Loop B and the N-terminal and C-terminal) of AQP1 (Figure 9A). Figure 9B shows the results for the binding of 22 sera from patients with known clinical characteristics as described in the next paragraph. Interestingly, with one exception (serum no 2, which did not bind to any peptide), all bound almost exclusively to a single peptide; specifically, 14 sera bound to extracellular peptides (11 to Loop-A, 1 to Loop-C, and 2 to Loop-E) and 7 to cytoplasmic peptides (5 to Loop-B and 2 to the N-terminal). No antibodies bound to the peptide corresponding to the C-terminal of AQP1. Loop A, which is the most antigenic, presents quite low identity (14%) and homology (43%) with the corresponding AQP4 sequences, whereas Loop E presents high similarity with the homologous sequence of AQP4 (67% identity and 89% homology); nevertheless, the two sera which bound to peptide Loop E (sera no. 8 and 18 in Figure 9B) did not bind to intact AQP4. As described below, peptide mapping correlated very well with the above immunoadsorption/RIPA results.

Retrospective analysis of MRI and clinical data for 22 anti-AQP1-seropositive patients

Magnetic resonance imaging (MRI) and clinical data were available for 22 anti-AQP1-seropositive, anti-AQP4-seronegative patients, consisting of 17 with suspected NMOsd and 5 MS patients (shown in Figure 1). In addition to the AQP1-RIPA, 19 of these sera were also tested by ELISA with intact AQP1 (included in Figure 5); only two of the 19 sera (sera no 10 and 21 in Table 1) did not bind in the ELISA. Anti-AQP4 seronegativity was determined by a two-step RIPA and confirmed by two CBA assays (in house according to [9] and commercial; see Methods), all found anti-AQP4 negative. Table 1 shows that 16 of the 17 patients with suspected NMOsd had LETM, five of which also had optic neuritis (i.e. typical NMO) and one had only transverse

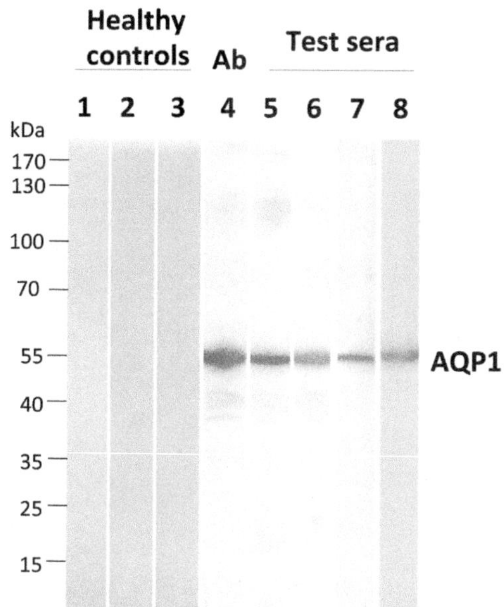

Figure 4. Detection of binding of anti-AQP1 antibody to denatured AQP1 by Western blotting. Yeast-expressed AQP1 was electrophoresed and transferred onto nitrocellulose membranes, which were then incubated with the test sera (see Methods). Lanes 1-3: Three sera from healthy controls at dilutions 1/250, 1/60, and 1/30, respectively; lane 4: commercial rabbit anti-AQP1 antibody; lanes 5-8: representative anti-AQP1-positive sera with titers of 132, 25, 10.5, and 5.8 nM, at dilutions of 1/500, 1/250, 1/30, and 1/30, respectively. None of the test anti-AQP1 sera bound to the control protein MuSK (not shown).

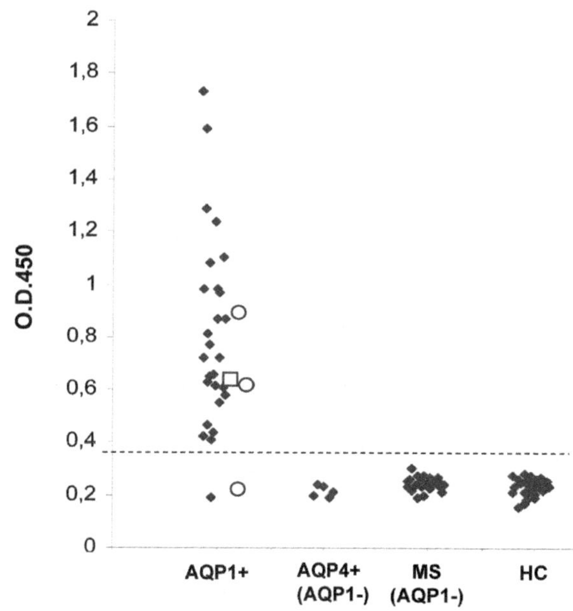

Figure 5. Detection of anti-AQP1 antibodies by an ELISA with immobilized purified human yeast-expressed AQP1. Sera previously tested by RIPA for AQP1 antibodies were tested for binding to immobilized AQP1 by ELISA. First column contains 31 sera found positive by the RIPA (including a double-positive anti-AQP1/AQP4, empty square, and 3 sera from anti-AQP1-positive MS patients, empty circles). The following 3 columns contain 5 anti-AQP4-positive/anti-AQP1-negative sera, 30 sera from anti-AQP1-negative MS patients and 44 sera from healthy controls). The dashed horizontal line denotes the cut-off (O.D.450: 0.36) for positive values.

myelitis. Fifteen sera from these 17 patients were also tested, in addition to RIPA, for antibodies binding to intact AQP1-expressing cells using immunoadsorption on intact AQP1-transfected HEK293 cells followed by RIPA (as described in the previous chapter and in Figure 8). The percentage of antibodies in each serum binding to the extracellular region of AQP1 was 79–99% in 10 sera, 13–30% in 3, and <10% in 2 sera (7th column in Table 1). These 22 sera were also tested for binding to synthetic peptides corresponding to the 6 AQP1 segments (Figure 9, described above). Interestingly, whether or not a high percentage of the antibodies in a serum sample bound to the intact AQP1-expressing cells correlated very well with whether or not they bound to an extracellular peptide (compare columns 7 and 9 in Table 1), further confirming the extracellular location of the antibody binding sites for most of these sera.

Of the 5 anti-AQP1-positive MS patients (no. 18–22 in Table 1), the first two (no. 18 and 19) also had predominant spinal cord lesions (though not LETM) and most (92–100%) of their anti-AQP1 antibodies bound to intact AQP1-expressing HEK293 cells. Interestingly, the anti-AQP1 antibodies in the remaining 3 patients (no. 20–22) with classic MS practically did not bind to the intact AQP1-expressing cells (0–6%) and bound to the cytoplasmic AQP1 Loop-B. Finally, 3 of the 22 anti-AQP1 positive patients also had neoplasms (nephroma, non-Hodgkin lymphoma, or mammary cancer).

Discussion

Many NMOsd patients are seronegative for anti-AQP4 antibodies [8–10], and, although AQP4 loss is usually observed in NMO brain lesions, no such loss has been observed in some NMO patients [27]. Recently, antibodies to the myelin-oligodendrocyte glycoprotein (MOG) have also been detected in seronegative NMO patients, however in limited cases (4/27) [28]. AQP1 might be a major candidate autoantigen for anti-AQP4-seronegative NMOsd, since, like AQP4, it is abundantly expressed in human astrocytes [13–16]. Interestingly, a recent study by Misu et al [17] showed that, in some NMO lesions, astrocyte AQP1 cell surface expression was selectively reduced or AQP1 was found in astrocytic intracellular granules, results similar to those observed for AQP4. However, in this study, we showed that anti-AQP1 autoantibodies would not be detected in the classic assay for NMO-IgG antibodies (Figure 2). Consequently, different approaches were necessary for the detection of anti-AQP1 autoantibodies.

We initially developed a RIPA for anti-AQP1 antibodies and screened sera from patients with suspected NMOsd. Interestingly, anti-AQP1 antibodies were more frequent than anti-AQP4 antibodies, with 16.7% of the 348 tested sera being anti-AQP1-positive and only 12% anti-AQP4-positive (Figure 1). Overall, the anti-AQP-seropositive patients were doubled, from 12% anti-AQP4 to 24.7% anti-AQPs (taking into account that 4% were double positive). It should be noted that, generally, only a small fraction (~9–12%) of patients referred for laboratory diagnosis of NMOsd are found to be anti-AQP4-positive [29]. The low female/male ratio of the anti-AQP1-seropositive patients (1.9:1) is identical to that determined for anti-AQP4 antibody-negative NMO patients in a recent large multicenter study [30], but much lower than that of 10.4:1 for the anti-AQP4-positive patients, determined in the same study or the 4.2:1 ratio we found in our anti-AQP4-seropositive patients. The relevance of the anti-AQP1 autoantibodies to NMOsd was supported by their absence in sera

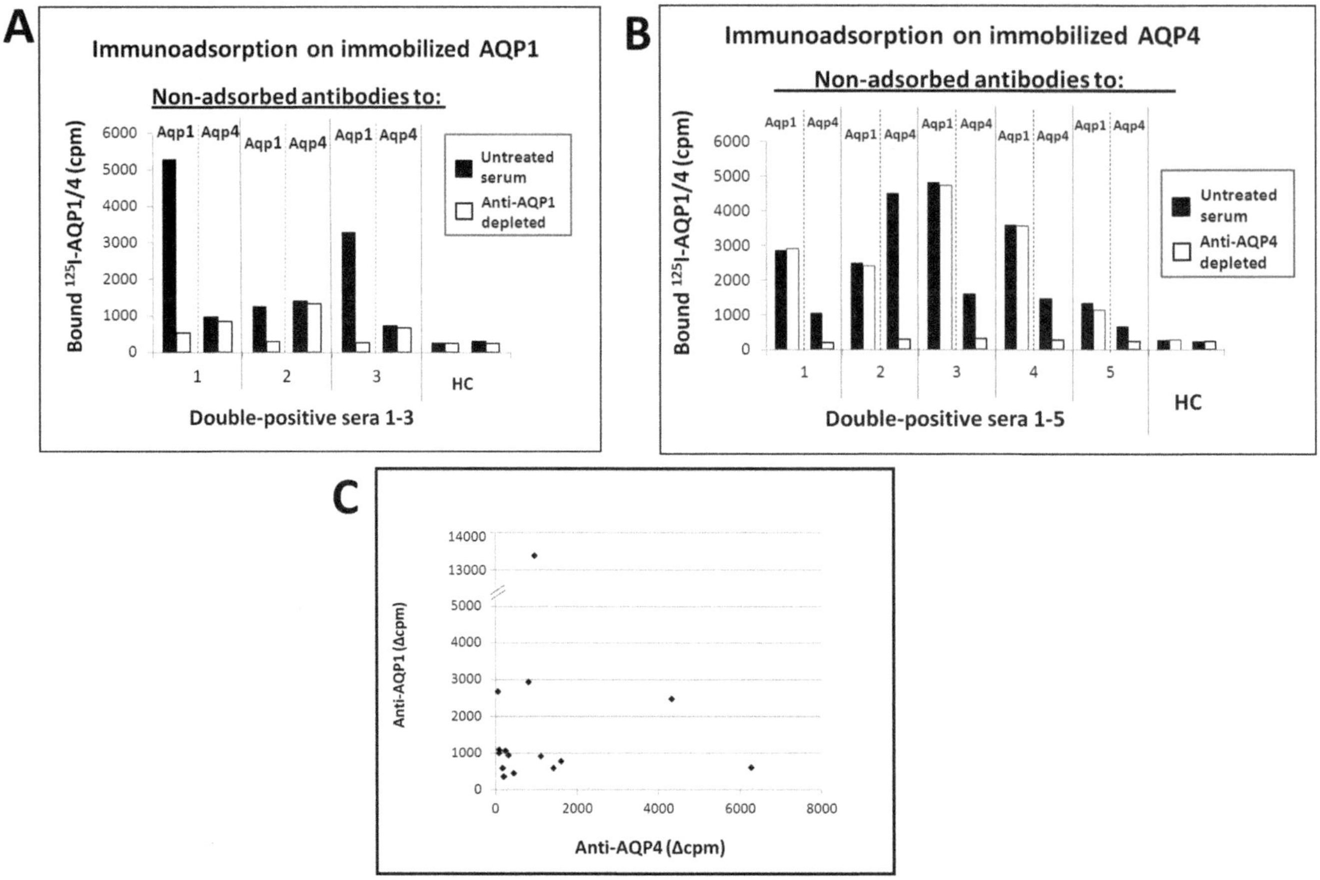

Figure 6. Anti-AQP1 antibodies and anti-AQP4 antibodies do not bind to the other antigen. (A and B). Search for cross-reactivity of anti-AQP1 and anti-AQP4 antibodies with AQP4 and AQP1, respectively. Three (A) or five (B) double-positive serum samples were incubated with immobilized AQP1 (A) or AQP4 (B) to immunoadsorb the corresponding antibodies, then were tested by RIPA for binding to ^{125}I-labeled AQP1 or AQP4 (white bars) in parallel with the untreated sera (black bars). HC, healthy control. (C). Lack of correlation of the amount of radiolabeled AQP1 precipitated by anti-AQP1 antibodies (y axis) or AQP4 (x axis) precipitated anti-AQP4 antibodies by double-positive sera in identical regular RIPAs performed using the two labeled antigens.

from 100 healthy individuals and 142 patients with non-demyelinating neuroimmune diseases (131 with MG) (Figure 1). The specificity of the anti-AQP1 antibodies in patients' sera was also confirmed by several competition experiments (Figure 3) and by the independent determination of autoantibody binding by Western blotting (Figure 4). Finally, although in general RIPAs are

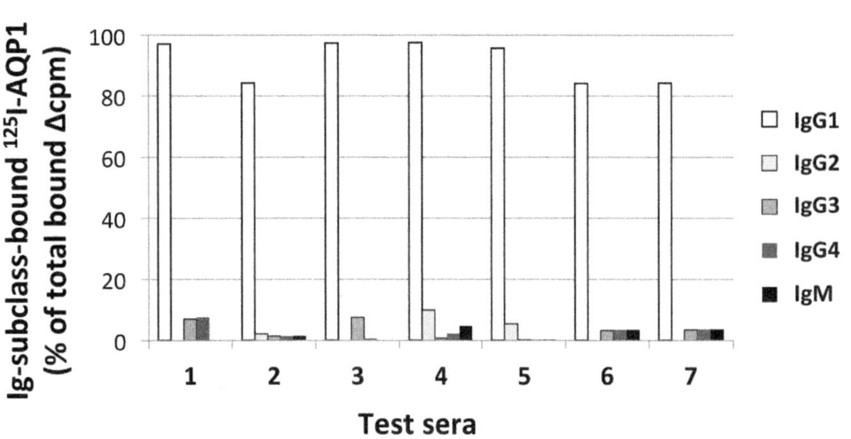

Figure 7. Ig class and IgG subclass of the anti-AQP1 antibodies. The Ig class and IgG subclass of the anti-AQP1 antibodies from 7 positive human sera was tested by RIPA using anti-class/subclass second antibodies. Sera no. 1-4 are from patients 11-14 in Table 1 and sera no. 5-7 are from the remaining anti-AQP1 antibody-positive sera.

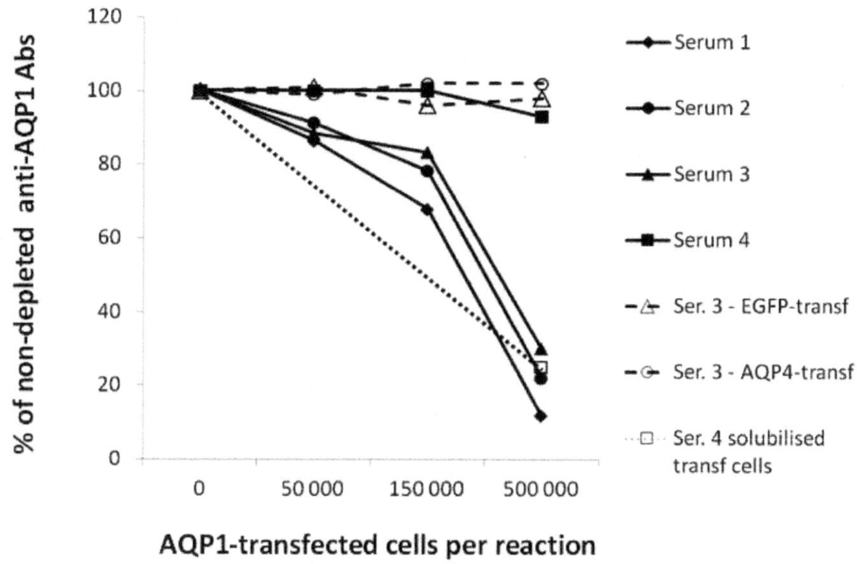

Figure 8. Determination of the percentage of antibodies directed against the extracellular domain of membrane-embedded AQP1. Four anti-AQP1-positive sera were left untreated or were preincubated with increasing numbers of AQP1-GFP- (filled symbols) or EGFP-transfected (shown only for serum 3; empty triangle) HEK293 cells treated with secretin to increase surface expression of AQP1; the untreated and treated samples were tested in the usual RIPA for anti-AQP1 antibodies. The sera were also treated with AQP4-transfected HEK293 cells (shown only for serum 3; open circle). Serum 4 was also treated with an extract of AQP1-transfected HEK293 cells (open square).

more sensitive and specific than ELISAs, because ELISAs are preferable in routine diagnosis we also developed a simple and sensitive ELISA with intact affinity-purified AQP1 which overall agrees well with the RIPA results (Figure 5). Only 2/31 (6.5%) sera previously found anti-AQP1 positive by the RIPA could not bind AQP1 in the ELISA. Therefore, AQP1-ELISA should prove useful in future routine diagnosis. These findings suggest that anti-AQP1 autoantibodies are at least as frequent as anti-AQP4 autoantibodies and thus anti-AQP1-seropositive patients may form a major subset of anti-AQP4-seronegative NMOsd patients.

Despite the fact that the two AQPs share considerable amino acid sequence identity (~51%), the two sets of antibodies in the double-positive sera showed no cross-reactivity (Figure 6A,B) or correlation of titers (Figure 6C). Nevertheless, the high frequency of double-positives (24% of the anti-AQP1-positives) and the observation that sera with a higher antibody titer for either AQP have a greater chance of being double-positive (data not shown) might indicate a related, or interdependent, immune system triggering mechanism. Since several human astrocytes express both AQPs [13], epitope spreading across the AQPs in the double-positive patients is a possibility.

Since the extracellular versus cytoplasmic location of the binding site for anti-AQP1 antibodies is potentially important for their potential pathogenic significance, we then developed assays to determine the location of binding of the anti-AQP1 antibodies. Despite many efforts and moderate success, we failed to develop a reproducible and reliable CBA using AQP1-transfected HEK293 cells, the main reason apparently being the observed low concentration of cell surface-expressed AQP1. Use of secretin increased surface expression of AQP1, but this was still much lower than the observed surface expression of AQP4 and it also dramatically increased background immunofluorescence. We therefore developed two alternative assays which quantitatively measure the location of the antibody binding sites on AQP1: (a) immunoadsorption of the serum antibodies on secretin-treated intact AQP1-transfected cells followed by RIPA of the unbound antibodies and (b) an ELISA to test serum binding to 3 synthetic peptides corresponding to the 3 extracellular loops and 3 peptides corresponding to most of the cytoplasmic domain of AQP1. The results of the two assays correlated very well, revealing which sera contained antibodies that bound mainly to the extracellular or the cytoplasmic domain of AQP1. In addition, peptide mapping revealed that the predominant extracellular epitope(s) of the tested sera were on Loop-A of AQP1.

The observed very good correlation between the immunoadsorption/RIPA and peptide mapping results was not necessarily expected, as often only a small minority of the antibodies to a protein antigen bind to linear peptides. Thus, in preliminary experiments, we tested whether the antibodies that bound to the synthetic peptides were the majority or a small minority; we performed competition RIPA experiments, in which 7 test sera were preincubated with 0.1–1.0 μg of individual peptides or with peptide mixtures (extracellular or cytoplasmic peptide groups) before incubation with labeled AQP1 in the regular RIPA. The results showed that, for each serum, ≥0.5 μg of the specific peptide, but not of the other peptides, inhibited antibody binding to AQP1 by >50% (not shown), demonstrating that the majority of these anti-AQP1 antibodies could bind to linear sequences of AQP1, and thus the correlation between the results for the two approaches (immunoadsorption on intact cells and peptide mapping) was well justified. Nevertheless, because often small peptides also bind antibodies to proteins with small homologies with the specific antigen, in order to minimize false-positive results, anti-AQP1 presence should be detected with the use of intact AQP1 (e.g. by RIPA or ELISA with intact AQP1) whereas peptide-ELISA should only be used for epitope mapping of the anti-AQP1 positive sera.

Retrospective analysis of the available MRI and clinical data was possible for 22 patients positive only for anti-AQP1 antibodies (17 suspected NMOsd patients among the seropositives in the 1st column of Figure 1 and the 5 seropositive MS patients in the 3rd column of Figure 1). This analysis showed that 19 patients had

A

```
AQP1 pept-Loop A     YKYPVGNNQTAVQDN                  (36-49)
AQP4                 G---GSEKPLPVDC
                        *.::.    *

AQP1 pept-Loop C     YTSSLTGNSLGRNDLADGVNSGQ          (116-137)
AQP4                  TPPSVVGGLGVTMVHGNLTAGH
                      *..  .  ..**  .  :  ..:.:*:

AQP1 pept-Loop E     YAIDYTG/SAVITHNFSNHW  (183-188 & 199-210)
AQP4                  AINYTG/PAVIMGNWENHW
                      **:***  .***   *:.***

AQP1 pept-N-term.    YMASEFK---KKLFWRAVVAEFLATT       (1-22)
AQP4                  M-VAFKGVWTQAFWKAVTAEFLAML
                      *    **    .: **:**.*****

AQP1 pept-Loop B     LGLLLSCQISIFRALMYIIA             (81-100)
AQP4                 VAMVCTRKISIAKSVFYIAA
                     :.:: : :*** ::::** *

AQP1 pept-C-term.    GQVEEYDLDAD-------DINSRVEMKPK  (248-270)
AQP4                 SQVETDDLILKPGVVHVIDVDRGEEKKGK
                      ***  **   .        *::    * * *
```

B

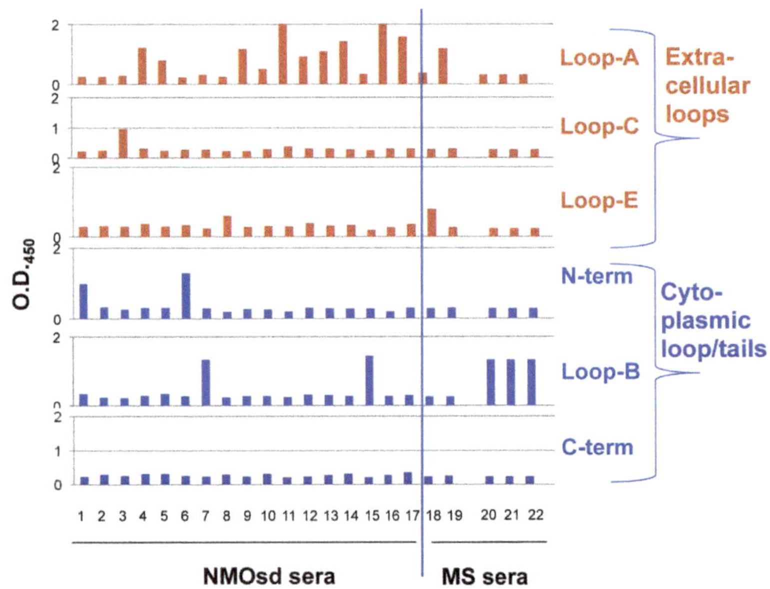

Figure 9. Anti-AQP1 antibody binding to synthetic peptides corresponding to AQP1 extracellular and cytoplasmic domains. A. Synthetic peptides corresponding to human AQP1 extracellular loops A (residues 36–49), C (116–137), and E (183–188, 199–210) or to the intracellular N-terminus (1–22), loop B (81–100), and the C-terminus (248–270) are shown (from Uniprot/Swissprot entry P29972, for the human AQP1). Extracellular residues are shown in bold and underlined. Loop-E is formed by two segments separated by a 14 residue gap; in "pept-Loop-E", 10 of these 14 residues (between TG and SE) were omitted (marked by:..). In the first 4 peptides, an extra tyrosine residue was added to the N-terminus for future radioiodination studies. Below the AQP1 peptide sequences are shown the corresponding AQP4 sequences; (*), identical amino acid residues; (:) and (.), homologous residues. **B.** Peptide mapping with anti-AQP1 positive sera**.** 22 sera from patients with known clinical characteristics (17 NMOsd and 5 MS) were tested for binding to 6 synthetic peptides corresponding to the 3 extracellular loops (loops A, C and E) and to the 3 cytoplasmic segments (N-terminal, Loop-B and C-terminal) by an ELISA assay. Positive values are considered those above $O.D._{450}$ 0.45 (see Methods). It is shown that most NMOsd sera bind to the Loop-A peptide.

spinal cord lesions (Table 1). Specifically, the vast majority (16) had LETM (5 also had optic neuritis, i.e. typical NMO), one had only transverse myelitis, and 2 were diagnosed with MS but also had a predominant spinal cord lesion load. Furthermore using intact AQP1-expressing cells we showed that the majority of the anti-AQP1 antibodies of the tested sera from the patients with spinal cord lesions of Table 1 bound to the extracellular domain of AQP1 (Table 1, column 7; average: 66.5%): 79–99% in 10 sera, 13–30% in 3, and only 2 sera had <10% of their antibodies to the extracellular domain of AQP1. In addition, peptide mapping experiments confirmed that the targets for these antibodies were predominantly extracellular and further restricted the major antigenic region to extracellular Loop-A (Figure 9B and Table 1). The cell-damaging factor in the patients with low concentrations of antibodies to the extracellular side of AQP1 may be different, as in the double-seronegative NMOsd patients, although these patients may also produce small quantities of antibodies to the extracellular side of AQP1, which were already immobilized on their tissue AQP1. Overall, these data strongly suggest that anti-AQP1 autoantibodies are a potential biomarker for NMOsd among anti-AQP4-seronegative patients.

The remaining 3 of the 22 patients of Table 1 (no. 20–22) had classic MS. However, the presence of anti-AQP1 autoantibodies in patients diagnosed with MS is not surprising, since NMO-IgG and anti-AQP4 autoantibodies have been detected in patients diagnosed with MS (~4% [18] to 25% [8]) and it was recently shown that sera from many MS patients bind to synthetic peptides corresponding to cytoplasmic epitopes of AQP4 [31]. Interestingly, almost none (0–6%) of the anti-AQP1 antibodies in the sera of the 3 patients with classic MS were able to bind to AQP1 on intact cells (Table 1, 7th column), nor to the extracellular loop peptides, suggesting that anti-AQP1 antibodies in classic MS may bind to the cytoplasmic domain of AQP1.

Importantly, 3 out of the 22 patients of Table 1 also had neoplasms, suggesting a paraneoplastic factor, similar to findings in NMO patients with anti-AQP4 antibodies [32]. Although the small numbers prohibit reliable conclusions for the frequency of neoplasms in anti-AQP1-seropositive patients, they suggest that this frequency may be roughly similar with that of the anti-AQP4-seropositive patients [32]. It is still unknown whether the anti-AQP1 autoantibodies play any pathogenic role in NMOsd. Since, in contrast to the situation in human astrocytes [13–16], AQP1 is not expressed on rodent astrocytes [16,19–21], rodents cannot be used as a model to study the pathogenic effect of anti-AQP1 autoantibodies on astrocytes and their role in NMOsd; therefore alternative models must be developed to investigate the possible pathogenic role of these antibodies. Nevertheless, the roughly similar range of anti-AQP1 and anti-AQP4 antibody titers (8,7±17,3 nM versus 7,8±12,5 nM, respectively) the similar IgG1 subclass predominance, and most antibodies binding to the extracellular domain of AQP1, suggest that a similar level of pathogenic potency of these antibodies is likely (possibly inducing local damage of astrocytes by a complement-mediated mechanism). In conclusion, we detected the presence of autoantibodies specific for AQP1 mainly in anti-AQP4-seronegative NMOsd patients, providing a promising diagnostic biomarker for this disease subtype, and observed similarities and differences to anti-AQP4-seropositive NMOsd. The low female-to-male ratio, the absence of coexisting autoimmune diseases (e.g. MG), and the predominance of LETM (with a low incidence of optic neuritis) in the anti-AQP1-seropositive NMOsd patients are similar to the characteristics of anti-AQP4-seronegative NMOsd [30,33]. These correlations and the high frequency of autoantibodies to the extracellular domain of AQP1 in patients with suspected NMOsd suggest that anti-AQP1-seropositivity may characterize a major subset of anti-AQP4-seronegative NMOsd patients.

Acknowledgments

We thank Jenny Katsoupi for technical assistance, and Kostas Lazaridis, Nikos Kouvatsos and Dafni Chroni-Tzartou for useful discussions.

Author Contributions

Conceived and designed the experiments: JST SJT. Performed the experiments: JST CS PZ. Analyzed the data: JST CS KK PZ TT SJT. Contributed reagents/materials/analysis tools: JST CS KK PZ TT SJT. Wrote the paper: JST SJT. Reviewed the manuscript, revising it critically for important intellectual content, and approved the final version: JST CS KK PZ TT SJT.

References

1. Miller DH, Weinshenker BG, Filippi M, Banwell BL, Cohen JA, et al. (2008) Differential diagnosis of suspected multiple sclerosis: a consensus approach. Mult Scler 14(9): 1157–74.
2. Wingerchuk DM, Lennon VA, Pittock SJ, Lucchinetti CF, Weinshenker BG (2006) Revised diagnostic criteria for neuromyelitis optica. Neurology. 66(10): 1485–9.
3. Wingerchuk DM, Lennon VA, Lucchinetti CF, Pittock SJ, Weinshenker BG (2007) The spectrum of neuromyelitis optica. Lancet Neurol. 6(9): 805–15.
4. Pittock SJ, Lennon VA, Krecke K, Wingerchuk DM, Lucchinetti CF, et al. (2006) Brain abnormalities in neuromyelitis optica. Arch Neurol. 63(3): 390–6.
5. Cabrera-Gomez JA, Kister I (2012) Conventional brain MRI in neuromyelitis optica. Eur J Neurol 19(6): 812–9.
6. Matiello M, Lennon VA, Jacob A, Pittock SJ, Lucchinetti CF, et al. (2008) NMO-IgG predicts the outcome of recurrent optic neuritis. Neurology 70(23): 2197–200.
7. Weinshenker BG, Wingerchuk DM (2010) Japanese optic-spinal MS: is it MS or neuromyelitis optica and does the answer dictate treatment? Neurology 75(16): 1404–5.
8. Siritho S, Nakashima I, Takahashi T, Fujihara K, Prayoonwiwat N (2011) AQP4 antibody-positive Thai cases: clinical features and diagnostic problems. Neurology 77(9): 827–34.
9. Leite MI, Jacob S, Viegas S, Cossins J, Clover L, et al. (2008) IgG1 antibodies to acetylcholine receptors in 'seronegative' myasthenia gravis. Brain 131(Pt 7): 1940–52.
10. Waters PJ, McKeon A, Leite MI, Rajasekharan S, Lennon VA, et al. (2012) Serologic diagnosis of NMO: a multicenter comparison of aquaporin-4-IgG assays. Neurology 78(9): 665–71; discussion 9.
11. Verkman AS, Mitra AK (2000) Structure and function of aquaporin water channels. Am J Physiol Renal Physiol 278(1): F13–28.
12. Benga O, Huber VJ (2012) Brain water channel proteins in health and disease. Mol Aspects Med 33(5–6):562–78.
13. Satoh J, Tabunoki H, Yamamura T, Arima K, Konno H (2007) Human astrocytes express aquaporin-1 and aquaporin-4 in vitro and in vivo. Neuropathology. 27(3): 245–56.
14. Moftakhar P, Lynch MD, Pomakian JL, Vinters HV (2010) Aquaporin expression in the brains of patients with or without cerebral amyloid angiopathy. J Neuropathol Exp Neurol. 69(12): 1201–9.
15. Perez E, Barrachina M, Rodriguez A, Torrejon-Escribano B, Boada M, et al. (2007) Aquaporin expression in the cerebral cortex is increased at early stages of Alzheimer disease. Brain Res 1128(1): 164–74.
16. Gao J, Tan M, Gu M, Marshall C, Ding J et al. (2012) Cellular localization of aquaporin-1 in the human and mouse trigeminal systems. PLoS One 7(9): e46379.
17. Misu T, Höftberger R, Fujihara K, Wimmer I, Takai Y et al. (2013). Presence of six different lesion types suggests diverse mechanisms of tissue injury in neuromyelitis optica. Acta Neuropathol. 125(6): 815–27.
18. Lennon VA, Wingerchuk DM, Kryzer TJ, Pittock SJ, Lucchinetti CF et al. (2004) A serum autoantibody marker of neuromyelitis optica: distinction from multiple sclerosis. Lancet 364(9451): 2106–12.
19. Lennon VA, Kryzer TJ, Pittock SJ, Verkman AS, Hinson SR (2005) IgG marker of optic-spinal multiple sclerosis binds to the aquaporin-4 water channel. J Exp Med 202(4): 473–7.
20. Nielsen S, Smith BL, Christensen EI, Agre P (1993) Distribution of the aquaporin CHIP in secretory and resorptive epithelia and capillary endothelia. Proc Natl Acad Sci U S A. 90(15): 7275–9.
21. Ribeiro Mde C, Hirt L, Bogousslavsky J, Regli L, Badaut J (2006) Time course of aquaporin expression after transient focal cerebral ischemia in mice. J Neurosci Res 83(7): 1231–40.
22. Dolman D, Drndarski S, Abbott NJ, Rattray M (2005) Induction of aquaporin 1 but not aquaporin 4 messenger RNA in rat primary brain microvessel endothelial cells in culture. J Neurochem 93(4): 825–33.
23. Talbot NC, Garrett WM, Caperna TJ. (2013) Analysis of the expression of aquaporin-1 and aquaporin-9 in pig liver tissue: comparison with rat liver tissue. Cells Tissues Organs. 174(3):117–28.
24. Polman CH, Reingold SC, Banwell B, Clanet M, Cohen JA, et al. (2011) Diagnostic criteria for multiple sclerosis: 2010 revisions to the McDonald criteria. Ann Neurol 69(2): 292–302.
25. Trakas N, Zisimopoulou P, Tzartos SJ (2011) Development of a highly sensitive diagnostic assay for muscle-specific tyrosine kinase (MuSK) autoantibodies in myasthenia gravis. J Neuroimmunol 240–1: 79–86.
26. Marinelli RA, Pham L, Agre P, LaRusso NF (1997) Secretin promotes osmotic water transport in rat cholangiocytes by increasing aquaporin-1 water channels in plasma membrane. Evidence for a secretin-induced vesicular translocation of aquaporin-1. J Biol Chem 272(20): 12984–8.
27. Matsuoka T, Suzuki SO, Suenaga T, Iwaki T, Kira J. (2011) Reappraisal of aquaporin-4 astrocytopathy in Asian neuromyelitis optica and multiple sclerosis patients. Brain Pathol. 21(5): 516–32.
28. Kitley J, Woodhall M, Waters P, Leite MI, Devenney E, et al. (2012) Myelin-oligodendrocyte glycoprotein antibodies in adults with a neuromyelitis optica phenotype. Neurology 79(12): 1273–7.

29. McKeon A, Fryer JP, Apiwattanakul M, Lennon VA, Hinson SR, et al. (2009) Diagnosis of neuromyelitis spectrum disorders: comparative sensitivities and specificities of immunohistochemical and immunoprecipitation assays. Arch Neurol 66(9): 1134–8.
30. Jarius S, Ruprecht K, Wildemann B, Kuempfel T, Ringelstein M, et al (2012) Contrasting disease patterns in seropositive and seronegative neuromyelitis optica: A multicentre study of 175 patients. J Neuroinflammation 9: 14. PMID:22260418
31. Iorio R, Fryer JP, Hinson SR, Fallier-Becker P, Wolburg H, et al. (2012) Astrocytic autoantibody of neuromyelitis optica (NMO-IgG) binds to aquaporin-4 extracellular loops, monomers, tetramers and high order arrays. J Autoimmun. Epub ahead of print. PMID:22906356.
32. Pittock SJ, Lennon VA (2008) Aquaporin-4 autoantibodies in a paraneoplastic context. Arch Neurol 65(5): 629–32.
33. Chang KH, Lyu RK, Chen CM, Wu YR, Chang HS, et al. (2012) Distinct features between longitudinally extensive transverse myelitis presenting with and without anti-Aquaporin 4 antibodies. Mult Scler. Epub ahead of print. PMID:22829325.

5

Improved Mass Spectrometry Assay For Plasma Hepcidin: Detection and Characterization of a Novel Hepcidin Isoform

Coby M. M. Laarakkers[1,2], Erwin T. Wiegerinck[1,2], Siem Klaver[1,2], Maria Kolodziejczyk[3], Hendrik Gille[3], Andreas M. Hohlbaum[3], Harold Tjalsma[1,2], Dorine W. Swinkels[1,2*]

1 Laboratory of Genetic, Endocrine and Metabolic Diseases, Department of Laboratory Medicine, Radboud University Medical Centre, Nijmegen, The Netherlands, **2** Hepcidinanalysis.com, Nijmegen, The Netherlands, **3** Pieris AG, Freising, Germany

Abstract

Mass spectrometry (MS)-based assays for the quantification of the iron regulatory hormone hepcidin are pivotal to discriminate between the bioactive 25-amino acid form that can effectively block the sole iron transporter ferroportin and other naturally occurring smaller isoforms without a known role in iron metabolism. Here we describe the design, validation and use of a novel stable hepcidin-25^{+40} isotope as internal standard for quantification. Importantly, the relative large mass shift of 40 Da makes this isotope also suitable for easy-to-use medium resolution linear time-of-flight (TOF) platforms. As expected, implementation of hepcidin-25^{+40} as internal standard in our weak cation exchange (WCX) TOF MS method yielded very low inter/intra run coefficients of variation. Surprisingly, however, in samples from kidney disease patients, we detected a novel peak (m/z 2673.9) with low intensity that could be identified as hepcidin-24 and had previously remained unnoticed due to peak interference with the formerly used internal standard. Using a cell-based bioassay it was shown that synthetic hepcidin-24 was, like the -22 and -20 isoforms, a significantly less potent inducer of ferroportin degradation than hepcidin-25. During prolonged storage of plasma at room temperature, we observed that a decrease in plasma hepcidin-25 was paralleled by an increase in the levels of the hepcidin-24, -22 and -20 isoforms. This provides first evidence that all determinants for the conversion of hepcidin-25 to smaller inactive isoforms are present in the circulation, which may contribute to the functional suppression of hepcidin-25, that is significantly elevated in patients with renal impairment. The present update of our hepcidin TOF MS assay together with improved insights in the source and preparation of the internal standard, and sample stability will further improve our understanding of circulating hepcidin and pave the way towards further optimization and standardization of plasma hepcidin assays.

Editor: Harpal Singh Randeva, University of Warwick - Medical School, United Kingdom

Funding: This work was in part funded by the European Union (grant agreement No. FP7-HEALTH-F4-2011-278408; http://www.eurocalin-fp7.eu). The funders had no role in study design, data collection and analysis, decision to publish, or preparation of the manuscript. No additional external funding was received for this study.

Competing Interests: CMML, ETW, SK, HT and DWS are employees of the Radboud University Medical Centre (Nijmegen, The Netherlands) that amongst others offers hepcidin measurements to the medical scientific and pharmaceutical communities through a service unit called "hepcidinanalysis.com" on a fee per sample basis. MK, HG and AMH are employees of Pieris AG (Freising, Germany), an independent, clinical stage biotechnology company advancing its proprietary Anticalin technology to create safer, more efficacious and more convenient protein therapeutics.

* E-mail: d.swinkels@labgk.umcn.nl

These authors contributed equally to this work.

Introduction

The peptide hormone hepcidin plays a central role in regulating dietary iron absorption and body iron distribution. Many human diseases are associated with alterations in hepcidin concentrations. The measurement of hepcidin in biological fluids is therefore a promising tool in the diagnosis and management of medical conditions in which iron metabolism is affected [1]. Hepcidin is a 25 amino-acid peptide hormone that is predominantly produced by hepatocytes and regulates systemic iron homeostasis. Under physiological conditions N−terminal truncated hepcidin−20 and −22 peptides have been observed in the urine, but not or at low concentrations, in plasma [2–7]. These smaller hepcidin isoforms mostly occur in plasma in diseases that are associated with significantly increased hepcidin concentrations, such as sepsis and kidney failure [6,8–10]. Much is still unknown about the origin of the smaller isoforms. Data suggest that a calcium-independent tissue activity present in pancreas extracts might be responsible for the systemic N-terminal truncation of hepcidin-25 to hepcidin-22, and that dipeptidylpeptidase 4 is involved in the processing of hepcidin-22 into hepcidin-20 [11,12]. It is however unknown whether hepcidin isoforms can also be the result of ex-vivo processing. Thus far, we and others measured hepcidin using a Weak Cation Exchange Time−of−Flight Mass Spectrometry (WCX−TOF MS) method, with the synthetic hepcidin−24 (desAsp-hepcidin-25) analogue spiked into the sample as an internal standard for quantification [5,6,10]. This hepcidin analogue was chosen because these assays were run on low/

medium resolution platforms such as surface-enhanced and matrix-assisted laser desorption/ionization (SELDI/MALDI) TOF MS platforms that need a relatively large mass difference to separately detect two peptide peaks. Despite the slightly different biochemical characteristics, the binding of desAsp-hepcidin and hepcidin-25 to WCX beads were similar [5], but different for IMAC-Cu^{2+} chips as applied in SELDI-TOF measurements (hepcidin-24/hepcidin-25 ratio's 0.71 and 0.93, as observed by Swinkels et al. [5] and Campostrini et al. [10], respectively). Importantly, the implementation of this internal standard to correct for inter-assay variation proved to be very useful and has successfully detected physiologic and pathologic changes in serum hepcidin in patients with various disorders of iron homeostasis [1,6,9,13–18]. However, the exact impact on the use of desAsp-hepcidin as an internal standard on the accurate quantification of hepcidin remained unclear, especially in disorders with increased concentrations of the smaller hepcidin isoforms.

Reliable quantitative hepcidin assays have been developed on mass spectrometry and immunochemical platforms. The mass spectrometry based methods can roughly be divided in those run on low/medium resolution platforms [5,6,10,19] and those run on high resolution platforms [20–25]. Most hepcidin assays on the latter platforms report the use of internal standards that comprise stable hepcidin isotopes causing mass shifts that vary between 8 and 14 Da [20–25].

At present there are considerable differences in hepcidin measurements using the various methodologies. Our two published international hepcidin round robins (distribution of identical samples to allow inter-laboratory assay comparisons) revealed i) generally high correlations between the participating methods and ii) similar between-sample and analytical variation of most methods [25,26]. However, absolute hepcidin concentrations differed widely between the assays. These discrepancies might be attributed to differences in the values that laboratories and companies assign to the internal and external standards used for the different methods, to impurities in these standards or to loss of the standard during storage, e.g. by aggregation [27] or differential adsorption of the synthetic hepcidins to tubes or other surfaces of laboratory materials. Hence, to bridge this gap there is a need for improved insights on accurate value assignment and optimal handling of synthetic hepcidin standards to allow the formulation of recommendations on this point.

In this article, we describe an update of our assay by replacing desAsp-hepcidin (hepcidin-24) as internal standard with a newly designed stable heavy isotope hepcidin-25 that has an increased molecular weight of 40 Da compared to native hepcidin-25. Notably, this relative large mass shift makes this isotope suitable for both low/medium resolution linear time-of-flight (TOF) platforms as well as high resolution mass spectrometry instruments. We validated this updated assay by assessing its reproducibility and its clinical relevance by measuring samples from various patients groups. We moreover investigated storage stability of hepcidin-25 and its smaller isoforms in samples derived from different disease entities under various conditions (anticoagulant, storage temperature and time). An unexpected finding concerned the identification of hepcidin-24 as a novel natural isoform in samples from kidney disease patients. We quantified the bioactivity of this hepcidin-24 isoform relative to bioactive hepcidin-25, and the -22 and -20 isoforms by assessing its ability to internalize and degrade ferroportin in a cell−based assay. Finally, we investigated the effect of the source and handling of different internal standards that are currently in use on the value assignment of the hepcidin results. These data aid in the understanding of circulating hepcidin-25 and its isoforms and the further optimization and standardization of plasma hepcidin assays throughout the world.

Materials and Methods

Sample Collection and Human Specimens

Ethics Statement. Use of serum and plasma from patients and healthy donors was approved by the Medical Ethical Committee of the district Arnhem-Nijmegen (The Netherlands) as its use in our studies conforms to the code for proper secondary use ("goed gebruik") of human tissue in the Netherlands and the declaration of Helsinki, respectively. Written informed consent was obtained from all healthy donors.

Blank plasma and serum. Hepcidin-25 blank serum was derived in April 2010 from an iron-depleted patient with juvenile hemochromatosis [28]; Hepcidin-25 blank heparin-plasma was obtained from a iron-deficient female donor in September 2012 and from the iron-depleted patient with juvenile hemochromatosis in February 2013.

Plasma from intensive care and Nephrology patients. The samples of patients from the intensive care unit (IC) and the nephrology department concerned leftovers from routine blood draws. Samples were used for measurement of hepcidin between 4–7 hrs after blood drawing and/or pooled and/or stored in aliquots at various temperatures for prolonged periods of time. EDTA (n = 10) and/or heparin plasma (n = 20) samples from IC patients were obtained in plastic tubes in October-November 2012. Heparin and EDTA samples of patients on hemodialysis (HD) were derived from aliquots (stored at −80°C) from a previous study collected between August and October 2010 [16]. For use in specific experiments, plasma pools were generated by collection of leftover samples within 24–48 hrs (at 4°C) after routine blood draw: i) heparin plasma from 30 different IC patients was collected in November-2011, ii) heparine plasma from 20 different nephrology patients was collected in November 2011 to obtain 2 different pools.

Serum and plasma from healthy controls. Serum samples from healthy controls that were used in this study concerned leftover aliquots (stored at −80°C) that were collected in February 2011 [18]. In addition, citrate plasma, heparin plasma, EDTA plasma and serum for stability studies was drawn in the afternoon from healthy volunteers (n = 5), known for having fasted morning hepcidin levels above the lower limit of detection (>0.5 nM) in September 2012. Fresh samples were measured within 2–4 hrs hours after blood drawing and stored in aliquots at various temperatures for prolonged incubation.

Quality Control (QC) samples. Heparin plasma QC samples (n = 2), high and low (HiQC2011 and LoQC2011), were obtained in November 2011 by pooling sample leftover samples from IC patients and iron-depleted HFE-hemochromatosis patients, respectively. Serum QC samples (n = 2), high and low (HiQC2009 and LoQC2009), were obtained in April 2009 from 2 blood donors from the local blood donation centre. To obtain the high QC sample, serum of one of the donors with a native hepcidin concentration <1 nM was spiked with synthetic hepcidin-25 from Peptide International (PI, Louisville, KY, USA) to a concentration of 9 nM.

Artificial Samples

Synthetic-hepcidin (100 nM in 20% acetonitril) is spiked into blank plasma or serum.

MALDI-TOF MS

WCX-TOF MS was performed between October 2011 and March 2013 as described previously by a combination of WCX bead-based hepcidin enrichment followed by TOF-MS [6]. As internal standards for quantification were used: i) synthetic hepcidin-24 (custom made Peptide International, 2673.9 Da; [5]); ii) novel custom made heavy hepcidin-25^{+40} (see below, Table 1). Mass-to-charge (*m/z*) spectra were generated by using MALDI-TOF MS (Microflex LT, Bruker Daltonics). Spectra were analyzed by using Bruker Daltonics FlexAnalysis software. A detailed protocol of this method is described in [6]. Total hepcidin concentration was defined as the sum of hepcidin-25, -24, -22, and -20 concentrations.

Novel Internal Standard Heavy Hepcidin-25^{+40}

Heavy hepcidin-25^{+40} (hepcidin-25^{+40} Da; DTHF($^{13}C_9$,^{15}N)P($^{13}C_5$,^{15}N)I($^{13}C_6$,^{15}N)CI($^{13}C_6$,^{15}N) F($^{13}C_9$,^{15}N)CCG(^{15}N)CCHRSKCGMCCKT disulfide bridged) was obtained from Peptides International (Table 1). Note that due to an isotope content of 98% the actual mass is 1 Da less than the theoretical mass of this peptide. Freeze dried hepcidin-25^{+40} was dissolved in H_2O in accordance with the manual. An 0.1 nmol/L solution of hepcidin-25^{+40} was aliquoted in 12 µL volumes in 200 µL polypropylene tubes and stored at −80°C until use.

Other Synthetic Hepcidin Peptides

In certain experiments, hepcidin-25 from Peptide International and Bachem LTD, were used (Table 1). Hepcidin-24 (custom), hepcidin-22 and hepcidin-20 were purchased from Peptide International. Confusingly, both Peptides International and Bachem differed in relative peptide content of the vials. Bachem provided the gross amount of hepcidin on the vial that (according to the package insert) consisted of 68.6% hepcidin-25 (as assessed by amino-acid analysis: the remaining being salts and water). In addition, peptide contents of catalog products of Peptides International manufactured by Peptides International itself (USA) were lower (65–75%) than those manufactured by Peptide Institute in Osaka, Japan (Table 1). According to manufacturers protocol both peptides had disulfide bonds between Cys1–Cys8, Cys2–Cys7, Cys3–Cys6, and Cys4–Cys5. Purity assessed by HPLC was >96.1–98% for Peptide International hepcidin (-isoforms) and >95% for Bachem hepcidin-25. Both companies did not determine the nature of the impurities, but in the case of peptides with multiple disulphide bridges as with hepcidin, they may concern misfolded peptides.

Pre-treatment with Anti-hepcidin Molecules

For *in vitro* immune-depletion studies, we used the Anticalin PRS-080, a highly specific and potent hepcidin antagonist exhibiting an affinity of 50 pM for hepcidin-25 and its amino−‴ terminal truncated versions including hepcidin-20 (Pieris AG, Freising, Germany; http://www.eurocalin-fp7.eu/; [29]). Anticalins are engineered human lipocalins where the natural ligand binding pocket is re-designed to bind therapeutically relevant targets in a monovalent fashion with a molar ratio of 1:1. This has also been experimentally confirmed for PRS-080 and other Anticalins through X-ray analysis of crystal structures of Anticalins in complex with their target ([30] and unpublished results). The exquisite binding specificity and selectivity of the Anticalin was investigated in detail with various methods including surface plasmon resonance in which the Anticalin did not exhibit any measurable affinity towards a range of structurally related and non related targets ([29]; and unpublished results).

Importantly, use of PRS-080 in a non-human primate model did effectively increase serum iron levels showing its ability to bind and inactivate hepcidin *in vivo* [29]. In the *in vitro* experiments described here, plasma and serum samples were incubated with known concentrations of PRS-080 (diluted in PBS) for half an hour at room temperature (RT), before employment in the WCX-TOF MS assay. Concentration of PRS-080 was determined by the UV280 method with a coefficient of variation of about 10% [31].

Hepcidin Bioactivity Assay

A stable cell line engineered for inducible expression of green fluorescent protein (GFP)-fused human ferroportin was used as a test system to determine hepcidin-24 activity relative to hepcidin-25, 22 and 20. In this cell system, ferroportin internalization and degradation provided a means to quantify hepcidin activity. The assay was implemented basically as described by Nemeth et al. [32,33] except that fluorescence was quantified in triplicate wells on a microplate reader. To determine EC_{50} values for hepcidin-25 and hepcidin-24, a curve was fitted by nonlinear regression with a four parameter logistic equation using Prism v5 (GraphPad).

Hepcidin Stability Experiments

Using the novel heavy hepcidin-25 internal standard, we assessed changes in hepcidin-25 and the smaller isoforms in various samples containing various anticoagulants *ex vivo* at room

Table 1. Characteristics of commercial hepcidin peptides used.

Hepcidin peptide	Commercial source	Product no	Lot number	MW (Da)	pI	Peptide content	Purity
Heavy hepcidin-25	PI*	n.a.	221-006221	2829.4†	8.22	100%^	>97.8%
Hepcidin-25	PI* B	PLP-4392-s	610310 3005438	2789.4	8.22	100%^ 68.6%	>98% >95%
Hepcidin-24	PI*	n.a.	740-609191	2673.9	8.51	100%^	96.3%
Hepcidin-22#	PI	PLP-3776 PI	001055V	2436.1	8.53	74.5%	96.8%
Hepcidin-20#	PI	PLP-3777-PI	001041V	2191.8	8.53	67.6%	96.1%

*catalog product of Peptide International but manufactured by Peptide Institute (Osaka, Japan);
^vialed at 100% peptide content;
†due to an isotope content of 98% the actual mass is 1 Da less than the theoretical mass of this peptide.
#delivered as 0.10 mg per vial; personal communication revealed that ≈ 0.11 mg was pipetted in each vial. PI, Peptide International (Louisville, KY, USA); B, Bachem LTD (St. Helens, UK); n.a., not applicable.

temperature (RT), 4°C, −20°C and −80°C. More specifically, changes in hepcidin concentrations relative to those found in freshly collected samples were determined i) at storage at RT and 4°C on day 0 (fresh samples), 1 (14–22 hrs), 2, 3, and 7 days after sampling/storage, ii) at −20°C after 1 week, and after 1, 4 and 6 months of storage and iii) at −80°C at after 6 months of storage. We also assessed the effect of the addition of protease inhibitors on the stability of hepcidin concentrations of heparin plasma of IC patients at RT during a 1 week storage. To this end, before storage we used 1 tablet of protease inhibitor (Complete, Mini, Roche Diagnostics, Mannheim, Germany) solved in 200 µL PBS, of which 20 µL was added to 1 ml plasma. Finally, we used our QC charts to assess the stability of hepcidin concentrations in heparin plasma and serum QC aliquots stored at −80°C for 2 years.

Results and Discussion

Design of a Heavy Hepcidin-25 Isotope for Linear TOF MS Application

Over the last years, we and others have quantified hepcidin-25 using affinity-enrichment followed by a linear TOF MS method with synthetic hepcidin−24 (desAsp-hepcidin-25) spiked into the sample as an internal standard for quantification [5,6,10]. The mass difference of 115.5 Da between synthetic hepcidin-24 and native hepcidin-25 makes resolution of both peaks highly convenient on low/medium resolution SELDI−/MALDI-TOF MS platforms. However, due to the slightly different biochemical characteristics (Table 1), the behavior of both peptides appeared different in the IMAC-Cu^{2+} SELDI-TOF approach, which motivated us to set-up hepcidin measurements using a weak cation exchange (WCX-)MALDI approach in which both peptides performed exactly the same. An important incentive of the present study was to design a broadly applicable heavy hepcidin-25 peptide that can be used as internal standard in different assay formats on various mass spectrometry platforms as this might contribute to the harmonization of the different hepcidin assays throughout the world [25]. Most assays run on high resolution MS platforms employ hepcidin-25 standards that cause a mass shift of hepcidin-25 between +8 and +14 Da [20–25]. As the methionine residue in hepcidin molecules is prone to oxidation [5](+16 Da), these relatively small mass shift internal standards would not be adequate on low/medium resolution platforms (SELDI/MALDI-TOF MS) due to peak interference between the internal standard and hepcidin-25-O_2. Therefore, we designed a heavy hepcidin-25 isotope with a relative large mass shift of +40 Da (molecular weight: 2829.4 Da; see Materials & Methods) because at this position we did not observe other interfering peaks in the WCX-TOF MS spectra of different patient populations (data not shown). The utility of the +40 Da hepcidin internal standard should be verified in high resolution MS platforms as well as in low/medium resolution TOF MS-based assays that employ other peptide enrichment approaches. These studies are foreseen to be part of a 3rd Round Robin for serum hepcidin assays, that we will initiate in 2014.

Using hepcidin-25^{+40} as a new internal standard, we performed validation experiments on different aspects of the WCX-TOF MS assay as outlined below.

Preparation of the Internal Standard and Effect on Hepcidin-25 Value Assignment

As quantification of native hepcidin in a sample is related to the known concentration of the spiked internal standard, the absolute levels of measured hepcidin are highly dependent on the value assignment of the used internal standard. One determining factor concerns proper dissolvent of the lyophilized peptide and whether or not a substantial amount of the internal standard is lost by aggregation or sticking to laboratory plastics during its preparation. Therefore, we compared the effects of three different solvents to prepare hepcidin-25^{+40} stock solutions on the measured hepcidin-25 concentration of a series of spiked blank serum samples to which known hepcidin-25 concentrations were added. These analyses clearly showed that significantly higher hepcidin-25 concentrations were measured when hepcidin-25^{+40} was dissolved in 5% phosphoric acid, compared to 20% acetonitril and water as dissolvent's (Figure 1). This finding implicates that use of 5% phosphoric acid leads to loss of about 50% of the peptide, while this seems to be the case for about 15% of the peptide when water is used. As this loss is not corrected for in the value assignment of the internal standard and the native hepcidin peak is related to the intensity of this standard peak (which is too low for its concentration), this results in measured values that are false high. Based on these observations, we defined 20% acetonitril as the best solvent to prepare stock solutions of the internal standard as this yielded a near 1:1 ratio between the measured hepcidin concentration and the nominal concentration (added concentration) hepcidin (dissolved in 20% acetonitril) in the measured samples. Therefore, 20% acetonitrile was implemented as solvent for the internal standard in our standard operating procedure, which was used throughout the remainder of this study. Notably, such differences in the preparation of internal standards could very well explain in part the high level of variation in absolute hepcidin levels measured by different assays throughout the world [25,26].

Quantification of Hepcidin Isoforms using Heavy Hepcidin-25^{+40}as Internal Standard

To assess whether hepcidin isoforms forms can also be accurately measured using the new internal standard, hepcidin-25^{+40} was spiked as internal standard into blank serum or heparin plasma samples to which synthetic hepcidin-24, -22 and -20 was added in known concentrations. As shown in Figure 2A, this approached showed good linearity in measurements of all three tested hepcidin isoforms, which was comparable to that of hepcidin-25. The small differences in the measured values of the

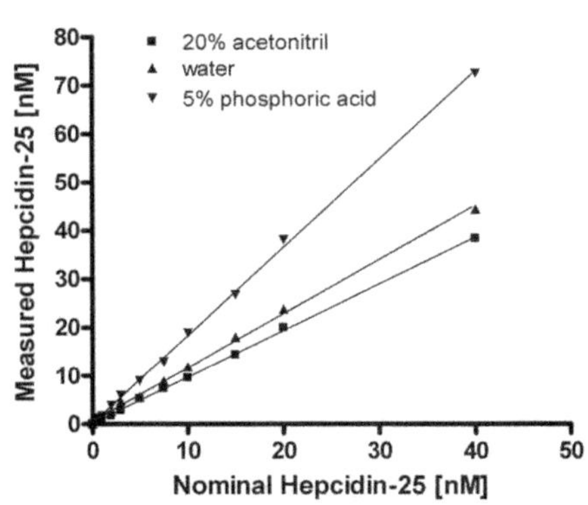

Figure 1. Effect on the solvent of hepcidin-25^{+40} on hepcidin-25 linearity and value assignment. Hepcidin-25^{+40} was dissolved in 20% acetonitril (acn), 5% phosphoric acid or water and used as internal standards in blank serum samples containing end concentrations of 0, 0.5,1, 2, 3, 5, 7.5, 10, 15, 20 and 40 nM synthetic hepcidin-25 purchased from Peptides International. Linearity of measured hepcidin-25 was determined by the formulas: Y = 1.822X+0.097 (5% phosphoric acid); Y = 1.119X+0.494 (water); Y = 0.964X+0.069 (20% acn).

different isoforms remained within the intra-run variation of the assay (Table 2). WCX-binding characteristics and flight behavior during TOF MS was assessed by spiking 10 nM of each of the different synthetic hepcidin peptides to the same blank serum prior to WCX-TOF MS. Figure 2B shows similar peak intensities of all hepcidin peptides, which confirms that hepcidin-25^{+40} internal standard is also an appropriate internal standard to quantification hepcidin-24, -22, and -20 isoforms in our assay.

Effect of Different Plasma Matrices on Measured Hepcidin-25 Concentration

To assess the effect of different plasma collection protocols on the measured hepcidin-25 concentration, hepcidin-25^{+40} was added to heparin plasma, EDTA plasma and citrate plasma that was simultaneously collected from the same patients or volunteers and frozen at −80°C until measurements. These comparisons showed good correspondence between hepcidin-25 levels in heparin plasma and citrate plasma (data not shown). However, we measured slightly higher hepcidin-25 concentrations in EDTA plasma compared to heparin plasma (Figure 3) and citrate (not shown). We have currently no clear explanation for this difference. Thus, to properly compare hepcidin data obtained within a study an identical plasma matrix should be used.

Identification of Hepcidin-24 as Native Hepcidin Isoform

To further test the applicability of hepcidin-25^{+40} as internal standard, it was used to quantify hepcidin-25 and its isoforms in samples from patients for which hepcidin isoforms were expected by WCX-TOF MS. To this purpose, we used samples from 2 heparin plasma pools that are representative for sample from intensive care and nephrology patients. As shown in Figure 4A and B, the resulting peptide profiles showed peaks at the m/z positions of hepcidin-25 and its expected hepcidin-22 and hepcidin-20 isoforms. Surprisingly, these peptide spectra did also contain a peak at m/z 2673.9 that matches the theoretical mass of hepcidin-24 (assuming that 4 intra-molecular disulphide bridges are present). To estimate the prevalence of this peak among individuals from various patients groups, we investigated samples from our bio-repository. Interestingly, the m/z 2673.9 peak was observed in none of 5 randomly selected healthy controls [18], in 7 out of 10 patients with chronic kidney disease [9], in 7 out of 10 patients on hemodialysis [16] and in 8 out of 10 IC patients (collected for the current study). In the patients (n = 20) with kidney failure (either chronic kidney disease or hemodialysis) we detected the putative hepcidin-24 peak in 3 out of 9 patients with ferritin <200 μg/L and in all 11 with a ferritin>200 μg/L, suggesting a relation with iron status and/or inflammation in these patients. To confirm the identity of the putative hepcidin-24

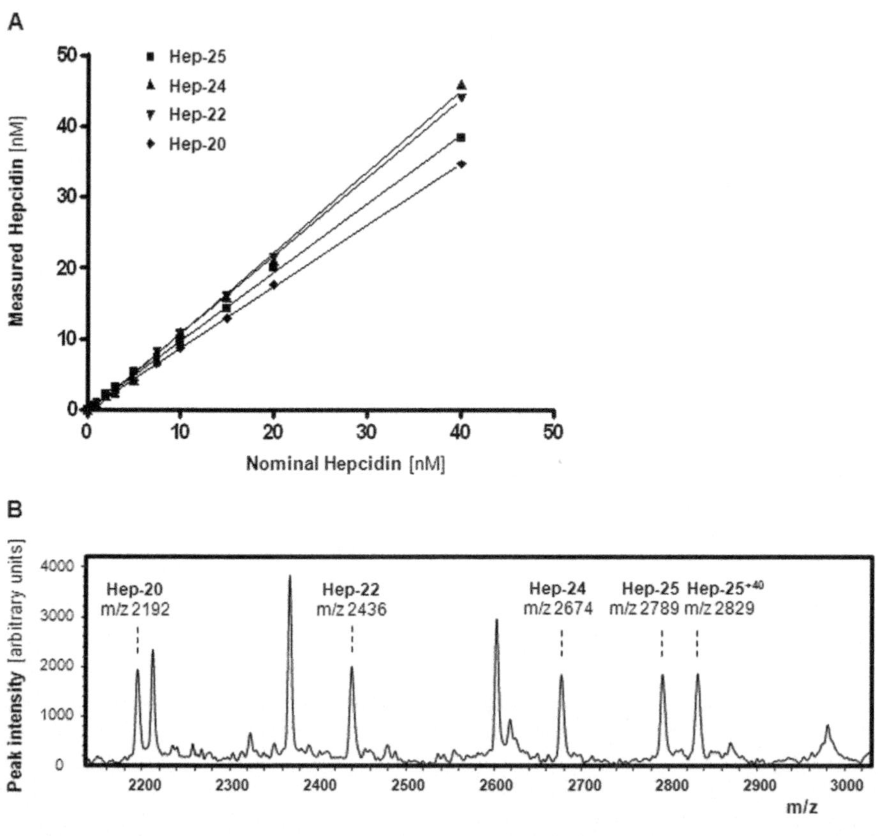

Figure 2. Quantification of hepcidin isoforms using hepcidin-25^{+40} as internal standard. A. Linearity (range 0–40 nM) for hepcidin-25, hepcidin-24, hepcidin-22, and hepcidin-20 as determined by hepcidin-25^{+40} as internal standard. Linearity curves are assessed in different runs. Blank serum (hep-24, hep-20 and hep-25) or heparin plasma (used for hep-22 as serum yields an interfering peak near the position of this isoform) was used as matrix for the addition of the synthetic hepcidin isoforms (PI) to end concentrations of 0, 0.5,1, 2, 3, 5, 7.5, 10, 15, 20 and 40 nM. Since there is a small interfering peak at 2191.8 Da in blank serum, the linearity curve of hepcidin-20 was corrected for the base line hepcidin-20 peak (data not shown). Description of the lines: hepcidin-25, Y = 0.964X+0.064 (R^2 = 0.9950); hepcidin-24, Y = 1.145X−0.767 (R^2 = 0.9975); hepcidin-22, Y = 1.100X−0.197 (R^2 = 0.9998); hepcidin-20, Y = 0.867X+0.055 (R^2 = 0.9998). **B.** WCX-TOF MS profile of blank plasma that was spiked with 10 nM of each of the synthetic hepcidin-20, -22, -24, -25, and -25^{+40} peptides, which illustrates that all these hepcidin analogues have the same WCX-binding characteristics and flight behavior during TOF MS.

Table 2. Reproducibility of the measurement of hepcidin-25 and its isoforms of the improved hepcidin assay with the novel hepcidin-25^{+40} as internal standard.*

	Intra-run (n = 8)		Inter-run (n = 8)	
Hepcidin form	**Level (nM)**	**CV(%)**	**Level (nM)**	**CV %**
Hepcidin-25	2.6	2.7	2.6	8.3
	3.5	3.3	10.5	4.6
	7.4	3.5		
	10.3	2.1		
Hepcidin-24	1.4	8.2	2.3	16.8
	2.3	4.0		
	2.4	2.9		
Hepcidin-22	1.4	8.6	1.4	11.6
	1.7	7.8		
Hepcidin-20	2.3	7.6	2.5	13.7
	4.9	4.2		
	5.8	6.0		

*Heparin samples used for intra-run experiments are the high and low QC samples and samples from nephrology patients (n = 2). Heparin samples used for inter-run experiments are the high and low QC samples.

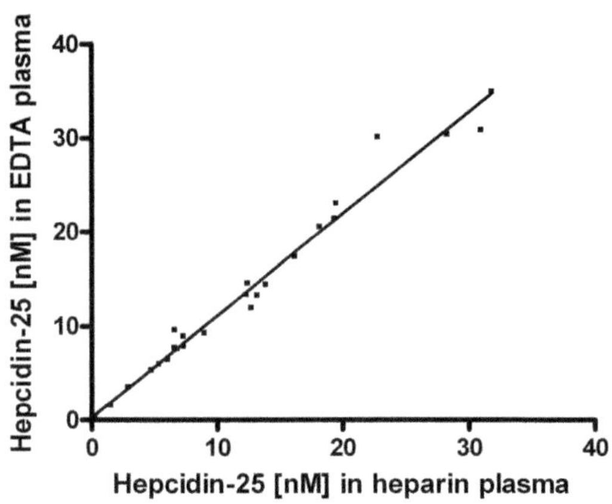

Figure 3. Comparison of hepcidin-25 concentrations measured in EDTA plasma and heparin plasma samples from healthy controls (n = 5), hemodialysis (HD) patients (n = 10) and intensive care (IC) patients (n = 10). Description of the line: Y = 1.083X+0.274 (R^2 = 0.9770).

peptide peak, we took advantage of the fact that pre-incubation of plasma samples with the anti-hepcidin molecule PRS-080 (Anticalin) prevents hepcidin binding to WCX beads due to a picomolar affinity of the Anticalin for hepcidin and thereby strong complex formation [29]. As shown in Figure 4C, hepcidin-24 disappeared, like hepcidin-25 and hepcidin-22, completely from the peptide profile after PRS-080 was added to the same plasma pool from nephrology patients that was used to generate the spectrum of panel B. This identity confirmation shows for the first time that hepcidin-24 can be a naturally occurring hepcidin isoform in the circulation of intensive care and nephrology patients. Another interesting observation concerned the fact that the intensity of the presumed peak of hepcidin-20 at m/z 2191.8 did not completely disappear from the spectrum after PRS-080 incubation. This finding suggests that another hepcidin-unrelated peptide is also present at this position as the *in vitro* affinity of PRS-080 is similar for both hepcidin-20 and hepcidin-25 (our unpublished observations) and hepcidin-25, hepcidin-24 and hepcidin-22 could be completely depleted with the employed excess of Anticalin. The latter idea is corroborated by the peptide profiles that belong to patients with iron deficiency anemia or iron-depleted juvenile hemochromatosis that lack hepcidin-25, but also do contain a detectable peak at position m/z 2191.8 (Figure 4D and E). Thus, it is highly unlikely that the peak that is observed at this position in the IC and nephrology samples is solely derived from the hepcidin-20 isoform and clearly points out that hepcidin isoform analysis on low/medium resolution platforms should be analyzed with great care. It advocates the regular validation of the peptide identity based on immune-depletion/capture on these platforms or MS/MS based peptide profiling of the involved peptide peaks when assessing different patient groups.

Bioactivity of the Newly Identified Hepcidin-24

To investigate possible clinical implications of the presence of circulating hepcidin-24 in certain patient groups, we quantified the ability of the newly detected hepcidin isoform to internalize and degrade GFP-fused ferroportin in a cell-based assay [32,33], relative to that of hepcidin-25, -22 and -20. As illustrated in Figure 5, the EC50 for hepcidin−24 was with 191.5 nM (95% confidence interval 126.8 nM - 289.2 nM) about 10-fold higher than that of hepcidin−25 with 14.7 nM (95% confidence interval 12.2 nM −17.8 nM), indicating a 10-fold lower activity of this newly identified hepcidin isoform. Notably, the activities of the shorter hepcidin-22 and -20 isoforms were at least 10-fold lower than that of hepcidin−24, which is in-line with the previously reported effects of truncation at the amino-terminus [33,34]. In view of the fact that the hepcidin-24 levels *in vivo* are at least 10-fold lower than those of hepcidin-25 (see Figure 4; data not shown), it is not anticipated that the residual bioactivity of hepcidin-24 contributes significantly to the ferroportin-regulating potential in these individuals with high circulating hepcidin-25/isoform levels.

Clinical Validation: Comparing IS Hepcidin-24 and Heavy Hepcidin

To validate the clinical use of our updated assay, we measured the hepcidin-25 concentrations in a variety of samples (n = 14) using either the novel hepcidin-25^{+40} isotope or synthetic hepcidin-24 as standard. We observed that values obtained by using the heavy isotope hepcidin-25 were slightly higher in samples that contained hepcidin isoforms than those obtained by our former hepcidin-24 standard, nevertheless the results of both assay had a strong correlation (Figure 6). Interestingly, upon correction of the hepcidin-25 values obtained by using the hepcidin-24 as internal standard, by the concentration of native hepcidin-24 in these samples measured by the use of the hepcidin-25^{+40} standard, both methodologies were nearly identical (Figure 6). This clearly shows that use of the novel hepcidin-25^{+40} internal standard leads to improved quantification in samples that contain hepcidin isoforms. On the other hand, it shows that hepcidin-24 is a convenient surrogate internal standard in samples that do not contain hepcidin isoforms.

Reproducibility of the Measurement of Hepcidin-25 and its Isoforms

To assess the robustness of the improved hepcidin assay with hepcidin-25^{+40} as internal standard, we determined coefficient of

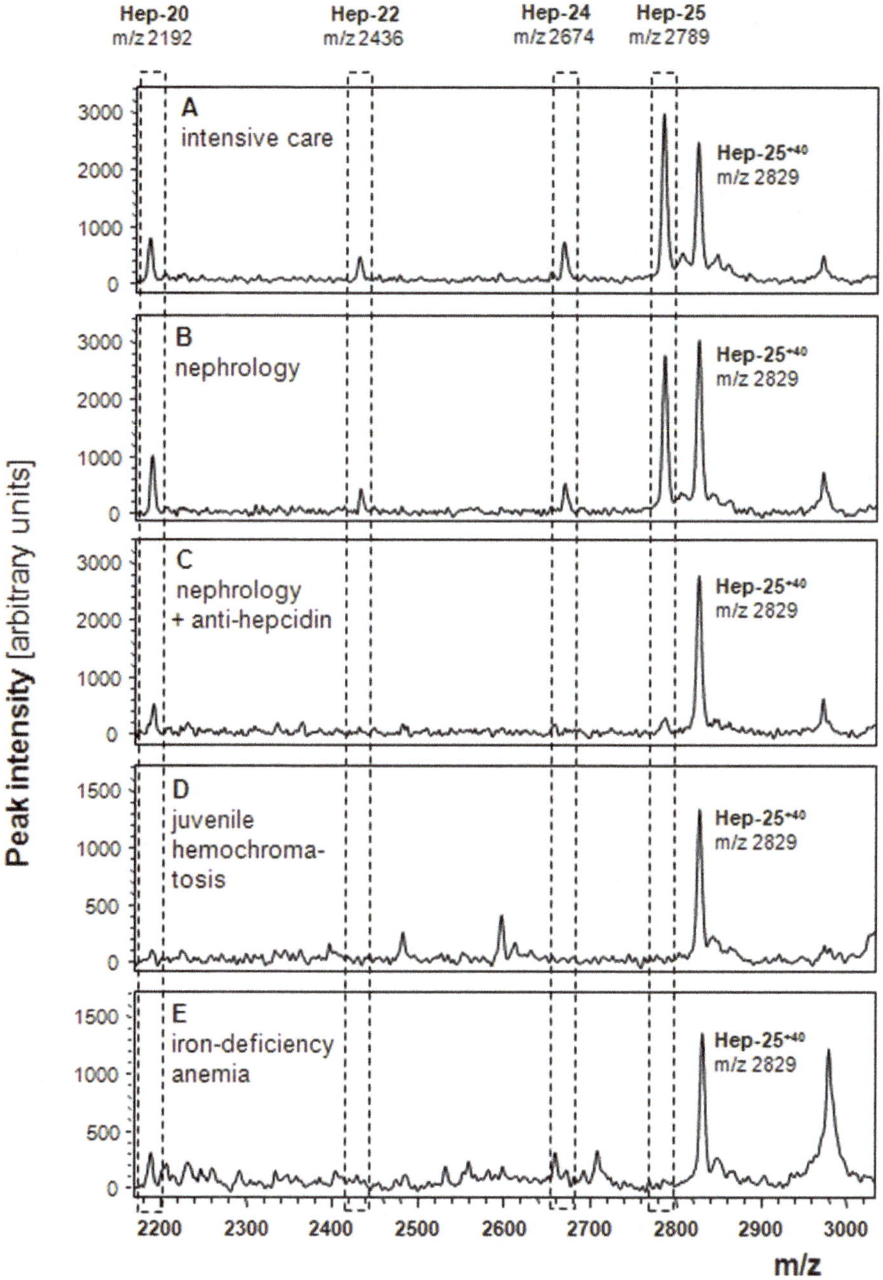

Figure 4. WCX-TOF MS profiles of sample pools of patients with presumed hepcidin isoforms. Panel **A**, peptide profile of a heparin plasma pool of IC patients; Panel **B/C**, peptide profile of heparin plasma pool of nephrology patients that were untreated (**B**) or pre-incubated with 1 molar excess of the anti-hepcidin molecule PRS-080 prior to WCX-TOF MS analysis (**C**); Panels **D/E**, control peptide profile of plasma from patients with juvenile hemochromatosis and iron deficiency anemia, respectively, that lack hepcidin-25. Positions in the spectrum: hepcidin-25^{+40} (internal standard), m/z 2829.4; hepcidin-25, m/z 2789.4; hepcidin-24, m/z 2673.9; hepcidin-22, m/z 2436.1; and hepcidin-20, m/z 2191.8. It should be noted that: i) profiles from IC and nephrology patients both clearly contain the m/z 2673.9 peak at the presumed position of hepcidin-24; ii) this peak disappears completely from the profile after incubation with PRS-080, similar to hepcidin-25/-22; iii) hepcidin-25^{+40} does not disappear from the profile as it was added after the PRS-080 incubation period, which limits complex formation; iv) the intensity of the presumed peak of hepcidin-20 at m/z 2191.8 after PRS-080 incubation decreases but does not disappear completely suggesting that another hepcidin-unrelated peptide is also present at this position; v) the peptide spectra of patient that lack hepcidin-25 also contain a peak at m/z 2191.8 (calculated between 1–2 nM), providing further evidence for the unlikeliness that this peak is solely derived from hepcidin-20.

variations (CV's) by repeated intra-run and inter-run measurements ($n = 8$) of samples with different concentrations of native hepcidin-25 and its isoforms (Table 2). These data are indicative of a reliable TOF MS assay for assessment of hepcidin-25 and its isoforms. They show higher CV's for the smaller hepcidin isoforms especially for the inter-run measurements and in the lower concentration ranges at which these smaller isoforms are generally observed.

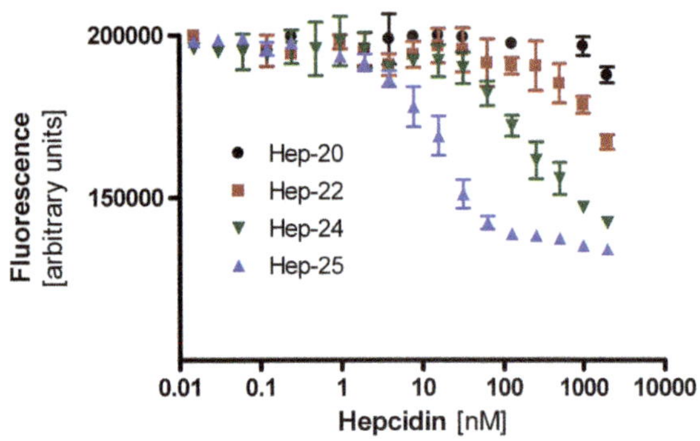

Figure 5. Hepcidin-mediated ferroportin internalization. Different concentrations of synthetic hepcidin-25, -24, -22 and -20 (indicated in nM on the horizontal axis) were added to the growth medium of a stable cell line that expresses green fluorescent protein-fused ferroportin (GFP-FPN). Hepcidin-mediated GFP-FPN internalization and degradation was quantified by measuring cellular fluorescence levels in arbitrary units.

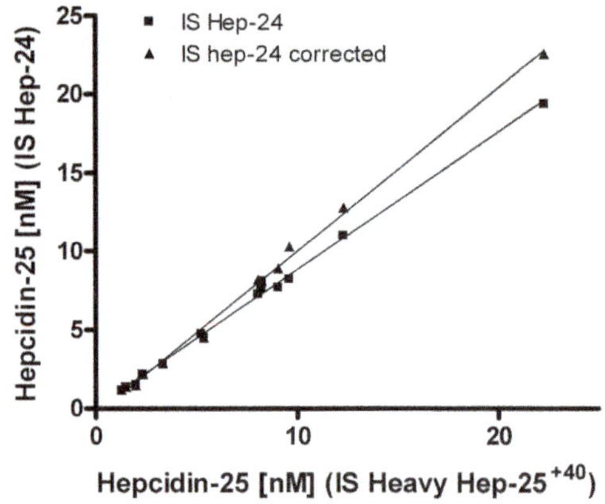

Figure 6. Comparison of hepcidin concentrations obtained by the respective internal standards hepcidin-24 and hepcidin-25^{+40}, with and without correction for native hepcidin-24 concentrations. Samples (n = 14) consisted of serum samples from healthy controls (n = 3), heparin plasma from nephrology patients (n = 7), heparin plasma high and low QC pools, serum high and low QC pools. Description of the lines: hepcidin-25 (IS HEP-24), Y = 0.878X+0.059 (R^2 = 0.9959); hepcidin-25 (IS HEP-24), with hep-24 correction, Y = 1.041X−0.425 (R^2 = 0.9960).

Value Assignment: Effect of Vendor of Synthetic Hepcidin-25

To better understand the cause of different absolute hepcidin concentrations that are measured by different hepcidin assays throughout the world [25], we used the hepcidin-25^{+40} standard to quantify synthetic hepcidin peptides from either Bachem or Peptides International that were spiked in blank serum with 10 nM of synthetic hepcidin-25 (according to the package inserts of the vendors). In the samples spiked with synthetic hepcidin from Peptides International we observed a mean concentration of 9.91 nM (n = 2) hepcidin-25, whereas this was 6.86 nM (n = 2) for the synthetic peptide from Bachem (Figure 7). These findings corroborate previous reported data and indicate that a significant difference in value assignment exists between the hepcidin-25 peptides from these two vendors, even when corrected towards a 100% peptide content of hepcidin-25 [25]. Consequently, if both would be used to quantify hepcidin, than the assay in which the Bachem peptide was used would give about 1.5-fold higher concentrations than when an internal standard from Peptides International was used.

Value Assignment Assessed by Immuno-depletion with Hepcidin Antagonists

To get an impression of the accuracy of the absolute hepcidin concentrations measured by our assay, we aimed to compare the decrease in hepcidin concentration in relation to the addition of known concentrations of the hepcidin-specific Anticalin PRS-080 that blocks binding of hepcidin to WCX beads during the affinity enrichment step of our assay. As shown in Figure 8, we observed a near equimolar decrease in hepcidin upon addition of increasing concentrations of PRS-080 for six separate native samples that did or did not contain hepcidin isoforms with concentrations between 10 nM and 40 nM of total hepcidin (sum of concentrations of hepcidin-25 and isoforms). More specifically, our observations were consistent with a molar ratio PRS-080 to hepcidin-25 of 1.53:1, suggesting that the absolute concentrations of hepcidin that are measured by our WCX-TOF MS assay are slightly too high assuming that all PRS-080 and hepcidin molecules have a confirmation that enables complex formation. Importantly, these data show that the hepcidin concentrations measured by the WCX-TOF MS assay come near to the real absolute values and are at least in the correct order of magnitude. The fact that we measure slightly higher levels than expected indicates that a minor fraction of the hepcidin-25^{+40} peptide is lost by sticking to pipette tips and/or tubes during preparation of the standard and does not end up in the stock solution of the internal standard, which thus is assigned a somewhat too high concentration. Based on our current experiments we predict that our thus far reported levels should be corrected by a factor 0.65 to approximate the absolute hepcidin levels in a sample. However, it should be emphasized that in biomedical studies the relative differences in hepcidin levels between study groups are more important than the absolute levels and that most currently available hepcidin assays are able to accurately do so, but do not allow inter-assay comparisons [25].

Effect of Sampling Material and Storage Conditions on Hepcidin-25 and its Isoforms

Storage at room temperature. To assess the stability of hepcidin in serum and plasma matrices, we kept samples from intensive care patients (n = 20) and healthy controls (n = 5) at room temperature for 0–7 days followed by WCX-TOF MS analysis using the hepcidin-25^{+40} internal standard. For heparin plasma samples, the results showed that hepcidin-25 gradually decreased at a rate that was highly sample dependent (Table 3, Figure S1A). For the samples at day 1 and day 7, the mean decrease in hepcidin-25 for the 5 controls and 20 intensive care patients was 90% (CV, 8%) and 55% (CV, 23%), respectively (Table 3).

Besides hepcidin-25 we also measured the smaller hepcidin-isoforms (Figure S1B-D, Table 3). In control samples, only the isoform hepcidin-20 emerged in time. However, the quantitative data of the isoform in time were considered not to be reliable since levels remained below 1 nM, and thus were associated with high inter-run CV's.

Interestingly, as illustrated by Figure 9, for samples from intensive care patients, the initial decrease in hepcidin-25 was paralleled by an increase in the smaller hepcidin isoforms during this period. For these patients, the respective mean increase of the hepcidin isoforms at day 7 were 116% (31%) for hepcidin-24, 160 (23%) for hepcidin-22, and 164% (41%) for hepcidin-20 (Table S1). However, the total amount of hepcidin decreased to −29% (17%) at day 7 (Table 3), suggesting that hepcidin-isoforms are

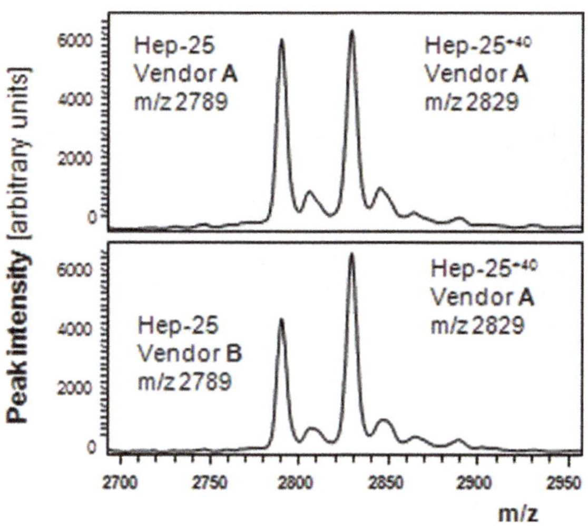

Figure 7. Peak Intensity of synthetic Hepcidin-25 of Peptides International (Vendor A; top panel) and Bachem (Vendor B; bottom panel) spiked to blank serum (to obtain a 10 nM concentration) and measured by WCX-TOF MS, using hepcidin-25^{+40} from Peptides International as internal standard. The theoretical concentrations of hepcidin-25 used in these experiments were adjusted towards 100% peptide content, based on the information provide in the package inserts of the respective Vendors (see Table 1). Hep-25, m/z 2789.4; Hep-25^{+40} (internal standard), m/z 2829.4.

degraded in smaller pieces that are not detected in our mass-spectrometry setup (Figure 9, Table 3).

For hepcidin-25 similar results were obtained for serum, EDTA plasma and citrate plasma of 5 healthy control subjects (Figure S2). Addition of protease inhibitors to heparin samples of IC patients (n = 10) prevented hepcidin-25 processing to a certain extend but did not block its decrease completely (Table S1).

Thus, since a 0–7 day stay of samples on the laboratory bench of samples with and without hepcidin isoforms results in *ex-vivo* changes in hepcidin-25 and its smaller isoforms, delays in measurements or in aliquoting for storage at lower temperatures should be prevented. These results differ from those of Itkonen et al. [35] who found hepcidin-25 levels of serum from 12 healthy controls to be stable for 1 day at room temperature.

Storage at 4°C. In heparin and EDTA plasma samples from intensive care patients (n = 10) and healthy controls (n = 5) and citrate-plasma and serum samples from healthy controls (n = 5) that were kept at 4°C for 0–7 days, concentrations of hepcidin−25, remained stable (Table S2/Figure S2). The same holds for total hepcidin and hepcidin isoforms of the intensive care patients (with the exception for hepcidin-20 in EDTA plasma, see Table S2). However, in these samples we cannot fully exclude small changes in hepcidin isoforms in time due to the relatively high inter-run CV's of their measurement. In controls concentrations of hepcidin isoforms remained <1 nM in time (data not shown).

Altogether, these data merely agree with the previously reported 6-day stability found for serum in 12 healthy controls [35] and indicate that samples can be shipped at 4°C in case this temperature can be guaranteed during transport of less than 7 days.

Storage at −20°C. Hepcidin-25 concentrations in i) serum and heparin plasma, EDTA plasma, and citrate plasma samples of healthy controls (n = 5) stored at −20°C did not change significantly for at least 6 months and ii) EDTA plasma and

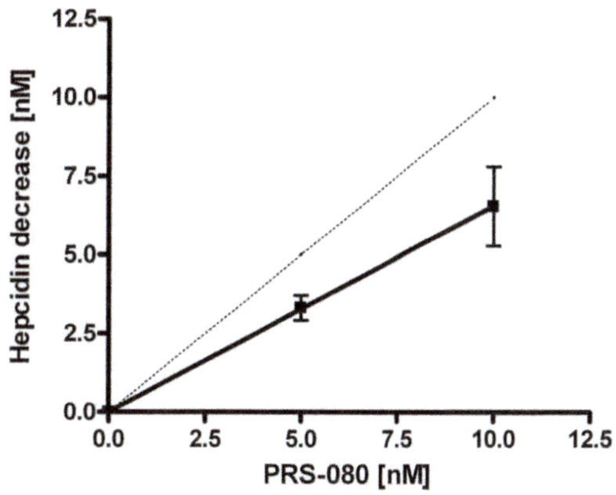

Figure 8. Hepcidin quantification by WCX-TOF MS in human samples that were pre-incubated with 0, 5 or 10 nM of PRS-080 to block binding of hepcidin to the WCX beads. Decrease in hepcidin concentration was determined by measurement of 6 different samples, with total hepcidin concentrations between 10 and 40 nM, incubated with PRS-080. Dotted line indicates the theoretically expected 1:1 ratio. Description of the line: Y = 0.655X+0.013 (R^2 = 0.9373).

heparin plasma samples of 10 intensive care patients and 5 controls stored at −20°C remained stable for at least 4 weeks (Tables S3 and S4). In time, concentrations of hepcidin isoforms remained <1 nM for controls (data not shown) and did not change in time for IC patients.

We observed no differences between the various anticoagulants used. We observed, however, a small non significant decrease of hepcidin-25 upon freezing (e.g. for heparin samples at 7 days: total intensive care and control (n = 15): 95% (CV 8%) relative to fresh

Table 3. Relative change of hepcidin at RT in heparin plasma*.

	Hepcidin level after one day at RT (%)				
	Hep-25	**Hep-24**	**Hep-22**	**Hep-20**	**Total-Hep**
average	90	120	127	115	98
CV	8	21	14	25	5
+2 SD	103	169	163	171	108
−2 SD	76	71	91	58	87
n =	**24**	**15**	**13**	**18**	**11**
	Hepcidin level after one week at RT (%)				
	Hep-25	**Hep-24**	**Hep-22**	**Hep-20**	**Total-Hep**
average	55	104	150	138	71
CV	23	29	29	38	17
+2 SD	80	164	237	242	94
−2 SD	30	44	62	34	47
n =	**24**	**15**	**13**	**18**	**11**

*Relative change in hepcidin-25, -24, -22 and -20 concentrations in heparin plasma samples expressed as percentage of the respective concentrations in fresh samples. Samples were from 19 intensive care (IC) patients and 5 controls and kept for 1 day and 7 days (1 week) at RT. Results <1.0 nM were removed from the calculations, among which are the results of the isoforms of 5 controls. Only results from samples with complete serial measurements are included.

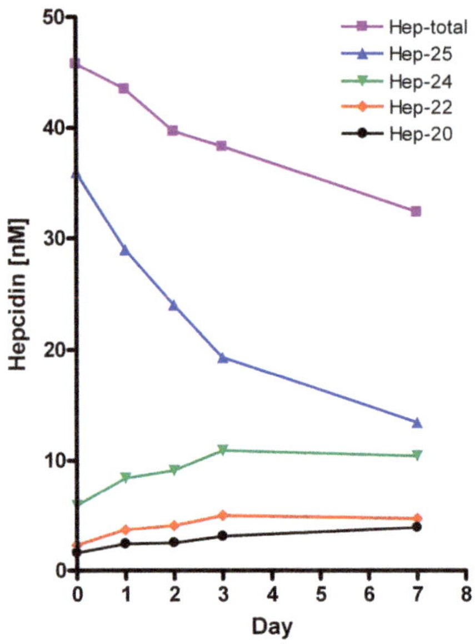

Figure 9. Changes in time at RT of hepcidin-25, 24, 22, and 20 and their sum (total hepcidin) in a representative heparin sample of an IC patient to illustrate that decrease in hepcidin-25 levels is accompanied by an increase in hepcidin-24, -22, and -20 levels. Similar observations for an extended set of samples are shown in Figure S1. Note that the total amount of hepcidin decreases despite the increase in hepcidin isoform levels.

samples, Table S3). Hepcidin isoforms did not change significantly in time. However, it should be noted that the reproducibility of the measurements of the smaller isoforms at low concentrations is low, especially the measurements of hepcidin-20 in EDTA plasma samples (Table 2, Table S3). Thus, hepcidin measurements by WCX-TOF MS are not influenced by storage at −20°C for at least 4 weeks, except for a possible small decrease in hepcidin-25 upon freezing. These data corroborate previous findings of a stability for serum at last 42 days in healthy controls [35]. Moreover, storage of EDTA plasma samples at −20°C does provide unreliable hepcidin-20 results with WCX-TOF MS.

Storage at −80°C. Hepcidin-25 results in QC serum samples measured by our former hepcidin-24 standard remained stable for 2 years at −80°C. Thereafter CV's increased, and control values regularly were out of range, necessitating re-analyses of the samples (Figure S3). This increase of CV's of hepcidin-25 measurements are most likely due to increasing noise in the MS profile upon prolonged storage.

Hepcidin-25 results in QC heparin plasma samples stored at −80°C and measured with the heavy hepcidin standard are stable for at least 1.5 years (Figure S3).

Storage of serum, heparin, EDTA and citrate plasma of 5 healthy volunteers results in a small but not significant average decrease of 4–6% in hepcidin-25 concentrations after 6 months (Table S5).

We conclude that hepcidin-25 results are not changed during 2 years of storage at −80°C, except for a possible small decrease upon freezing. However, after 2 years storage at −80°C, hepcidin results become less precise and individual measurements might become less reliable. Average results for a population may not be affected, but existing differences in hepcidin levels between groups, or correlations of hepcidin with other parameters might be more difficult to assess. These data extend previous reports describing hepcidin stability at −80°C for at least 6–8 months [23,36,37].

Concluding Remarks

Our WCX-TOF MS assay has proven to be an accurate and reproducible methodology to quantify hepcidin-25 and provides physiologically relevant results that were published in more than 60 biomedical studies over the last five years (see: www.hepcidinanalysis.com). Here we show that the implementation of the hepcidin-25^{+40} isotope, which replaces hepcidin-24 as a internal standard, further improves the test characteristics of this methodology, especially with respect to samples from patients with hepcidin isoforms. Our data on the stability of hepcidin-25 and its isoforms under various conditions will be helpful in the design of clinical studies. The implementation of the hepcidin-25^{+40} now also allows the identification and quantification of hepcidin-24 as a novel isoform in specific patient groups. Importantly, we have pinpointed critical steps in the handling and preparation of the internal standards in general to provide clues for a more accurate determination of the absolute hepcidin concentration. We strongly believe that different sources and protocols for preparation of hepcidin standards is most strongly contributing to the differences in values between methodologies as observed in the world-wide round robins [25,26]. The difference in hepcidin results obtained with standards from different vendors suggests issues with value assignment of these materials, while the choice of dissolvent significantly affects the recovery of the reference peptide. Suboptimal materials (lower starting amount) and protocols (lost of standard due to sticking to laboratory plastics) all lead to a lower actual concentration of the hepcidin standard than expected and used to calculate the measured hepcidin concentration. This leads to false high measured hepcidin levels and this observation implies that the assays that report the lowest hepcidin concentration may be most accurate in terms of absolute quantification. Interestingly, in the Round Robins for serum hepcidin our WCX-TOF MS (Method II in [26]; MS1 in Table 2 and onwards in [25]), was among the assays that measured lowest absolute hepcidin concentrations. Our titration experiments with the hepcidin-specific Anticalin PRS-080 seems to indicate that even our optimized protocols do not allow full recovery of the internal standard and that we may still measure about 1.53-fold higher hepcidin levels than actually present in a sample. Nevertheless, the hepcidin-25^{+40} peptide could aid in the desired harmonization of hepcidin throughout the world. As the relative large mass shift of 40 Da makes this isotope suitable for easy-to-use medium resolution linear TOF platforms, but can also be used on high-resolution mass spectrometry platforms as well as immunoassays. Therefore, this peptide is a strong candidate to be introduced as higher order reference materials to validate the different reference peptides that are in use throughout the world. Taken together, our present study aids in the understanding of circulating hepcidin-25 and its isoforms and provide important leads for the further harmonization of plasma hepcidin assays.

Supporting Information

Figure S1 Changes in concentrations of hepcidin-25 (**A**), hepcidin isoforms (**B, C, D**) in heparin samples from IC patients (n = 19 in black) and healthy controls (n = 5, in red) kept for 0–7 days at room temperature. Only samples with complete serial measurements are included; results <1 nM were excluded. None of the freshly collected samples from healthy control subjects contained hepcidin isoforms levels >1 nM.

Figure S2 Changes in concentrations of hepcidin-25 in heparin, EDTA and citrate plasma as well as serum in samples from the same 5 healthy controls (collected at the same moment) kept for 0–7 days at room temperature (RT) or 4°C.

Figure S3 Control charts of results from HiQC samples measured throughout time. A, serum; B, plasma. QC samples were aliquoted and stored at −80°C, a fresh aliquot was used for each measurement. Deviation starts to increase after 2 years for the serum QC sample, the plasma QC sample is still stable after 1.5 years.

Table S1 Relative change of hepcidin-25, -24, -22, -20 concentrations in heparin samples from 10 intensive care (IC) patients after 1 day (**A**) and 1 week (7 days; **B**) at room temperature (RT) with and without addition of protease inhibitors.

Table S2 Relative change of hepcidin-25, -24, -22 and -20 concentrations in heparin and EDTA plasma samples from 10 intensive care (IC) patients and 5 controls after 1 day (**A**) and after 1 weeks (7 days; **B**) at 4°C. *Results of samples with measured hepcidin <1.0 nM were deleted from the calculations. Samples from controls did not contain hepcidin isoform levels >1.0 nM; #, Measurements of hepcidin-20 in stored EDTA plasma proved unreliable due to the variable and unexplained appearance of an additional peak of similar mass in the WCX-TOF MS profile.

Table S3 Relative change of hepcidin-25, -24, -22 and -20 concentrations in heparin and EDTA plasma samples from 10 intensive care (IC) patients and 5 controls after 1 week (**A**) or 1 month (4 weeks; **B**) at −20°C. *Results of samples with measured hepcidin <1.0 nM were deleted from the calculations. Samples from controls did not contain hepcidin isoform levels >1.0 nM; #, Measurements of hepcidin-20 in stored EDTA plasma proved unreliable due to the variable and unexplained appearance of an additional peak of similar mass in the WCX-TOF MS profile, e.g. for EDTA plasma samples in 7 out of 10 patients, intra-individual changes in hepcidin-20 in time (fresh, week 1, week 4) showed an outlier (defined as 1 out of 3 serial measurements >100% different from other 2 measurements).

Table S4 Relative change of hepcidin-25 concentration in heparin -, EDTA-, and citrate plasma and serum from 5 controls after 1 week (wk), 1 month (mnt), 4 months (mnts) and 6 mnts at −20°C.

Table S5 Relative change of hepcidin-25 concentration in heparin -, EDTA-, and citrate plasma and serum from 5 controls after 6 months at −80°C.

Acknowledgments

We thank Hilde Peters and Jack Wetzels for collection of samples of patients on hemodialysis and Charlotte Schaap for the samples of healthy donors.

Author Contributions

Conceived and designed the experiments: CMML HT DWS. Performed the experiments: CMML ETW SK MK. Analyzed the data: CMML ETW SK HT DWS. Contributed reagents/materials/analysis tools: HG AMH HT DWS. Wrote the paper: CMML HG AMH HT DWS.

References

1. Kroot JJ, Tjalsma H, Fleming RE, Swinkels DW. (2011) Hepcidin in human iron disorders: diagnostic implications. Clin Chem 257: 1650–1669.
2. Park CH, Valore EV, Waring AJ, Ganz T. (2001) Hepcidin, a urinary antimicrobial peptide synthesized in the liver. J Biol Chem 276: 7806–7810.
3. Kemna E, Tjalsma H, Laarakkers C, Nemeth E, Willems H, et al. (2005) Novel urine hepcidin assay by mass spectrometry. Blood 106: 3268–3270.
4. Kemna EH, Tjalsma H, Podust VN, Swinkels DW. (2007) Mass spectrometry-based hepcidin measurements in serum and urine: analytical aspects and clinical implications. Clin Chem 53: 620–628.
5. Swinkels DW, Girelli D, Laarakkers C, Kroot J, Campostrini N, et al. (2008) Advances in quantitative hepcidin measurements by time-of-flight mass spectrometry. PLoS ONE 3: e2706.
6. Kroot JJC, Laarakkers CM, Geurts-Moespot AJ, Grebenchtchikov N, Pickkers P, et al (2010) Immunochemical and Mass-Spectrometry-Based Serum Hepcidin Assays for Iron Metabolism Disorders. Clin Chem 256: 1570–1579.
7. Campostrini N, Traglia M, Martinelli N, Corbella M, Cocca M, et al. (2012) Serum levels of the hepcidin-20 isoform in a large general population: The Val Borbera study. J Proteomics 76: 28–35.
8. Tomosugi N, Kawabata H, Wakatabe R, Higuchi M, Yamaya H, et al. (2006) Detection of serum hepcidin in renal failure and inflammation by using ProteinChip System. Blood 108: 1381–1387.
9. Peters HP, Laarakkers CM, Swinkels DW, Wetzels JF. (2010) Serum hepcidin-25 levels in patients with chronic kidney disease are independent of glomerular filtration rate. Nephrol Dial Transplant 25: 848–853.
10. Campostrini N, Castagna A, Zaninotto F, Bedogna V, Tessitore N, et al. (2010) Evaluation of hepcidin isoforms in hemodialysis patients by a proteomic approach based on SELDI-TOF MS. J Biomed Biotechnol 2010: 329646.
11. Schranz M, Bakry R, Creus M, Bonn G, Vogel W, et al. (2009) Activation and inactivation of the iron hormone hepcidin: Biochemical characterization of prohepcidin cleavage and sequential degradation to N-terminally truncated hepcidin isoforms. Blood Cells Mol Dis 43: 169–179.
12. Valore EV, Ganz T. (2008) Posttranslational processing of hepcidin in human hepatocytes is mediated by the prohormone convertase furin. Blood Cells Mol Dis 40: 132–138.
13. van Santen S, van Dongen-Lases EC, de Vegt F, Laarakkers CM, van Riel PL, et al. (2011) Hepcidin and hemoglobin content parameters in the diagnosis of iron deficiency in rheumatoid arthritis patients with anemia. Arthritis Rheum. 63: 3672–3680.
14. Santini V, Girelli D, Sanna A, Martinelli N, Duca L, et al. (2011) Hepcidin levels and their determinants in different types of myelodysplastic syndromes. PLoS One 6: e23109.
15. Nai A, Pagani A, Silvestri L, Campostrini N, Corbella M, et al. (2011) TMPRSS6 rs855791 modulates hepcidin transcription in vitro and serum hepcidin levels in normal individuals. Blood 118: 4459–4462.
16. Peters HP, Rumjon A, Bansal SS, Laarakkers CM, van den Brand JA, et al. (2012) Intra-individual variability of serum hepcidin-25 in haemodialysis patients using mass spectrometry and ELISA. Nephrol Dial Transplant 27: 3923–3929.
17. Ambaglio I, Malcovati L, Papaemmanuil E, Laarakkers CM, Della Porta MG, et al. (2013). Inappropriately low hepcidin levels in patients with myelodysplastic syndrome carrying a somatic mutation of SF3B1. Haematologica 98: 420–342.
18. Schaap CC, Hendriks JC, Kortman GA, Klaver SM, Kroot JJ, et al. (2013) Diurnal rhythm rather than dietary iron mediates daily hepcidin variations. Clin Chem 59: 527–535.
19. Ward DG, Roberts K, Stonelake P, Goon P, Zampronio CG, et al. (2008) SELDI-TOF-MS determination of hepcidin in clinical samples using stable isotope labelled hepcidin as an internal standard. Proteome Sci;6: 28.
20. Murao N, Ishigai M, Yasuno H, Shimonaka Y, Aso Y. (2007) Simple and sensitive quantification of bioactive peptides in biological matrices using liquid chromatography/selected reaction monitoring mass spectrometry coupled with trichloroacetic acid clean-up. Rapid Commun Mass Spectrom 21: 4033–4038.
21. Kobold U, Dulffer T, Dangl M, Escherich A, Kubbies M, et al. (2008) Quantification of hepcidin-25 in human serum by isotope dilution micro-HPLC-tandem mass spectrometry. Clin Chem 54: 1584–1586.
22. Li H, Rose MJ, Tran L, Zhang J, Miranda LP, et al. (2009) Development of a method for the sensitive and quantitative determination of hepcidin in human serum using LC-MS/MS. J Pharmacol Toxicol Methods 59: 171–180.
23. Bansal SS, Abbate V, Bomford A, Halket JM, Macdougall IC, et al. (2010) Quantitation of hepcidin in serum using ultra-high-pressure liquid chromatography and a linear ion trap mass spectrometer. Rapid Commun Mass Spectrom 24: 1251–1259.
24. Anderson DS, Kirchner M, Kellogg M, Kalish LA, Jeong JY, et al. (2011) Design and validation of a high-throughput matrix-assisted laser desorption ionization time-of-flight mass spectrometry method for quantification of hepcidin in human plasma. Anal Chem 83: 8357–8362.
25. Kroot JJ, van Herwaarden AE, Tjalsma H, Jansen RT, Hendriks JC, et al. (2012) Second round robin for plasma hepcidin methods: first steps toward harmonization. Am J Hematol 87: 977–983.

26. Kroot JJ, Kemna EH, Bansal SS, Busbridge M, Campostrini N, et al. (2009) Results of the first international round robin for the quantification of urinary and plasma hepcidin assays: need for standardization. Haematologica. 94: 1748–1752.
27. Hunter HN, Fulton DB, Ganz T, Vogel HJ. (2002) The solution structure of human hepcidin, a peptide hormone with antimicrobial activity that is involved in iro uptake and hereditary hemochromatosis. J Biol Chem 277: 37597–37603.
28. van Dijk BA, Kemna EH, Tjalsma H, Klaver SM, Wiegerinck ET, et al (2007) Effect of the new HJV-L165X mutation on penetrance of HFE. Blood 109: 5525–5526.
29. Hohlbaum AM, Trentman S, Gille H, Allersdorfer A, Belaiba RS, et al. (2011) Discovery and Preclinical Characterization of a Novel Hepcidin Antagonist with Tunable PK/PD Properties for the Treatment of Anemia in Different Patient Populations Blood 118: Abstract 687.
30. Schönfeld D, Matschiner G, Chatwell L, Trentmann S, Gille H, et al. (2009) An engineered lipocalin specific for CTLA-4 reveals a combining site with structural and conformational features similar to antibodies. Proc Natl Acad Sci USA 106: 8198–8203.
31. Pace CN, Vajdos F, Fee L, Grimsley G, Gray T. (1995) How to measure and predict the molar absorption coefficient of a protein. Protein Sci 4: 2411–2423.
32. Nemeth E, Tuttle MS, Powelson J, Vaughn MB, Donovan A, et al. (2004) Hepcidin regulates cellular iron efflux by binding to ferroportin and inducing its internalization. Science 306: 2090–2093.
33. Nemeth E, Preza GC, Jung CL, Kaplan J, Waring AJ, et al. (2006) The N-terminus of hepcidin is essential for its interaction with ferroportin: structure-function study. Blood 107: 328–333.
34. Rivera S, Nemeth E, Gabayan V, Lopez MA, Farshidi D, et al. (2005) Synthetic hepcidin causes rapid dose-dependent hypoferremia and is concentrated in ferroportin-containing organs. Blood 106: 2196–2199.
35. Itkonen O, Parkkinen J, Stenman UH, Hämäläinen E. (2012) Preanalytical factors and reference intervals for serum hepcidin LC-MS/MS method. Clin Chim Acta 413: 696–701.
36. Kroot JJ, Hendriks JC, Laarakkers CM, Klaver SM, Kemna EH, et al. (2009) (Pre)analytical imprecision, between-subject variability, and daily variations in serum and urine hepcidin: implications for clinical studies. Anal Biochem 389: 124–129.
37. Ganz T, Olbina G, Girelli D, Nemeth E, Westerman M. (2008) Immunoassay for human serum hepcidin. Blood 112: 4292–4297.

6

Strategy for Identifying Dendritic Cell-Processed CD4$^+$ T Cell Epitopes from the HIV Gag p24 Protein

Leonia Bozzacco[1]*, Haiqiang Yu[2¤a], Jörn Dengjel[3,4], Christine Trumpfheller[1], Henry A. Zebroski III[2], Nawei Zhang[2], Victoria Küttner[3,4], Beatrix M. Ueberheide[5¤b], Haiteng Deng[2¤c], Brian T. Chait[5], Ralph M. Steinman[1†], Svetlana Mojsov[1], David Fenyö[6]

1 Laboratory of Cellular Physiology and Immunology and Chris Browne Center, The Rockefeller University, New York, New York, United States of America, **2** Proteomics Resource Center, The Rockefeller University, New York, New York, United States of America, **3** Freiburg Institute for Advanced Studies, University of Freiburg, Freiburg, Germany, **4** Center for Biological Systems Analysis, University of Freiburg, Freiburg, Germany, **5** Laboratory of Mass Spectrometry and Gaseous Ion Chemistry, The Rockefeller University, New York, New York, United States of America, **6** Laboratory of Computational Proteomics, Center for Health Informatics and Bioinformatics, New York University Medical Center, New York, New York, United States of America

Abstract

Mass Spectrometry (MS) is becoming a preferred method to identify class I and class II peptides presented on major histocompability complexes (MHC) on antigen presenting cells (APC). We describe a combined computational and MS approach to identify exogenous MHC II peptides presented on mouse spleen dendritic cells (DCs). This approach enables rapid, effective screening of a large number of possible peptides by a computer-assisted strategy that utilizes the extraordinary human ability for pattern recognition. To test the efficacy of the approach, a mixture of epitope peptide mimics (mimetopes) from HIV gag p24 sequence were added exogenously to Fms-like tyrosine kinase 3 ligand (Flt3L)-mobilized splenic DCs. We identified the exogenously added peptide, VDRFYKTLRAEQASQ, and a second peptide, DRFYKLTRAEQASQ, derived from the original exogenously added 15-mer peptide. Furthermore, we demonstrated that our strategy works efficiently with HIV gag p24 protein when delivered, as vaccine protein, to Flt3L expanded mouse splenic DCs *in vitro* through the DEC-205 receptor. We found that the same MHC II-bound HIV gag p24 peptides, VDRFYKTLRAEQASQ and DRFYKLTRAEQASQ, were naturally processed from anti-DEC-205 HIV gag p24 protein and presented on DCs. The two identified VDRFYKTLRAEQASQ and DRFYKLTRAEQASQ MHC II-bound HIV gag p24 peptides elicited CD4$^+$ T-cell mediated responses *in vitro*. Their presentation by DCs to antigen-specific T cells was inhibited by chloroquine (CQ), indicating that optimal presentation of these exogenously added peptides required uptake and vesicular trafficking in mature DCs. These results support the application of our strategy to identify and characterize peptide epitopes derived from vaccine proteins processed by DCs and thus has the potential to greatly accelerate DC-based vaccine development.

Editor: Silke Appel, University of Bergen, Norway

Funding: This work was supported by grants from the National Institute of Allergy and Infectious Diseases, AI13013, 081677 and AI051573 to RMS and RR000862 to BTC. The funders had no role in study design, data collection and analysis, decision to publish, or preparation of the manuscript.

Competing Interests: The authors have declared that no competing interests exist.

* E-mail: lbozzacco@rockefeller.edu

¤a Current address: Thermo Fisher Scientific, San Jose, California, United States of America
¤b Current address: Proteomics Core Research Lab, Department of Biochemistry, New York University Langone Medical Center, New York, New York, United States of America
¤c Current address: School of Life Sciences, Tsinghua University, Beijing, China

† Deceased

Introduction

Cellular immune responses against any foreign non-self antigens depend upon recognition of peptides presented on the surface of antigen presenting cells (APCs), such as dendritic cells (DCs) [1]. Antigenic peptides are derived from the processing of exogenous proteins and are presented on MHC II molecules that are abundantly expressed on DCs [2]. Studies analyzing antigen presentation generally require the use of antigen-specific T cells isolated from immunized animals or human volunteers. Indeed, T cell epitopes are defined by screening sets of partially overlapping synthetic peptides covering the entire sequence of the known antigenic protein [3], or covering the entire proteome if no protein antigen is known. However, disadvantages of using T cell based-assays are the difficulty in quantitating peptides presented on the MHC complex by an absolute measurement and the large number of peptides that need to be assayed [4].

Other approaches for identification of T cell epitopes use predictive computational models of MHC binding affinity to obtain candidate peptides [5]. To date many epitopes have been identified using HLA binding motifs and peptide prediction algorithms. For example, Assarsson et al. [6] identified 54 novel epitopes from more than 4000 computer-defined peptides with HLA binding motifs, including 38 that were CD8$^+$ T cell specific. However, it is not known whether these are immunodominant or subdominant within any given CD8$^+$ T cell response.

Mass spectrometry (MS) provides an effective means to sample and quantitate MHC II-associated self- peptides from DCs for many MHC II proteins [7,8]. Here we seek to identify among the total mixture of MHC II peptides the specific antigenic peptides of low abundance derived from exogenous proteins. Due to the limited dynamic range of MS instrumentation, typically only a small fraction of the observed peptides in a complex sample is successfully sequenced, even with the fastest current instruments [9]. Targeted data acquisition approaches have been shown to achieve much lower detection limits [10], but only a limited number of peptides can be targeted in any given analysis. Thus, current data-dependent analysis will not consistently capture the peptides of interest, and targeted data acquisition cannot effectively cover the large number of possibilities.

Here, we solved this problem by a four step-approach (Figure 1B). First, we enumerate all theoretical possible MHC-II peptides from the exogenously added protein sequence. Second, we obtain mass spectra of the enriched MHC II peptides from DCs. Third, we search these spectra for mass evidence of the theoretical peptides. Fourth, if a potential match is found, this is tested by targeted tandem mass spectrometry.

To evaluate this strategy we chose HIV gag p24 peptide sequences because the gag protein plays an important role in inducing protective T cell immunity to HIV [11] and is consequently a component of several experimental vaccines [3,12,13,14]. Indeed, identification of the $CD4^+$ T cell epitopes presented on class II MHC complexes are of particular interest because of their role in the development of protective antibody responses induced by vaccination [15]. We used our strategy to identify two MHC II-bound HIV gag p24 peptides, which elicited strong T cell responses *in vitro*. We also demonstrate that presentation of MHC II-associated HIV gag peptides by mature DCs takes place intracellularly.

Results

EpiSifter for Rapid Detection of Specific MHC II Peptides

To detect, identify and measure exogenous MHC II-bound peptides presented by DCs among the abundant and complex mixture of endogenous peptides, Flt3L-mobilized DCs are pulsed *in vitro* with synthetic peptides or with a protein [1,7,16,17]. DC MHC II-complexes are then enriched by immunoprecipitation, whereupon bound peptides are eluted and purified for MS analysis (Figure 1A) [7]. However, it is difficult to detect these specific low abundance peptides by conventional data-dependent acquisition. In data-dependent acquisition, the masses of all peptides eluting at a particular time are first measured. Then, based on this measurement, starting typically from the highest intensity peak species, a number of peptide ions are sequentially isolated and their fragmentation spectra (MS/MS) acquired. In highly complex samples, as is the case for DC MHC II peptide mixture, this method often fails to detect the less abundant species of interest [18].

To overcome this limitation we developed a four step strategy (Figure 1B): (I) the program EpiSifter generates all theoretical possible MHC II peptides, ranging from 9 up to 25 amino acid in length from the amino acid sequence of the known antigen (peptides or protein). Their possible mass to charge (m/z) ratios are calculated along with their isotope ratios. Typical modifications, like oxidized methionine, are also considered for calculation of the theoretical m/z values. (II) The masses of the enriched MHC II peptides isolated from DCs are measured with high accuracy by LC-MS analysis. (III) Mass chromatograms of the m/z ratios for all the theoretically possible peptides are visualized as color-coded plots, allowing a fast and efficient checking of the spectra. (IV) Peptides, where signals are observed at the correct m/z ratio, have the expected chromatographic peak width and have

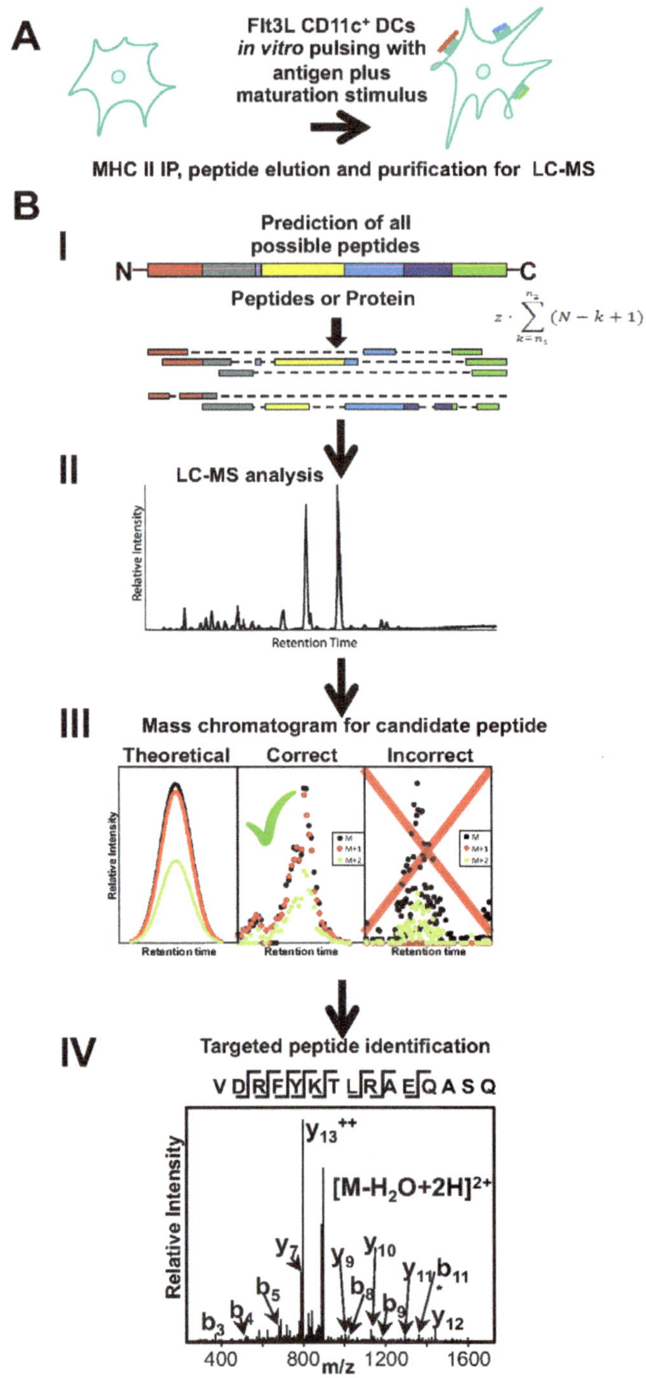

Figure 1. A schematic overview of the strategy to identify and analyze MHC II peptides presented by DCs from exogenous protein and peptide sequences. (A) MHC II peptide-complexes are isolated by immunoprecipitation from DCs pulsed *in vitro* with antigen. (B, I) A list of all possible theoretical peptides is generated from the amino acid sequence of the known antigen. (B, II) The masses of the enriched DC MHC II peptides are measured by LC-MS analysis. (B, III) The mass chromatograms for the isotopic peaks of all possible theoretical peptides are visualized as color-coded plots. (B, IV) The observed peptide masses are selected for targeted MS/MS analysis.

the correct isotope distribution, are selected for targeted MS/MS analysis in a separate run to test their sequence and identity.

Using this strategy we can rapidly find mass evidence for putative exogenous MHC II peptides and then test their identity by targeted MS/MS. An additional advantage of this strategy is that it identifies the most abundant charge state for subsequent targeted analysis, further reducing the list of candidate masses dramatically.

Identification of DC MHC II-bound Peptides from an Exogenous HIV gag p24 Peptide Mixture

To test the strategy, DCs were loaded with four HIV gag p24 15-mer peptides at concentrations of 1 µM each. These were previously identified by ELISPOT assay as mimetopes after anti-DEC gag p24 antibody vaccine immunization in two distinct mouse haplotypes (Table 1) [3]. EpiSifter generated a list of all 321 theoretically possible peptides, 9 to 15 amino acids in length (Table S1), for the four HIV gag 15 mer mimetope sequences by applying the following rules: a) removing up to four amino acids from both termini; b) allowing charge states from +1 to +4; c) considering methionine in its oxidized and unoxidized form.

Following LC-MS analysis of DC MHC II-bound peptides [7], the mass chromatograms were visualized with EpiSifter for each theoretically predicted peptide (Table S1). Figure 2 and Figure S1 show the mass chromatograms for DC MHC II-bound HIV gag p24 peptides from one representative experiment of three biological replicates. By manual inspection of the 321 mass chromatograms (one for each predicted peptide) we observed for two peptide sequences, VDRFYKTLRAQASQ (D6 in Table 1) and DRFYKTLRAQASQ (designated D6.1), a region in the mass chromatogram where the first three isotopes of the peptide were present at approximately the expected intensity ratios (Figure 2A and 2B) [19]. There was no evidence in the mass chromatograms for the other possible versions of the other three HIV gag p24 peptides added to DCs (Figure S1A, S1B, S1C). By conventional database search of the original data, MHC II HIV gag p24 peptides could not be identified among the endogenous mouse MHC II peptides. (Table S2) [7]. Next, we performed targeted MS/MS analysis for the two peptides shown in Figure 2A and 2B. Figure 2C and 2D show the MS/MS spectra of the double charged peptides VDRFYKTLRAQASQ and DRFYKTLRAQASQ at m/z 906.48 and m/z 856.94 respectively. The corresponding MS/MS spectra of MHC II-bound HIV gag p24 peptide sequences (Figure 2C and 2D) were identical to the ion product spectra of the synthetic isotope labeled peptides VDRFYKT***L**RAQASQ (m/z 909.99, z = 2, $*^{13}C_6N_1$ **L**) and DRFYKT***L**RAQASQ (m/z 860.45, z = 2, $*^{13}C_6N_1$ **L**) respectively (Table S3), confirming unambiguously the validity of the identification (Figure S2).

Table 1. HIV gag p24 immunoreactive peptides previously identified by ELISPOT assay.

BALB/c(I-A/E^d)	A9	aa 165–179	SPEVIPMFSALSEGA
	C8	aa 257–271	PVGEIYKRWIILGLN
C57BL/6 (I-A^b)	A4	aa 145–159	QAISPRTLNAWVKVV
	D6	aa 297–311	VDRFYKTLRAEQASQ

HIV gag p24 mimetope peptides identified previously from 15-mer peptide pools to stimulate IFNγ secretion by CD4$^+$ T cells in two MHC haplotypes after anti-DEC-HIV gag p24 immunization [3].

We concluded that by using EpiSifter strategy we quickly and successfully identified the exogenously added peptide, VDRFYKTLRAEQASQ, and a second peptide, DRFYKLTRAEQASQ, derived from the original added HIV gag p24 15-mer peptide by cleavage of the N-terminal amino acid (Table 1) (see below).

The Two MHC II-bound HIV gag p24 Peptides Presented by DCs are Derived from a Single Exogenous Peptide Sequence

To test that the DRFYKLTRAEQASQ peptide is indeed derived from the original exogenously added VDRFYKLTRAEQASQ peptide sequence by an active physiological process, we performed a set of experiments, the results of which are provided in Figure 3. Figures 3A and 3B show details of the mass spectra of VDRFYKTLRAQASQ (m/z 906.48, z = 2) and DRFYKTLRAQASQ (m/z 856.94, z = 2) recorded from the MHC II eluted peptide mixture. Figure 3C and 3D show mass spectra of the isotope pairs corresponding to VDRFYKTLRAQASQ (Figure 3C) and DRFYKTLRAQASQ (Figure 3D) peptides when DCs were pulsed with an equimolar mixture of heavy and light isotope forms of HIV gag p24 VDRFYKTLRAQASQ peptide (Table S3). Observation of these pairs supports the hypothesis that the 14-mer peptide DRFYKTLRAQASQ is generated from the original 15-mer peptide sequence VDRFYKTLRAQASQ added to DCs *in vitro*.

To examine in more detail the processing of the exogenously added peptides by DCs, we repeated the same experiment in the presence of 20 mM ammonium chloride (NH_4Cl) as an inhibitor of endo-lysosome compartment acidification. In the presence of 20 mM NH_4Cl, we did not detect any signal at the expected m/z values (Figure S3A and S3B), suggesting that binding to MHC II complexes of the HIV gag p24 peptides VDRFYKTLRAQASQ and DRFYKTLRAQASQ occurred intracellularly and was inhibited by 20 mM NH_4Cl.

Exogenously Added HIV gag p24 Peptides are Presented on DC MHC II Molecules in a Similar Concentration Range as the Endogenous Self-MHC II Peptides

To evaluate the absolute abundance of MHC II-bound HIV gag VDRFYKTLRAQASQ and DRFYKTLRAQASQ peptides presented on DCs, we performed quantitative analysis by using isotope labeled peptide standards [7]. Figure S4 shows the mass chromatogram of the VDRFYKTLRAQASQ and the DRFYKTLRAQASQ peptide pairs measured in one representative experiment. We extrapolated from three biological replicates that the amount of VDRFYKTLRAQASQ presented on MHC II molecules of DCs was 19±4 fmol/µL, corresponding to 590±120 copies per cell. Quantitation of DRFYKTLRAQASQ peptide showed that it was presented on DC MHC II molecules at 1800±120 copies per cell, calculated from 60±4 fmol/µL.

In conclusion, we observed that DCs presented exogenous HIV gag p24 peptides on MHC II complexes in the range of concentration detected previously for the endogenous MHC II peptide repertoire, i.e. 2.5 fmol/µL to 12 pmol/µL or from approximately 13 to 2×10^5 copies per DC [7].

Identification of VDRFYKTLRAEQASQ and DRFYKTLRAEQASQ as MHC II-bound Peptides Derived from Processing of Anti-DEC 205-HIV gag p24 Protein by DCs

To extend our strategy to a relevant vaccine protein delivered as exogenous antigen, we used anti-DEC 205 HIV gag p24 protein,

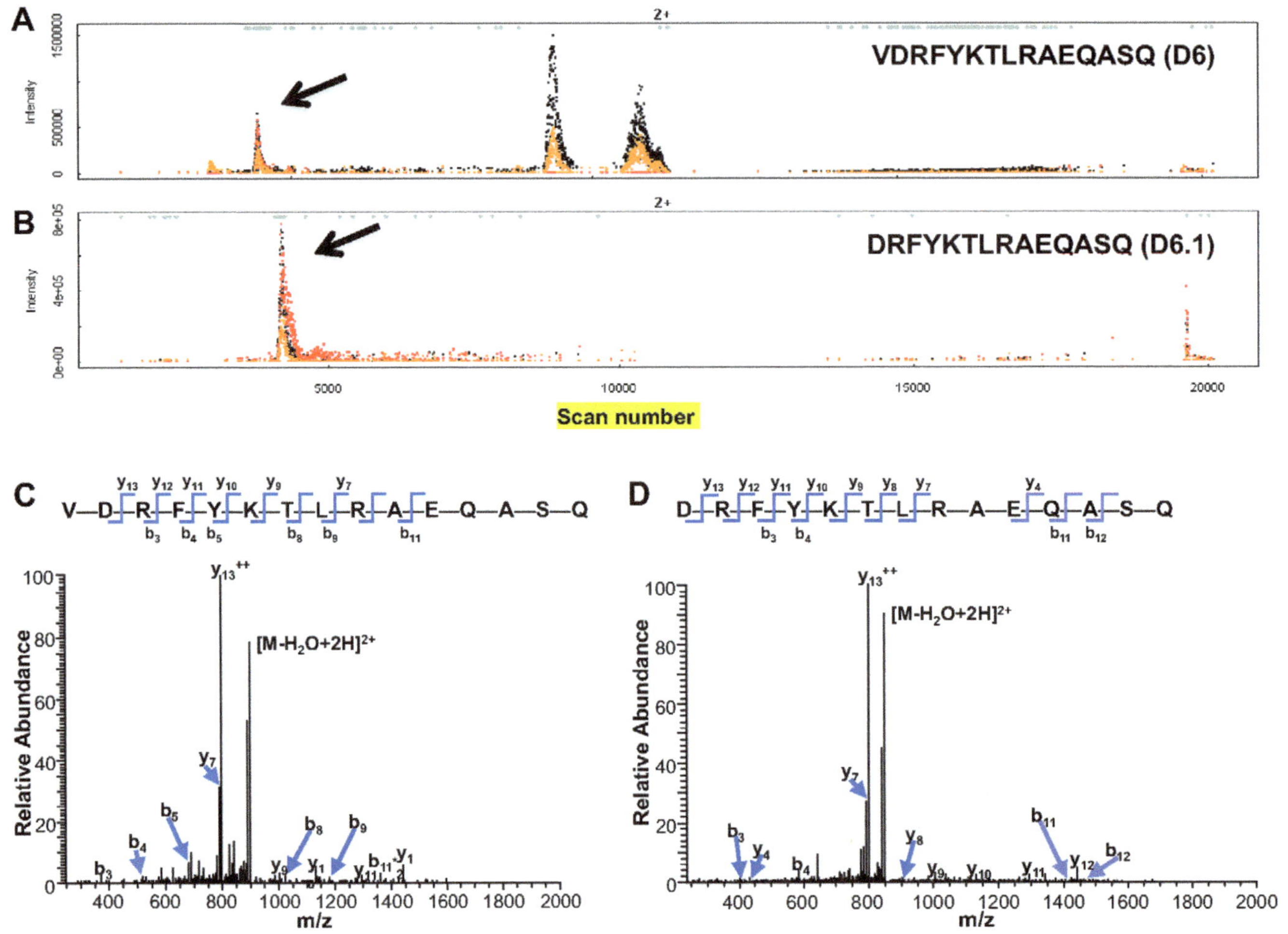

Figure 2. Query of DC MHC II- bound HIV gag p24 peptides derived from an exogenous peptide mixture. Full scan profile of two predicted peptides generated by EpiSifter. The black histogram indicates the monoisotopic peak, the red one corresponds to the first ^{13}C isotope, while the grey-coded one represents the second. The isotope envelopes marked with arrows correspond to the peptide sequences: (A) VDRFYKTLRAEQASQ (designated as D6) and (B) DRFYKTLRAEQASQ (designated as D6.1). (C, D) MS/MS spectra of the query peptides, VDRFYKTLRAEQASQ (m/z 906.48, z = 2) (C) and DRFYKTLRAEQASQ (m/z 856.94, z = 2) (D).

delivered specifically to DCs *in vitro* through the cell-specific DEC-205 receptor [3,12,13,14]. We previously demonstrated that introduction of a protein within anti-DEC-205 antibody greatly enhances the capacity of Flt3L DCs to present an antigen, such as for example HIV gag p24 protein [1,7,16,17]. Here, we first pulsed mouse spleen Flt3L-DCs *in vitro* for 5 h with 1 µg/mL (5 pM) of anti-DEC-205 HIV gag p24 protein and then we eluted MHC II peptides as described above. In this case, EpiSifter generated a list of all 8189 theoretically possible peptides, 7–27 amino acids in length (Table S4), for the HIV gag p24 protein sequence (aa 133–363 derived from HIV isolate BH10), considering charge states from +1 to +4 and methionine oxidation.

Following LC-MS analysis of eluted DC MHC II-bound peptides, the mass chromatograms were visualized with EpiSifter for all the theoretically predicted peptides. Figure 4 shows the mass chromatograms of the previously identified (Figure 2) HIV gag p24 peptides VDRFYKTLRAEQASQ (D6) and DRFYKTLRAEQASQ (D6.1), as DC MHC II-bound peptides naturally processed from the anti-DEC-205 HIV gag p24 protein added exogenously to splenic Flt3L-DCs *in vitro*. We observed for the two peptides, VDRFYKTLRAEQASQ (D6) and DRFYKTLRAEQASQ (D6.1) identified above, a region in the mass chromatogram where the first three isotopic peaks of the peptides were present with similar predicted intensities, in two different charge states, +3 and +4, (Figure 4A and 4B). By EpiSifter we did not detect the other previously described HIV gag p24 mimetopes (Table 1), in agreement with our initial observations with exogenously added 15-mer peptides (Figure S1).

We performed MS/MS analysis on the experiment shown in Figure 5 but we did not obtain the sequences of VDRFYKTLRAQASQ and DRFYKTLRAQASQ peptides. We hypothesized that the sequencing could not be performed because they were present in low amounts, thus preventing successful fragmentation of the peptide ions. Then, to prove the identity of VDRFYKTLRAQASQ and DRFYKTLRAQASQ peptides, we introduced their corresponding synthetic isotopically labeled peptides VDRFYKT***L**RAQASQ (*$^{13}C_6N_1$ **L**) and DRFYKT***L**RAQASQ (*$^{13}C_6N_1$ **L**) (Table S3) into the sample prior to LC-MS run. This pair of peptides (light and heavy isotopes) is chemically identical, with the same amino acid sequence and easily detectable because it coelutes during LC separation. In Figure 5 we show the chromatographic profile of a representative example of LC-MS experiment for VDRFYKTLRAEQASQ (Figure 5A) and DRFYKTLRAE-

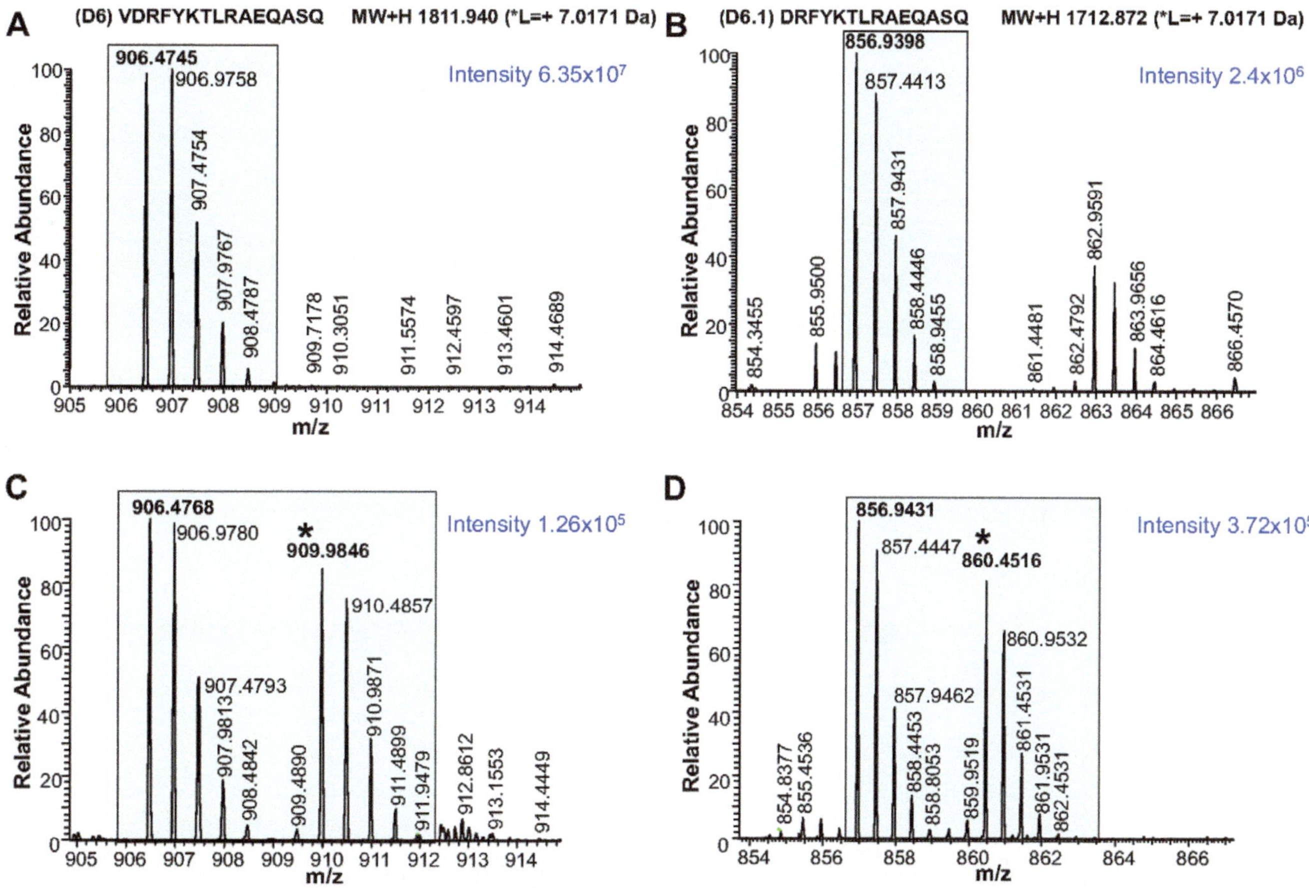

Figure 3. Detection of VDRFYKTLRAEQASQ and DRFYKTLRAEQASQ as DC MHC II-bound peptides after DCs were pulsed *in vitro* with VDRFYKTLRAEQASQ HIV gag p24 peptide sequence. (A) MS profile of the VDRFYKTLRAEQASQ peptide identified in the MHC II peptide mixture eluted from one representative sample. (B) MS profile of the DRFYKTLRAEQASQ identified from the same sample as in A. (C) MS profile of the VDRFYKTLRAEQASQ isotope pair eluted from DCs pulsed with an equimolar ratio of heavy and light VDRFYKTLRAEQASQ peptide (Table S3). (D) MS profile of the DRFYKTLRAEQASQ peptide pair identified from the same sample as in (C). Isotope peaks corresponding to m/z (z = 2) values for the detected peptides are shown in the shaded area. Heavy isotope peaks are marked with a (*).

QASQ (Figure 5B) isotope peptide pair. We enlarged the area of the spectrum comprising the masses of interest to show the extracted ion chromatographic profile of the triple charged peptides VDRFYKTLRAEQASQ (Figure 5A) and DRFYKTL-RAEQASQ (Figure 5B) at m/z 604.6532 (z = 3) and m/z 571.6297 (z = 3), respectively. The simultaneous detection of peaks at m/z 604.6532 and 606.9917 (z = 3, Figure 5A) indicated the co-elution of VDRFYKTLRAQASQ and VDRFYKT***L**RAQASQ peptides during LC separation (retention time = 50.19 minute, data not shown). We calculated as 2.1 and 1.98 ppm the mass deviation between the measured m/z values (604.6532 and 606.9917 for VDRFYKTLRAEQASQ and VDRFYKT***L**RAQASQ, respectively) and the theoretical values, 604.6519 and 606.9905, of the VDRFYKTLRAEQASQ peptide pair. In other words, the exact mass measurement of the VDRFYKTLRAEQASQ peptide pair and co-elution during LC separation led to the unambiguous peptide assignment.

The same method was applied to confirm the presence of peptide DRFYKTLRAQASQ in the sample. The measured m/z of DRFYKTLRAEQASQ and DRFYKT***L**RAQASQ ($*^{13}C_6N_1$ **L**) peptides were 571.6297 and 573.9688 respectively with only 1.75 ppm and 1.91 ppm mass deviation (Figure 5 B), compared to the theoretical mass, 571.6287 and 573.9677, calculated respectively for DRFYKTLRAEQASQ and DRFYKT***L**RAQASQ peptide sequences.

Last, the MS/MS spectra obtained for the synthetic isotope labeled peptide VDRFYKT***L**RAQASQ at m/z 606.9917, z = 3 and for DRFYKT***L**RAQASQ at m/z 573.9688, z = 3 are shown in Figure 5C and 5D, respectively. Their MS/MS sequences further confirmed the validity of the identification of VDRFYKTLRAEQASQ at m/z 604.6532 (Figure5A and 5C) and DRFYKTLRAEQASQ at m/z 571.6297 (Figure 5B and 5D) as DC MHC II-bound peptides derived from anti-DEC-205 HIV gag p24 protein exogenously added to mouse splenic DCs. Taken together the above data provide strong support for the presence of naturally processed VDRFYKTLRAEQASQ (D6) and DRFYKTLRAEQASQ (D6.1) MHC II-associated peptides derived from anti-DEC 205 HIV gag p24 protein.

We concluded that by using EpiSifter strategy we quickly and successfully identified two peptides, VDRFYKTLRAEQASQ and DRFYKLTRAEQASQ as naturally processed DC MHC II-bound peptides derived from the exogenously added anti-DEC-205 HIV gag p24 protein. In addition, these data validated our findings with exogenously added VDRFYKTLRAEQASQ (D6) HIV gag p24 peptide identified in the first part of our work (Figure 2).

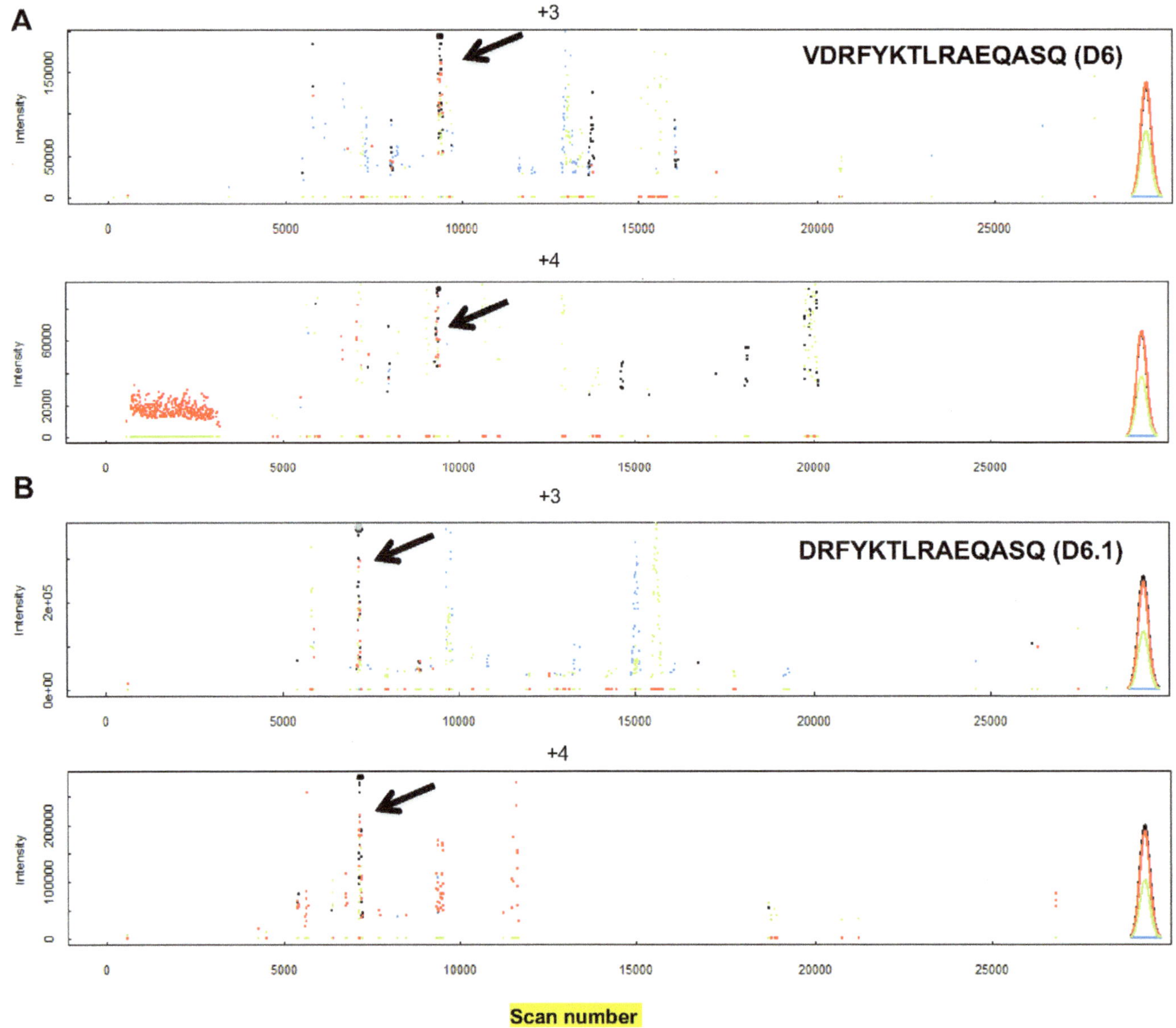

Figure 4. Query of DC MHC II- bound peptides derived from exogenously added anti-DEC-205 HIV gag p24 protein. (A, B) Full scan profile of VDRFYKTLRAEQASQ (D6) and DRFYKTLRAEQASQ (D6.1) peptides generated by EpiSifter. The isotope envelopes marked with arrows correspond to the previously identified peptide sequences: (A) VDRFYKTLRAEQASQ (D6) and (B) DRFYKTLRAEQASQ (D6.1), both found in the +3 and +4 charge states.

DCs Display Low Amounts of Processed MHC II Peptides from Anti-DEC-205 HIV gag p24 Protein

We extracted quantitative data for the HIV gag p24 VDRFYKTLRAEQASQ (D6) and DRFYKLTRAEQASQ (D6.1) peptides derived from anti-DEC-205 HIV gag p24 protein in the experiment described in Figure 5. By using a known amount (1 ng) of isotopically labeled peptides standards [7], from a single quantitation analysis (Figure 5), we measured 1.4 fmol/μL of VDRFYKTLRAEQASQ (D6) and 0.14 fmol/μL of DRFYKTL-RAEQASQ (D6.1), corresponding approximately to 17 and 1.6 copies/cell, respectively. As a control in the same experiment, we quantitated the amount of KELEEQLGPVAEETR MHC II peptide derived from endogenous apolipoprotein E [7], to be 180 copies/cell (15 fmol/μL) (Figure S5). This measurement is in good agreement with our previous quantitation of 548 copies per cell (42 fmol/μL) [7]. It falls within the 30% range of variability tolerated when different biological samples and different analytical runs are compared [20]. However, the values obtained for the VDRFYKTLRAEQASQ (D6) and of DRFYKTLRAEQASQ (D6.1) peptides presented on MHC II molecules on DCs may be an underestimation due to large differences in the peak intensities between the MHC II-bound peptides and the isotopically labeled standards added in the sample prior to the MS run.

Ag-specific CD4$^+$ T Cells Recognize MHC II-bound HIV gag Peptides Identified by LC-MS/MS

To determine the reactivity of VDRFYKTLRAQASQ and DRFYKTLRAQASQ MHC II- bound HIV gag p24 peptides identified with the present strategy, we performed 6-carboxy-fluorescein succinimidyl ester (CFSE) T cell proliferation assays.

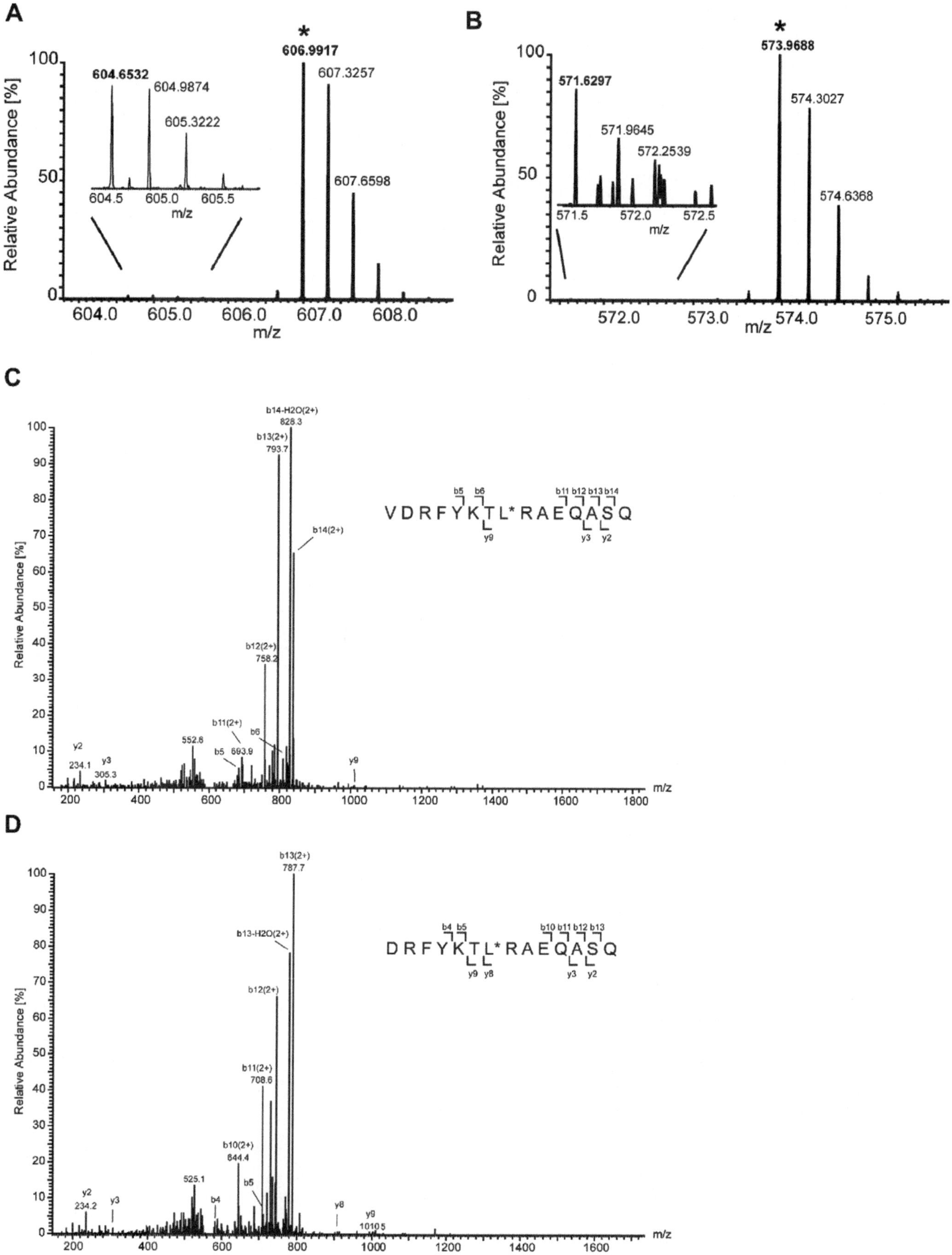
A
606.9917
607.3257
607.6598
604.6532
604.9874
605.3222
Relative Abundance [%]
m/z
B
573.9688
574.3027
574.6368
571.6297
571.9645
572.2539
Relative Abundance [%]
m/z
C
VDRFYKTL*RAEQASQ
b14-H2O(2+) 828.3
b13(2+) 793.7
b14(2+)
b12(2+) 758.2
b11(2+)
b6
b5 693.9
552.8
y2 234.1
y3 305.3
y9
Relative Abundance [%]
m/z
D
DRFYKTL*RAEQASQ
b13(2+) 787.7
b13-H2O(2+)
b12(2+)
b11(2+) 708.6
b10(2+) 644.4
525.1
b5
b4
y2 234.2
y3
y8
y9 1010.5
Relative Abundance [%]
m/z

Figure 5. Detection of VDRFYKTLRAEQASQ (D6) and DRFYKTLRAEQASQ (D6.1) as DC MHC II-bound peptides after DCs were pulsed *in vitro* with anti-DEC-205 HIV gag p24 protein. One representative experiment of two biological replicates is shown. MS profile of VDRFYKTLRAEQASQ (D6) (A) and DRFYKTLRAEQASQ (D6.1) (B) isotope peptide pair identified as MHC II-bound peptides derived from anti-DEC-205 HIV gag p24 protein, using synthetic isotopically labeled peptides as internal standards, marked with (*). For clarity, the restricted mass range of interest was zoomed in. (C, D) Identification of MHC II-bound peptides from anti-DEC-205 HIV gag p24 protein. MS/MS spectra were obtained for the synthetic isotope labeled peptides at m/z 606.9917, z = 3 for VDRFYKT*LRAQASQ (C) and at m/z 573.9688, z = 3 for DRFYKT*LRAQASQ (D) recorded in the LC-MS analysis shown in Figure 5A and 5B. The isotopically labeled amino acid is labeled with a (*). The corresponding y and b series are marked.

Synthetic VDRFYKTLRAQASQ and DRFYKTLRAQASQ HIV gag p24 peptides were added to bulk splenocytes isolated from HIV gag p24 immunized mice [21]. In Figure 6A, we measured proliferation and IFNγ production in the presence of HIV gag p24 VDRFYKTLRAQASQ (D6) and DRFYKTLRAQASQ (D6.1) peptides. We observed that CD4$^+$ T cells proliferated actively after restimulation with VDRFYKTLRAQASQ and DRFYKTLRAQASQ peptides as well as in presence of the 56-peptide mixture spanning the entire HIV gag p24 sequence (Figure 6A). No reactivity was measured with the negative control HIV gag p17 peptide mix [3] and with the endogenous MHC II apolipoprotein E (ApoE) peptide, which sequence is shown in Figure S5 [7].

We also compared the reactivity of VDRFYKTLRAQASQ peptide with the other three HIV gag p24 mimetopes (Table 1), which were not detected by LC-MS as MHC II-bound peptides (Figure S1). Only the HIV gag p24 VDRFYKTLRAQASQ peptide stimulated efficiently proliferation and IFNγ secretion of HIV p24 specific CD4$^+$ T cells, comparably to the response with the 56-peptide mix HIV gag p24 (Figure S6). None of the other HIV gag p24 mimetopes were reactive in expanding antigen-specific bulk splenocytes.

Next, to directly assess the function of DCs in presenting the above-identified HIV gag p24 VDRFYKTLRAQASQ (D6) and DRFYKTLRAQASQ (D6.1) peptides to T cells, peptide pulsed DCs were cocultured with a fixed number of T cells isolated from HIV gag p24 immunized mice [21]. T cell proliferation and IFNγ

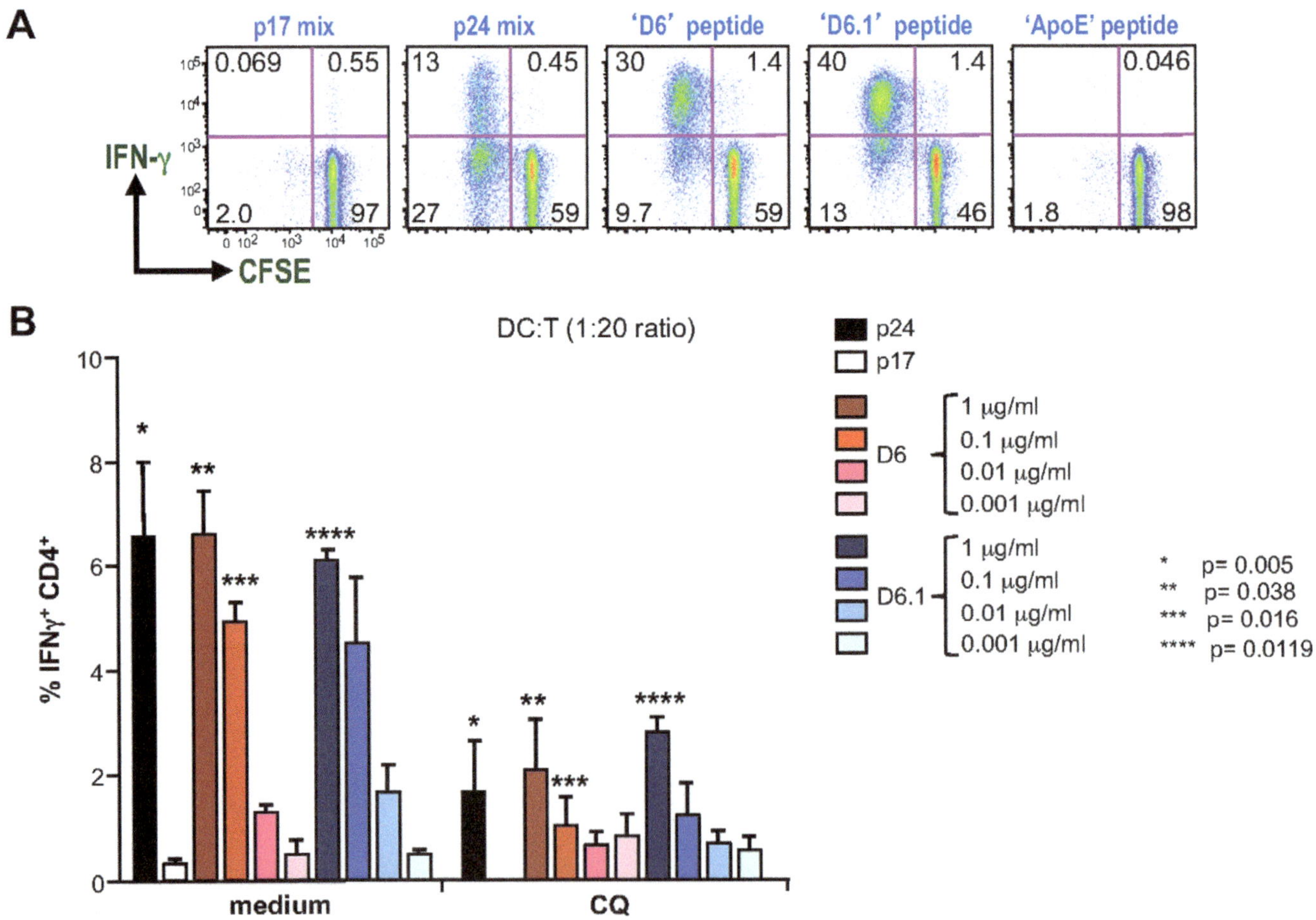

Figure 6. Immuno-reactivity of the MHC II-bound HIV gag p24 peptides measured by FACS analysis. (A) CFSE-labeled bulk splenocytes from HIV gag p24 immunized mice were stimulated with negative control HIV gag p17 peptide mix, positive control HIV gag p24 peptide mix and HIV gag VDRFYKTLRAEQASQ (D6), DRFYKTLRAEQASQ (D6.1) peptides and unrelated ApoE peptide (Figure S5). (B) Antigen presentation assay of MHC II–bound HIV gag p24 peptides in DC:T cell cocultures. Untreated (medium) and CQ-treated DCs were loaded with graded doses of peptides and added to CFSE-labeled purified T cells. Values represent mean ± SD of the percentage of CD4$^+$ IFNγ$^+$ T cells pooled from four independent experiments. Paired Student's t-test was performed for untreated- *versus* CQ-treated DCs.

secretion were measured as above. We used a T cell: DC ratio of 20:1, at which only minimal level of unspecific T cell proliferation was detected. We observed a robust proliferation and cytokine secretion of $CD4^+$ T cells when DCs were pulsed with graded dose of VDRFYKTLRAEQASQ (D6) and DRFYKTLRAEQASQ (D6.1) peptides (Figure 6B, left "medium" chart). And as expected no T cell activation was induced when DCs were pulsed with unspecific peptides (HIV gag p17).

Several reports have suggested that binding of exogenous mimetope peptides to MHC molecules may occur at the cell surface, without the need for intracellular processing [22]. Other reports have indicated that binding of exogenous peptides may take place intracellularly upon uptake of the peptides [23]. To begin to understand the mechanism of processing and presentation of exogenous peptides by DCs, we inhibited endosomal-lysosomal system acidification with chloroquine (CQ). In control experiments we first determined the optimal concentrations of CQ by using DC viability as an endpoint. Treatment of DCs with 100 μM CQ did not alter cell viability compared to untreated cells (1.8×10^6 and 1.7×10^6, respectively, recovered from initial 2×10^6 DC/well). Since treatment with CQ distinctly inhibited T cell-activation by DCs in response to HIV gag p24 peptides, we inferred that intracellular processing of exogenously added VDRFYKTLRAQASQ was required, (Figure 6B, right "CQ" chart). In four independent experiments $CD4^+$ T cell activation to peptide-loaded DCs was decreased by a median of 26% for VDRFYKTLRAQASQ ($p = 0.038$ compared to peptide alone at 1 μg/ml) and 46% for DRFYKTLRAQASQ peptide ($p = 0.012$ compared to peptide alone at 1 μg/ml). In line with other reports [24,25], we found that inhibition of peptide presentation by NH_4Cl was reversible after removal of NH_4Cl by extensive washing of DCs with PBS (Figure S7). All together these results demonstrate that low concentrations of MHC II-associated HIV gag p24 VDRFYKTLRAQASQ and DRFYKTLRAQASQ peptides induce expansion of antigen-specific $CD4^+$ T cells and that MHC II presentation of 15- or 14-mer peptides by mature DCs occurs through intracellular pathways.

Finally, we assessed the functional avidity of T cell responses following VDRFYKTLRAEQASQ and DRFYKTLRAEQASQ peptides presentation by DCs [26]. The magnitude of the $CD4^+$ T cell response measured for each peptide concentration from Figure 6B was expressed as a percentage of the maximum response observed with 1×10^{-6} M peptide (Figure S8). The EC_{50}, calculated as estimation of functional avidity indicated that VDRFYKTLRAEQASQ and DRFYKTLRAEQASQ peptides, presented on MHC II molecules of DCs, have similar sensitivity for antigen-specific T cells (EC_{50} 3.7 and 3.6×10^{-8} M, respectively). For comparison, a 30 fold higher concentration of Her2 derived peptide, PDSLRDLSVF, amino acid 420–429, from the Her2/Neu breast cancer antigen [27] is required to induce antigen-specific T cell responses in a vaccine model for breast cancer [28]. The measured high avidity of T cells specific to HIV gag p24 VDRFYKTLRAEQASQ and DRFYKTLRAEQASQ peptides reinforces the use of the HIV gag p24 protein as a vaccine target in clinical studies.

Discussion

We have described a strategy, which combines the use of a computational tool, EpiSifter, with LC-MS analysis, to identify antigen-specific MHC II peptides eluted from DCs. To test the strategy, we first used DCs that were pulsed *in vitro* with HIV gag p24 peptides and we identified two MHC II-bound HIV gag p24 sequences, VDRFYKTLRAQASQ and DRFYKTLRAQASQ, present at concentrations within the range of previously identified MHC II self-peptides on DCs [7]. For the reasons we discussed earlier, traditional data-dependent LC-MS/MS analysis was unable to identify (Table S2), these sequences and this limitation was overcome by the present strategy (Figure 1B).

There have been only few reports of exogenous MHC-bound peptides identified by LC-MS/MS analysis, and only following exhaustive manual interrogation of thousands of MS/MS spectra or by prior peptide fractionation of complex mixture followed by screening of antigenic peptide fractions in T cell assays [4,29,30,31,32]. The benefit of our strategy is that it depends in the first instance on the acquisition of accurate mass measurements that narrow the potential m/z values, which are compatible with predicted peptides. The EpiSifter predicts all possible candidate peptides and uses an effective graph representation to target only those peptide masses that are actually detected. Then, targeted MS sequencing is performed for the candidate peptides that are initially best revealed by the program (Figure 1B). This strategy is particularly useful for the analysis of unanticipated peptide species where other techniques such as selected reaction monitoring (SRM) or targeted analyses are not feasible.

In the initial experiments (Figure 2) we used VDRFYKTLRAQASQ (D6) and other three HIV gag p24 mimetope sequences (Table 1) as a proof of concept. However, this strategy is applicable to any other protein sequence. Indeed, we extended our approach and evaluated antigen processing following exogenous pulsing of mouse spleen DCs with a relevant vaccine protein, such as anti-DEC-205 HIV gag p24. We successfully demonstrated that VDRFYKTLRAQASQ (D6) and DRFYKTLRAQASQ (D6.1) peptides are two naturally processed epitopes from the entire HIV gag p24 protein (Figure 4 and 5), when delivered to DCs through the DEC-205 receptor [3,12,13,14]. We found that DCs display relatively small amounts of naturally processed VDRFYKTLRAQASQ (D6) and DRFYKTLRAQASQ (D6.1) MHC II peptides derived from anti-DEC-205 HIV gag p24 protein (Figure 5). Nonetheless, the identified MHC II-associated HIV gag p24 peptides elicited strong dose-dependent responses from antigen specific $CD4^+$ T cells, which have acquired high avidity for these MHC II-bound peptides as a result of the immunization with anti-DEC 205-HIV gag p24 protein (Figure 6 and Figure S8) [33]. These observations are consistent with previously published work suggesting that T cell activation correlates inversely with the number of MHC complexes of cognate peptide per DC and that very few MHC-peptide complexes are necessary for the induction of full-fledged T cell activation [34,35,36].

As demonstrated above, a major advantage of the described strategy is the ability to identify distinct MHC-bound peptides from antigenic, vaccine proteins. In the experiments presented in this manuscript we focused our efforts on identifying the two immunodominant (D6) and DRFYKTLRAQASQ (D6.1) peptides derived from the HIV gag p24 protein. Future experiments will examine whether additional peptides derived from the exogenously added DEC-205 HIV gag p24 protein are presented on mouse spleen DCs following the delivery of the HIV gag p24 protein through the DEC-205 receptor.

When VDRFYKTLRAQASQ peptide was added exogenously to DCs *in vitro* its presentation required intracellular processing (Figure 3C and 6B), suggesting an active role of DCs in generating MHC II-bound peptides from an exogenous peptide source [23,37] and not a direct binding of peptides to HLA molecules at the cell surface [22,38]. Three out of the four HIV gag p24 mimetopes, added exogenously, were not detected by LC-MS/MS (Figure S1) and did not elicit proliferation of $CD4^+$ T cells measured by a CFSE dilution assay (Figure S6).

Application of MS analysis for identification of antigen-derived peptides presented on HLA molecules is becoming a major focus of vaccine development [32]. Naturally processed and presented peptides can be used as vaccine candidates - either as multi epitope-based vaccine [11] or DC-based vaccine candidates [12,13,39] – which offer the advantages of high immunogenicity and valuable efficacy [40]. Recently, Ranasinghe et al. reported that an extended HIV gag D6 peptide sequence, YVDRFYKTL-RAQASQEV (amino acid 296–313), was frequently recognized among HIV infected subjects (40% of responders) and associated with spontaneous control of disease progression, independently of the patient HLA haplotype [41]. Furthermore, it has been demonstrated that sequence conservation in this region results in a high degree of cross-clade $CD4^+$ T cell recognition [42] and elicits high avidity $CD4^+$ T cells [33].

In conclusion, the present strategy promises to assist vaccine research by facilitating the identification and selection of vaccine-derived epitopes. The direct characterization of T cell epitopes opens up a powerful new approach to assess the efficacy of different vaccination strategies, formulations, antigenic inserts and adjuvants as well as new robust method for clinical trial monitoring. This advance will allow us to correlate the peptide repertoire presented on DCs with the repertoire of T cell responses elicited in vaccine recipients [12]. In turn, this understanding should lead to an accelerated rational design of clinical HIV vaccine.

Materials and Methods

Detailed materials and methods are provided in **Materials and Methods S1.**

Ethics Statement

All procedures were approved by the Institutional Animal Care and Use Committee (IACUC) at The Rockefeller University, New York (Approval ID numbers 11478 and 11460). Balb/c×C57Bl/6 (I-$A^{b,d}$/Ed) F1 mice were used in this study. Animals were maintained under specific pathogen-free conditions and used at 6–8 wk of age in accordance with the IACUC guidelines.

Cell Preparation

$CD11c^+$ DCs were obtained from Flt3L-treated Balb/c×C57Bl/6 (I-$A^{b,d}$/E^d) F1 mice [7,16]. Total $20–30\times10^6$ purified DC per well were pulsed with 1 µM of synthetic 15-mer peptides or 1 µg/ml (5 pM) of anti-DEC-205 HIV gag p24 protein for 5–6 h. Twenty five µg/mL polyinosinic:polycytidylic acid (poly IC) was also added as a maturation stimulus. In some experiments, 20 mM of NH_4Cl or 100 µM CQ was used to preincubate DCs. Treated and untreated DCs were loaded with peptides.

Affinity Purification of MHC II-peptide Complexess and Peptides Purification

MHC II molecules were purified by immunoaffinity from cell lysate [7]. The MHC II peptide complexes were eluted with 10% acetic acid [7]. Purified MHC II peptides were dried and reconstituted in 20 µL 0.1% TFA/water. Half of the peptide mixture ($3.5–4\times10^8$ cell equivalents) was injected for LC-MS analysis.

LC-MS Analysis

We used a Dionex U3000 capillary/nano-HPLC system (ThermoFisher Scientific, San Jose, CA) interfaced with the Thermo Fisher LTQ-Orbitrap mass spectrometer, operated in a data-dependent acquisition mode cycling through a single MS full-scan (620–1200 m/z, 30,000 resolution) followed by 6 MS/MS scans in the ion trap [7].

EpiSifter

EpiSifter was implemented using Perl (www.perl.org) and R (www.r-project.org). First, sequences of the proteins of interest are cut to generate all possible MHC II peptides, and corresponding m/z's are calculated for charge states +1 to +4, considering oxidized and unoxidized methionine. Second, the mass chromatograms for the possible peptides, modifications, charge states, and isotopic peaks, and interfering peaks are extracted from the data within the mass accuracy of the mass spectrometer (10 ppm). Third, the mass chromatograms are plotted together with the expected intensity ratio's using R.

Quantitation of MHC II Peptides

MHC II peptide mixture (10 µL) in 0.1% TFA/water was mixed with 10 µL of a solution containing known amounts (1 ng) of the two synthetic isotope labeled peptides dissolved in 50% acetonitrile/0.1% TFA/water. The entire peptide mixture was subjected to LC-MS analysis [7].

T Cell Cultures

Antigen-specific bulk splenocytes were isolated from F1 mice primed intraperitoneally with 5 µg anti-DEC-205 HIV gag p24 protein and 50 µg of polyinosinic-polycytidylic acid stabilized with polylysine and carboxymethylcellulose (poly ICLC), as adjuvant, and boosted under the same condition 4–6 weeks later [21]. Spleens from immunized mice (2–4 mice) were collected 2 weeks later after boost. CFSE-labeled bulk splenocytes were cultured with either 0.05 µg/ml of HIV gag p24 peptide mix, HIV gag D6 and D6.1 peptides, HIV gag p17 or ApoE peptide (KE-LEEQLGPVAEETR) [7]. For DC:T cell cocultures, 2×10^6 splenic $CD11c^+$ DCs were pulsed with poly IC and different HIV gag p24 peptide concentrations. NH_4Cl-, CQ-treated or untreated DCs were loaded with peptides, washed and added to CFSE labeled HIV gag p24 primed T cells. After 4 days of culture, cell proliferation and cytokine production were determined by intracellular cytokine staining and FACS analysis.

Supporting Information

Figure S1 Query of DC MHC II- bound HIV gag p24 peptides after DCs were pulsed *in vitro* with a peptide mixture. Panels A, B and C show the full scan generated by EpiSifter for the 3 HIV gag mimetopes (Table 1). Note that no signals with correct m/z and isotope distribution ratios were detected (compare to Figure 2).

Figure S2 Identification of MHC II-bound HIV gag p24 peptides. Comparison of MS/MS spectra of eluted HIV gag p24 peptides VDRFYKTLRAEQASQ (m/z 906.4745, z = 2, Figure 3A) (A) and DRFYKTLRAEQASQ (m/z 856.9398, z = 2, Figure 3B) (B) with MS/MS spectra of the corresponding synthetic isotopically labeled peptides (m/z 909.9822, z = 2, Table S3) (C) and (m/z 860.4479, z = 2, Table S3) (D). The isotopically labeled amino acid is labeled with a (*). The corresponding y and b series are marked.

Figure S3 NH_4Cl inhibits MHC II- binding of HIV gag p24 peptides VDRFYKTLRAQASQ and DRFYKTLRA-QASQ. (A, B) The same experiment as in Figure 3A and 3B was performed in the presence of 20 mM of NH_4Cl. No peptide ions (m/z = 906.4736 and m/z = 856.9394) corresponding to the peptides above could be detected.

Figure S4 Quantitation of MHC II-associated VDRFYKTLRAEQASQ and DRFYKTLRAEQASQ HIV gag p24 peptides. Heavy isotopes peaks are indicated with a (*). MS profile of VDRFYKTLRAEQASQ (A) and DRFYKTL-RAEQASQ (B) isotope peptide pair identified as MHC II-bound peptides. The detected m/z values are indicated in bold letters.

Figure S5 Quantitation analysis of a selected endogenous MHC II peptide. Heavy isotopes peaks are indicated with a (*). MS profile of the ApoE isotope peptide pair (KE-LEEQL*GPVAEETR, mass shift 1727.88+7 Da) identified in the MHC II peptide mixture eluted from the experiment shown in Figure 5. The detected m/z values are indicated in bold letters.

Figure S6 Immuno-reactivity of HIV gag p24 mimetopes measured by FACS analysis. CFSE-labeled bulk splenocytes from HIV gag p24 immunized mice were stimulated with negative control HIV gag p17 peptide mix, positive control HIV gag p24 peptide mix and with the four HIV gag p24 mimetopes listed in Table 1.

Figure S7 Presentation of HIV gag p24 VDRFYKTL-RAEQASQ peptide in DC:T cell cocultures. Untreated (medium) and NH_4Cl-treated DCs, prepared from the experiment described in Figure 3A and 3C, were loaded with graded doses of VDRFYKTLRAEQASQ peptide, washed with PBS and added to CFSE-labeled purified T cells. Values represent mean ± SD of the percentage of $CD4^+$ $IFN\gamma^+$ T cells pooled from two independent experiments.

Figure S8 Functional avidity of MHC II–bound HIV gag p24 peptides in DC:T cell cocultures. For each peptide concentration (−logM), the frequencies of $CD4^+$ $IFN\gamma^+$ T cells measured in (Figure 6B) were expressed as percentage of the maximum response obtained at saturation (1×10^{-6} M). The indicated EC_{50} represents the peptide concentration of VDRFYKTLRAEQASQ and DRFYKTLRAEQASQ peptides, respectively, required to attain 50% of maximal IFNγ production by HIV gag p24 specific $CD4^+$ T cells.

Table S1 List of all theoretically predicted peptides, m/z ratios and relative intensities for the isotope distributions, calculated by EpiSifter from the HIV gag p24 mimetope sequences shown in Table 1.

Table S2 List of identified DC MHC II-bound peptides. Peptide sequences were analyzed by tandem MS and identified by searching the IPI database for mouse protein sequences plus the HIV gag p24 protein sequence using the X! Tandem algorithm (see Materials and Methods), allowing for enzymatic cleavage between any pair of amino acids. Only peptides with mass accuracy better than 20 ppm and 0.4 Da for precursors and fragments, respectively, were included.

Table S3 List of isotope labeled peptides synthesized for quantitation analysis and their MS parameters. [a]Mass corresponds to the monoisotopic form of the selected peptide. [b]Predicted mass shift due to the incorporation of the isotopically labeled amino acid.

Table S4 List of all theoretically predicted peptides, m/z ratios and relative intensities for the isotope distributions, calculated by EpiSifter from the HIV gag p24 protein sequence (aa 133–363 derived from HIV isolate BH10).

Materials and Methods S1 Detailed materials and methods.

Acknowledgments

All the authors read and approved the final manuscript - with the exception of Dr. Ralph Steinman, who passed away untimely on September 30th 2011. Most importantly, Dr. Steinman actively contributed to the design and realization of this work. We thank Judy Adams for graphics, and Marguerite Nulty for help with the references.

Author Contributions

Conceived and designed the experiments: LB BTC RMS DF. Performed the experiments: LB HY JD CT NZ VK. Analyzed the data: LB HY JD CT BMU DF. Contributed reagents/materials/analysis tools: HAZ HD BMU BTC SM DF. Wrote the paper: LB CT BMU BTC SM DF. Designed and implemented the software used in analysis: DF.

References

1. Mellman I, Steinman RM (2001) Dendritic cells: specialized and regulated antigen processing machines. Cell 106: 255–258.
2. Neefjes J, Jongsma ML, Paul P, Bakke O (2011) Towards a systems understanding of MHC class I and MHC class II antigen presentation. Nat Rev Immunol 11: 823–836.
3. Trumpfheller C, Finke JS, Lopez CB, Moran TM, Moltedo B, et al. (2006) Intensified and protective $CD4^+$ T cell immunity in mice with anti-dendritic cell HIV gag fusion antibody vaccine. J Exp Med 203: 607–617.
4. Tan CT, Croft NP, Dudek NL, Williamson NA, Purcell AW (2011) Direct quantitation of MHC-bound peptide epitopes by selected reaction monitoring. Proteomics 11: 2336–2340.
5. Rammensee H-G, Friede T (1995) MHC ligands and peptide motifs: first listing. Immunogenetics 41: 178–228.
6. Assarsson E, Bui HH, Sidney J, Zhang Q, Glenn J, et al. (2008) Immunomic analysis of the repertoire of T-cell specificities for influenza A virus in humans. J Virol 82: 12241–12251.
7. Bozzacco L, Yu H, Zebroski HA, Dengjel J, Deng H, et al. (2011) Mass spectrometry analysis and quantitation of peptides presented on the MHC II molecules of mouse spleen dendritic cells. J Proteome Res 10: 5016–5030.
8. Luber CA, Cox J, Lauterbach H, Fancke B, Selbach M, et al. (2010) Quantitative proteomics reveals subset-specific viral recognition in dendritic cells. Immun 32: 279–289.
9. Eriksson J, Fenyo D (2007) Improving the success rate of proteome analysis by modeling protein-abundance distributions and experimental designs. Nat Biotech 25: 651–655.
10. Kalkum M, Lyon GJ, Chait BT (2003) Detection of secreted peptides by using hypothesis-driven multistage mass spectrometry. Proc Natl Acad Sci USA 100: 2795–2800.
11. Rolland M, Heckerman D, Deng W, Rousseau CM, Coovadia H, et al. (2008) Broad and Gag-biased HIV-1 epitope repertoires are associated with lower viral loads. PloS One 3: e1424.
12. Bozzacco L, Trumpfheller C, Siegal FP, Mehandru S, Markowitz M, et al. (2007) DEC-205 receptor on dendritic cells mediates presentation of HIV gag protein to $CD8^+$ T cells in a spectrum of human MHC I haplotypes. Proc Natl Acad Sci USA 104: 1289–1294.
13. Trumpfheller C, Longhi MP, Caskey M, Idoyaga J, Bozzacco L, et al. (2012) Dendritic cell-targeted protein vaccines: a novel approach to induce T-cell immunity. J Intern Med 271: 183–192.

14. Rerks-Ngarm S, Pitisuttithum P, Nitayaphan S, Kaewkungwal J, Chiu J, et al. (2009) Vaccination with ALVAC and AIDSVAX to prevent HIV-1 infection in Thailand. N Engl J Med 361: 2209–2220.
15. Goenka R, Barnett LG, Silver JS, O'Neill PJ, Hunter CA, et al. (2011) Dendritic cell-restricted antigen presentation initiates the follicular helper T cell program but cannot complete ultimate effector differentiation. J Immunol 187: 1091–1095.
16. Bozzacco L, Trumpfheller C, Huang Y, Longhi MP, Shimeliovich I, et al. (2010) HIV gag protein is efficiently cross-presented when targeted with an antibody towards the DEC-205 receptor in Flt3 ligand-mobilized murine DC. Eur J Immunol 40: 36–46.
17. Blander JM, Medzhitov R (2006) Toll-dependent selection of microbial antigens for presentation by dendritic cells. Nature 440: 808–812.
18. Deutsch EW, Lam H, Aebersold R (2008) Data analysis and bioinformatics tools for tandem mass spectrometry in proteomics. Physiol Genomics 33: 18–25.
19. Zhang G, Ueberheide BM, Waldemarson S, Myung S, Molloy K, et al. (2010) Protein quantitation using mass spectrometry. Methods Mol Biol 673: 211–222.
20. Callister SJ, Barry RC, Adkins JN, Johnson ET, Qian WJ, et al. (2006) Normalization approaches for removing systematic biases associated with mass spectrometry and label-free proteomics. J Proteome Res 5: 277–286.
21. Trumpfheller C, Caskey M, Nchinda G, Longhi MP, Mizenina O, et al. (2008) The microbial mimic poly IC induces durable and protective $CD4^{+}$ T cell immunity together with a dendritic cell targeted vaccine. Proc Natl Acad Sci USA 105: 2574–2579.
22. Santambrogio L, Sato AK, Carven GJ, Belyanskaya SL, Strominger JL, et al. (1999) Extracellular antigen processing and presentation by immature dendritic cells. Proc Natl Acad Sci USA 96: 15056–15061.
23. Day PM, Yewdell JW, Porgador A, Germain RN, Bennink JR (1997) Direct delivery of exogenous MHC class I molecule-binding oligopeptides to the endoplasmic reticulum of viable cells. Proc Natl Acad Sci USA 94: 8064–8069.
24. Grant KI, Casciola LA, Coetzee GA, Sanan DA, Gevers W, et al. (1990) Ammonium chloride causes reversible inhibition of low density lipoprotein receptor recycling and accelerates receptor degradation. J Biol Chem 265: 4041–4047.
25. Trombetta ES, Ebersold M, Garrett W, Pypaert M, Mellman I (2003) Activation of lysosomal function during dendritic cell maturation. Science 299: 1400–1403.
26. Slifka MK, Whitton JL (2001) Functional avidity maturation of CD8+ T cells without selection of higher affinity TCR. Nat Immunol 2: 711–717.
27. Ercolini AM, Machiels JP, Chen YC, Slansky JE, Giedlen M, et al. (2003) Identification and characterization of the immunodominant rat HER-2/neu MHC class I epitope presented by spontaneous mammary tumors from HER-2/neu-transgenic mice. J Immunol 170: 4273–4280.
28. Wang B, Zaidi N, He L-Z, Zhang L, Kuroiwa JMY, et al. (2012) Targeting of the non-mutated tumor antigen HER2/neu to mature dendritic cells induces an integrated immune response that protects against breast cancer in mice. Breast Cancer Res 14: R39.
29. Hogan KT, Sutton JN, Chu KU, Busby JA, Shabanowitz J, et al. (2005) Use of selected reaction monitoring mass spectrometry for the detection of specific MHC class I peptide antigens on A3 supertype family members. Cancer Immunol Immunother 54: 359–371.
30. Tsai S-L, Chen M-H, Yeh C-T, Chu C-M, Lin A-N, et al. (1996) Purification and characterization of a naturally processed hepatitis B virus peptide recognized by CD8+ cytotoxic T lymphocytes. J Clin Invest 97: 577–584.
31. Herr W, Ranieri E, Gambotto A, Kierstead LS, Amoscato AA, et al. (1999) Identification of naturally processed and HLA-presented Epstein-Barr virus peptides recognized by CD4(+) or CD8(+) T lymphocytes from human blood. Proc Natl Acad Sci USA 96: 12033–12038.
32. Ovsyannikova IG, Johnson KL, Bergen HR 3rd, Poland GA (2007) Mass spectrometry and peptide-based vaccine development. Clin Pharmacol Ther 82: 644–652.
33. Vingert B, Perez-Patrigeon S, Jeannin P, Lambotte O, Boufassa F, et al. (2010) HIV controller CD4+ T cells respond to minimal amounts of Gag antigen due to high TCR avidity. PLoS Pathog 6: e1000780.
34. Henrickson SE, Mempel TR, Mazo IB, Liu B, Artyomov MN, et al. (2008) T cell sensing of antigen dose governs interactive behavior with dendritic cells and sets a threshold for T cell activation. Nat Immunol 9: 282–291.
35. Zheng H, Jin B, Henrickson SE, Perelson AS, von Andrian UH, et al. (2008) How antigen quantity and quality determine T-cell decisions in lymphoid tissue. Mol Cell Biol 28: 4040–4051.
36. Cochran JR, Cameron TO, Stern LJ (2000) The relationship of MHC-peptide binding and T cell activation probed using chemically defined MHC class II oligomers. Immunity 12: 241–250.
37. Huang XL, Fan Z, Borowski L, Mailliard RB, Rolland M, et al. (2010) Dendritic cells reveal a broad range of MHC class I epitopes for HIV-1 in persons with suppressed viral load on antiretroviral therapy. PloS One 5: e12936.
38. Sherman MA, Weber DA, Spotts EA, Moore JC, Jensen PE (1997) Inefficient peptide binding by cell-surface class II MHC molecules. Cell Immunol 182: 1–11.
39. Niu L, Termini JM, Kanagavelu SK, Gupta S, Rolland MM, et al. (2011) Preclinical evaluation of HIV-1 therapeutic ex vivo dendritic cell vaccines expressing consensus Gag antigens and conserved Gag epitopes. Vaccine 29: 2110–2119.
40. Sette A, Newman M, Livingston B, McKinney D, Sidney J, et al. (2002) Optimizing vaccine design for cellular processing, MHC binding and TCR recognition. Tissue Antigens 59: 443–451.
41. Ranasinghe S, Flanders M, Cutler S, Soghoian DZ, Ghebremichael M, et al. (2012) HIV-specific CD4 T cell responses to different viral proteins have discordant associations with viral load and clinical outcome. J Virol 86: 277–283.
42. Ondondo BO, Rowland-Jones SL, Dorrell L, Peterson K, Cotten M, et al. (2008) Comprehensive analysis of HIV Gag-specific IFN-gamma response in HIV-1- and HIV-2-infected asymptomatic patients from a clinical cohort in The Gambia. Eur J Immunol 38: 3549–3560.

7

High-Resolution Mass Spectrometry Driven Discovery of Peptidic Danger Signals in Insect Immunity

Arton Berisha[1], Krishnendu Mukherjee[2], Andreas Vilcinskas[2], Bernhard Spengler[1], Andreas Römpp[1]*

1 Institute of Inorganic and Analytical Chemistry, Justus Liebig University, Giessen, Germany, **2** Institute of Phytopathology and Applied Zoology, Justus Liebig University, Giessen, Germany

Abstract

The 'danger model' is an alternative concept for immune response postulating that the immune system reacts to entities that do damage (danger associated molecular patterns, DAMP) and not only to entities that are foreign (pathogen-associated molecular patterns, PAMP) as proposed by classical immunology concepts. In this study we used *Galleria mellonella* to validate the danger model in insects. Hemolymph of *G. mellonella* was digested with thermolysin (as a representative for virulence-associated metalloproteinases produced by humanpathogens) followed by chromatographic fractionation. Immune-stimulatory activity was tested by measuring lysozyme activity with the lytic zone assays against *Micrococcus luteus* cell wall components. Peptides were analyzed by nano-scale liquid chromatography coupled to high-resolution Fourier transform mass spectrometers. Addressing the lack of a genome sequence we complemented the rudimentary NCBI protein database with a recently established transcriptome and *de novo* sequencing methods for peptide identification. This approach led to identification of 127 peptides, 9 of which were identified in bioactive fractions. Detailed MS/MS experiments in comparison with synthetic analogues confirmed the amino acid sequence of all 9 peptides. To test the potential of these putative danger signals to induce immune responses we injected the synthetic analogues into *G. mellonella* and monitored the anti-bacterial activity against living *Micrococcus luteus*. Six out of 9 peptides identified in the bioactive fractions exhibited immune-stimulatory activity when injected. Hence, we provide evidence that small peptides resulting from thermolysin-mediated digestion of hemolymph proteins function as endogenous danger signals which can set the immune system into alarm. Consequently, our study indicates that the danger model also plays a role in insect immunity.

Editor: Joy Sturtevant, Louisiana State University, United States of America

Funding: This work was supported by the Hessian ministry of Science and Arts (HMWK) through LOEWE focus "Ambiprobe" and LOEWE focus "Insect biotechnology", by the Deutsche Forschungsgemeinschaft DFG through project Sp314/13-1, and by the European Research Council ERC-STG 210356. The funders had no role in study design, data collection and analysis, decision to publish, or preparation of the manuscript.

Competing Interests: The authors have declared that no competing interests exist.

* E-mail: andreas.roempp@anorg.chemie.uni-giessen.de

Introduction

A prerequisite for the evolution of Metazoa was the ability to discriminate between self and nonself. This power of distinction is mediated by a functioning immune system which encompasses all endogenous mechanisms and molecules contributing to host defense against microbes and parasites.

Vertebrate and invertebrate animals share the ancient innate immunity mediating discrimination between infectious nonself and noninfectious self by a limited number of germ line encoded receptors which bind to molecules common among pathogens, but absent from the host such as microbial cell wall components, particularly, bacterial lipopolysaccharides (LPS), peptidoglycans, and fungal ß-1,3-glucans [1]. This binding of receptor proteins to pathogen-associated molecular patterns (PAMPs) differs from the highly specific recognition of a wide range of antigens by a large and diverse spectrum of somatically rearranged receptors of T and B cells which represent an evolutionary novelty of vertebrates. Their adaptive or acquired immunity relies on antibody-based immunological memory of infection [2]. Janeways model claiming that activation of the immune system requires recognition of PAMPs has become a paradigm, particularly in insect immunity [3]. Binding of PAMPs to corresponding pattern recognition receptors results in activation of immune responses including both cellular mechanisms such as phagocytosis or multicellular encapsulation of microbes, and humoral defences such as the rapid synthesis of antimicrobial peptides and proteins [4].

Matzinger has challenged immunological research by introducing her danger model, which is based on the idea that the immune system is more concerned with entities that do damage than with those that are foreign [5]. Her alternative model explains recognition of foreign threat by molecules released from damaged cells or wounded tissues whose binding to corresponding pattern recognition receptors results in activation of immune responses. Such damage-associated molecular patterns (DAMPs) function as danger signals which can set the immune system into alarm [6]. First evidence for the presence of immunity-related danger signals has been reported from insects, particularly in the larvae of the greater wax moth *Galleria mellonella* [7]. The latter have in between emerged as powerful model hosts for pathogens infecting insects or humans [8]. This study elucidated that activation of innate immune responses in *G. mellonella* does not necessarily require PAMPs, the presence of microbial enzymes, particularly metalloproteinases of the M4 family with thermolysin as the prototype, in the hemolymph is sufficient to produce DAMPs [7]. Members of the thermolysin-family encompass prominent virulence factors and toxins of human pathogens such as aureolysin, bacillolysin and pseudolysin which have been implicated to be responsible for increase of vascular permeability, hemorrhagic edema, sepsis and necrotic tissue destruction [9]. Thermolysin-mediated digestion of hemolymph proteins in *G. mellonella* results in formation of small peptidic fragments smaller than 3 kDa which elicit innate immune responses that are qualitatively (spectrum of immune-related proteins secreted within the hemolymph) and quantitatively (expression levels of antimicrobial peptides) comparable with the response to injected bacterial lipopolysaccharide (LPS), a widely used microbial elicitor of innate immune responses [10].

Mass spectrometry (MS) is a powerful and universal tool for the analysis and identification of proteins and peptides. Complex biological samples are usually separated by liquid chromatography or gel electrophoresis prior to MS analysis. The MS-based identification of peptides and proteins from *G. mellonella* is hindered by fact that the genome is not sequenced and thus protein databases are only rudimentary. Traditional database search approaches using SEQUEST [11] or Mascot [12] do not result in satisfactory results. To compensate for this impediment we have recently subjected the immunity-related transcriptome of *G. mellonella* to next generation sequencing using the Roche 454-FLX platform combined with traditional Sanger sequencing to obtain a comprehensive immune gene repertoire [13]. In addition, we have developed a method for *de novo* sequencing of peptides isolated from *G. mellonella* hemolymph [14]. The reliability of peptide identification can be significantly improved by using high resolution and accurate measurements [15,16]. The highest mass resolution is achieved by Fourier Transform mass spectrometers based on orbital trapping [17] or ion cyclotron resonance [18,19]. A method that simplifies *de novo* sequencing is the composition based sequencing (CBS) approach which takes advantage of accurate mass measurements [20]. An example for CBS analysis of the tree frog H. savigni with unknown genome sequence was recently reported [21].

The application of high-resolution mass spectrometry enabled the identification of a number of hemolymph protein fragments resulting from thermolysin-mediated hydrolysis of hemolymph. Their putative function as danger signals was verified using synthetic analogues which were injected into *G. mellonella* larvae in order to test their capacity to elicit humoral immune responses. As in our previous studies [7,10] we used freeze-dried or living *Micrococcus luteus* bacteria as indicator organisms in lytic zone or inhibition zone assays, respectively.

Materials and Methods

Insect rearing and sample preparation

G. mellonella larvae were reared on an artificial diet (22% maize meal, 22% wheat germ, 11% dry yeast, 17.5% bee wax, 11% honey and 11% glycerin) at 32°C in darkness. For collecting the hemolymph the larvae were first cooled at 4°C for 15 min and the prolegs were pierced with a sterile needle. Hemolymph melanisation was prevented by the use of few crystals of phenylthiourea. Cell-free hemolymph was degraded with thermolysin (*Sigma*, Taufkirchen, Germany) with an end concentration of 1 mg/mL in sterile bidistilled water. The mixture was incubated for 1h at 36°C on a rotary shaker. These parameters (moderate enzyme concentration and short incubation time) were chosen in order to mimic the *in vivo* conditions. Thermolysin-degraded fragments smaller than 3 kDa were obtained gradually by centrifugation at 7000 x g for 4-6h at 4°C using Centricon centrifugal concentrators Centricon-30, Centricon-10 and Centricon-3 (*Millipore*, Billerica, Massachusetts, USA), respectively.

Fractionation of hemolymph sample

Separation of the thermolysin-digested hemolymph sample (< 3 kDa) was realized by an *Agilent 1200 RP* HPLC system (*Agilent*, Waldbronn, Germany) with a *Symmetry* C18 4.6×250 mm column at a flow-rate of 1 mL/min. The injection volume of the sample was 100 µL. Separation was carried out using a 40 min gradient starting at 5% acetonitrile in water (v/v). After 15 min the solvent concentration of acetonitrile was raised up to 50% (v/v). After 40 min the concentration of acetonitrile was increased to 95% (v/v). The resulting fractions were collected manually in 30sec intervals, lyophilized and dissolved in 50 µL H_2O. The fractions were injected in *Galleria* larvae and immune stimulation was determined by lytic zone assay, subsequently.

Mass spectrometry

NanoHPLC-ESI FTMS measurements. The measurements of the samples were performed on a binary nanoHPLC system consisting of the units Switchos, Famos and Ultimate (*LCPackings/Dionex*, Idstein, Germany). Solvent A was water (HPLC grade, *Fluka*, Neu-Ulm, Germany) containing 2% acetonitrile (v/v) (Uvasol® grade, *Merck KGaA* Darmstadt, Germany) and 0.1% formic acid (v/v) (puriss p.a. for ms, *Fluka* Neu-Ulm, Germany). Solvent B was acetonitrile containing 20% water (v/v) and 0.08% formic acid (v/v). Separation of the hemolymph sample was carried out using a 77 min gradient. First B was increased from 0% to 10% in 10 min, subsequently increased to 30% in 30 min and then increased to 100% in 5 min and maintained for 11 min. The injection volumes were 0.3 µL for the hemolymph bulk sample and 1.2 to 5 µL for hemolymph fractions, respectively. After pre-concentration on a C18 PepMap trap column (5 mm×300 µm i.d.) the samples were separated on a fused silica C18 PepMap100 capillary column (150 mm×75 µm i.d.) (*Dionex*, Idstein, Germany) at 200 nL/min. The separation was monitored by a UV detector at 214 nm.

The nanoHPLC system was coupled to the mass spectrometer by a nanospray source. Pico-Tip® Emitter (*New Objective*, Woburn, MS, *USA*) were used as nanospray needles. The separated peptides were measured on a tandem mass spectrometer (*LTQ FT Ultra*, *Thermo Fisher Scientific GmbH*, Bremen, Germany) consisting of a linear ion trap (IT) and a Fourier Transform Ion Cyclotron Resonance (FTICR) mass spectrometer. For database searches high-resolution (R = 100.000 at m/z 400) survey spectra were measured on the FTICR with high mass accuracy (< 2 ppm). Fragment ion spectra were measured both with high resolution and in the ion trap. For analyses with manual *de novo* and the composition-based sequencing method precursor and fragment ions were solely measured with high resolution and high mass accuracy. *Collision induced dissociation* (CID) was used for fragmentation. MS/MS data was obtained on the FT also using the wide scan range (WSR) method with three microscans in order to compensate the loss of ions due to the time-of-flight effect of the FTICR mass spectrometer.

In addition, coupling of nanoHPLC with a linear ion trap / Fourier transform orbital trapping (IT-FTOT MS) mass spectrometer (*LTQ Orbitrap Discovery*, *Thermo Scientific GmbH*, Bremen, Germany) equipped with a nanospray ion source, was used. This mass spectrometer allowed, in addition to CID-fragmentation, the detection of low mass fragments by *higher-energy collisional dissociation* (HCD), providing additional information for peptide sequencing. Full scan and MS/MS spectra on the FTOT instrument were acquired with high resolution (R = 30.000 at m/z 400) and high mass accuracy (< 4 ppm). External calibration of both FT instruments was performed immediately before measurements according to the instructions of the manufacturer. All MS measurements were performed using the positive ion mode.

Offline ESI FTMS measurements. Offline nano ESI MS measurements of synthetic peptides were performed for validation of identified peptides using the orbitral trapping mass spectrometer. Pico tip emitters (5 μm i.d., *DNU-MS*, Berlin, Germany) were used for this task. The synthesized peptides were dissolved in 200 μL water/acetonitrile 50:50 (v/v) with 0.1% formic acid and diluted 1:10 before measurement.

MALDI FTMS measurements. For matrix-assisted laser desorption/ionization (MALDI) measurements the commercial MALDI source of the orbital trapping mass spectrometer, equipped with a N_2-Laser (337 nm), was used. Alpha-cyano-4-hydroxycinnamic acid (10 mg/mL) dissolved in acetonitrile/water 70:30 (v/v) with 0.1% trifluoroacetic acid was used as matrix. Two μL matrix solution were mixed with 1 μL sample on a stainless steel sample plate. The laser energy was set to 20 μJ and the crystal positioning system (CPS) was activated. MS and MS/MS experiments were performed with collision induced dissociation (CID) and higher-energy collisional dissociation (HCD) to obtain low-mass fragment ions.

Data analysis / Identification of peptides

Database search with standard database. Data obtained from HPLC-MS measurements were analyzed with SEQUEST database search using the BioWorks V3.3.1 software (*Thermo Fisher Scientific*, Bremen, Germany). For bulk samples 5 HPLC-MS measurements were carried out and at least 2 measurements for each bioactive fraction. Protein sequence database entries for *G. mellonella* are very limited. Therefore all available database entries for *Galleria* were downloaded from the server of the "*National Center of Biotechnology Information*" NCBI (http://www.ncbi.nlm.nih.gov/). This database contained 162 protein sequence database entries. The database searches of all HPLC-MS measurements were carried out with following settings: precursor ion tolerance 5 ppm; fragment ion tolerance 1 Da. The following variable modifications were selected: phosphorylation on serine, threonine, and tyrosine, hydroxylation on proline, threonine and methionine and carbamidomethylation on cysteine. These modifications are considered in the software with a mass increment of the respective amino acid, for hydroxylation $\Delta m = +15.99492$, for phosphorylation $\Delta m = +79.96633$ and for carbamidomethylation $\Delta m = +57.02146$, respectively. Thermolysin cleaves at the N-terminal side of amino acids with hydrophobic or bulky side chains [22]. The amino acids alanine, isoleucine, leucine, phenylalanine, valine and methionine were selected as cleavage sites. The number of allowed missed cleavage sites was set to 10. Peptides with a *Peptide Probability* below 0.001 were considered as identified. False discovery rates (FDR) were calculated according to Nesvizhskii et al. [23].

Database search with experimental database. An additional database search was applied with a newly implemented *G. mellonella* database obtained from transcriptome analysis [13]. The transcriptomic data were translated into the single letter amino acid code, manually implemented in a FASTA file and used the same way as standard database entries for the SEQUEST database searches with BioWorks. The database included 96756 entries, mainly short sequences (64% of all entries had a chain length of less than 50 amino acids).

***De novo* sequencing.** Manual *de novo* sequencing was performed with accurately measured masses. Additionally, the composition-based *de novo* sequencing approach was applied using the computer program *Peptide Composer 1.0* [20,24]. Calculations of amino acid compositions of peptides were performed using accurately measured mass spectra with a tolerance of 2 ppm for FTICR and 4 ppm for FTOT mass spectra, both for precursor ions and fragment ions. (The isobaric peptides leucine and isoleucine could not be differentiated by exact mass measurements due to their identical elemental composition.)

Synthesized peptides

After sequence determination, peptides were custom-synthesized by *GL Biochem (Shanghai Ltd.,* China*)* with a purity of over 85%.

Bioactivity tests

Last instar *G. mellonella* larvae, weighing between 250–350 mg, were used in this study for determining the immune stimulation. TenμL of collected fractions or 20 μL of synthesized peptide solutions (20 mM), respectively, were injected dorsolaterally into the hemocoel of the larva using 1 mL disposable syringes and 0.4×20 mm needles mounted on a microapplicator. After injection, the larvae were incubated at 32°C in darkness. 24 hours post injection, hemolymph was collected to investigate the immune stimulation.

As a simple read-out system to determine the immune-stimulatory activity of samples, we used the inhibition zone assay using freeze-dried *Micrococcus luteus* [7,25]. The inhibition zone assay against living bacteria such as *Micrococcus luteus* is a standard assay which is widely used in microbiology to qualitatively and quantitatively determine the activity of soluble antimicrobial compounds. The latter are pipetted into wholes which have been punched in the agar inoculated with bacteria. The sample defund into the agar around thereby inhibiting the growth of the bacteria. The diameter of the clear zones without growing bacterial colonies can be measured and subsequently quantified by using a calibration curve with a standard antibiotic such as gentamycin. Accordingly, the anti-bacteria activity in samples can by quantified in gentamycin equivalents. Briefly, seven milliliter of *M. luteus* agar consisting of 1% high-purity agar-agar (*Carl Roth*, Karlsruhe, Germany), 5 mg/mL freeze-dried *M. luteus* (*Sigma*, Taufkirchen, Germany), and 67 mM potassium phosphate buffer (PBS) pH 6.4 were placed onto petri dishes (Ø100 mm). Holes with a diameter of 4 mm were filled with 3 μL hemolymph and incubated for 24h at 37°C. Lysozyme activity was quantitatively determined (units/mL) by establishing a calibration curve from standard chicken lysozyme (*Sigma*, Taufkirchen, Germany).

For validating the potency of identified peptidic danger signals to elicit immune responses we used living *M. luteus* as an indicator organism in the inhibition zone assay because bacterial growth inhibition indicates the presence of induced antibacterial peptides, whereas the lytic zone assay using freeze-dried *M. luteus* indicates enzymatic digestion by lysozyme. Synthetic peptides were dissolved in sterile saline solution (120 mM NaCl) with an end concentration of 20 mM and 20 μL were injected into larvae. For each peptide 6 individual larvae were used. The anti-*M. luteus*-activity was calculated to gentamicin equivalents (inμg/mL) using a calibration curve with gentamicin as previously described [10,26]. Saline solution (peptide solvent) was used as control, and injected in the larvae following the same procedure as for the tested peptides. Average value and standard deviation for 6 individual animals were calculated for each peptide. Statistically significant differences between activities of larvae injected with peptides and control were determined using Students *t*-test.

Results and Discussion

Figure 1 shows the workflow for the combined approach used for the identification of bioactive peptides. The thermolysin-digested hemolymph sample was measured directly by nanoLC-FTMS. Second, the hemolymph sample was fractionated and resulting fractions showing lysozyme inducing activity were analyzed in more detail by nanoLC-FTMS and MALDI MS. Data analysis was performed by SEQUEST database search and *de novo* sequencing. Identified peptides were tested individually for immune stimulation by synthetic analogues.

Fractionation of hemolymph sample and initial bioactivity tests

Initial HPLC-MS measurements revealed that the thermolysin-digested sample (< 3 kDa) was too complex for the direct determination of bioactive compounds. For this reason the bulk sample was fractionated by HPLC and collected fractions were tested with a *lytic zone assay* to screen for elevated bioactivity. Details on chromatographic fractionation and activity tests can be found in the supporting information (File S1). Two fractions, termed A and B, in which the highest immune-stimulatory activity was detected were subjected to a second round of HPLC fractionation in order to further decrease the complexity. The resulting sub-fractions of these separation steps were tested again for immune-stimulatory activity. This procedure resulted in 4 sub-fractions for fraction A and 2 sub-fractions for fraction B. The tests of these sub-fractions showed that sub-fraction A1 and B1 had the highest activity and were therefore subjected to more detailed analysis.

Peptide identification by mass spectrometry

The hemolymph samples were measured by nanoHPLC coupled to FTMS instruments. The acquired data was analyzed using standard database searches as well as experimental databases derived from transcriptome data. Bioactive fractions were analyzed in detail by nanoHPLC-FTMS and additionally by MALDI Orbitrap measurements. Peptide sequence analysis was finally extended to database-independent *de novo* sequencing methods. Identification results are reported for the bulk sample and for bioactive fractions separately, in the following. A list of all identified peptides can be found in the supporting information (Table S1).

Database search with standard database. One of the standard procedures in proteomics for characterization of peptides is high performance liquid chromatography coupled to mass spectrometry followed by protein database searches. In this work HPLC-FTMS measurements of the thermolysin-digested hemolymph sample were performed with database search using SEQUEST. For this purpose all rudimentary available protein sequence entries from the public database NCBI (162 protein entries) of *G. mellonella* were used.

Database search of the bulk sample with the NCBI database led to the identification of 75 peptides. The majority of these identified peptides were obtained by HPLC-MS/MS measurements where the precursor ion was measured in the FT cell with high resolution, and the five most intense ions of this survey scan were fragmented and detected in the ion trap.

Most of the identified peptides could be assigned to abundant hemolymph proteins such as hexamerin, apolipophorin and transferrin precursor. The number of identified peptides was

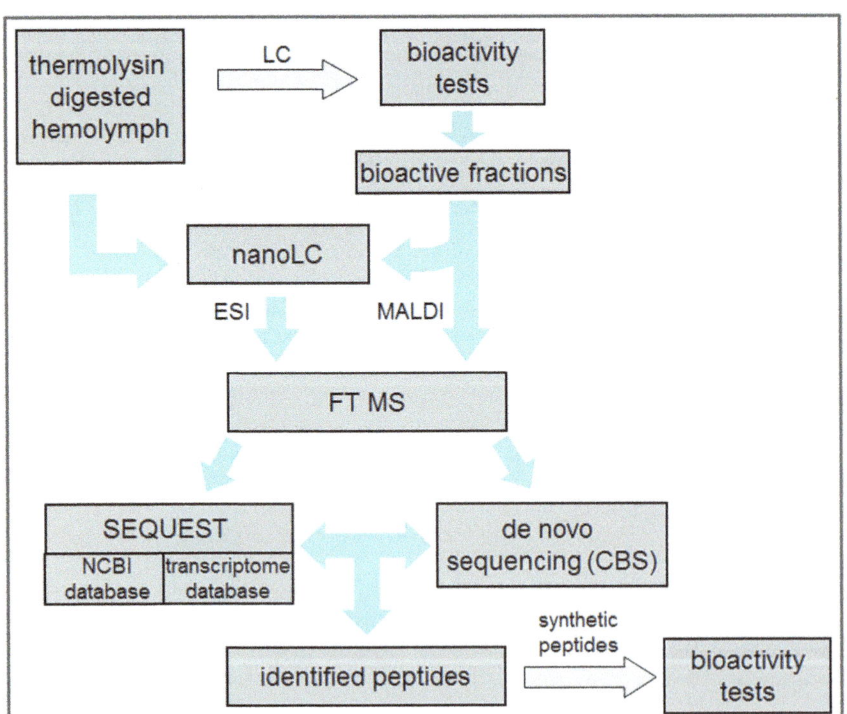

Figure 1. Workflow for the identification of bioactive peptides. On the one hand the thermolysin-digested hemolymph sample was analyzed directly by nanoHPLC-FTMS. On the other hand the thermolysin-digested sample was pre-fractionated using a HPLC-system and the collected samples were tested for bioactivity for a more detailed analysis of immuno-relevant peptides.

relatively low, especially when compared with database searches of fully-sequenced species where in one run several hundred peptides can be identified. This was due to the limited number of protein entries in the database for *G. mellonella* and showed the necessity of further identification methods. The determination of a *false discovery rate* (FDR) that is common in database search studies was only partly possible, because of the small size of the used databases. An FDR of 2,7% was determined by searching a decoy database that was generated by reversing the sequences of the original NCBI database [27].

SEQUEST database search was also performed with nanoHPLC-MS measurements of the bioactive fractions A, B and the sub-fractions A1 and B1 (which had the highest bioactivity in the initial activity tests). Five peptides could be identified in this manner in the fraction B. These peptides were VV-9, IE-8, FN-9, IN-10 and LY-11 (see Table 1 for sequence information). Of these peptides IE-8 and LY-11 were also identified in the hemolymph bulk measurements discussed above. The NCBI database searches of fraction A and sub-fractions A1 and B1 showed no peptide identification.

Database search with experimental database. A recently established comprehensive transcriptome of *G. mellonella* [13] was used as a database in order to identify additional peptides. Database search of the bulk sample with this experimental database resulted in 38 identified peptides. The decoy database searches of three HPLC-MS runs using the reversed database resulted in a FDR rate of 29%. These high values are most likely due to the fact that most of the database entries in the experimental database were short and therefore tended to lead to more hits with the decoy database.

The HPLC-MS measurements of the bioactive fractions were also analyzed by database search using the experimental database. In fraction B1 only one peptide could be identified by this approach: APPSGPAAPPAKTP (AP-14). In fraction B peptide VV-9 could be identified that was also identified with the standard database approach. No peptides were identified in fraction A and sub-fraction A1. The total number of peptides identified with the experimental database was 40.

***De novo* sequencing.** *De novo* sequencing of peptides allows identifying peptides without prior knowledge about possible amino acid sequences. Therefore this approach was used in order to identify peptides which are not included in the NCBI or transcriptome database. Data analysis by *de novo* sequencing was always performed using high resolution and high mass accuracy measurements, both for precursor and fragment ions. In doing so, it was possible to determine amino acid-specific increment masses within the MS/MS spectra. For example lysine and glutamine have the same nominal mass, but can easily be differentiated by accurate mass measurements due to their mass difference of 36 mDa.

Twenty-two peptides were identified using the *de novo* sequencing approach in the bulk sample. Nine of this set were identified by manual *de novo* sequencing of non-significant SEQUEST database search hits (Peptide Probability above threshold of 0.001).

The *de novo* sequencing approach led to no identification of peptides in the bioactive fractions. Therefore additional experiments were performed that implemented matrix-assisted laser desorption/ionization (MALDI) and the higher-energy collisional dissociation device (HCD) as a fragmentation technique. This fragmentation method allows the fragmentation in a dedicated quadrupolar collision cell, which results in a lower mass cut-off for fragmentation ions. An example of a peptide identified by this approach is discussed in the following. In sub-fraction A1 two peaks, $m/z = 517.28360$ and $m/z = 631.38722$, could be sequenced by the composition based sequencing approach (CBS) (see [20,24] for more details). The MS/MS spectrum of precursor ion $m/z = 631.38722$ using HCD fragmentation is shown in Figure 2. For the characterization of this peptide the first step was the determination of the amino acid composition. The detected mass of $m/z = 631.38722$ corresponds to 11 possible amino acid compositions within an accuracy of 3 ppm. By checking against the fragment ions this value could be reduced to 4 possible amino acid compositions (A,A,S,V,K,R), (G,A,S,L,K,R), (A,A,V,T,K,R) and (A,K,K,E,R). In the MS/MS spectrum two neutral mass losses indicated the loss of glutamic acid (single letter code: E) with $\Delta m/z = 129.0425$ Da (between fragment ions $m/z = 432.2558$ and $m/z = 303.2133$) and $\Delta m/z = 129.0421$ Da (between $m/z = 329.1812$ and $m/z = 200.1391$), respectively. On the other hand, there was no indication for a loss of serine, threonine, valine or leucine. This information was used to determine the amino acid composition, as only one of the 4 remaining possibilities contains glutamic acid. The only remaining amino acid composition could

Table 1. Peptides identified in the bioactive fractions.

Fraction	Sequence	$[M+H]^+_{exp.}$	Abbr.	Used method	Ionization/ Fragmentation
B	**IYHKPTTE**	988.50713	IE-8	NCBI DB	ESI/CID
B	**IKANAPQAEN**	1055.54729	IN-10	NCBI DB	ESI/CID
B	**FQATNDNKN**	1051.47905	FN-9	NCBI DB	ESI/CID
B	**LKTKNPSPDTY**	1263.65666	LY-11	NCBI DB	ESI/CID
B	**VDGKSAPNV**	886.46198	VV-9	NCBI & transcriptome DB	ESI/CID
B1	**APPSGPAAPPAKTP**	1258.67646	AP-14	transcriptome DB	ESI/CID
B1	**SRPSPNYP**	917.44540	SP-8	*de novo*	ESI/HCD
A1	**ERRG**	517.28360	EG-4	*de novo*	MALDI/HCD
A1	**KAERK**	631.38722	KK-5	*de novo*	MALDI/HCD

Listed are: the amino acid sequence of the peptides, the fraction in which the peptide was identified, and the method that was used for identification.
CID: Collision induced dissociation.
HCD: Higher-energy collisional induced dissociation.
DB: Database.

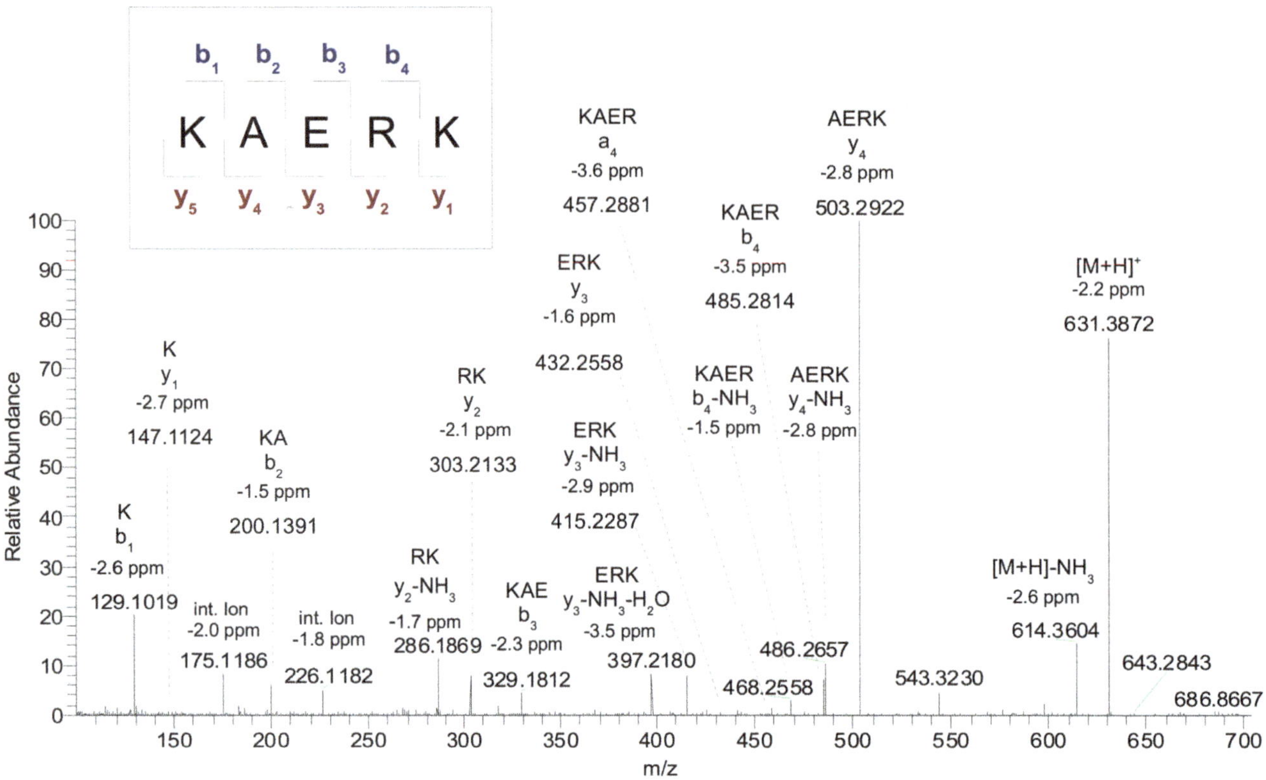

Figure 2. MALDI-Orbitrap MS/MS measurement with higher-energy collisional dissociation mode (HCD). The $[M+H]^+$ precursor ion at m/z 631.3872 was detected in the bioactive fraction A1. The presented peptide sequence KAERK was determined by *de novo* sequencing (CBS).

therefore be determined as (A,K,K,E,R). In the second step of the composition-based *de novo* sequencing the peptide sequence was determined by scoring the agreement between observed and expected fragment ions of permuted sequence propositions. For that all fragment ion signals of the MS/MS spectrum (Figure 2) were used. The results of the *de novo* (CBS) sequencing showed that the sequence KAERK (abbreviation 'KK-5' in Table 1) was matched with the highest CBS score value. The peptide sequence of precursor ion m/z = 517.28360 was identified using the same approach as ERRG (EG-4). Three peptides were identified by *de novo* sequencing in the bioactive sub-fractions. Peptide SP-8 was identified in sub-fraction B1 in addition to the two peptides EG-4 and KK-5 discussed above. In total 25 peptides were identified by *de novo* sequencing.

Comparison of identification methods. The different approaches for the identification of peptides were summarized in a Venn diagram shown in Figure 3. Nineteen peptides were exclusively identified using the *de novo* sequencing (CBS) approach, 63 peptides could only be identified using the standard NCBI protein sequence databases of *G. mellonella.* The database search with the experimental database resulted in the identification of 30 additional peptides. Nine peptides were found in the database searches both with the NCBI and experimental databases. Five peptides were identified with *de novo* and NCBI database search. Consequently most peptides were identified by only one method. Only one peptide (IKIPAPYE) was identified with all three approaches. These results show that each sequencing approach has specific advantages and limitations. The combination of the used methods led to the identification of 127 peptides in total. Most of these peptides could be identified using the NCBI database search. In contrast, the number of peptides identified with the *de novo* approach in the bulk sample was lower. It has to be mentioned, however, that identification with the *de novo* (CBS) sequencing approach is not finally completed. Due to the fact that highly resolved MS spectra were analyzed manually, there is still the possibility of identifying additional peptides. Furthermore, reliability of the *de novo* sequencing results was higher compared to the database search results, because fragment ions were all

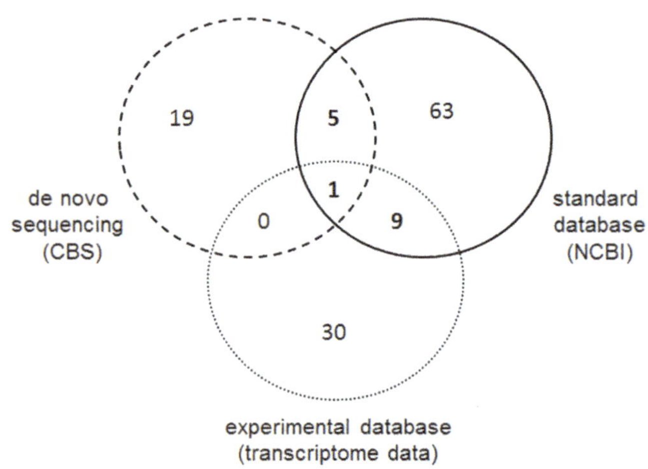

Figure 3. Venn diagram comparing identified peptides obtained by using different approaches. In total, 127 peptides were identified. 25 peptides were identified with the *de novo* approach, 78 with the standard database search and 40 with the experimental database.

assigned based on accurate mass measurements. The FDR rate of the search with the experimental database was high, indicating that there were also false positives among these.

Of the 127 peptides that were identified in total, 9 peptides were identified in the fractions with increased bioactivity. These peptides are summarized in Table 1. Within this set, 4 peptides were identified in fraction B using the standard NCBI database search, whereas one peptide (VV-9) was identified both with the standard and experimental database. Two peptides each were identified in sub-fraction A1 and B1. Three peptides (EG-4, KK-5 and SP-8) were identified by the *de novo* approach, whereas one peptide (AP-14) was determined by database search with the experimental database. Using the standard database search alone would have led to no results in the sub-fractions. Inclusion of MALDI and HCD was also necessary in order to identify the peptides in the sub-fractions. These results show that different ionization techniques and different approaches for peptide analysis had to be used for the identification of a broad set of peptides. All of the identified peptides in the bioactive fractions were of particular interest and had to be validated and tested for bioactivity as discussed in the following section.

Confirmation by synthetic peptides. To validate the sequence of the peptides identified in the bioactive fractions summarized in Table 1, synthetic analogs were synthesized. Figure 4 shows the comparison of the FT-MS/MS spectra of peptide VV-9 detected in the hemolymph fraction and its synthetic analogue. Fragmentation experiments for validation were performed the same way as for the hemolymph sample. The fragment ions of the identified peptide and of the synthetic analogue, matched all with high accuracy thus confirming the initial identification.

Seven peptides could be validated by accurate mass measurements for precursor ion and fragment ions. Two peptides identified by the database search, FN-9 and AP-14, were confirmed by accurate precursor ion mass and fragment ions measured in the ion trap. MS/MS spectra of these peptides can be found in the supporting information (File S2). Consequently all peptide sequences identified in bioactive fractions shown in Table 1 were confirmed by MS/MS measurements of synthesized peptides.

Bioactivity tests of identified peptides

The peptides identified by our MS approaches have been postulated to represent putative danger signals resulting from digestion of hemolymph by microbial metalloproteinases. In order to validate their immune-stimulatory activity we produced synthetic analogues which were injected into last instar larvae of *G. mellonella* to determine their capacity to induce immune response in vivo. As a simple read-out system we used the inhibition zone assay against living *M. luteus* to determine antibacterial activity in the hemolymph [25].

Interestingly, not all synthetic peptides turned out to exhibit immune-stimulatory activity when injected (Figure 5). In comparison with control injections with saline alone, the peptides VV-9, IE-8, FN-9, IN-10, LY-11 and AP-14 induced significantly higher anti-*M. luteus*-activity. On the other hand peptides KK-5, EG-4 and SP-8 did not induce a significant response. Therefore, we conclude that only particular peptidic hemolymph protein fragments can function as danger signals when released upon metalloproteinase-mediated hydrolysis.

The larvae of the lepidopteran *G. mellonella* have been established as a powerful model host for pathogens of insects or humans and as a source for novel anti-infective therapeutics [28]. In addition, they were used to demonstrate for the first time that the presence of microbial metalloproteinases alone is sufficient to induce potent innate immune responses [7]. Thermolysin is the prototype metalloproteinase belonging to the M4 family which encompasses many virulence factors of human pathogens [9]. Thermolysin-mediated hydrolysis of hemolymph proteins from *G. mellonella* results in formation of peptidic fragments which, when injected, can elicit immune responses that are qualitatively and quantitatively comparable with those observed upon injection of bacterial cell wall components such as LPS [10].

Peptide analysis and identification in our study was more complicated in comparison to standard proteomics experiments due to several factors. Firstly, lacking a sequenced genome of *G. mellonella* it was difficult to characterize these putative danger signals by standard proteomics approaches. However, fully sequenced insect systems such as *Drosophila* and *Tribolium* are very small and therefore it is difficult to obtain hemolymph samples in sufficient quantities (for mass spectrometric analysis and bioactivity tests). Furthermore we have discovered the first specific peptidic inhibitor of metalloproteinases in *G. mellonella*. This so-called IMPI is currently being developed as new second generation antibiotic to cure symptoms caused by thermolysin-like metalloproteinases during infections in humans [28]. Secondly, peptides produced by thermolysin are more difficult to ionize than those resulting from other enzymes. Trypsin, as a digestion enzyme which produces ions which are much easier to ionize, would not be representative for the activity of microbial pathogens. It was shown previously that trypsin does not induce immune response in *Galleria mellonella* [7]. Finally, thermolysin digestion was performed at moderate enzyme concentration and for 1 hour only, which results in incomplete digestion of hemolymph proteins. A longer incubation time and/or higher thermolysin concentrations would result in a higher number of peptides. But this situation would not be comparable with in vivo conditions during infection of *G. mellonella* where only limited amounts of thermolysin-like metalloproteinases are produced, plausibly to avoid activation of host immune responses [29]. In addition it is known from our previous studies that stronger or total digestion of hemolymph proteins does not necessarily increases their immune-stimulatory activity [7].

To account for these specific challenges, we applied in this study high resolution mass spectrometry combined with *de novo* sequencing of peptides generated by metalloproteinase-mediated digestion of hemolymph proteins from *G. mellonella*. This approach led to the identification of several peptides in bioactive fractions. Following a manual inspection of LC-MS/MS data, we estimate that the two sub-fractions A1 and B1 contain about 10 peptides each (see supporting information File S3 for details). The full sequence of 2 peptides each was identified in these fractions and partial sequences were obtained for other peptides. In order to validate the postulated function of the *de novo* sequenced peptides as danger signals we produced synthetic analogues which were injected into *G. mellonella* larvae to assess their capacity to induce innate immune responses. The latter are characterized by the release of antimicrobial peptides among which some exhibit activity against living *Micrococcus luteus* [*28*]. Strikingly but in accordance with our expectation, it turned out that not all peptidic fragments exhibit immune-stimulatory activity. Only the peptides designated as VV-9, IE-8, FN-9, IN-10, LY-11 and AP-14 were confirmed as elicitors of innate immune responses when injected.

We therefore postulate that the release of metalloproteinases produced by invading pathogens and parasites results in generation of these danger signals which, in turn, induce innate immune responses [29]. The formation of these peptidic danger signals can only be detected by direct (mass spectrometric) analysis on the peptide level. Transcriptomic approaches would only detect the endogenous precursor peptides/proteins of these digestion prod-

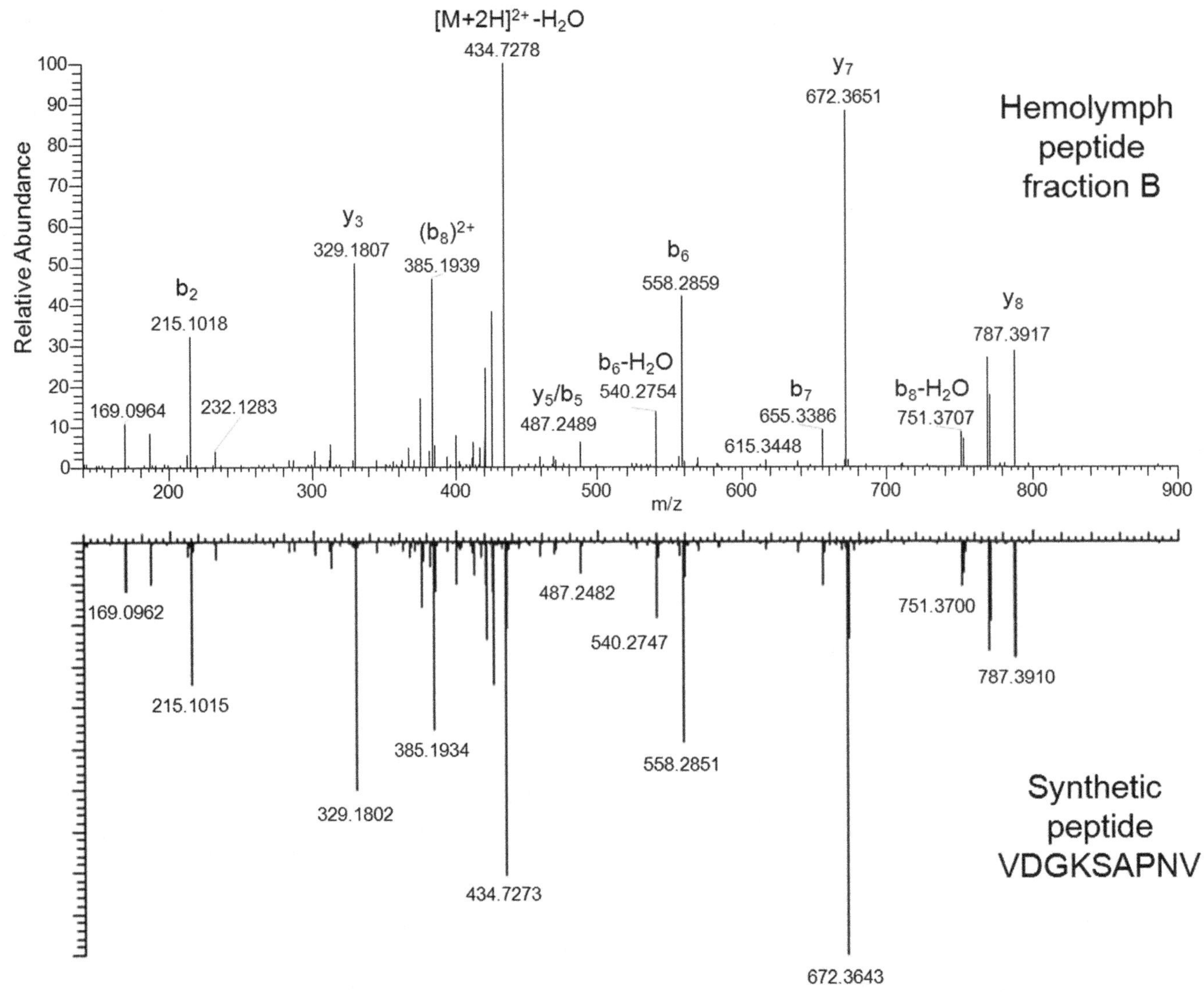

Figure 4. Validation of the identified peptide VV-9 in hemolymph fraction B. The MS/MS spectrum of the precursor ion m/z 886.46 was acquired with accurate mass using CID(35) as fragmentation technique. The sequence of this peptide was determined as VDGKSAPNV by database search. The fragmentation pattern of the synthetic peptide is in good accordance with the MS/MS spectrum of the identified peptide in the bioactive fraction.

ucts. In this context it is important to note that *G. mellonella* is the only known animal which is capable to produce a specific peptidic inhibitor against microbial metalloproteinases [30]. This insect metalloproteinase inhibitor (IMPI) was originally discovered in and purified from hemolymph of immune-stimulated larvae [31], and mediates reportedly feedback-loop regulation of harmful microbial metalloproteinases [32].

Consequently, we have identified peptides which are generated in *G. mellonella* under presence of microbial metalloproteinases, and which mediate the activation of immune responses encompassing the synthesis and release into the hemolymph of both antimicrobial peptides and the IMPI. Regarding the high *in vivo* toxicity of thermolysin-like metalloproteinases, the characterized peptidic products of their activity can be considered as danger signals because they can set the immune system into alarm. This efficient mechanism providing the ability to sense invading pathogens or parasites by the activity of their virulence-associated enzymes [29] adds to the sophisticated strategies determining the tremendous evolutionary plasticity of insect immunity [33].

Conclusions

In this study we used high-resolution mass spectrometry to identify peptidic hemolymph protein fragments which are generated in the hemolymph of the model host *G. mellonella* when harmful microbial metalloproteinases belonging to the M4 family, with thermolysin as the prototype, are present. Addressing the lack of a genome sequence we complemented the rudimentary NCBI database with a transcriptome database and *de novo* sequencing approaches for peptide identification. This approach led to the identification of 9 potentially bioactive peptides. These (tentative) results were validated by studying synthesized peptide analogues. Detailed MS/MS experiments confirmed the amino acid sequence of all 9 peptides. Six out of 9 peptides identified in the bioactive

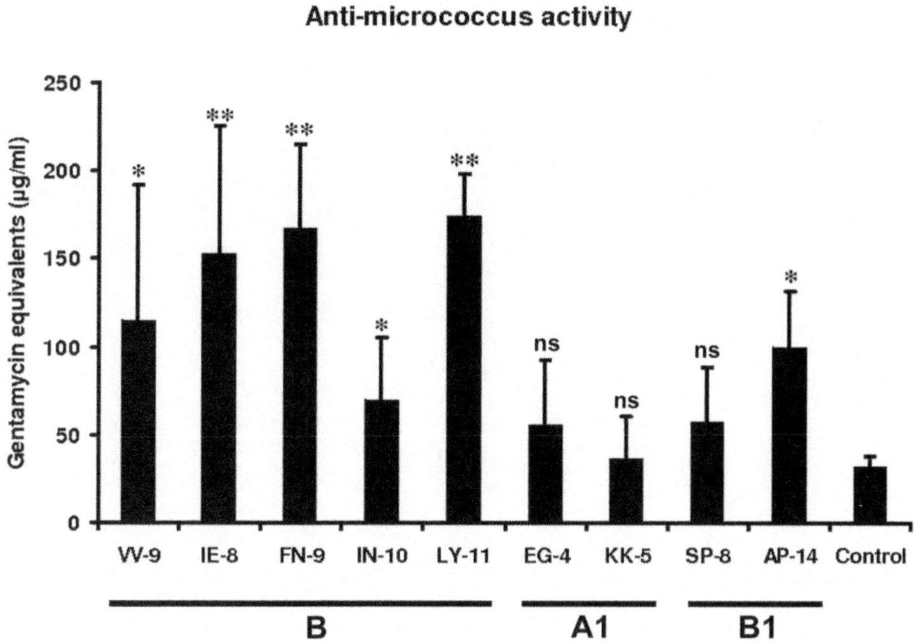

Figure 5. Bioactivity test of the identified peptides against (living) Micrococcus bacteria. The activity is shown in Gentamycin equivalents inµg/mL. Peptides VV-9, IE-8, FN-9, IN-10, LY-11 and AP-14 exhibit significantly higher immune-stimulatory activity than the used solvent control. Peptide KK-5 shows no activity compared to the solvent control, in contrast. EG-4 and SP-8 show slightly elevated immune activity. Statistically significant differences between activities of larvae injected with peptides and control were determined using Students *t*-test and are indicated by * ($p<0.05$) and ** ($p<0.005$), ($n=4$-6 for each peptide).

fractions exhibited immune-stimulatory activity when injected in *Galleria* larvae. These six peptidic elicitors of immune responses are postulated to function as danger signals indicating the presence of microbial metalloproteinases and mediating their regulation by specific inducible metalloproteinase inhibitors. Our results suggest that the validity of the immunity-related danger model proposed by Matzinger can be expanded beyond mammals to include insects. The detailed molecular pathways and importance of this process will have to be investigated in follow-up studies.

Acknowledgments

We thank Hala Mahfoud for HPLC separation of the hemolymph bulk sample and for the lytic zone assays of the fractions.

Author Contributions

Conceived and designed the experiments: AR AV BS. Performed the experiments: AB KM. Analyzed the data: AB AR KM. Contributed reagents/materials/analysis tools: BS AV AR. Wrote the paper: AR AB AV. Provided unique instrumentation and data analysis capabilities: BS.

References

1. Janeway CA, Medzhitov R (2002) Innate immune recognition. Annual Review of Immunology 20: 197–216.
2. Akira S, Uematsu S, Takeuchi O (2006) Pathogen recognition and innate immunity. Cell 124: 783–801.
3. Lazzaro BP, Rolff J (2011) Immunology. Danger, microbes, and homeostasis. Science 332: 43–44.
4. Wiesner J, Vilcinskas A (2010) Antimicrobial peptides The ancient arm of the human immune system. Virulence 1: 440–464.
5. Matzinger P (2002) The danger model: A renewed sense of self. Science 296: 301–305.
6. Matzinger P (2007) Friendly and dangerous signals: is the tissue in control? Nature Immunology 8: 11–13.
7. Griesch J, Wedde M, Vilcinskas A (2000) Recognition and regulation of metalloproteinase activity in the haemolymph of Galleria mellonella: a new pathway mediating induction of humoral immune responses. Insect Biochemistry and Molecular Biology 30: 461–472.
8. Mylonakis E, Casadevall A, Ausubel FM (2007) Exploiting amoeboid and non-vertebrate animal model systems to study the virulence of human pathogenic fungi. Plos Pathogens 3: 859–865.
9. Adekoya OA, Sylte I (2009) The Thermolysin Family (M4) of Enzymes: Therapeutic and Biotechnological Potential. Chemical Biology & Drug Design 73: 7–16.
10. Altincicek B, Linder M, Linder D, Preissner KT, Vilcinskas A (2007) Microbial metalloproteinases mediate sensing of invading pathogens and activate innate immune responses in the lepidopteran model host Galleria mellonella. Infection and Immunity 75: 175–183.
11. Eng JK, McCormack AL, Yates JR (1994) An Approach to Correlate Tandem Mass-Spectral Data of Peptides with Amino-Acid-Sequences in a Protein Database. Journal of the American Society for Mass Spectrometry 5: 976–989.
12. Perkins DN, Pappin DJC, Creasy DM, Cottrell JS (1999) Probability-based protein identification by searching sequence databases using mass spectrometry data. Electrophoresis 20: 3551–3567.

13. Vogel H, Altincicek B, Gloeckner G, Vilcinskas A (2011) A comprehensive transcriptome and immune-gene repertoire of the lepidopteran model host Galleria mellonella. Bmc Genomics 12: 1–19.
14. Altincicek B, Berisha A, Mukherjee K, Spengler B, Roempp A, et al. (2009) Identification of collagen IV derived danger/alarm signals in insect immunity by nanoLC-FTICR MS. Biological Chemistry 390: 1303–1311.
15. Römpp A, Dekker L, Taban I, Jenster G, Boogerd W, et al. (2007) Identification of leptomeningeal metastasis-related proteins in cerebrospinal fluid of patients with breast cancer by a combination of MALDI-TOF, MALDI-FTICR and nanoLC-FTICR MS. Proteomics 7: 474–481.
16. Schober Y, Schramm T, Spengler B, Römpp A (2011) Protein identification by accurate mass matrix-assisted laser desorption/ionization imaging of tryptic peptides. Rapid Communications in Mass Spectrometry 25: 2475–2483.
17. Scigelova M, Makarov A (2006) Orbitrap mass analyzer—overview and applications in proteomics. Proteomics 6 Suppl 2: 16–21.
18. Marshall AG, Hendrickson CL (2002) Fourier transform ion cyclotron resonance detection: principles and experimental configurations. International Journal of Mass Spectrometry 215: 59–75.
19. Römpp A, Taban IM, Mihalca R, Duursma MC, Mize TH, et al. (2005) Examples of Fourier transform ion cyclotron resonance mass spectrometry developments: from ion physics to remote access biochemical mass spectrometry. European Journal of Mass Spectrometry 11: 443–456.
20. Spengler B (2004) De novo sequencing, peptide composition analysis, and composition-based sequencing: A new strategy employing accurate mass determination by Fourier transform ion cyclotron resonance mass spectrometry. Journal of the American Society for Mass Spectrometry 15: 703–714.
21. Langsdorf M, Ghassempour A, Römpp A, Spengler B (2010) Characterization of a peptide family from the skin secretion of the Middle East Tree Frog Hyla savignyi by composition-based de novo sequencing. Rapid Communications in Mass Spectrometry 24: 2885–2899.
22. Keil B (1992) Specificity of Proteolysis. Berlin-Heidelberg-New York: Springer. 336 p.
23. Nesvizhskii AI, Vitek O, Aebersold R (2007) Analysis and validation of proteomic data generated by tandem mass spectrometry. Nature Methods 4: 787–797.
24. Spengler B (2007) Accurate mass as a bioinformatic parameter in data-to-knowledge conversion: Fourier transform ion cyclotron resonance mass spectrometry for peptide de novo sequencing. European Journal of Mass Spectrometry 13: 83–87.
25. Vilcinskas A, Matha V (1997) Effect of the entomopathogenic fungus Beauveria bassiana on the humoral immune response of Galleria mellonella larvae (Lepidoptera: Pyralidae). European Journal of Entomology 94: 461–472.
26. Altincicek B, Vilcinskas A (2006) Metamorphosis and collagen-IV-fragments stimulate innate immune response in the greater wax moth, Galleria mellonella. Developmental and Comparative Immunology 30: 1108–1118.
27. Reidegeld KA, Eisenacher M, Kohl M, Chamrad D, Korting G, et al. (2008) An easy-to-use Decoy Database Builder software tool, implementing different decoy strategies for false discovery rate calculation in automated MS/MS protein identifications. Proteomics 8: 1129–1137.
28. Vilcinskas A (2011) ANTI-infective Therapeutics from the Lepidopteran Model Host Galleria mellonella. Current Pharmaceutical Design 17: 1240–1245.
29. Vilcinskas A (2010) Coevolution between pathogen-derived proteinases and proteinase inhibitors of host insects. Virulence 1: 206–214.
30. Vilcinskas A, Wedde M (2002) Insect inhibitors of metalloproteinases. IUBMB Life 54: 339–343.
31. Wedde M, Weise C, Nuck R, Altincicek B, Vilcinskas A (2007) The insect metalloproteinase inhibitor gene of the lepidopteran Galleria mellonella encodes two distinct inhibitors. Biological Chemistry 388: 119–127.
32. Clermont A, Wedde M, Seitz V, Podsiadlowski L, Lenze D, et al. (2004) Cloning and expression of an inhibitor of microbial metalloproteinases from insects contributing to innate immunity. Biochemical Journal 382: 315–322.
33. Vilcinskas A (2013) Evolutionary plasticity of insect immunity. Journal of Insect Physiology 59: 123–129.

8

Novel Processed Form of Syndecan-1 Shed from SCC-9 Cells Plays a Role in Cell Migration

Annelize Z. B. Aragão[1,9], Marília Belloni[1,9], Fernando M. Simabuco[1], Mariana R. Zanetti[1], Sami Yokoo[1], Romênia R. Domingues[1], Rebeca Kawahara[1], Bianca A. Pauletti[1], Anderson Gonçalves[1], Michelle Agostini[2], Edgard Graner[2], Ricardo D. Coletta[2], Jay W. Fox[3], Adriana F. Paes Leme[1]*

1 Mass Spectrometry Laboratory, Brazilian Biosciences National Laboratory – CNPEM, Campinas, Brazil, **2** Piracicaba School of Dentistry, University of Campinas (UNICAMP), Piracicaba, Brazil, **3** School of Medicine, University of Virginia, Charlottesville, Virginia, United States of America

Abstract

The extracellular milieu is comprised in part by products of cellular secretion and cell surface shedding. The presence of such molecules of the sheddome and secretome in the context of the extracellular milieu may have important clinical implications. In cancer they have been hypothesized to play a role in tumor growth and metastasis. The objective of this study was to evaluate whether the sheddome/secretome from two cell lines could be correlated with their potential for tumor development. Two epithelial cell lines, HaCaT and SCC-9, were chosen based on their differing abilities to form tumors in animal models of tumorigenesis. These cell lines when stimulated with phorbol-ester (PMA) showed different characteristics as assessed by cell migration, adhesion and higher gelatinase activity. Proteomic analysis of the media from these treated cells identified interesting, functionally relevant differences in their sheddome/secretome. Among the shed proteins, soluble syndecan-1 was found only in media from stimulated tumorigenic cells (SCC-9) and its fragments were observed in higher amount in the stimulated tumorigenic cells than stimulated non-tumorigenic cells (HaCaT). The increase in soluble syndecan-1 was associated with a decrease in membrane-bound syndecan-1 of SCC-9 cells after PMA stimuli. To support a functional role for soluble syndecan-1 fragments we demonstrated that the synthetic syndecan-1 peptide was able to induce cell migration in both cell lines. Taken together, these results suggested that PMA stimulation alters the sheddome/secretome of the tumorigenic cell line SCC-9 and one such component, the syndecan-1 peptide identified in this study, was revealed to promote migration in these epithelial cell lines.

Editor: Effie C. Tsilibary, National Center for Scientific Research Demokritos, Greece

Funding: This study was supported by grants from Conselho Nacional de Desenvolvimento Científico e Tecnológico (470567/2009-0) and grants and fellowships from Fundação de Amparo à Pesquisa do Estado de São Paulo (2009/52833-0; 2009/18301-1; 2010/09642-7; 2010/19278-0). The funders had no role in study design, data collection and analysis, decision to publish, or preparation of the manuscript.

Competing Interests: The authors have declared that no competing interests exist.

* E-mail: adriana.paesleme@lnbio.org.br

9 These authors contributed equally to this work.

Introduction

Oral cancer is one of the most common malignancies worldwide and despite improvements in diagnosis and treatment, the overall survival rate for advanced patients has not been significantly improved over the last three decades [1]. Indeed, the lack of biomarkers avoids prognostic prediction or specific treatment for oral squamous cell carcinomas (OSCC), the most common presentation of oral cancer.

New approaches on clinical proteomics, such as secretome-based analysis, have been developed to identify novel biomarkers. Secretome/sheddome is a proteomic area that allows the analysis of a dynamic extracellular environment including secreted, released, degraded or shed proteins [2–4]. Soluble proteins in the extracellular milieu can have specific functions and can induce a variety of responses that are still not predictable, for instance, notch, E-cadherin and CD44 are known candidates for potential outside-in signal transduction [5–8]. These fragments can carry over conserved sequences that can regulate autocrine and paracrine targets [9].

In order to evaluate the differences between the secretome/sheddome of normal and tumorigenic cells, two epithelial cell lines, HaCaT and SCC-9, were treated with phorbol-ester (PMA). Here we showed that PMA stimulation induced distinct migration, adhesion and gelatinase activity as well as differences in the secretome/sheddome. Components in the media such as soluble and fragments of syndecan-1 were found mainly in stimulated tumorigenic cells. Syndecans are known family of cell surface proteoglycans that play regulatory roles in many biological processes, including migration, proliferation, wound healing, inflammation, angiogenesis and tumorigenesis [10,11]. The role of syndecans in tumor progression may vary with tumor stage and type [10]. In squamous cell carcinoma, the reduction of syndecan-1 expression is correlated with the progression of carcinogenesis [12], histological grade of malignancy [13], tumor size and the mode of invasion [14]. Furthermore, we also demonstrated evidence that the fragment of syndecan-1 identified was able to induce cell migration.

Results

Analysis of secretome/sheddome in tumorigenic and non-tumorigenic cells

Secretome/sheddome composition is distinct in tumorigenic and non-tumorigenic cells. Fifty-three proteins were identified in the extracellular media and classified as extracellular matrix proteins, secreted proteins, membrane-bound proteins, and intracellular proteins that have a membrane projection. Differences between the cells either treated or not with PMA were observed, and based on the ratio of quantitative values, proteins with changes higher than 1.5-fold (i.e. >1.5 or <0.66) were considered as significantly regulated by PMA treatment (Table 1, Fig. 1).

Non-tumorigenic cells stimulated with PMA showed exclusively 14 up-regulated proteins, including agrin, laminin subunit alpha-5, cathepsin H, collagenase 3, interstitial collagenase, EGF-containing fibulin-like extracellular matrix protein 1, proprotein convertase subtilisin, serpin B5, amyloid-like protein 2, glypican-1, isoform 1 of beta-enolase, junction plakoglobin, plakophilin-3, tripeptidyl-peptidase 1 and vinculin. Tumorigenic cells also showed exclusively up-regulated proteins, such as fibronectin, annexin A1, annexin A5, 4F2 cell-surface antigen heavy chain, syndecan-1, syndecan-4, ezrin, fascin, clathrin and talin-1. However, there were commonly up-regulated proteins in response to PMA activation, namely laminin subunit alpha-3, laminin subunit beta-3, laminin subunit gamma-2, kallikrein-10, thrombospondin-1, CD44 antigen and moesin.

Peptidomic analysis reveals a potential candidate for cancer regulation. The analysis of peptides in the media showed that the treatment with PMA induced the increase of syndecan-1 fragments (Table 2). Table 2 shows the number of unique peptides and number of spectral counts for each peptide sequence identified by LC-MS/MS. The syndecan-1-derived peptides increased mainly in tumorigenic cells stimulated with PMA (Mann-Whitney test, $p = 0.0273$). The higher number of spectral counts for syndecan-1-derived peptides was observed for

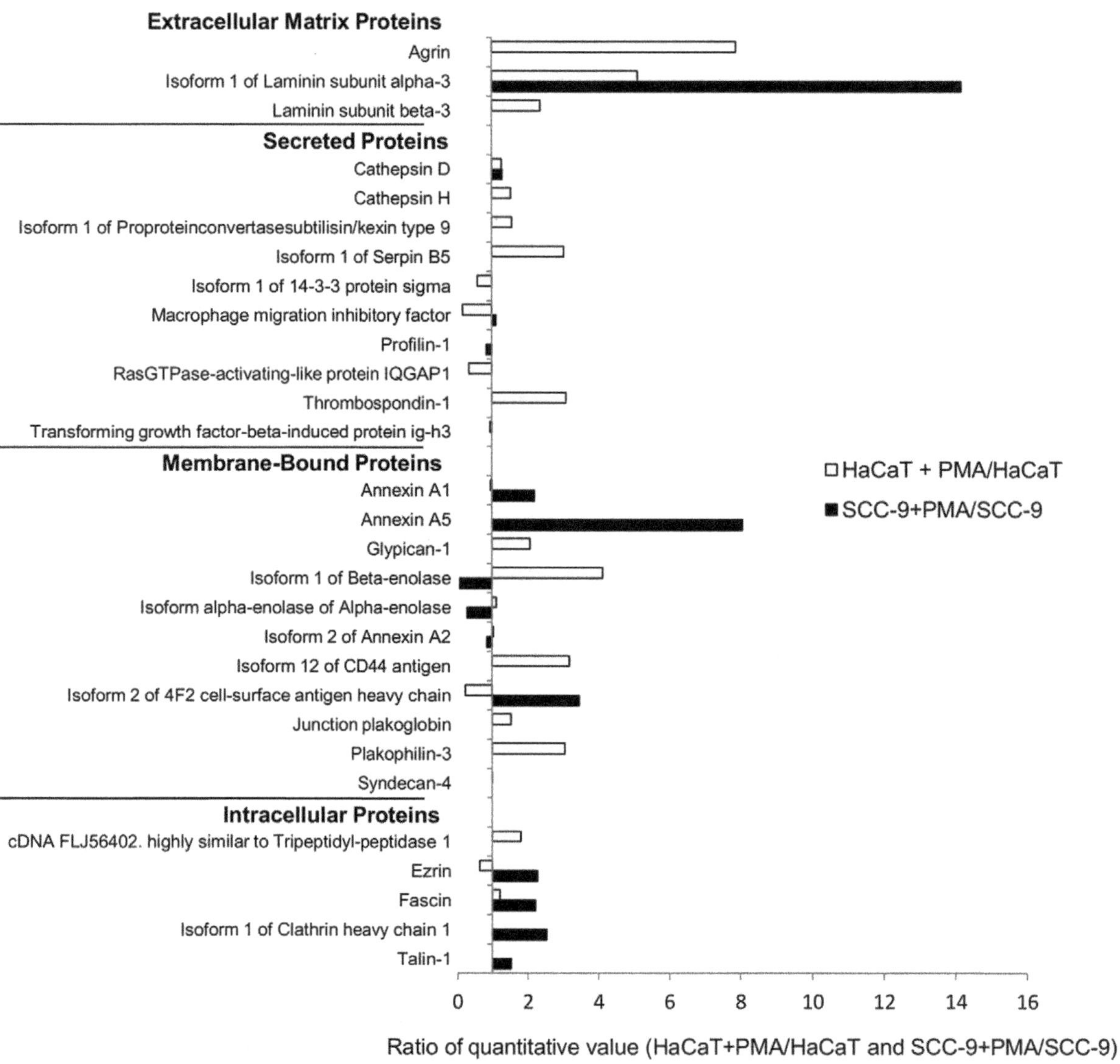

Figure 1. Identification of proteins in the conditioned media by LC-MS/MS according to the ratio of quantitative value (normalized spectral counts), as indicated in Table 1.

Table 1. Identification of proteins in the conditioned media by LC-MS/MS according to the ratio of quantitative value.

Accession Number	Protein Identification	HaCat + PMA/HaCaT	SCC-9+ PMA/SCC-9	SCC-9/HaCaT	SCC-9+PMA/ HaCaT+PMA
	Extracellular Matrix Proteins				
IPI00374563	Agrin	7.88	absent	only in HaCaT	only in HaCat+PMA
IPI00022418	Isoform 1 of Fibronectin	absent	only in SCC+PMA	absent	only in SCC+PMA
IPI00003951	Isoform 1 of Laminin subunit alpha-3	5.09	14.17	0.46	1.29
IPI00015117	Isoform Long of Laminin subunit gamma-2	only in HaCaT + PMA	only in SCC+PMA	absent	0.70
IPI00783665	Laminin subunit alpha-5	only in HaCaT + PMA	absent	absent	only in HaCat+PMA
IPI00299404	Laminin subunit beta-3	2.36	Only in SCC+PMA	only in HaCaT	0.39
	Secreted Proteins				
IPI00011229	Cathepsin D	1.28	1.30	0.67	0.68
IPI00297487	Cathepsin H	1.54	absent	only in HaCaT	only in HaCat+PMA
IPI00021738	Collagenase 3 (MMP-13)	only in HaCaT + PMA	absent	absent	only in HaCat+PMA
IPI00023673	Galectin-3-binding protein	absent	only in SCC	only in SCC	absent
IPI00023728	Gamma-glutamyl hydrolase	only in HaCaT	absent	only in HaCaT	absent
IPI00008561	Interstitial collagenase	only in HaCaT + PMA	absent	absent	only in HaCat+PMA
IPI00026314	Isoform 1 of Gelsolin	only in HaCaT	absent	only in HaCaT	absent
IPI00291262	Isoform 1 of Clusterin	only in HaCaT	absent	only in HaCaT	absent
IPI00029658	Isoform 1 of EGF-containing fibulin-like extracellular matrix protein 1	only in HaCaT + PMA	absent	absent	only in HaCat+PMA
IPI00387168	Isoform 1 of Proprotein convertase subtilisin/kexin type 9	1.57	absent	only in HaCaT	only in HaCat+PMA
IPI00783625	Isoform 1 of Serpin B5	3.03	absent	only in HaCaT	only in HaCat+PMA
IPI00013890	Isoform 1 of 14-3-3 protein sigma	0.59	absent	only in HaCaT	only in HaCat+PMA
IPI00480121	Kallikrein-10	only in HaCaT + PMA	only in SCC+PMA	absent	10.43
IPI00293276	Macrophage migration inhibitory factor	0.16	1.12	0.86	5.97
IPI00216691	Profilin-1	absent	0.84	only in SCC	only in SCC+PMA
IPI00013895	Protein S100-A11	only in HaCaT	absent	only in HaCaT	absent
IPI00294879	Ran GTPase-activating protein 1	only in HaCaT	absent	only in HaCaT	absent
IPI00009342	Ras GTPase-activating-like protein IQGAP1	0.34	absent	only in HaCaT	only in HaCat+PMA
IPI00296099	Thrombospondin-1	3.09	only in SCC+PMA	only in HaCaT	0.20
IPI00018219	Transforming growth factor-beta-induced protein ig-h3	0.95	1.00	0.39	0.41
	Membrane-bound proteins				
IPI00218918	Annexin A1	0.96	2.19	0.85	1.93
IPI00024095	Annexin A3	absent	only in SCC	only in SCC	absent
IPI00329801	Annexin A5	absent	8.06	only in SCC	only in SCC+PMA
IPI00414320	cDNA FLJ55482. highly similar to Annexin A11	Only in HaCaT	absent	only in HaCaT	absent
IPI00015688	Glypican-1	2.07	absent	only in HaCaT	only in HaCat+PMA
IPI00010271	Isoform A of Ras-related C3 botulinum toxin substrate 1	absent	only in SCC	only in SCC	absent
IPI00218474	Isoform 1 of Beta-enolase	4.12	0.07	74.08	1.25
IPI00465248	Isoform alpha-enolase of Alpha-enolase	1.12	0.28	7.44	1.86
IPI00031030	Isoform 1 of Amyloid-like protein 2	only in HaCaT + PMA	absent	absent	only in HaCat+PMA
IPI00418169	Isoform 2 of Annexin A2	1.04	0.85	0.92	0.75
IPI00871158	Isoform 2 of Annexin A8-like protein 2	absent	only in SCC	only in SCC	absent
IPI00297160	Isoform 12 of CD44 antigen	3.18	only in SCC+PMA	only in HaCaT	1.31
IPI00027493	Isoform 2 of 4F2 cell-surface antigen heavy chain	0.23	3.46	1.58	24.32
IPI00554711	Junction plakoglobin	1.53	absent	only in HaCaT	only in HaCat+PMA
IPI00026952	Plakophilin-3	3.05	only in SCC	1.18	only in HaCat+PMA
IPI00002441	Syndecan-1	absent	only in SCC+PMA	absent	only in SCC+PMA

Table 1. Cont.

Accession Number	Protein Identification	HaCat + PMA/HaCaT	SCC-9+ PMA/SCC-9	SCC-9/HaCaT	SCC-9+PMA/ HaCaT+PMA
IPI00011564	Syndecan-4	1.01	only in SCC+PMA	only in HaCaT	8.18
	Intracellular proteins				
IPI00020599	Calreticulin	absent	only in SCC	only in SCC	absent
IPI00298237	cDNA FLJ56402. highly similar to Tripeptidyl-peptidase 1	1.81	absent	only in HaCaT	only in HaCat+PMA
IPI00843975	Ezrin	0.64	2.27	0.34	1.19
IPI00163187	Fascin	1.22	2.21	0.39	0.71
IPI00024067	Isoform 1 of Clathrin heavy chain 1	absent	2.53	only in SCC	only in SCC+PMA
IPI00014898	Isoform 1 of Plectin-1	absent	only in SCC	only in SCC	absent
IPI00291175	Isoform 1 of Vinculin	only in HaCaT + PMA	absent	absent	only in HaCat+PMA
IPI00219365	Moesin	only in HaCaT + PMA	only in SCC+PMA	absent	1.01
IPI00915869	Putative uncharacterized protein MDH1	absent	only in SCC	only in SCC	absent
IPI00298994	Talin-1	Only in HaCaT	1.53	24.56	only in SCC+PMA

The HaCaT and SCC-9 cells were treated with PMA for 24 h, the media were collected, the proteins were digested with trypsin and analyzed by LC-MS/MS. The data were submitted to Mascot search engine and the .dat files from Mascot output were analyzed in Scaffold Q+, which calculates the quantitative value by normalizing spectral counts across the experiments. The ratio of quantitative value of the PMA-stimulated cells/DMSO-treated cells for each protein is shown in the table. The proteins exclusively found in one condition are indicated as "only" and the proteins that are not present are indicated as "absent".

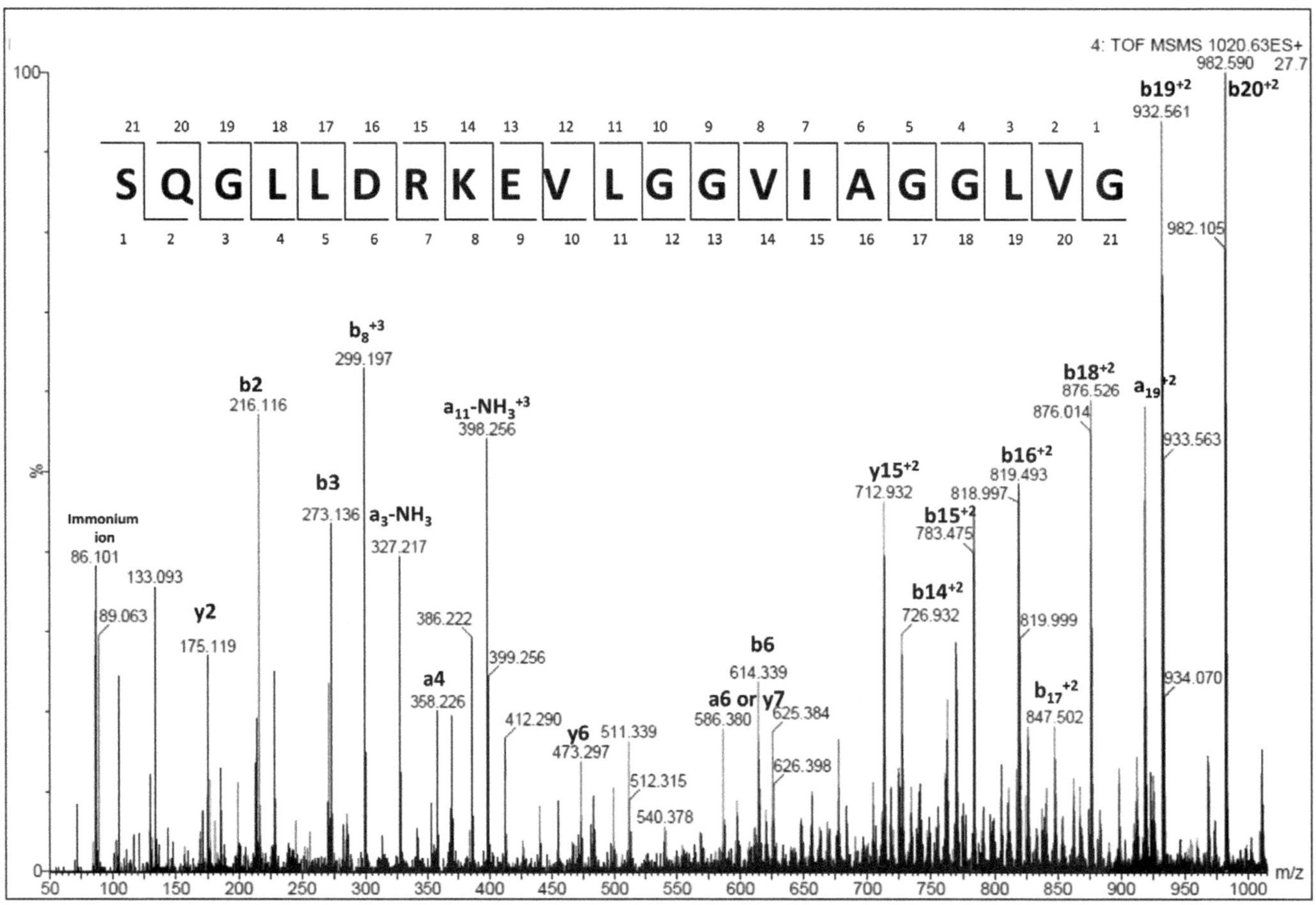

Figure 2. CID spectrum of a syndecan-1 peptide identified by LC-MS/MS. Endogenous peptides were identified by LC-MS/MS in the media after 24 h of PMA-treatment. The spectrum of syndecan-1 peptide (m/z 1019.5598, +2) was manually validated for b and y ion series.

Table 2. Identification of endogenous peptides in the conditioned media by LC-MS/MS according to the total number of unique peptides and spectral counts.

Accession number	Protein Identification	Experiment 1								Experiment 2							
		HaCaT		HaCaT+ PMA		SCC-9		SCC- 9+PMA		HaCaT		HaCaT+ PMA		SCC-9		SCC-9+ PMA	
		Run 1	Run 2	Run 1	Run 2	Run 1	Run 2	Run 1	Run 2	Run 1	Run 2	Run 1	Run 2	Run 1	Run 2	Run 1	Run 2
IPI00002441	**Syndecan-1**	**2/5**	**2/6**	**2/9**	**2/9**	**3/8**	**4/12**	**6/23**	**4/20**			**1/3**	**1/2**	**2/8**	**3/8**	**3/13**	**2/10**
	SQGLLDRKEVLGGVIAGGLVG	3	4	6	6	4	5	8	9			3	2	3	6	8	7
	SQGLLDRKEVLGGVIAGGLVGLI	2	2	3	3	3	4	5	4					5	1	3	3
	SQGLLDRKEVLGGVIA					1	2	5	5								
	LDRKEVLGGVIAGG							2							1		
	SQGLLDRKEVLGGVIAGG							1									
	RNQSPVDQGATGASQGLLDRKE							2	2							2	
	VLGGVIAGGLVG																
	DLHTPHTED						1										

The HaCaT and SCC-9 cells were treated with PMA for 24 h, the media were collected and the endogenous peptides were analyzed by LC-MS/MS. The total numbers of the unique peptides and the number of spectral counts are shown in bold and the number of spectral counts is shown for each sequence. There is a statistically significant difference in the number of spectral counts of syndecan-1 fragments between SCC-9 cells and SCC-9 cells treated with PMA (Mann-Whitney test, p = 0.0273).

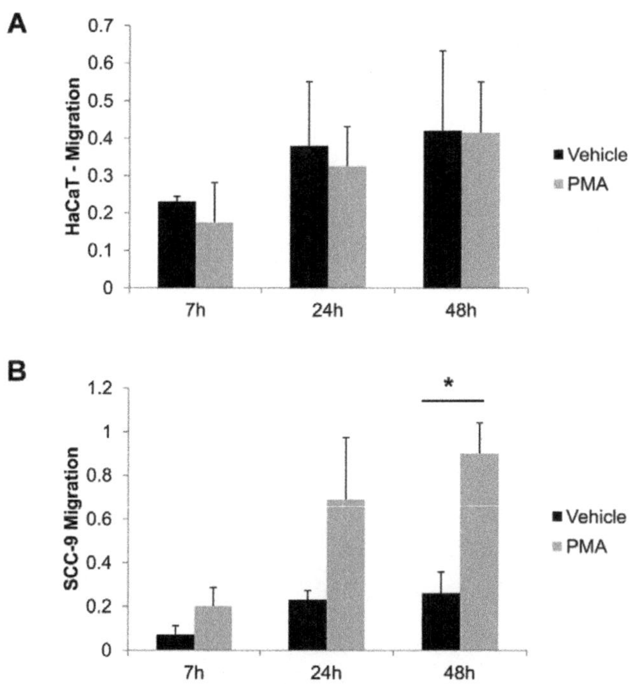

Figure 3. PMA increased migration only in SCC-9 cells by scratch assay. PMA and vehicle effect on the migration of HaCaT (A) and SCC-9 (B) cells after 7, 24 and 48 h. Cells grown in 12-well plates to confluence were scraped with p200 pipette tip. The closure was measure after 0, 7, 24 and 48 h and normalized to the control (vehicle: DMSO). Two independent experiments were performed with triplicates. Columns represent mean ± SD (n = 2) and * indicates p<0.01, compared to control (vehicle).

the sequence SQGLLDRKEVLGGVIAGGLVG (SYN-1), for which the CID spectra were manually validated (Fig. 2).

Analysis of the effect of PMA treatment on tumorigenic and non-tumorigenic cells

PMA induced cell migration in tumorigenic cells. The effect of PMA on migration was evaluated by scratch assays (Fig. 3) and it was observed that PMA induced migration in SCC-9 cells, exclusively, upon 48 h-PMA treatment supplemented of 1% FBS (p = 0.01, Students' *t*-test).

PMA promoted lower cell adhesion in HaCaT and SCC-9 cells. HaCaT (p = 0.002, Students' *t*-test) and SCC-9 (p<0.001, Students' *t*-test) cells showed reduced ability to adhere in Matrigel^{TM} upon 24 h-PMA treatment, with a higher effect on SCC-9 cells (Fig. 4).

Gelatinase activity was increased with PMA stimuli. HaCaT and SCC-9 cells were treated with PMA and after 24 h the gelatinase activity was evaluated in the conditioned media. There was an increase of activity upon PMA treatment in both cell lines for ~72 kDa gelatinase (p = 0.01 and p = 0.02 for HaCaT and SCC-9 cells, respectively, Students' *t*-test). The ~95 kDa gelatinase showed higher activity upon PMA treatment only in SCC-9 cells (p = 0.04, Students' *t*-test). The activity of MMP-2 immunoprecipitated from the media confirmed the gelatinase activity (Fig. 5).

Membrane syndecan-1 decreased after 30 min and 24 h upon PMA treatment. The loss of membrane-bound syndecan-1 in SCC-9 cells was confirmed by immunofluorescence in a time-course experiment performed after 5 min, 30 min and 24 h upon PMA treatment (Fig. 6). The decrease of membrane-bound syndecan-1 was statistically significant after 30 min (Students' *t*-test, p = 0.02) and 24 h (Students' *t*-test, p = 0.001).

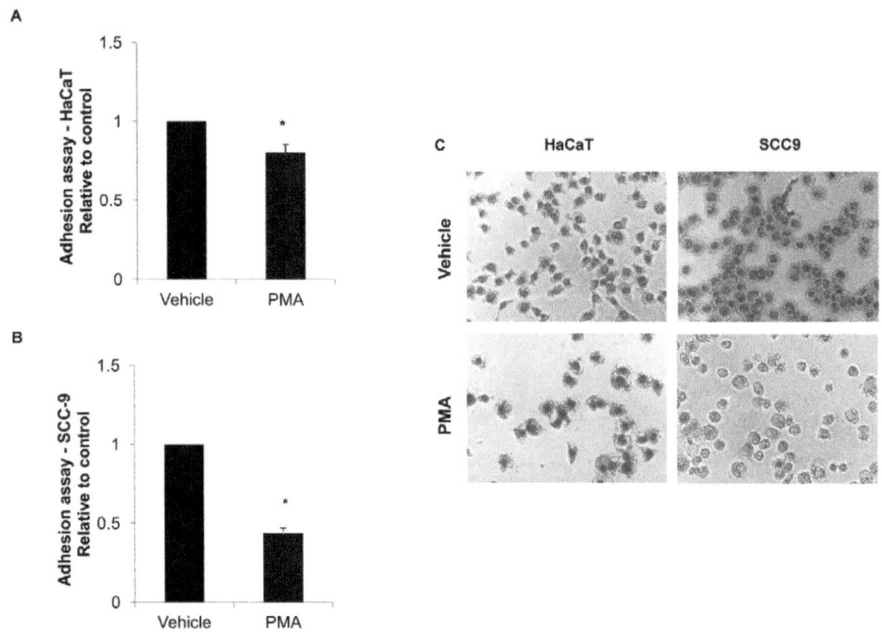

Figure 4. PMA promoted the loss of adhesion in Matrigel™ in HaCaT cells and SCC-9 cells. Adhesion assay showed that the cells stimulated by PMA decreased the ability to adhere to extracellular matrix proteins. HaCaT (A) and SCC-9 (B) cells had the ability to adhere in Matrigel™ diminished after 24 h of PMA treatment. Three independent experiments were performed with triplicates. Columns represent mean ± SD (n = 3) and * indicates $p<0.01$, normalized with the control (vehicle: DMSO). (C) Representative micrographs (magnification 40×) of adherent cells after PMA and vehicle (DMSO) treatments.

Analysis of the effect of syndecan-1 synthetic peptide on tumorigenic and non-tumorigenic cells

Syndecan-1 synthetic peptide induced migration in tumorigenic and non-tumorigenic cells by scratch assay. The synthetic peptide of syndecan-1 (SYN-1) and its scrambled were evaluated at the concentrations of 1 μM, 10 μM and 100 μM by scratch assay. In HaCaT cells, 10 μM of SYN-1 peptide induced migration at 48 h in the absence of FBS (p = 0.02, Students' *t*-test) (Fig. 7B). In the presence of 1% FBS, it induced migration at the concentration of 1 μM, at 24 (p = 0.03, Students' *t*-test) and 48 h (p = 0.02, Students' *t*-test) (Fig. 7C and 7D). With respect to SCC-9 cells, SYN-1 peptide induced migration with statistical significance only at the concentration of 1 μM in the absence of FBS at 24 h (p = 0.04, Students' *t*-test) and 48 h (p = 0.01, Students' *t*-test) (Fig. 7E and 7F).

Syndecan-1 synthetic peptide induced migration in tumorigenic cells by transwell assay. The effect of SYN-1 peptide on migration was also evaluated by transwell assay at concentration of 10 μM in absence of FBS. The results showed an increase of migration only in SCC-9 cells (p = 0.02, Students' *t*-test) (Fig. 8B).

Analysis of the effect of PMA treatment on tumorigenic and non-tumorigenic cells by qRT-PCR

mRNA levels of syndecan-1 did not change in a time-course experiment, but they showed higher expression in

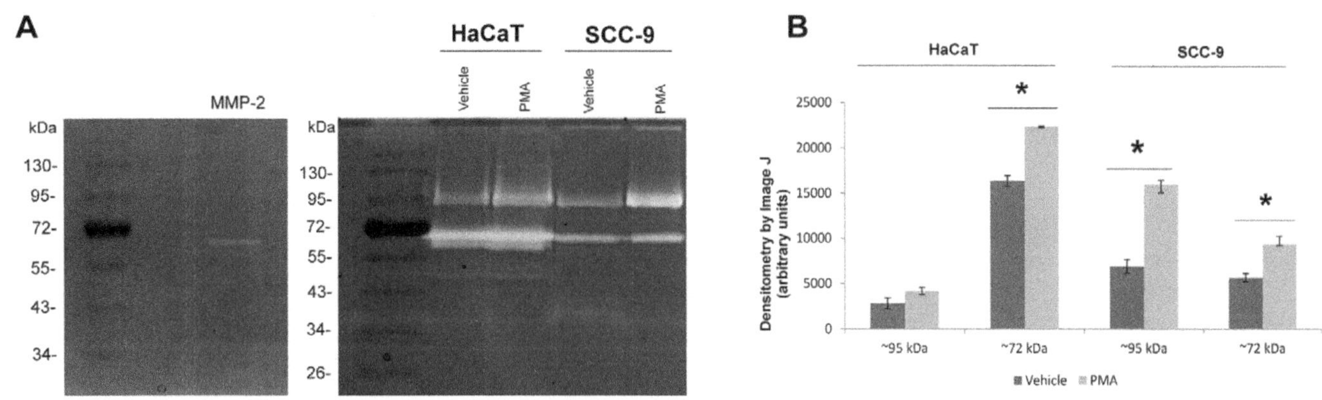

Figure 5. Gelatinase activity was increased with PMA stimuli. Representative 1D-zymography of conditioned media (12 μg of proteins) of HaCaT and SCC-9 cells treated or not with 50 ng/ml PMA two times with 12 h interval, collected after 24 h. The gelatinase activity of immunoprecipitated MMP-2 was used as positive control. Numbers on the left indicate the molecular mass marker mobility (A). The densitometry of clear areas showed the increase of activity after PMA treatment (B). Columns represent mean ± SD (n = 3) and * indicates $p<0.05$.

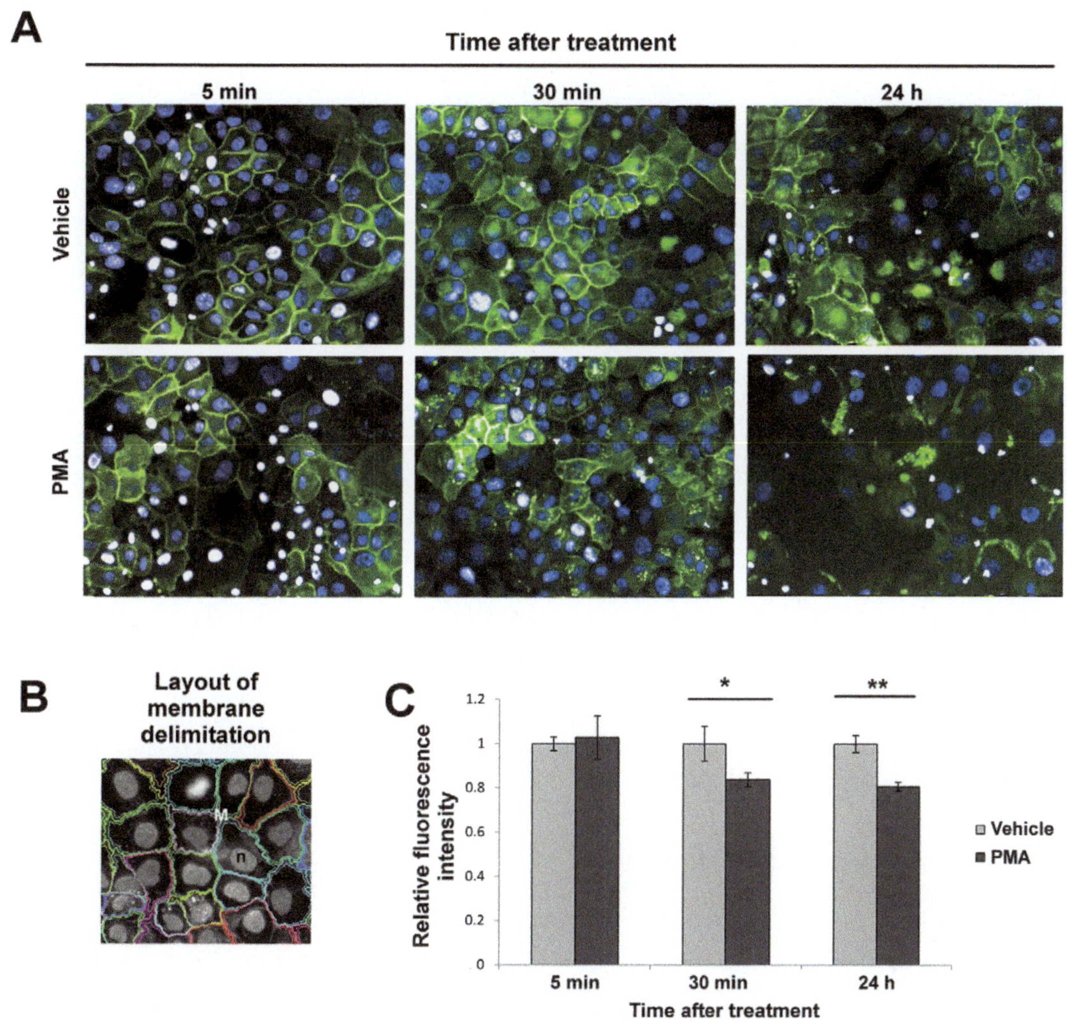

Figure 6. Loss of membrane-bound syndecan-1 localization after PMA treatment in SCC-9 cells. SCC-9 cells were treated with PMA for 5 min, 30 min and 24 h, fixed, and labeled for syndecan-1 with goat anti-syndecan-1 antibody. (A) Immunofluorescence images revealed diminished syndecan-1 in membrane localization after 30 min and 24 h of PMA treatment. Green: anti-goat antibody conjugated with Alexa Fluor 488. Blue: DAPI. (B) Example of masks used for quantification of syndecan-1 in the cell surface membrane. (C) Quantification performed by the Operetta high content image system showed loss of cell membrane of syndecan-1 after 30 min and 24 h of PMA treatment. Three independent experiments were performed. Columns represent mean ± SD (n = 3) and * and ** indicate $p<0.05$ and $p<0.01$, respectively.

SCC-9 cells compared with HaCaT cells. One-way ANOVA analysis shows that PMA treatment did not change the mRNA expression levels of syndecan-1 in HaCaT and SCC-9 cells according to the time-course experiment ($p>0.05$). However, there were statistically significant differences in the expression levels of syndecan-1 between the cell lines at 0, 5 and 30 min upon PMA treatment ($p = 0.009$, 0.01 and 0.004, respectively) (Fig. 9).

Discussion

Ectodomain shedding of cell surface proteins, secreted proteins and proteolysis-derived fragments are soluble protein candidates to contribute to a diverse extracellular milieu and subsequently certain biological events related to cancer [15,16]. One percent of the extracellular ectodomains of membrane-anchored proteins are released from the cell surface by endogenous proteolytic cleavage [17], but the novel biological functions in which some of these molecules can be involved in cancer remain unclear. Here, we analyzed the secretome/sheddome of two epithelial cells lines in attempt to understand how they may affect tumor development. The non-tumorigenic (HaCaT) and tumorigenic (SCC-9) cells were stimulated with PMA that activates proteases and promotes protein shedding, secretion and proteolysis [2].

Differences between these cell lines, with or without stimulation by PMA, were observed in migration, adhesion as well as in gelatinase activity (Fig. 3, 4, 5, respectively). In agreement with these data, the secretomes/sheddomes were also different following stimulation with PMA (Fig. 1, Tables 1 and 2). From the proteomic analysis, seven proteins were up regulated by the effect of PMA treatment in both cell lines, while 14 and 10 proteins were exclusively found in HaCaT and SCC-9 cells, respectively (Fig. 1, Table 1). Consistent with these results, the secretome of human mammary epithelial (HMEC) cells, a non-tumorigenic cell line, also showed the increase of 36 extracellular proteins after PMA stimulation [2]. Similar to their findings, a significant number of PMA-stimulated proteins were related to proteolytic and extracellular matrix function. Interestingly, we found in the PMA-stimulated HaCaT cells, which are also non-tumorigenic, the increase of proteases expression, named cathepsin, kallikrein and

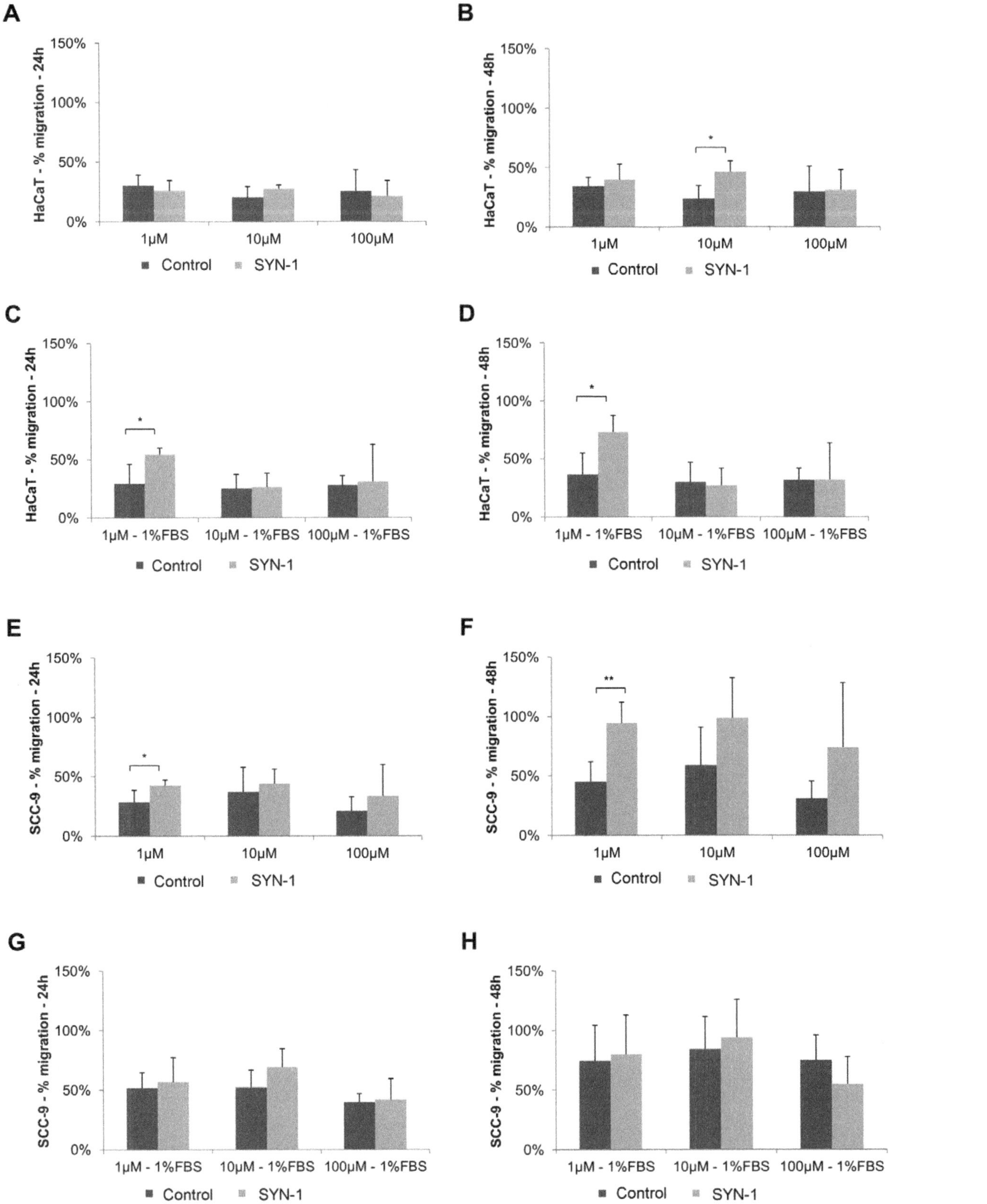

Figure 7. Synthetic syndecan-1-derived peptide promoted migration in HaCaT and SCC-9 cells by scratch assay. Migration of HaCaT and SCC-9 cells treated with synthetic syndecan-1 peptide (SYN-1) and its scrambled peptide (control) in the concentrations of 1 μM, 10 μM and 100 μM were evaluated by scratch assay. SYN-1 did not induce migration in HaCaT cells at 24 h (A), but 10 μM of SYN-1 promoted migration at 48 h in absence of FBS (B), 1 μM of SYN-1 induced migration in HaCaT cells at 24 h (C) and 48 h (D), both in the presence of 1% FBS. SCC-9 cell migration was statistically significant at the concentration of 1 μM of SYN-1 in the absence of FBS at 24 h (E) and 48 h (F), but it was not statistically significant when

compared to the scrambled peptide in the presence of 1% FBS at 24 h (G) and 48 h (H). Three independent experiments were performed with duplicates. Columns represent mean ± SD (n = 3) and * and ** indicate $p<0.05$ and $p<0.01$, respectively.

collagenases. Besides, tumorigenic cells exclusively showed higher expression of membrane-bound proteins, such as syndecans. Furthermore, several PMA-stimulated proteins identified in the secretome of SCC-9 cells were also found in the secretome of other cancer cells [3,4,18], in which many of them represent potential biomarkers for cancer.

Among the identified proteins, heparan sulphate proteoglycans, such as syndecan-1, were found only in the media of stimulated SCC-9 cells (Table 1, Fig. 1). The increase of the soluble syndecan-1 in the media of stimulated SCC-9 cells was associated with the loss of syndecan-1 from the cell surface evaluated by immunofluorescence assay (Fig. 6). In fact, in a specific experiment to evaluate the syndecan-1 cleavage, previous study showed the increase of syndecan-1 cleavage in PMA-stimulated GAC myeloma cells compared with DMSO treatment, which was also enhanced or decreased, respectively, after Heparinase III and inhibitor (BB-94) treatments [19]. Interestingly, another large-scale secretome study to find cancer biomarkers also observed syndecan-1 in the extracellular media in three different types of cancer cell lines originated from oral cancer, colorectal carcinoma and lung cancer [18]. Furthermore, the loss of syndecan-1 from the membrane surface was previously correlated with carcinogenesis, suggesting that it could be a useful marker of malignant transformation or prognostic factor in oral squamous carcinoma [20,21].

Interestingly, fragments of syndecan-1 were also found in higher abundance in stimulated SCC-9 cells (Table 2). We found seven peptides derived from the syndecan-1 and most of which have a leucine in the P1' position. Possible candidate proteinases responsible for such cleavage include ADAM-17 protease, which has valine or leucine residues as preferential sites at P1' position [22], and MMP-7 and MT1-MMP, which show a strong preference for leucine at P1' position [23,24]. However, in this case it is not possible to predict which protease is responsible for the cleavage, since the media was collected 24 h after cell stimulation and more than one protease could had been involved in this process.

To support a functional role for soluble syndecan-1 fragments, we demonstrated that the synthetic syndecan-1 peptide (SYN-1) was able to induce cell migration in both cell lines, as observed by a scratch assay (Fig. 7), and also induced migration, by transwell assay, in tumorigenic cells, SCC-9 cells (Fig. 8). A recent study has demonstrated that both full-length and truncated syndecan-1 can modulate fibrosarcoma cell migration and adhesion, in which the extracellular domain is more important for promoting cell adhesion and the transmembrane and cytoplasmic domains in inhibiting cell migration [11]. In addition, a study using breast cancer cells observed that membrane-bound syndecan-1 increased cell proliferation, whereas soluble syndecan-1 increased cell invasion [25].

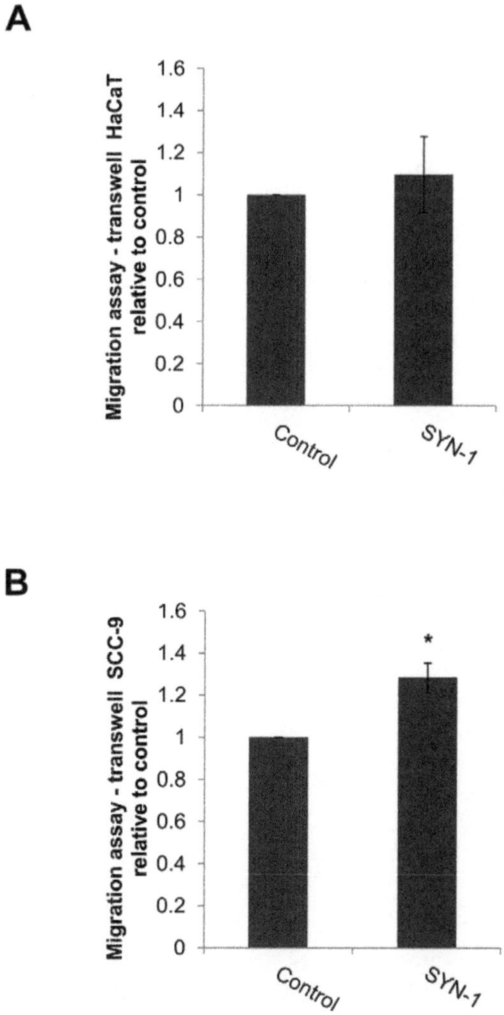

Figure 8. Synthetic syndecan-1-derived peptide promoted migration in SCC-9 cells by transwell migration assay. The migration of HaCaT and SCC-9 cells treated with synthetic syndecan-1-derived peptide (SYN-1) and its scrambled in a concentration of 10 µM in the absence of FBS was evaluated by transwell assay. The measurements were normalized with the control (scrambled peptide). SYN-1 did not induce migration in HaCaT cells (A). SCC-9 cell migration was statistically significant when compared to scrambled peptide (B). Three experiments were performed with triplicates. Columns represent mean ± SD (n = 3) and * indicate $p<0.05$.

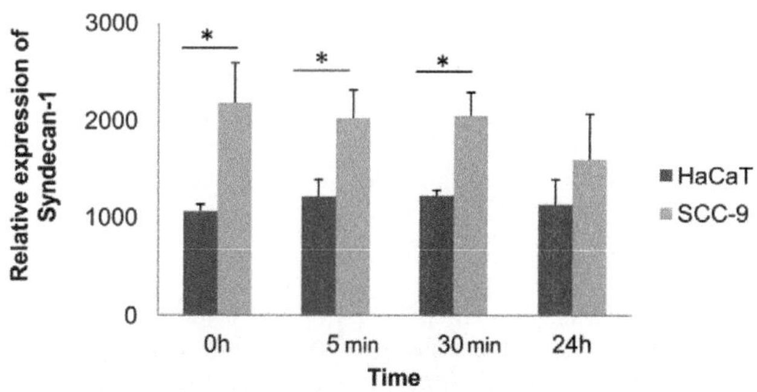

Figure 9. qRT-PCR analysis of syndecan-1 in HaCaT and SCC-9 cells treated with PMA at 0, 5 min, 30 min and 24 h. Relative mRNA expression levels of syndecan-1 were measured by the real-time quantitative PCR. The data were normalized with glyceraldehyde-3-phosphate dehydrogenase gene. One-way ANOVA analysis shows that PMA treatment did not increase the expression levels of syndecan-1 in each cell line in the time-course experiment. Student's *t*-test indicated differences in the expression level between the HaCaT and SCC-9 cells. Syndecan-1 showed higher expression in tumorigenic SCC-9 cells. Three independent experiments were performed with triplicates. Columns represent mean ± SD (n = 3) and * indicates $p\leq0.01$.

Another study showed that soluble syndecan-1 promoted growth of myeloma tumors [26]. Several studies have shown the involvement of syndecan-1 in cell migration and its regulation can occur through the regulation of the activities of integrins, including αvβ3, αvβ5, β4 and α2β1 [27–29] as well as by the cleavage of syndecan-1 by MMP-7, which could release the restrictions to migration [27]. However, here we showed the first example where a fragment of syndecan-1 was demonstrated to potentially be involved in cell migration. Based on previously investigations, it is possible that the soluble syndecan-1 also has an effect on carcinoma cells and it will be investigated in future studies.

The analysis of mRNA of syndecan-1 excluded the possibility that the increase of syndecan-1 in the media would be influenced by the increase of gene expression. We showed here that syndecan-1 mRNA expression levels did not change in a time-dependent manner (Fig. 9).

In summary, this study demonstrated that the repertoire of secreted, shed and degraded proteins in the extracellular milieu of PMA-stimulated non-tumorigenic HaCaT and tumorigenic SCC-9 cells could be involved in fundamental cell processes and reveals that a fragment of syndecan-1 was able to induce cell migration.

Materials and Methods

Cell Culture

The human OSCC cell line SCC-9 was obtained from American Type Culture Collection (ATCC, Manassas, VA, USA), and cultured as recommended. SCC-9 cells originated from human squamous carcinoma from the tongue. The HaCaT cells, an immortalized but not transformed epithelial cell line [30], was maintained in DMEM containing 10% fetal bovine serum (FBS) and antibiotics at 37°C in a 5% CO_2 air atmosphere. HaCaT cells are human keratinocytes originated from skin.

Mass Spectrometry Analysis: PMA Stimulation

The HaCaT and SCC-9 cells were cultured until 80% of confluence in 500-cm^2 plates, washed with serum-free media and then stimulated two times, with 12 h interval, with 50 ng/ml PMA (Sigma) diluted in DMSO. The same concentration of DMSO was used as control in the time 0 and after 12 h. After 24 h, the media were collected and the proteins and peptides were analyzed as described below.

Protein and Peptide extraction, Protein Digestion

The media were collected and the final concentration of 1 mM EDTA and 0.5 mM PMSF were added to the media. The protocol to obtain peptides and proteins was performed with few modifications [31]. Briefly, cell debris were eliminated by centrifugation at 4,000×*g* during 20 min at 4°C and the supernatants were heated in a water bath for 20 min at 80°C to inactivate proteases. After cooling, pH was adjusted to 2–3 by adding 10 M HCl to precipitate the proteins. After centrifugation at 10,000×*g* for 1 h at 4°C, the protein pellet was ressuspended in 200 mM ammonium bicarbonate and the peptides in the supernatant were concentrated in Sep-Pak® Vac tC18 cartridge 6cc/500 mg (Waters) and dried in a vaccum. The proteins in the extracellular media (50 µg) were treated with the final concentration of 4 M urea, following reduction, alkylation and digestion with trypsin (1:50, w/w) [32].

Mass spectrometry analysis

For protein analysis, an aliquot of 4.5 µl containing 15 µg of proteins of the resulting peptide mixture was evaluated as previously described [33] and for endogenous peptide analysis, we based on the intracellular protein concentration to inject similar concentration of peptides. Peptides (4.5 µl) were separated by C18 (100 µm×100 mm) RP-nanoUPLC (nanoAcquity, Waters) coupled with a Q-Tof Ultima mass spectrometer (Waters) with nanoelectrospray source at a flow rate of 0.6 µl/min. The gradient was 2–90% acetonitrile in 0.1% formic acid over 45 min for the digested proteins, and 60 min for endogenous peptides. The nanoelectrospray voltage was set to 3.5 kV, a cone voltage of 30 V and the source temperature was 100°C. The instrument was operated in the 'top three' mode, in which one MS spectrum is acquired followed by MS/MS of the top three most-intense peaks detected. After MS/MS fragmentation, the ion was placed on exclusion list for 60 s and for the analysis of endogenous cleavage peptides, a real time exclusion was used.

Data Analysis

The spectra were acquired using software MassLynx v.4.1 and the raw data files were converted to a peak list format (mgf) without summing the scans by the software Mascot Distiller v.2.3.2.0, 2009 (Matrix Science Ldt.) allowing the label-free analysis, and searched against Human International Protein Database (IPI) v. 3.72 (86392 sequences, 35093930 residues; release date April, 2010) using Mascot engine v.2.3.01 (Matrix Science Ltd.), with carbamidomethylation as fixed modifications, oxidation of methionine as variable modification, one trypsin missed cleavage and a tolerance of 0.1 Da for both precursor and fragment ions. For the protein quantitation, the .dat files from Mascot output were analyzed in Scaffold Q+ (version 3_00_03, Proteome Software) and the quantitative value (normalized spectral counts) was obtained [34,35]. For endogenous peptide identification, oxidation of methionine was set as variable modification, and a tolerance of 0.1 Da for both precursor and fragment ions. For label-free quantitation of endogenous peptides, the spectral counts [36] and the number of unique peptides were assessed. The peptide was considered as unique when it differs in at least 1 amino acid residue, covalently modified peptides, including N- or C-terminal elongation (i.e. missed cleavages) count as unique and different charge states of the same peptide and modifications were not count as unique. Only peptides with a minimum of five amino acid residues which showed significant threshold ($p<0.05$) in Mascot-based score were considered in the results. Two independent experiments were performed for proteins and peptides analysis. MS/MS spectrum of the most abundance endogenous peptide of syndecan-1 was manually validated for b and y ion series.

Cell Migration Assay

The migration of HaCaT and SCC-9 cells was investigated through an in vitro monolayer assay. Cells grown in 12-well plates to confluence were scraped with a p200 pipette tip to create a cell-free area [37]. The cells were washed three times to remove cell debris. The same PMA treatment described above was performed for this assay (50 ng/ml PMA two times, with 12 h interval). The experiments were performed in serum-free media and 1% FBS supplemented media. The closure was evaluated after 0, 7, 24 and 48 h and analyzed by ImageJ software (http://rsb.info.nih.gov/ij/). Two independent experiments were performed in triplicates.

Cell adhesion assay

The ability of HaCaT and SCC-9 cells to adhere to extracellular matrix proteins was evaluated in the adhesion assay [38]. First, the cells were plated in 100 mm dishes (Corning) at the density of 2×10^5 cells for the PMA treatment (50 ng/ml, two times, with 12 h interval) and the another 96-well plate was coated with Matrigel™ (2 μg per well). After 24 h, these cells were trypsinised and seeded in this coated 96-well plate, previously washed three times with PBS and blocked with 3% BSA (bovine serum albumin) during 2 h in serum-free media supplemented with 3% BSA. The adhesion was evaluated for 1 h, and then the cells were washed and fixed with 10% formaldehyde. Briefly, the cells were stained with 1% toluidine blue containing 1% borax for 5 min. The dye was eluted using 1% SDS and the absorbance was measured at 620 nm. Three independent experiments were performed in triplicates.

Zymography of conditioned media

For 1-D zymography, the proteins (12 μg) in the conditioned media collected after cell treatments with 50 ng/ml PMA two times with 12 h interval were submitted to 1-D electrophoresis on 12% SDS-polyacrylamide gels containing 1 mg/ml gelatin under nonreducing conditions, and gelatinolytic activity was performed as previously described [39]. We immunoprecipitated MMP-2, as positive control, from the conditioned media of HaCaT cells. Briefly, anti-MMP-2 antibody (1 μg) (Santa Cruz) was added to conditioned media containing 5 μg of proteins and incubated for 16 h at 4°C. The immunocomplexes were collected by incubating with 20 μl of protein-A Sepharose for 1 h at 4°C followed by three washed with 1% Tween in PBS. After centrifugation at 2,000×*g* for 2 min at 4°C, the beads were resuspended in sample buffer under nonreducing conditions as described before. Gels were stained with Coomassie blue and destained. Gelatin digestion was identified as clear bands against a blue background. Three independent experiments were performed.

Immunofluorescence Assay

For immunofluorescence assays, SCC-9 cells were cultivated in a CellCarrier (Perkin Elmer) 384 well plate. After PMA stimulation as described above, the cells were analyzed after 5 min, 30 min and 24 h. The cells were washed with PBS, fixed with 4% formaldehyde for 10 min, washed again and permeabilized with 0.5% Triton X-100 for 10 min. The cells were blocked with blocking solution (PBS containing 0.2% Triton and 3% non-fat dry milk) for 30 min and then incubated with the goat anti-syndecan-1 antibody (R&D Systems) diluted in blocking solution for 1 h. Cells were washed with PBS and incubated with an anti-goat Alexa Fluor 488-conjugated antibody (Invitrogen), diluted in blocking solution for 1 h. Finally, the cells were washed with PBS, incubated with DAPI (4′,6-diamidino-2-phenylindole) solution for 10 min, washed again and analyzed by the Operetta high content image system (Perkin Elmer). All pictures were acquired with the same contrast and brightness parameters. For automatic fluorescence quantification in the Operetta platform, DAPI staining was used for cell counting. Syndecan-1 Alexa Fluor 488 staining, restricted to the cell membranes, was defined by the most appropriate mask. Three independent experiments were performed.

Synthetic peptides

The peptide of syndecan-1 identified by MS and manually validated (H-SQGLLDRKEVLGGVIAGGLVG-OH), named SYN-1, and the scrambled control (H-IGVGGLRELVK-QLGDLGGVSA-OH) were chosen for functional experiments. The peptide was synthesized (Proteimax, São Paulo, Brazil) with the same sequence identified by mass spectrometry.

Cell Migration Assay with Syndecan-1-Derived Peptide

Migration of HaCaT and SCC-9 cells was performed as described before. Cells grown in 12-well plates to confluence were scraped with p200 pipette tip to create a cell-free area. The cells were washed three times with serum-free media to remove cell debris and incubated with serum-free media or 1% FBS containing the synthetic peptides SYN-1 and its scrambled in the concentrations of 1 μM, 10 μM and 100 μM. The migration was evaluated after 0, 24 and 48 h and analyzed by ImageJ software. Three independent experiments were performed with duplicates.

Transwell migration assay with Syndecan-1-Derived Peptide

HaCaT and SCC-9 cells (7.5×10^4 cells) were plated in the top chambers of 8 μm pore transwells (HTS Transwell-96 Well Plate, Corning) in the serum-free culture medium after a starvation period of 18 h. The cells were allowed to migrate towards serum-free medium supplemented with 10 μM SYN-1 peptide or its scrambled over a period of 6 h. At the end of the assay, cells at the top chamber were removed with a cotton swab and the cells at the bottom of the filter were fixed with 10% formaldehyde for 10 min, washed with PBS and stained with 1% toluidine blue solution in 1% borax for 5 min. The dye was eluted using 1% SDS and the absorbance was measured at 620 nm. Three independent experiments were performed in triplicates.

Real-time quantitative PCR

In order to analyze the expression of syndecan-1, HaCaT and SCC-9 cells were cultured for 24 h in serum-free medium and treated with PMA during 5, 30 min and 24 h as described before. Total RNA was obtained using the TRIzol reagent (Invitrogen Corporation) and 2 μg of total RNA were used for retro-transcription using the First-Strand cDNA Synthesis Kit (GE Healthcare). Real-time quantitative PCR for syndecan-1 was performed using SYBR® Green PCR Master Mix (Applied Biosystems), and the dissociation curves were performed to confirm the specificity of products. The syndecan-1 forward primer was 5′- AGAAGAAGGACGAAGGCAGCTACT- 3′ and reverse primer was 5′- ATTCCTCCTGTTTGGTGGGCTTCT- 3′. The threshold cycles (CT) values of target gene were normalized relative to glyceraldehydes-3-phospate dehydrogenase gene, and relative expression ratios were calculated by the $2-\Delta\Delta$ Ct method. Three-independent experiments were performed with duplicates.

Statistical analysis

For the functional experiments, the assumptions of adherence of the errors to the Gaussian distribution were tested, for all response variables, using the Shapiro-Wilk Test. The GLIMMIX procedure of the SAS System (SAS Institute Inc. The SAS System, release 9.12. SAS Institute Inc., Cary:NC, 2008) were used to calculate the analysis of variance followed Student's *t*-test or Tukey test. For the analysis of mRNA expression of syndecan-1, the analysis of variance was followed by Tukey test. For the statistical analysis of syndecan-1 fragments, it was used Mann-Whitney test to compare the effect of PMA on HaCaT and SCC-9 cells. The level of significance was stated at 0.05 in all statistical tests applied.

Acknowledgments

We thank Dr. Sandra Dias from Brazilian Biosciences National Laboratory for helping with immunofluorescence experiments. We thank Dr. Marcelo Corrêa Alves from "Luiz de Queiroz" College of Agriculture for helping with statistical analysis.

Author Contributions

Conceived and designed the experiments: AZBA MB AFPL. Performed the experiments: AZBA MB FS MZ SY RRD BP RK MA AFPL. Analyzed the data: AZBA MB AG AFPL. Contributed reagents/materials/analysis tools: MA EG RDC AFPL. Wrote the paper: EG RDC JF AFPL.

References

1. Jemal A, Siegel R, Ward E, Hao Y, Xu J, et al. (2009) Cancer Statistics, 2009. CA Cancer J Clin 59: 225–249.
2. Jacobs JM, Waters KM, Kathmann LE, Camp II DG, Wiley HS, et al. (2008) The mammary epithelial cell secretome and its regulation by signal transduction pathways. J Proteome Res 2: 558–569.
3. Faça VM, Ventura AP, Fitzgibbon MP, Pereira-Faça SR, Pitteri SJ, et al. (2008) Proteomic analysis of ovarian cancer cells reveals dynamic processes of protein secretion and shedding of extra-cellular domains. PLoS One 3(6): e2425. Available: http://www.plosone.org/article/info%3Adoi%2F10.1371%2Fjournal.pone.0002425.
4. Yao L, Zhang Y, Chen K, Hu X, Xu L X (2011) Discovery of IL-18 as a novel secreted protein contributing to doxorubicin resistance by comparative secretome analysis of MCF-7 and MCF/Dox. PLoS One 6:e24684. Available: http://www.plosone.org/article/info%3Adoi%2F10.1371%2Fjournal.pone.0024684. Accessed 2012 Jul 26.
5. Brou C, Logeat F, Gupta N, Bessia C, LeBail O, et al. (2000) A novel proteolytic cleavage involved in Notch signaling: the role of the disintegrin–metalloprotease TACE. Mol Cell 5:207–16.
6. Okamoto I, Kawano Y, Murakami D, Sasayama T, Araki N, et al. (2001) Proteolytic release of CD44 intracellular domain and its role in the CD44 signaling pathway. J Cell Biol 155: 755–62.
7. Hartmann D, de Strooper B, Serneels L, Craessaerts K, Herreman A, et al. (2002) The disintegrin/metalloprotease ADAM 10 is essential for Notch signaling but not for alpha-secretase activity in fibroblasts. Hum Mol Genet 11: 2615–24.
8. Marambaud P, Shioi J, Serban G, Georgakopoulos A, Sarner S, et al. (2002) A presenilin-1/gamma-secretase cleavage releases the E-cadherin intracellular domain and regulates disassembly of adherens junctions. EMBO J 21: 1948–56.
9. Demidova-Rice TN, Wolf L, Deckenback J, Hamblin MR, Herman IM (2012) Human platelet-rich plasma- and extracellular matrix-derived peptides promote impaired cutaneous wound healing in vivo. PLoS One 7(2): e32146. Available: http://www.ncbi.nlm.nih.gov/pmc/articles/PMC3285658/pdf/pone.0032146.pdf. Accessed 2012 Jul 26.
10. Manon-Jensen T, Itoh Y, Couchman JR (2010) Proteoglycan in health and disease: the multiple roles of syndecan shedding. FEBS Journal 277: 3876–3889.
11. Zong F, Fthenou E, Mundt F, Szatmáril T, Kovalszky I, et al. (2011) Specific syndecan-1 domains regulate mesenchymal tumor cell adhesion, motility and migration. Available: http://www.plosone.org/article/info%3Adoi%2F10.1371%2Fjournal.pone.0014816. Accessed 2012 Jul 26.
12. Martínez A, Spencer ML, Brethauer U, Grez P, Marchesani FJ, et al. (2009) Reduction of syndecan-1 expression during lip carcinogenesis. J Oral Pathol Med 38(7): 580–583.
13. Kurokawa H, Zhang M, Matsumoto S, Yamashita Y, Tanaka T, et al. (2006) Reduced syndecan-1 expression is correlated with the histological grade of malignancy at the deep invasive front in oral squamous cell carcinoma. J Oral Pathol Med. 35(5): 301 306.
14. Ro Y, Muramatsu T, Shima K, Yajima Y, Shibahara T, et al. (2006) Correlation between reduction of syndecan-1 expression and clinico-pathological parameters in squamous cell carcinoma of tongue. Int J Oral Maxillofac Surg. 35(3): 252–257.
15. Doucet A, Butler GS, Rodríguez D, Prudova A, Overall CM (2008) Metadegradomics: toward in vivo quantitative degradomics of proteolytic post-translational modifications of the cancer proteome. Mol Cell Proteomics 7(10): 1925–1951.
16. Karagiannis GS, Pavlou MP, Diamandis EP (2010) Cancer secretomics reveal pathophysiological pathways in cancer molecular oncology. Mol Oncol 4 (6): 496–510.
17. Cauwe B, Van den Steen PE, Opdenakker G (2007) The biochemical, biological, and pathological kaleidoscope of cell surface substrates processed by matrix metalloproteinases. Crit Rev Biochem Mol Biol 42(3): 113–185.
18. Wu CC, Hsu CW, Chen CD, Yu CJ, Chang KP, et al. (2010) Candidate serological biomarkers for cancer identified from the secretomes of 23 cancer cell lines and the human protein atlas. Mol Cell Proteomics 9: 1100–1117.
19. Ramani VC, Pruett PS, Thompson CA, DeLucas LD, Sanderson RD (2012) Heparan sulfate chains of syndecan-1 regulate ectodomain shedding. J Biol Chem 287: 9952–9961.
20. Martínez A, Spencer ML, Brethauer U, Grez U, Marchesani FJ, et al. (2009) Reduction of syndecan-1 expression during lip carcinogenesis. J Oral Pathology & Medicine 38 (7): 580–583.
21. Kurokawa H, Zhang M, Matsumoto S, Yamashita Y, Tanaka T, et al. (2006) Reduced syndecan-1 expression is correlated with the histological grade of malignancy at the deep invasive front in oral squamous cell carcinoma. J Oral Pathology & Medicine 35 (5): 301–306.
22. Caescu CI, Jeschke GR, Turk BE (2010) Active site determinants of substrate recognition by the metalloproteinases TACE and ADAM10. Biochem J 424 (1): 79–88.
23. Endo K, Takino T, Miyamori H Kinsen H, Yoshizaki T, et al. (2003) Cleavage of syndecan-1 by membrane type matrix metalloproteinase-1 stimulates cell migration. J Biol Chem 278 (42): 40764–40770.
24. Heinz A, Jung M C, Duca L, Sippl W, Taddese S, et al. (2010) Degradation of tropoelastin by matrix metalloproteinases –cleavage site specificities and release of matrikines. FEBS J 277 (8): 1939–1956.
25. Nikolova V, Koo CY, Ibrahim SA, Wang Z, Spillmann D, et al. (2009) Differential roles for membrane-bound and soluble syndecan-1 (CD138) in breast cancer progression. Carcinogenesis. 30(3): 397–407.
26. Yang Y, Yaccoby S, Liu W, Langford JK, Pumphrey CY, et al. (2002) Soluble syndecan-1 promotes growth of myeloma tumors in vivo. Blood. 100(2): 610–617.
27. Chen P, Abacherli LE, Nadler ST, Wang Y, Li Q, et al. (2009) MMP7 Shedding of syndecan-1 facilitates re-epithelialization by affecting $\alpha 2\beta 1$ integrin activation. Available: http://www.plosone.org/article/info%3Adoi%2F10.1371%2Fjournal.pone.0006565. Accessed 2012 Jul 26.
28. Beauvais DM, Rapraeger AC (2003) Syndecan-1-mediated cell spreading requires signaling by $\alpha v\beta 3$ integrins in human breast carcinoma cells. Exp Cell Res 286 (2): 219–232.
29. McQuade KJ, Beauvais DM, Burbach BJ, Rapraeger AC (2006) Syndecan-1 regulates $\alpha v\beta 5$ integrin activity in B82L fibroblasts. J Cell Sci 119 (12): 2445–2456.
30. Boukamp P, Petrussevska RT, Breitkreutz D, Hornung J, Markham A, et al. (1988) Normal keratinization in a spontaneously immortalized aneuploid human keratinocyte cell line. J Cell Biol 106 (3): 761–771.
31. Berti DA, Morano C, Russo LC, Castro LM, Cunha FM, et al. (2009) Analysis of Intracellular substrates and products of thimet oligopeptidase in human embryonic kidney 293 cells. J Biol Chem 284 (21): 14105–14116.
32. Villén J, Gygi SP (2008) The SCX/IMAC enrichment approach for global phosphorylation analysis by mass spectrometry. Nat Protocol 3 (10): 1630–1638.
33. Paes Leme AF, Sherman NE, Smalley DM, Sizukusa LO, Oliveira AK, et al. (2012) Hemorrhagic activity of HF3, a snake venom metalloproteinase: insights from the proteomic analysis of mouse skin and blood plasma. J Proteome Res 11(1): 279–291.
34. Escalante T, Rucavado A, Pinto AFM, Terra RMS, Gutirrez JM, et al. (2009) Wound exudate as a proteomic window to reveal different mechanisms of tissue damage by snake venom toxins. J Proteome Res 8 (11): 5120–5131.
35. Eming SA, Koch M, Krieger A, Brachvogel B, Kreft S, et al. (2010) Differential proteomic analysis distinguishes tissue repair biomarker signatures in wound exudates obtained from normal healing and chronic wounds. J Proteome Res 9 (9): 4758–4766.
36. Liu H, Sadygov RG, Yates JR III (2004) a model for random sampling and estimation of relative protein abundance in shotgun proteomics. Anal Chem 76 (14): 4193–4201.
37. Liang CC, Park AY, Guan JL (2007) In vitro scratch assay: a convenient and inexpensive method for analysis of cell migration in vitro. Nat Protocol 2(2): 329–33.
38. Humphries MJ (2001) Cell-Substrate Adhesion Assays. Current Protocols in Cell Biology. 2001 John Wiley & Sons, Inc. DOI 10.1002/0471143030.cb0901s00.
39. Paes Leme AF, Kitano E, Furtado MF, Valente R, Camargo ACM, et al. (2009) Analysis of the subproteomes of proteinases and heparin-binding toxins of eight bothrops venoms. Proteomics 9: 733–745.

Potent Neutralization of Botulinum Neurotoxin/B by Synergistic Action of Antibodies Recognizing Protein and Ganglioside Receptor Binding Domain

Changchun Chen[1]⁹, Shuhui Wang[1]⁹, Huajing Wang[2,3]⁹, Xiaoyan Mao[4], Tiancheng Zhang[2,3], Guanghui Ji[2,3], Xin Shi[2,3], Tian Xia[2,3], Weijia Lu[4], Dapeng Zhang[2,3], Jianxin Dai[2,3,5]*, Yajun Guo[2,3,5]*

1 School of Pharmacy, The Center for Antibody Medicine of Ministry of Education, Shanghai Jiao Tong University, Shanghai, People's Republic of China, **2** International Joint Cancer Institute, The Second Military Medical University, Shanghai, People's Republic of China, **3** National Engineering Research Center for Antibody Medicine, State Key Laboratory of Antibody Medicine and Targeting Therapy and Shanghai Key Laboratory of Cell Engineering, Shanghai, People's Republic of China, **4** Lanzhou Institute of Biological Products, Lanzhou, Gansu, People's Republic of China, **5** PLA General Hospital Cancer Center, Beijing, People's Republic of China

Abstract

Background: Botulinum neurotoxins (BoNTs), the causative agents for life-threatening human disease botulism, have been recognized as biological warfare agents. Monoclonal antibody (mAb) therapeutics hold considerable promise as BoNT therapeutics, but the potencies of mAbs against BoNTs are usually less than that of polyclonal antibodies (or oligoclonal antibodies). The confirmation of key epitopes with development of effective mAb is urgently needed.

Methods and Findings: We selected 3 neutralizing mAbs which recognize different non-overlapping epitopes of BoNT/B from a panel of neutralizing antibodies against BoNT/B. By comparing the neutralizing effects among different combination groups, we found that 8E10, response to ganglioside receptor binding site, could synergy with 5G10 and 2F4, recognizing non-overlapping epitopes within Syt II binding sites. However, the combination of 5G10 with 2F4 blocking protein receptor binding sites did not achieve synergistical effects. Moreover, we found that the binding epitope of 8E10 was conserved among BoNT A, B, E, and F, which might cross-protect the challenge of different serotypes of BoNTs *in vivo*.

Conclusions: The combination of two mAbs recognizing different receptors' binding domain in BoNTs has a synergistic effect. 8E10 is a potential universal partner for the synergistical combination with other mAb against protein receptor binding domain in BoNTs of other serotypes.

Editor: Michel R. Popoff, Institute Pasteur, France

Funding: This work was supported by grants from National Nature Science Foundation for China, Ministry of Science & Technology of China (973 and 863 Program Projects), Shanghai Commission of Science & Technology (Key Laboratory and projects) as well as special program projects for infectious diseases and new drug development from Ministry of Science & Technology of China (2010ZX09401-407). Dr. Jianxin Dai is recipient of Pujiang Scholar Award from Shanghai Commission of Education. The funders had no role in study design, data collection and analysis, decision to publish, or preparation of the manuscript.

Competing Interests: The authors have declared that no competing interests exist.

* E-mail: yjguo@smmu.edu.cn (YJG); jxdai@smmu.edu.cn (JXD)

⁹ These authors contributed equally to this work.

Introduction

Botulinum neurotoxins (BoNTs) comprise a group of highly lethal toxins consisting of 7 serotypes (BoNT/A-G) produced by the anerobic bacteria, Clostridium botulinum [1,2]. Four of the BoNT serotypes (A, B, E, and F) cause human botulism, a neuroparalytic disease which results from ingestion of pre-formed toxin present in contaminated food and from toxin produced *in vivo* from infected wounds [3]. Owing to their extreme potency and lethality, BoNTs are included in the list of category A select agents and toxins [4]. Each BoNT isoform is synthesized as a single polypeptide chain with a molecular mass of ~150 kDa. The inactive precursor protein is cleaved either by clostridial or tissue proteases into a 50-kDa light chain (LC) and a 100 kDa heavy chain (HC) linked by an essential interchain disulfide bridge and by the belt, a loop from the HC that wraps around the LC [5]. The LCs act as zinc metallopeptidases, which solely hydrolyze one of three SNARE proteins depending on the serotype: BoNT A and E cleave synaptosomal-associated protein of 25 kDa (snap-25) and BoNT/B and F cleave the vesicle associated membrane protein (VAMP) [6,7,8,9]. resulting in a blockade of neurotransmission and flaccid paralysis [10]. The heavy chain is divided into two functionally distinct regions: a C terminal binding domain (Hc) and a N terminal translocation domain (H_N) [11]. The binding domain initially interacts with low affinity to a group of gangliosides on the presynaptic plasma membrane [12], after which it binds to a protein acceptor. Interestingly, the BoNTs' serotypes that exhibit highest sequence similarity share the same protein receptor, i.e., BoNT types A, E, and F bind SV2 [13,14,15], whereas BoNT types B bind SytI and II [16]. The existence of two classes of binding sites distinguished by different affinities and the discovery of protease-sensitive binding to neurons resulted in a double-receptor concept. In a first step complex polysialogangliosides accumulate BoNTs on the plasma membrane surface; and in a second step, protein receptors mediate their endocytosis.

Monoclonal antibodies (mAbs) have been intensely explored as inhibitors of the recognition step between BoNTs and their cellular receptors. [17] However, the two-receptor model makes it difficult to development of a mAb-based antitoxin for botulism. Single mAb recognizing only one epitope can hardly block the binding between BoNT and cell completely. It might be the reason why single mAb can only neutralize at most 10 to 100 times the 50% lethal dose (LD50) of toxin in mice [17]. Combination of two or three mAbs recognizing nonoverlapping epitopes can neutralize BoNT A very potently [18]. Although the binding areas of antibodies were mapped, however, the epitopes which these antibodies bind to are not finely defined. However, not all randomly paired mAbs have the potent synergistical neutralizing function. It is more difficult to select mAb pairs with desirable synergistical function from a panel of neutralizing mAbs whose epitopes are not clear. Because both the protein receptor and ganglioside receptor are essential for the entrance of BoNTs into neurons [19], we predicted that two monoclonal antibodies, which recognized protein and ganglioside receptor binding domain respectively, can neutralize synergistically against BoNTs.

In this study, we selected three neutralizing mAbs, which recognize different non-overlapping epitopes of BoNT/B, and compare the neutralizing effects among different antibody combinations. We found that 8E10, response to ganglioside receptor binding site, could collaborate with 5G10 and 2F4, recognizing non-overlapping epitopes within Syt II binding sites. However, the combination between two mAbs, 5G10 and 2F4, blocking protein receptor binding sites, did not achieve synergistical effects. We also found that the binding epitope of 8E10 was conserved among BoNT A, B, E and F. It could cross-protect the challenge of different serotypes of BoNTs *in vivo*. The results provide a potential universal partner for the synergistical combination with other mAb against protein receptor binding domain in BoNTs of other serotypes.

Results

1. A panel of neutralizing antibodies specific for BoNT/B-Hc

To generate mAbs capable of neutralizing BoNT/B, gene encoding protective antigen Hc fragment of botulism neurotoxin serotype B (BoNT/B-Hc) was optimized and synthesized. The reconstructed Hc proteins were expressed in *E.coli* BL21 (DE3) in soluble form and purified for preparation of a toxoid immunogen. Sixteen hybridomas were obtained and screened for the production of anti-BoNT/B Hc mAbs, using ELISA with purified BoNT/B Hc (Fig. 1A). Cell culture medium corresponding to the positive wells was subjected to western blot assay for testing the binding specificity of these mAbs (Fig. 1B). The results showed that all of the positive hybridomas were able to bind to native or denatured BoNT/B Hc specifically. The supernatants of these clones were then tested for their ability to protect mice against 4 LD50 of BoNT/B for 4 days. Among them, 7 mAbs (8D1, 8E10, 1F4, 2F4, 5G10, 2H12 and 2B2) could neutralize BoNT/B *in vivo* (Fig. 1C).

2. Blocking activities of the neutralizing mAbs

The first step in BoNT attachment to nerve membranes starts with C-terminus of Hc binding to gangliosides of neuron membranes. Ganglioside binding could either bring the toxin and the protein receptor into close proximity or cause a conformational change at the protein receptor binding site to enhance interactions [2]. To identify binding domain which was blocked by the seven neutralizing mAbs, we used ELISA-based analysis of GT1b or SytII binding with BoNT/B Hc pre-incubated with or without mAbs. The results indicated that mAb 8E10, 1F4 and 2B2 blocked the interaction between BoNT/B Hc and GT1b (Fig. 2A), whereas 8D1, 2F4, 5G10 and 2H12 interrupted the binding between BoNT/B Hc and Syt II (Fig. 2B).

To further investigate if the epitopes recognized by the antibodies were overlapped, we tested mAbs using competitive binding ELISAs. The optimal dilution of each HRP-conjugated mAb was determined by testing serial two-fold dilution in BoNT B Hc-coated ELISA plates and selecting this as the working dilution to generate an optical density of around 1.5. The selected working dilution was 1/8000 for mAbs 2B2 and 2F4, 1/16000 for mAbs 8D1, 1F4, 8E10, 5G10 and 1/4000 for 1F4, 2H12. The results of competitive ELISAs were reported in Table 1.

HRP-conjugated 8E10 inhibited the binding of 1F4 or 2B2 to the antigen and vice versa, showing that 8E10, 1F4 and 2B2 recognized overlapping epitopes. HRP-conjugated 5G10 inhibited the binding of 2H12 to the antigen and vice versa, indicating that 5G10 and 2H12 recognized overlapping epitopes. The same competition was also observed between 2F4 and 8D1. MAbs were thus categorized into 3 groups. MAbs binding to the same epitopes or overlapping epitopes were put into the same group (Table 2). Group 1 consisted of mAbs 8E10, 1F4 and 2B2, group 2 included mAbs 5G10 and 2H12 and group 3 included mAbs 2F4 and 8D1. Because the antigen binding affinity of the mAbs 8E10, 5G10 and 2F4 was higher than that of other 4 neutralizing mAbs (Table 1), they were selected for fine epitope mapping based on phage-display.

3. Epitope mapping of mAb 8E10, 5G10 and 2F4

After three successive rounds of panning on 8E10, 5G10 and 2F4 respectively, we randomly selected and purified 100 positive phages. DNA sequence analysis of the phage-encoded peptides revealed 4–6 amino-acid consensus sequences, which was aligned with the sequence of BoNT/B displayed in Table 3. All phages selected by 8E10 have a common motif of SXWY. Alignment of this sequence on the sequence of BoNT/B provides a best fit with 1259SKWY1262. 5G10 and 2F4 bind to SDXFY and KXSP, which correspond exactly to residues 1201SDEFY1205 and 1114KDSP1117 of BoNT/B respectively. The peptides containing 1259SKWY1262, 1201SDEFY1205 and 1114KDSP1117 motif in the BoNT/B were synthesized and denoted as P1, P2 and P3 respectively. The ELISA assay results showed that mAb 8E10, 5G10 and 2F4 bound to KLH-conjugated P1, P2 and P3 peptides (P1, P2, P3-KLH) specifically since these mAbs did not cross-react with the other peptides recognized by the other two mAbs and control peptide-KLH (CP-KLH)(Fig. 3A). The capability of the synthesized peptides to block the binding of mAb 8E10, 5G10 and 2F4 to BoNT/B Hc was also determined. The results revealed that peptide P1, P2, P3 effectively inhibited the binding of mAb 8E10, 5G10, 2F4 to BoNT/B Hc respectively. In contrast, control peptide had not inhibitory effect on their interaction (Fig. 3B). These results further demonstrated that the motifs listed above were the epitopes of 8E10, 5G10 and 2F4 respectively.

BoNT/B Hc has been crystalized in complex with trisaccharide sialyllactose (a mimic of GT1b) (PDB 1F31) and with SytII peptide (PDB 2NM1). The crystal structure reveals two distinguished domains which bind to GT1b and SytII respectively. Therefore, it is possible to compare the binding site of protein and ganglioside receptor to BoNT/B Hc with the localization of the mAb 8E10, 5G10 and 2F4 epitope identified in this study. The illustration of

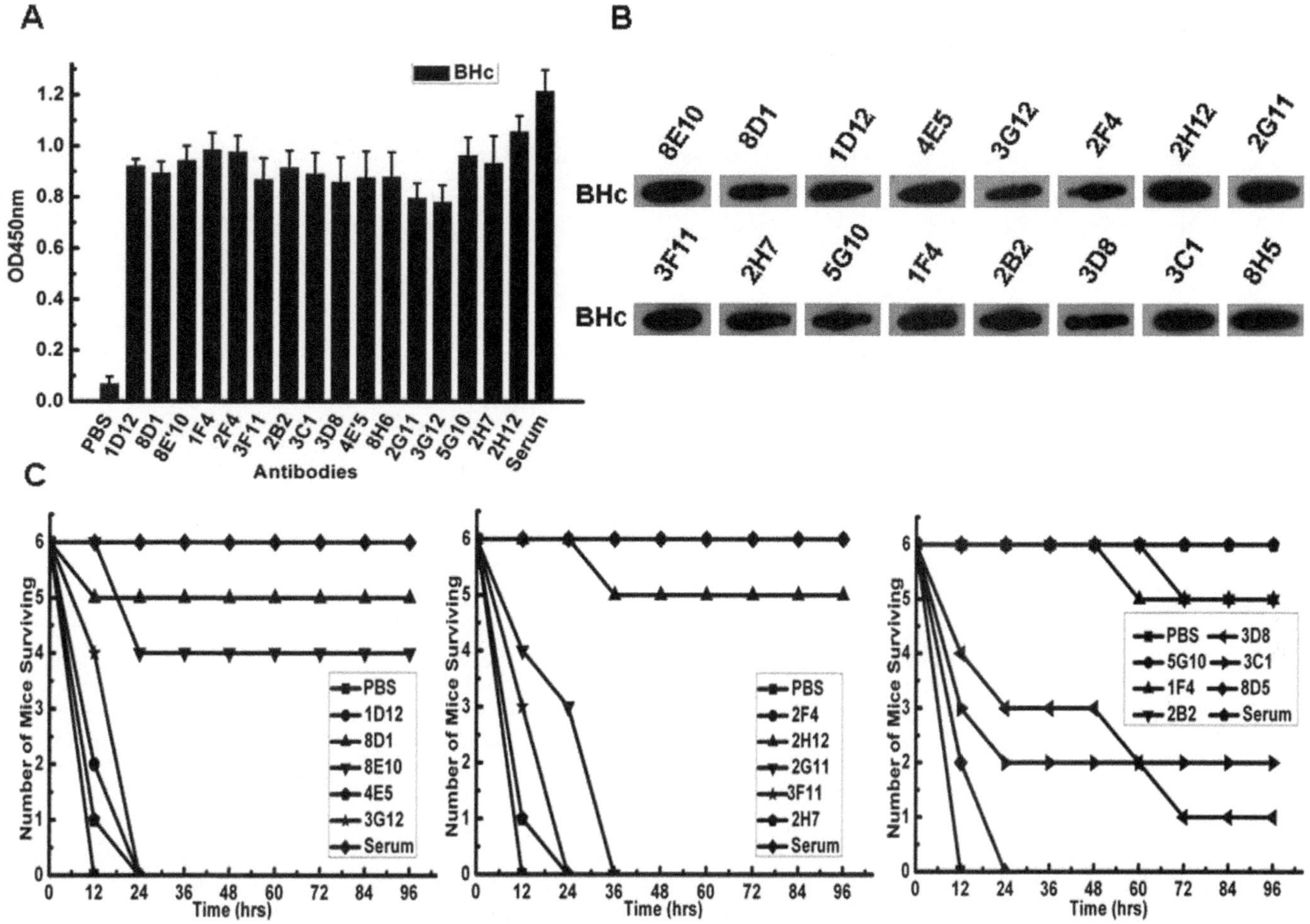

Figure 1. Characterization of mAbs specific for BoNT/B Hc. The binding activity of mAb was determined by ELISA (A) and Western blot analysis (B). Neutralizing mAbs were selected by *in vivo* neutralization assay(C). For the initial screening of the positive hybridoma that produced the neutralizing antibody, 200 µl of culture supernatant from hybridomas grown in a 24-well tissue culture flask for 3–5 days, was pre-incubated with 4 LD50 of BoNT/B for 1 hour at 37°C, and the reaction mixtures intraperitoneally injected to each of six mice. More than 4 survivals in four days were considered as positive. Serum of the immunized mice in the dilution of 1:100 was set as control. Left: PBS, 1D12, 8D1, 8E10, 4E5, Serum. Middle: PBS, 2F4, 2H12, 2G11, 3F11, 2H7, Serum. Right: PBS, 5G10, 1F4, 2B2, 3D8, 3C1, 8H5, Serum. The error bars represented standard deviations.

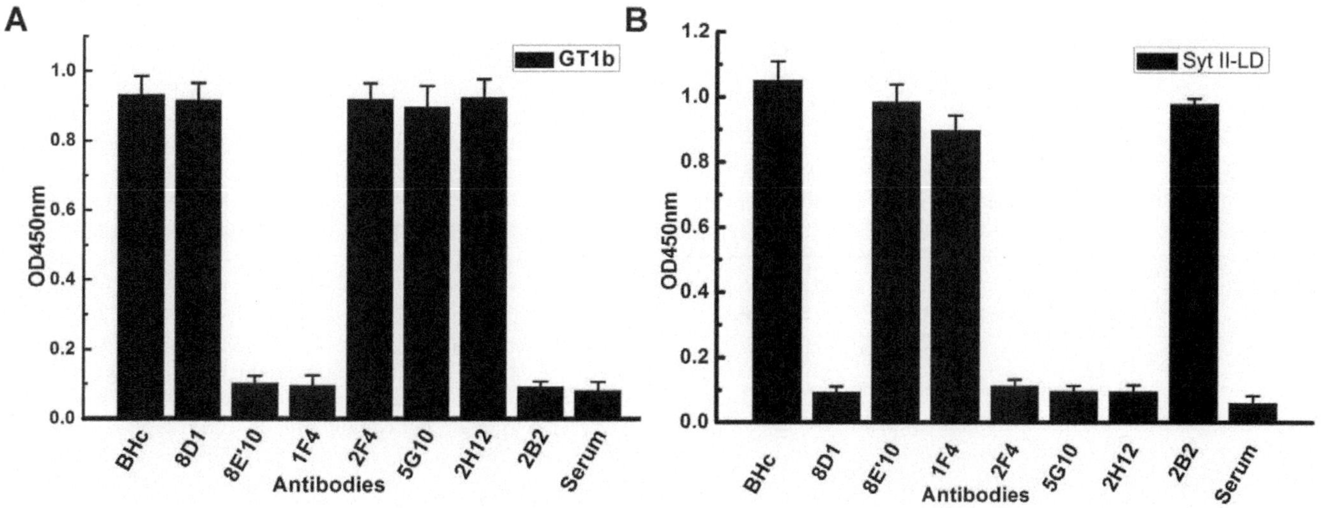

Figure 2. Binding domains recognized by neutralizing mAbs. Plates were coated with GT1b (A) or recombinant syt II-LD (B) respectively. HRP-conjugated BoNT/B Hc was pre-incubated with neutralizing mAbs at molar ratio of 1:100. Bound BoNT/B Hc was detected by adding TMB as a substrate.

Table 1. Results of competition binding ELISAs for MAb epitope studies.

	unlabelled mAbs	HRP-conjugated mAbs						
		8E10-HRP	1F4-HRP	2B2-HRP	5G10-HRP	2H12-HRP	2F4-HRP	8D1-HRP
Group 1	8E10	9	7	8	90	92	90	92
	1F4	8	5	7	88	91	91	93
	2B2	6	9	8	92	88	95	93
Group 2	5G10	91	90	85	5	6	78	80
	2H12	87	90	89	7	4	81	79
Group 3	2F4	90	87	91	82	79	3	5
	8D1	88	93	87	80	78	6	4
	PBS	100	100	100	100	100	100	100

Results were expressed as percent binding of HRP-conjugated mAbs. The amount of binding obtained in the absence of unlabeled antibody was set at 100% for each HRP conjugated mAb. First column indicates the group assignments of mAbs based on competition binding assay. mAbs binding to overlapping epitopes are grouped together, while non-competing mAbs are grouped individually.

structure of BoNT/B Hc is showed in Fig. 4. The epitopes of mAbs were colored and overlaid on the three-dimensional structure of BoNT/B. The epitope of 8E10 (colored in red) is just located inside the GT1b-binding domain (colored in yellow) (Fig. 4 A). The residues "1114KDSP1117" recognized by 2F4 (Fig. 4B, blue) cover part of the Syt II binding pocket formed by residues K1114, S1116, P1117, V1118, Y1183, E1191, K1192, F1194 and F1204 (Fig. 4B, green). The 5G10 epitope comprises 5 amino acids "1201SDEFY1205" (Fig. 4B, cyan), which overlaps another side of Syt II binding site formed by residues P1117, W1178, Y1181, F1194, A1196, P1197 and F1204 (Fig. 4B, green). However, the epitopes of 5G10 and 2F4 are not overlapping with each other. These results provide the structure basis of neutralizing functions of these mAbs. The mAb 8E10 neutralizes BoNT/B through blocking the interaction between BoNT/B Hc and its ganglioside receptor on neurons. 5G10 and 2F4 inhibit the binding of BoNT/B Hc with protein receptor SytII.

4. Binding inhibition assay *in vitro*

In order to determine the toxin neutralization by IgG *in vitro*, we used SytII transfected and ganglioside treated PC12 cells, a neuroendocrine cell line of which wild type is lack of functional toxin receptors [20], to serve as a target cell model. Flow cytometry results showed that FITC-labeled recombinant BoNT/B Hc could bind to SytII transfected and ganglioside treated PC12. 8E10 could completely block the interaction between BoNT/B Hc and ganglioside treated PC12 (Fig. 5A). 5G10 and 2F4 could inhibit BoNT/B Hc to bind to the SytII transfected PC12 effectively (Fig. 5B). However, as to the SytII transfected and ganglioside treated PC12, none of the mAbs above could completely block its binding to BoNT/B Hc. In contrast with the single mAb, the combination of 8E10 with 5G10 or 2F4 could completely inhibit the binding of BoNT/B Hc to the double treated PC12, while the combination of 5G10 with 2F4 could not block the binding of PC12 to BoNT/B Hc (Fig. 5C). The results of confocal microscope showed that the combination of 5G10 and 2F4 could only decrease the fluorescence signal on the membrane of double treated PC12 similar to the effects of single antibody treatment. In contrast, the mixture of 8E10 and 5G10 or 2F4 abrogated the fluorescence signal from the cell membrane completely (Figure S1). These results demonstrated that mAbs, which recognized ganglioside or SytII binding domain respectively, had the potent to neutralize BoNT/B synergistically.

Table 2. Kinetic constants and isotypes of mAbs binding to BoNT/B-Hc.

mAb	*k*off (M-1 S-1)	*k*on (S-1)	*KD* (M)	Subtype
8E10	4.070E−4	1.838E+5	2.214E−9	IgG1/κ
2B2	5.941E−4	7.572E+4	7.854E−9	IgG1/κ
1F4	2.674E−4	4.605E+4	5.806E−9	IgG1/κ
5G10	4.146E−5	1.374E+4	2.975E−9	IgG1/κ
2H12	9.390E−3	6.631E+5	1.301E−8	IgG1/κ
2F4	1.085E−4	5.425E+4	2.000E−9	IgG1/κ
8D1	2.335E−3	2.468E+4	9.460−8	IgG1/κ

5. Protection of mice from BoNT/B toxicity using neutralizing mAbs

In vivo toxin neutralization was studied using a standard mouse protection assay. We incubated 100 μg mAbs with BoNT/B for 1 hour prior to intraperitoneal injection into BALB/c mice and the groups of mice with 6 animals for each group were used to test dose level. Mice were observed for morbidity and mortality over 30 days. Complete protection was observed with doses up to 20 LD50s for 8E10, and 40 LD50s for 5G10 and 2F4. Partial protection, as indicated by increased survival compared with antibody-free control mice, was afforded with higher doses (Table 4). At 80 LD50s, mice receiving 5G10 and 2F4 survived 72 hours, compared to 6 hours for the control mice. As combination of mAbs have demonstrated synergy in BoNT/A neutralization [18], the different pairs of mAb were studied at increasing doses of toxin to explore potent combination. We mixed 50 μg each of the mAbs (total of 100 μg) with different LD50s BoNT/B and tested the combination by intravenous injection. At 640 LD50s, only 1 of 6 mice receiving the pair of 5G10+2F4 survived whereas none of mice receiving two of the pairs of mAbs (8E10+5G10 or 8E10+2F4) died. 3 of 6 mice that received the pair of 8E10+5G10, and 2 of 6 mice that received the pair of 8E10+2F4 survived for 96 hours in the group challenged with 1,280 LD50s. In contrast, the control mice died in less than

Table 3. Alignment of phage-displayed peptide sequences selected by anti-BoNT/B neutralizing antibodies.

Phage clone	Phage-displayed peptide sequence	Absolute and relative frequency
5G10 selected		
5G10-A1	--AGQ**SDS**Y**Y**LYHP--	12 (24%)
5G10-G4	--SF**ADYYY**QYSPR--	4 (8%)
5G10-C10	---T**SD**E**WY**NLHFLR	6 (12%)
5G10-F6	-----TYK**SDHFY**HWQP--	20 (40%)
5G10-D1	-----TPT**SD**R**FF**NLWP--	8 (16%)
Consensus	--**SD**X**FY**--	
BoNT/B	-^{1201}**SD**E**FY**^{1205}-	
8E10 selected		
8E10-B9	--RTMKSM**SDWY**QK--	3 (4.7%)
8E10-D6	---TNSD**SEWF**GHYM--	4 (6.3%)
8E10-E3	--TKDAE**SRWY**HAF--	24 (37.5%)
8E10-F2	---SV**TDYF**NLTILQ-	5 (7.8%)
8E10-F10	--FS**TRWY**ESWLNP	12 (18.8%)
8E10-H1	--SMLG**SEYY**MMVM--	16 (25%)
Consensus	---**SXWY**---	
BoNT/B	--^{1259}**SKWY**^{1262}---	
2F4 selected		
2F4-A9	---MN**RDSP**EHIVAL--	3 (5.9%)
2F4-C11	--EALT**KHTF**TQLV---	7 (13.7%)
2F4-E8	--TMI**KHSET**FNQL---	9 (17.6%)
2F4-2A1	--IPR**KESP**YQRIW--	21 (41.2%)
2F4-2G4	---M**RQCP**KEHIVAL-	11 (21.6%)
Consensus	---**KXSP**--	
BoNT/B	--^{1114}**KDSP**^{1117}--	

Phage-displayed consensus amino acids are shown in boldface.

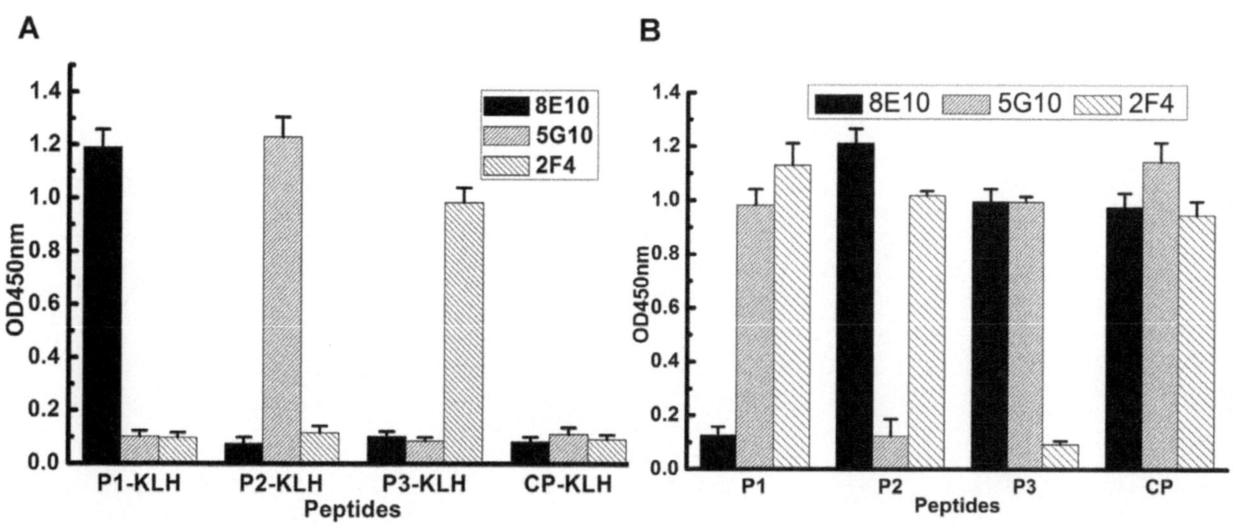

Figure 3. Reactivity of synthetic epitope peptides with neutralizing mAbs. Binding of synthetic peptide with mAb. Peptides P1, P2, P3, containing the motifs recognized by 8E10, 5G10 and 2F4 respectively, were synthesized and conjugated with KLH, which were indicated as P1-, P2-, P3-KLH. The binding abilities of mAbs were detected by ELISA. A non-associated peptide was set as control peptide (CP), and conjugated with KLH as CP-KLH. (B) Competitive inhibition of BoNT/B Hc binding with mAbs by synthetic peptides. CP-KLH: synthetic peptide THPHLPRALMRS was coupled to KLH (bearer protein) to generate CP-KLH; P1-KLH: synthetic peptide YFCISKWYLKEV was coupled to KLH (bearer protein) to generate P1-KLH. P2-KLH: synthetic peptide PISDSDEFYNTI was coupled to KLH (bearer protein) to generate P2-KLH; P3-KLH: synthetic peptide IKLKKDSPVGEI was coupled to KLH(bearer protein) to generate P3-KLH.

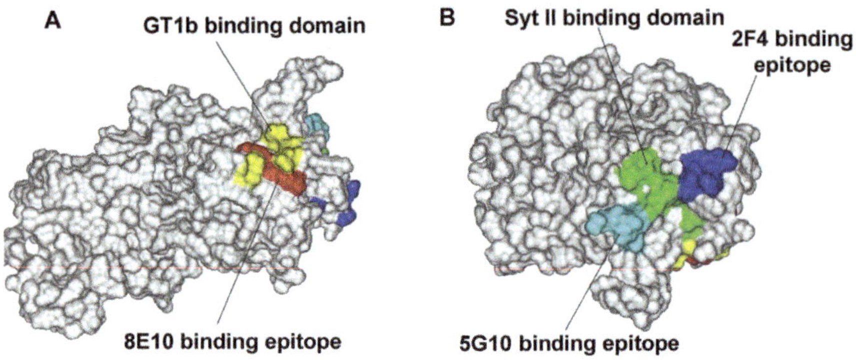

Figure 4. Molecular model overlay of neutralizing epitopes within the BoNT/B Hc binding domain. The model was established using the software Discovery Studio 2.0 (Accelrys, San Diego, CA) based on the crystal structure of BoNT/B Hc (PDB 1F31) from the Protein Data Bank. BoNT/B Hc is shown in a surface representation. (A) The residues reported as GT1b-binding site are colored yellow (*Nat. Struct. Biol. 1998*), and the residues recognized by 8E10 are colored red. (B) The SytII-binding site residues are colored green, and the residues of 5G10 and 2F4 are indicated in cyan and blue.

3 hours. The survival rate of the group received 8E10+5G10+2F4 is the same as that of the group received the pair of 8E10+5G10. No mice survived in the group challenged with 2,560 LD50s (Data not shown).

6. Neutralizing potency of mAb 8E10 against different serotypes of BoNTs

Given the binding epitope of 8E10 is located at the conservative motif of BoNT A, B, E, F (Fig. 6A), we wondered whether it could bind and neutralize across these serotypes. To evaluate this, we first expressed the Hc domain of BoNT A, E and F, and determined the binding kinetics of the 8E10 antibody to the BoNT/A, B, E, and F Hc by surface plasmon resonance (SPR) and showed the results in Table 5 and Figure S2. The results show that the binding affinity between 8E10 and BoNT/A or F is a little bit lower than the binding affinity between 8E10 and BoNT/E, whereas the affinity between 8E10 and BoNT/B is the highest. The fact that all four serotypes of BoNTs bind 8E10 with nanomolar efficiency demonstrates that all serotypes tested are of high affinity. The results of ELISA (Fig. 6B) and western blotting (Fig. 6C) showed that 8E10 could bind with native and denatured Hc of BoNT A, B, E, F, indicating that 8E10 recognized the conserved linear epitope among these BoNTs. In addition, we compared the potency of BoNT/A, B, E and F neutralization *in vivo* by mAbs 8E10. All of 6 mice challenged with 20 LD50s of BoNT/B were protected by 100 μg of 8E10. Same dosage of 8E10 protected 5/6 of mice challenged with 20 LD50s of BoNT/A or F, respectively and 4 of 6 mice also survived under the challenge of 20 LD50s of BoNT/E (Fig. 6D). In contrast, none of the mice survived longer than 24 hours after challenged with BoNTs without the injections of mAbs.

Discussion

For the last two decades, the relationship between the structure and function of BoNT has been studied intensively. With the complex structure of BoNT/B and its receptors, protein receptor Syt-II [21] and its ganglioside receptor GT1b [22], we can combine these data to provide the structure basis of the "double receptor" interaction proposed by Montecucco et al. [23]. Bio-functional assays also provided the evidence that both protein receptor SytII and ganglioside co-receptor were necessary for the infective process of BoNT/B [24]. This led us to the hypothesis that a combination comprised of a pair of neutralizing antibodies that bound to different receptors binding domains of the toxin would be more effective neutralizers than either member of the pair alone. We generated three mAbs, 8E10, 5G10 and 2F4 which could neutralize the challenge of 20 LD50 of BoNT/B *in vivo*. Finely epitope mapping revealed that 8E10 recognized ganglioside binding domain in BoNT/B, 5G10 and 2F4 bound with 2 non-overlapping epitopes surround the SytII binding domain respectively. In addition, we found that the combination between 8E10 (recognizing ganglioside binding domain) and 5G10 or 2F4 (recognizing protein receptor binding domain) could result in a more than 30–60 folds increase in potency compared with that of any the single mAb. However, for the combination of 5G10 and 2F4, which both blocked protein receptor binding sites by non-overlapping epitopes, no synergistic effect was observed. This indicates that double blocking the ganglioside and protein receptor binding domain simultaneously with two mAbs is helpful for the synergistic effects of the two antibodies' combination against BoNT/B.

The combinations of antibodies, which recognize nonoverlapping epitopes, synergistically cooperate in neutralization potency have been reported previously [18]. In that study, random combinations between 2 of the 3 neutralizing mAbs significantly prolonged the time to neuroparalysis compared with single mAbs. However, in our study, 2 mAbs, which bind to the same function domain of BoNT/B, had much less synergistic effects than those whose epitopes located in different function domains of BoNT/B. In addition, a BoNT/B-specific triplex antibody combination exhibited cooperative neutralizing effects to the toxin in vivo that were no better than those of the pairs of antibodies (Table 4). These findings are different from the conclusion of the previous report. Fine epitope mapping showed that the two mAbs (3D12 and S25) which bind the BoNT/A H_{CC} (C terminal of Hc) overlap the putative sialoganglioside binding site and cover a large portion of H_{CC}. The other mAb C25 bound a conformational epitope that consisted of the sequence from the N- and C-terminal subdomains of BoNT/A Hc, which overlapped with a putative inositol phosphate binding site that may be important for attachment to the lipid membrane. This function domain is essential for the binding to anionic lipid in the environment of lipid raft, which is important for the translocation processing of the toxin. [25,26]. Interestingly, the neutralizing potency of the pair of mAbs (C25+3D12), which bind to translocation domain and sialoganglioside binding domain respectively, is 10 times higher than that of S25+3D12, whose epitopes are located at the same function

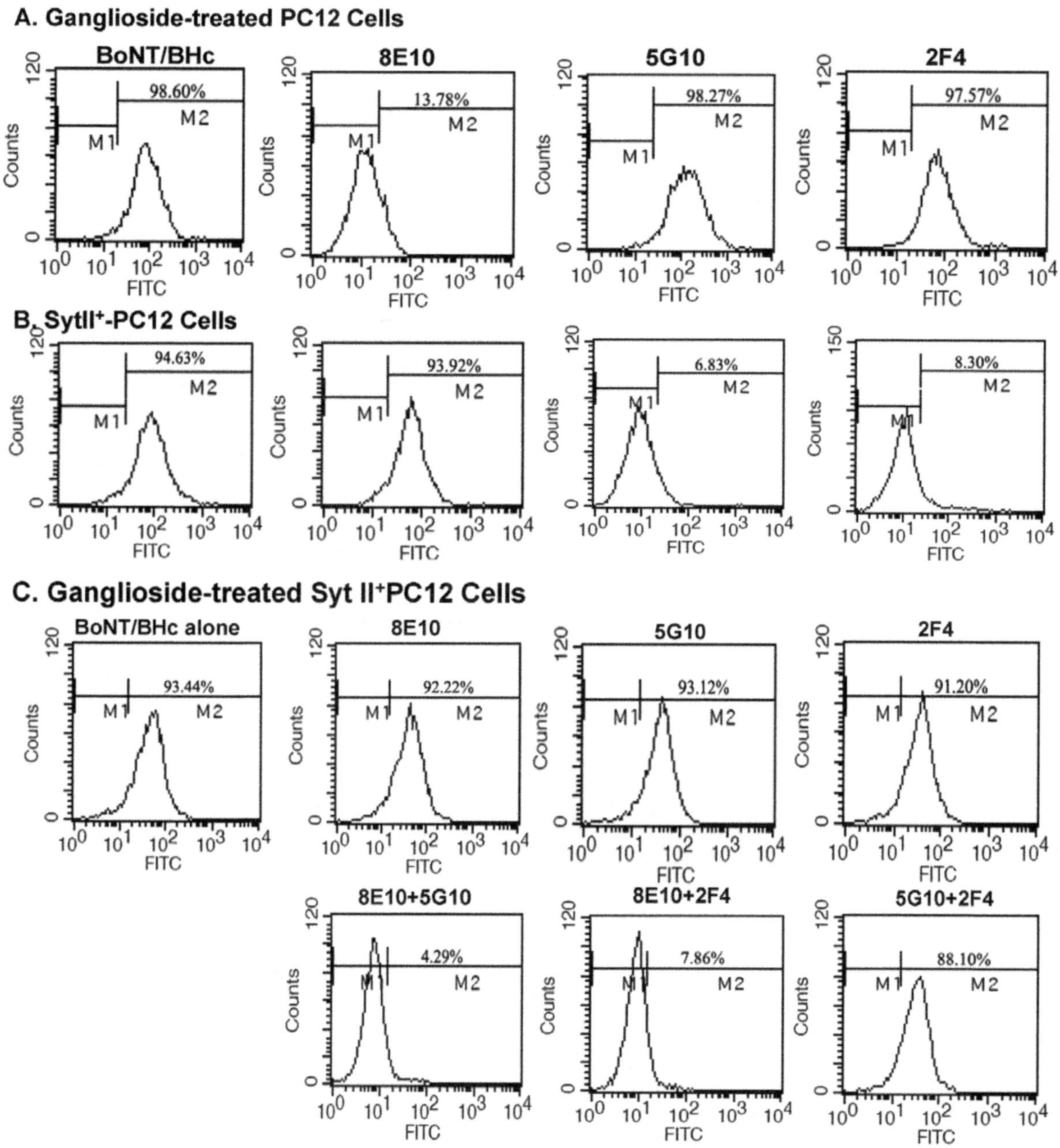

Figure 5. Inhibition of mAbs and pairs of mAbs to the interaction between BoNT/B and neurons by flow cytometer. The single or pairs of mAbs were mixed with FITC-BoNT/B Hc at a molar ratio of 1:100 before staining PC12 cells treated as indicated. Inhibition of mAbs for neuron binding activity of BoNT/B was indicated by the flow cytometer. (A) ganglioside-treated cells and BoNT/B Hc with or without single mAb (8E10, 5G10 or 2F4) (B) the binding of SytII+ cells and BoNT/B Hc with or without single mAb (8E10, 5G10 or 2F4), (C) the binding of ganglioside-treated syt II+PC12 cells and BoNT/B Hc with or without single mAb (8E10, 5G10 or 2F4) or pairs of mAbs (8E10+5G10, 8E10+2F4, 5G10+2F4). For pairs of mAbs, the molar ratio between BoNT/B Hc and mAbs is 1:50:50.

domain. This finding provides evidence that the synergetic effects of mAbs, which recognize different function domains of BoNT/A, are better than the effects of those recognizing the same function domain. Although we didn't observe the cooperative effects between the mAbs binding to SytII binding domain of BoNT/B, we did prove that the combination of two mAbs recognizing different receptors' binding domain in BoNT/B has a synergistic effect. Accordingly, the data indicate that the combination between 2 mAbs recognizing different function domains in BoNTs could be the general principle for the potent synergistic effect.

Neutralizing mAbs binding multiple serotypes of botulinum neurotoxin are rare have been reported previously [27]. The cross-reactive mAbs bound to a relatively conserved epitope at the tip of the BoNT H_N. This is a functionally important epitope for intoxication, as mAb binding leads to potent BoNT neutralization. In this study, a neutralizing mAb 8E10 binding the conservative domain on Hc was reported. The structure of Hc from BoNT/B (PDB 2NM1) [21], BoNT/E (PDB 3FFZ) [28] and BoNT/F (PDB 3FUQ) [15] supports the view that the Hc fold is highly conserved. GT1b binds on a cleft formed by W1266 and Y1267 on one face

Table 4. *In vivo* toxin neutralization by mAb alone, pairs or triplex of mAbs.

mAbs	BoNT/B LD_{50}	Survival time (hours)	Survival (alive/tested)
Control	20	<12	0/6
5G10	20	96	6/6
8E10	20	96	6/6
2F4	20	96	6/6
Control	40	<8	0/6
5G10	40	96	6/6
8E10	40	72	0/6
2F4	40	96	6/6
Control	80	<6	0/6
5G10	80	72	0/6
8E10	80	48	0/6
2F4	80	72	0/6
Control	640	<4	0/6
5G10+2F4	640	60	1/6
8E10+5G10	640	96	6/6
8E10+2F4	640	96	6/6
8E10+2F4+5G10	640	96	6/6
Control	1280	<3	0/6
5G10+2F4	1280	24	0/6
8E10+5G10	1280	96	3/6
8E10+2F4	1280	96	2/6
8E10+2F4+5G10	1280	96	3/6

and E1203, H1253, and F1252 on the other for BoNT/A [29]. The structure of Hc from BoNT/B in complex with the trisaccharide sialyllactose, a mimic of GT1b (PDB 1F31), displays a similar binding cavity with corresponding residues W1261, Y1262, and H1240 [22]. In addition, the crystal structure of Hc from BoNT/F displays a ganglioside-binding pocket with corresponding residues W1250, Y1251, and H1241 [15]. Amino acid sequence alignment shows that S1264, W1266 and Y1267, conserved among all BoNT serotypes [30], constitute key residues of a lactose-binding motif (H. . .SXWY. . .G) that contribute to the crevice binding with GT1b. The implication is that the GT1b-binding pocket for all these BoNTs is similar. We found in this study that the motif 1259SKWY1262 is the recognizing epitope of 8E10, which could also cross react with the Hc of BoNT A/B/E/F (Fig. 6). This conserved epitope may partially explain how 8E10 could cross-protect the challenge of different serotype of BoNTs *in vivo*. Although its synergistic effect with the mAbs blocking protein receptor binding domain in BoNT A, E or F has not been detected, we can predict that 8E10 could act synergistically with the mAbs recognizing protein receptor binding domain against BoNT A, E or F. This study provides a potential universal partner for the synergistical combination with other mAb against protein receptor binding domain in BoNTs of other serotypes.

Materials and Methods

1. Ethics Statements

Mice were purchased from Animal Center of Chinese Academy of Sciences and maintained under pathogen-free conditions. All animal experimental procedures were carried out in strict accordance with the guidelines of the Animal Experiment Committee of the International Joint Cancer Institute, and were approved by the Animal Experiment Committee of the International Joint Cancer Institute.

2. Holotoxin, Antigens, protein and peptides

BoNT/A, BoNT/B, BoNT/E and BoNT/F were provided by Lanzhou Institute of Biological Products (Lanzhou, China). DNA encoding BoNT/A Hc (residues 868 through 1296), BoNT/B Hc (residues 853 through 1291), BoNT/E Hc (817 through 1255), BoNT/F Hc (847 through 1280), and Syt II-LD fragment (residue 37 through 86) were synthesized and subcloned into a pET expression vector, expressed in E. coli BL21(DE3). The recombinant histidine tagged BoNT/A, B, E, F Hc and Syt II-LD fragment were isolated and purified by nickel affinity gel column chromatography and their molecular weight and purity were verified by gel electrophoresis [31,32].

Peptides used in this study, P1 (YFCISKWYLKEV), P2 (PISDSDEFYNTI), P3 (IKLKKDSPVGEI), P4 (THPHLPRALMRS), were synthesized by Yeli Bio-Scientific Inc. (Shanghai, China).

3. Animals and cell lines

6- to 8-week-old female BALB/c mice were purchased from Animal Center of Chinese Academy of Sciences and maintained under pathogen-free conditions. PC12 (rat adrenal pheochromocytoma cells) was purchased from Cell Bank of Chinese Academy of Science (Shanghai, China). To generate cells that express syt II (syt II), full-length mouse syt II (Genechem. Shanghai, China) was subcloned into pCDNA3.1 (CLONTECH, Mountain View, CA) and transfected into PC12 cells via electroporation. Transfected

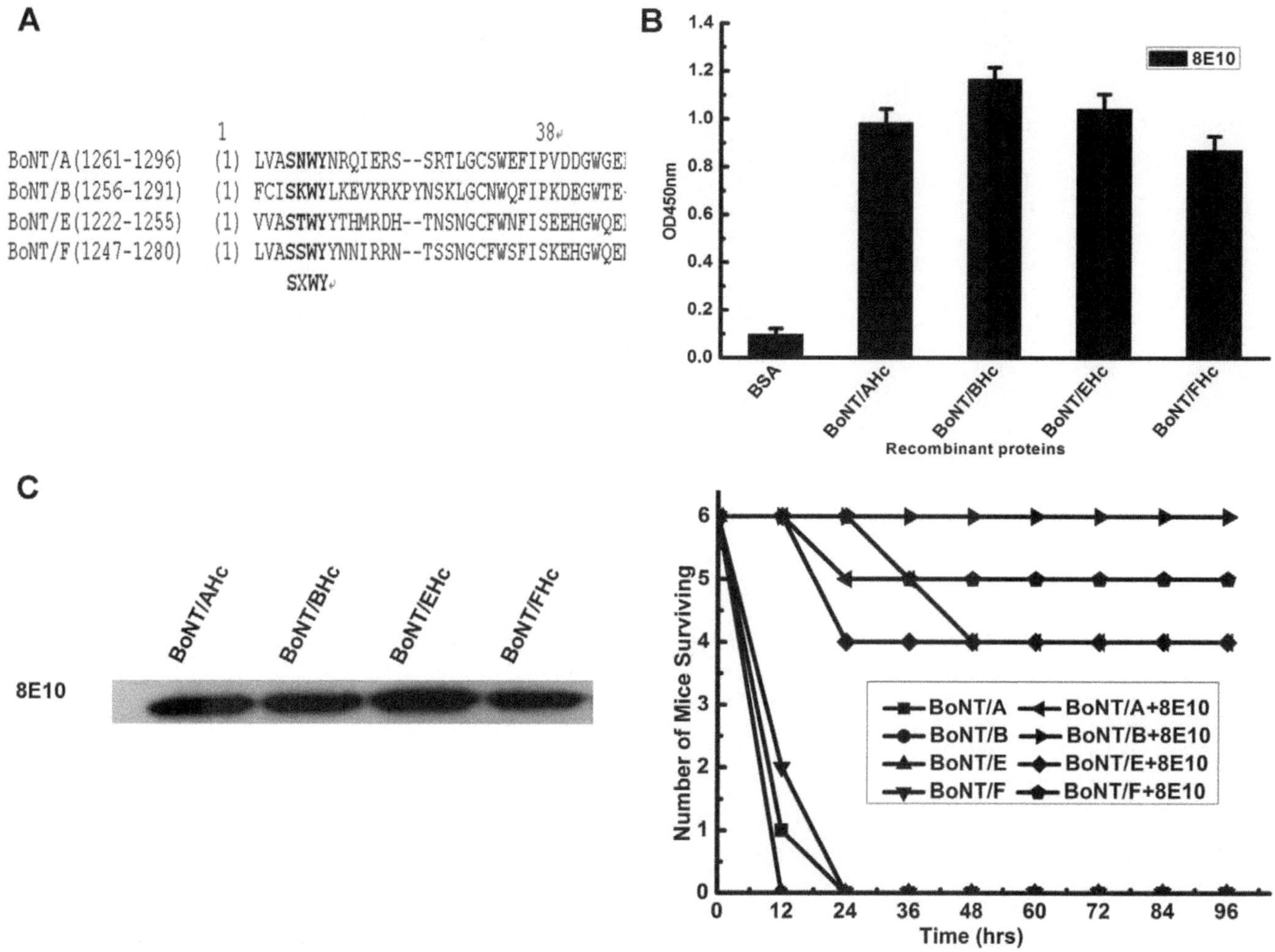

Figure 6. Cross-reactive and cross-neutralizing effects of 8E10. (A) Amino acid sequence alignment of the carboxyl-terminal region of BoNT (A, B, E, F). The residues, which are conservative among BoNTs/A, B, E and F are presented as bold letters. (B) ELISA of 8E10 to BoNT Hcs of different serotypes A, B, E and F. (C) BoNT Hcs (A, B, E, F) were separated by SDS-PAGE under reducing conditions and visualized by western blotting with 8E10. (D) Cross-neutralizing effects *in vivo*. 100 μg of 8E10 was pre-incubated with 20 LD50s of BoNT/A, B, E, F for 1 hour respectively. The mixtures were injected into mice, and the final death tally was determined 4 days after injection.

cells were selected with 1 mg/ml G418, and several independent monoclonal cell lines were established. For experiment in which cells were preloaded with gangliosides (Merk Chemicals, Darmstadt, Germany), cells were grown to 80% confluence followed by incubation in serum-free media plus 250 μg/ml gangliosides. 24 h later, the serum-free/ganglioside media was replaced with complete media, and the cells were incubated with FITC labeled BoNT/B Hc.

Table 5. Kinetic constants of mAb 8E10 binding to BoNT/A/B/E/F-Hc.

Serotype	*k*off (M-1 S-1)	*k*on (S-1)	*KD* (M)
BoNT A	1.077E−4	1.442E+4	7.468E−9
BoNT B	3.270E−5	1.261E+4	2.592E−9
BoNT E	6.209E−5	1.189E+4	5.223E−9
BoNT F	1.030E−4	1.339E+4	7.690E−9

4. mAb preparation

6-week-old female BALB/c mice were subcutaneously immunized twice at 3-week intervals with 10 ìg of BoNT/B Hc emulsified in Freund's complete or incomplete adjuvant (Sigma-Aldrich, Shanghai, China). Three days after a final immunization with BoNT/B Hc antigen alone, spleen cells from the mice and mouse myeloma NS1 cells (Cell Bank, Chinese Academy of Science, Shanghai, China) were fused and maintained according to the standard procedure [33]. The hybridomas producing anti-BoNT/B antibodies were screened by an indirect enzyme-linked immunosorbent assay (ELISA), using the purified recombinant BoNT/B Hc as a coated antigen. For ascites production, 5×10^6 hybridoma cells were injected intraperitoneally into BALB/c mice, which had been primed with 0.5 ml pristane (Sigma-Aldrich, Shanghai, China). Ascites fluid was collected, and the mAbs purified, using HiTrap Protein G coupled to Sepharose 4B (Amersham Bioscience, Uppsala, Sweden). mAb isotypes were determined by the Mouse Typer Isotyping Panel Kit (Bio-Rad, Hercules, CA). The binding specificities of the mAbs were determined by western blotting. The binding strength of neutralizing mAbs to the BoNT/B Hc was analyzed by surface plasmon resonance technology using a Biacore T100 instrument (Amersham Bioscience; Uppsala, Sweden) [34].

5. Binding specificity and cross-reactivity assay by western blot and ELISA

0.1 μg of purified BoNT/B Hc were mixed with 2× SDS sample buffer (125 mM Tris pH 6.8, 4% SDS, 10% glycerol, 0.006% bromophenol blue, 1.8% ß-mercaptoethanol), boiled and loaded onto 10% polyacrylamide gels. After electrophoresis, the samples were transferred on nitrocellulose sheets, probed with all of these 16 positive binding mAbs respectively (1:1000), and stained with a goat anti-mouse IgG coupled to horseradish peroxidase (HRP) (BD Bioscience, San Jose, CA) and developed with ECL Plus.

To determine the cross-reactivity of 8E10, Easy Wash 96-well plates (Corning, Corning, NY) were coated at 4°C overnight with 100 μl/well purified BoNT A/B/E/F Hc (at 5 μg/ml) respectively (BSA was set as negative control). After blocked with blocking buffer (5% non fat dry milk in PBS) for 2 hours at room temperature, 1 ng of 8E10 in 100 ul PBS was added and incubated for 2 hour at 37°C. The wells were washed and incubated at 37°C for 1 h with HRP-conjugated goat anti-mouse IgG at a 1:3000 dilution. After extensive washing, 3, 3, 5, 5-tetramethylbenzidine (TMB) was added (50 ul/well) and incubated for 10 min at room temperature in the dark and the reaction was stopped by the addition of 2 M H2SO4 (50 ul/well). Absorbance of the samples in the plates was read in an automated ELISA microplate reader at 450 nm. For western blot, purified BoNT A/B/E/F Hc were subjected to SDS PAGE, transfer to nitrocellulose membrane and probed with 8E10 as described above.

6. Kinetic analysis

BIAcore measurements were performed with the Biacore T200 instrument (GE Healthcare) at 25°C in running buffer (10 mM Tris pH 8.0, 100 mM NaCl, 0.005% surfactant P20, 50 μM NiCl). CM5 chips were coated with BoNT/B Hc. 7 neutralizing mAbs were passed over the chip surface at concentrations ranging from 4.4 nM to 22.2 nM for 240 seconds at a flow rate of 30 ul/min and dissociation was recorded during 60 minutes. The chip was regenerated with 20 μl of 35 mM EDTA at 50 μl/min. Binding kinetics were evaluated using the BiaEvaluation software package (GE Healthcare) using a Langmuir model 1:1.

To determine the binding abilities of mAb 8E10 and BoNT/A/B/E or F Hc, Protein A was cross-linked to the dextran surface of a CM5 sensor chip. Protein A was immobilized using amine coupling with 1-ethyl-3-[3-dimethylaminopropyl] carbodiimide hydrochloride and N-hydrosuccinimide to a density of 1000–2000 response units (RU). mAb 8E10 was captured to approximately 100 RU. The analyte BoNT/A/B/E or F Hc was passed over the chip surface at concentrations ranging from 4.4 nM to 22.2 nM as indicated for 240 seconds at a flow rate of 30 μl/min and dissociation was recorded during 60 minutes. Binding kinetics were evaluated as described above.

7. Competition ELISAs

To identify the binding domain which was blocked by the neutralizing mAbs, BoNT/B Hc was conjugated with horseradish peroxidase (HRP) using HRP Plus Activated Conjugation Kit (Thermo, Rockford, IL) according to the instruction of the kit. Easy Wash 96-well plates (Corning, Corning, NY) were coated at 4°C overnight with 100 μl/well recombinant syt II-LD (at 5 μg/ml) or GT1b (at 10 μg/ml) in PBS, washed with PBS/0.05% Tween-20 (Sigma-Aldrich, Shanghai, China) and then blocked for 1 hour at 37°C with PBS/0.05% Tween-20/5% bovine calf serum/3% goat serum (Sigma-Aldrich, Shanghai, China). HRP-conjugated BoNT/B Hc was pre-incubated with neutralizing mAbs at molar ratio of 1:100 for 1 hour at 37°C before added into coated plate. 3, 3, 5, 5-tetramethylbenzidine (TMB) was used as a substrate (Sigma-Aldrich, Shanghai, China) to detect bound BoNT/B Hc.

To determine the binding epitope of the mAbs were overlapping or not, competition was tested by the competition-ELISA [35] with some modifications. Mixtures of constant quantities of the HRP-mAbs (8D1, 8E10, 1F4, 5G10, 2H12, 2B2 and 2F4) conjugate in the optimal working dilutions with dilutions of the competing non-labelled mAb were incubated on BoNT/B Hc coated plates for 1 hour at 37°C. Follow a final wash cycle, 50 ul of substrate and stopping solution were added and absorbance values were determined as previously described. A control without competing antibody (PBS) and a control containing the same mAb (self-competition) were included in the test. Results were expressed as percent binding of HRP-conjugated mAb. The amount of binding obtained in the absence of unlabeled antibody was set at 100% for each HRP-conjugated mAb.

8. Epitope mapping by biopanning

Anti-BoNT/B antibodies (8E10, 5G10 and 2F4) were separately immobilized on 96-well plate. Phages (1.5×10^{11} pfus) from Ph.D. -12 Phage Display Peptide Library (New England Biolabs, Beijing, China) were pre-absorbed by immobilized mouse IgG. The phages, diluted in Tris-buffered saline, were incubated at 4°C for 1 h with immobilized mAbs. The wells were washed for five times with TBST (0.1% Tween 20/Tris-buffered saline). Then, the phages that bound with mAbs were amplified by direct infection with E. coli ER2738. The amplified phages were purified by precipitation with 20% PEG8000, 2.5 M NaCl and used in the next cycle. Three rounds of selection were routinely performed.

The indirect binding assay to test the reactivity of anti-BoNT/B Hc mAbs with synthetic peptides was performed in 96-well plates as described previously [36] with minor modifications. The plates were coated with KLH-peptide 1, 2, 3 respectively. Following 2 washings and blockade of free protein-binding sites, mAbs were added to each well respectively. Mabs binding to peptide were detected by sequential addition of an appropriate dilution of HRP-conjugated goat anti-mouse IgG. To confirm the specific of the epitopes identified, anti-BoNT/B mAbs were incubated with synthetic peptides (P1, P2, P3) respectively, and then the mixture was respectively added to plates precoated with BoNT/B Hc. After 2 h incubation, the wells were washed and detected with an appropriate dilution of HRP-conjugated secondary antibody. After the addition of TMB and stop solution, absorbance was read at 450 nm by a microplate reader.

9. Flow cytometry assay

The inhibition of mAbs for cell binding activity of BoNT/B was examined by performing flow cytometry [37]. BoNT/B Hc was labeled with the FITC labeling kit according to the instruction (Pierce Biotechnology, Rockford, IL).The single or pairs of mAbs were mixed with FITC-BoNT/B Hc at 37°C for 1 h. Ganglioside-treated or Syt II transfected (Syt II+) or double positive PC12 cells were labeled by overnight incubation with the mixture at 4°C and washed with cold PBS, and examined by flow cytometry (BD Biosciences, San Jose, CA). The PC12 cells alone and FITC-labeled Hc fragment of BoNT/B were used as systemic control and negative control, respectively.

10. Confocal microscopy

FITC labeled BoNT/B Hc (1 μg) was incubated with control medium, monoclonal antibody (100 μg) or pairs of mAbs (50 μg of each Ab) for one hour at room temperature. Sub-confluent Ganglioside-treated SytII$^+$-PC12 cells, plated on glass cover slips, were incubated with the BoNT/B Hc-mAbs mixtures at 4°C for one hour and then for 2 hours at 37°C. Cells were then washed with PBS for 15 minutes, fixed with 4% paraformaldehyde for 15 mins, and washed again before mounting. Confocal laser scanning was performed on a Olympus Fluoview 500 system using the 40×objective, and Fluoview software was used for image analysis.

11. Neutralization of BoNT/B activity *in vivo*

For the initial screening of the positive hybridoma that produced the neutralizing antibody, 0.5 ml of culture supernatant from hybridomas grown in a 24-well tissue culture flask for 3–5 days, was pre-incubated with 4 LD50 of BoNT/B for 1 h at room temperature, and the reaction mixtures intraperitoneally (i.p.) injected to each of six mice. More than four survivals in four days were considered as positive. In order to define the activity of the positive neutralizing mAbs and synergistic effect of pairs of mAbs, 100 μg of the positive neutralizing mAbs were added to the indicated number of mouse LD50 of BoNT/B neurotoxin in a total volume of 0.5 ml of gelatin phosphate buffer and incubated at RT for 60 min. For pairs of mAbs, 50 μg of each mAb was added. The mixture was then injected i.p. into female BALB/c mice (16–22 g). Mice were studied in groups of six and were observed at least daily. The morbidity and mortality of mice were determined over 30 days. For the cross-protection assays, 100 μg of 8E10 was pre-incubated with 20 LD50s of BoNT/A, B, E, F for 1 hour respectively. The mixtures were injected into mice, and the final death tally was determined 4 days after injection.

Supporting Information

Figure S1 The effect of BoNT/B-specific antibodies on binding of BoNT/B Hc with ganglioside-treated Syt II+PC12 cells. Ganglioside-treated Syt II+PC12 cells were cultured with FITC-labeled BoNT/B Hc (green), with or without mAbs as indicated and visualized by confocal microscopy.

Figure S2 Binding affinity of mAb 8E10 and BoNT A/B/E/F Hc. The 8E10 was immobilized on the surface of a Protein A cross-linked CM5 chip. Purified BoNT/A/B/E/F Hc at concentrations of 22.2 nM, 14.8 nM, 9.9 nM, 6.6 nM, 4.4 nM were directly injected over the target and the reference surface. The signal obtained from the reference surface was subtracted to avoid the binding of non-specificity.

Author Contributions

Conceived and designed the experiments: JXD YJG CCC HJW SHW. Performed the experiments: CCC SHW HJW XYM GHJ. Analyzed the data: JXD CCC HJW SHW YJG TCZ. Contributed reagents/materials/analysis tools: CCC SHW XYM TCZ GHJ XS TX WJL DPZ. Wrote the paper: JXD YJG CCC.

References

1. Johnson EA, Bradshaw M (2001) Clostridium botulinum and its neurotoxins: A metabolic and cellular perspective. Toxicon 39: 1703–1722.
2. Singh BR (2000) Intimate details of the most poisonous poison. Nature structural biology 7: 617.
3. West M (2005) DATE OF ISSUE 02/02/2005 REVIEW INTERVAL Five Yearly AUTHORISED BY Malcolm West.
4. Hicks RP, Hartell MG, Nichols DA, Bhattacharjee AK, Van Hamont JE, et al. (2005) The medicinal chemistry of botulinum, ricin and anthrax toxins. Current medicinal chemistry 12: 667–690.
5. Rossetto O, Montecucco C (2008) Presynaptic neurotoxins with enzymatic activities. Pharmacology of Neurotransmitter Release: 129–170.
6. Schiavo G, Shone CC, Rossetto O, Alexander F, Montecucco C (1993) Botulinum neurotoxin serotype F is a zinc endopeptidase specific for VAMP/synaptobrevin. Journal of Biological Chemistry 268: 11516.
7. Schiavo G, Rossetto O, Catsicas S, Polverino de Laureto P, DasGupta B, et al. (1993) Identification of the nerve terminal targets of botulinum neurotoxin serotypes A, D, and E. Journal of Biological Chemistry 268: 23784.
8. Blasi J, Chapman ER, Link E, Binz T, Yamasaki S, et al. (1993) Botulinum neurotoxin A selectively cleaves the synaptic protein SNAP-25. Nature 365: 160–163.
9. Schiavo G, Santucci A, Dasgupta BR, Mehta PP, Jontes J, et al. (1993) Botulinum neurotoxins serotypes A and E cleave SNAP-25 at distinct COOH-terminal peptide bonds. FEBS letters 335: 99–103.
10. Darios F, Niranjan D, Ferrari E, Zhang F, Soloviev M, et al. (2010) SNARE tagging allows stepwise assembly of a multimodular medicinal toxin. Proceedings of the National Academy of Sciences 107: 18197–18201.
11. Lacy DB, Tepp W, Cohen AC, DasGupta BR, Stevens RC (1998) Crystal structure of botulinum neurotoxin type A and implications for toxicity. Nature Structural & Molecular Biology 5: 898–902.
12. Simpson L, Rapport M (1971) The binding of botulinum toxin to membrane lipids: sphingolipids, steroids and fatty acids. Journal of neurochemistry 18: 1751–1759.
13. Dong M, Yeh F, Tepp WH, Dean C, Johnson EA, et al. (2006) SV2 is the protein receptor for botulinum neurotoxin A. Science 312: 592.
14. Dong M, Liu H, Tepp WH, Johnson EA, Janz R, et al. (2008) Glycosylated SV2A and SV2B mediate the entry of botulinum neurotoxin E into neurons. Molecular biology of the cell 19: 5226–5237.
15. Fu Z, Chen C, Barbieri JT, Kim JJP, Baldwin MR (2009) Glycosylated SV2 and gangliosides as dual receptors for botulinum neurotoxin serotype F. Biochemistry 48: 5631–5641.
16. Nishiki T, Tokuyama Y, Kamata Y, Nemoto Y, Yoshida A, et al. (1996) The high-affinity binding of Clostridium botulinum type B neurotoxin to synaptotagmin II associated with gangliosides GT1b/GD1a. FEBS letters 378: 253–257.
17. Amersdorfer P, Wong C, Smith T, Chen S, Deshpande S, et al. (2002) Genetic and immunological comparison of anti-botulinum type A antibodies from immune and non-immune human phage libraries. Vaccine 20: 1640–1648.
18. Nowakowski A, Wang C, Powers D, Amersdorfer P, Smith T, et al. (2002) Potent neutralization of botulinum neurotoxin by recombinant oligoclonal antibody. Proceedings of the National Academy of Sciences 99: 11346.
19. Dong M, Tepp WH, Liu H, Johnson EA, Chapman ER (2007) Mechanism of botulinum neurotoxin B and G entry into hippocampal neurons. The Journal of cell biology 179: 1511–1522.
20. Lomneth R, Martin TFJ, DasGupta BR (1991) Botulinum neurotoxin light chain inhibits norepinephrine secretion in PC12 cells at an intracellular membranous or cytoskeletal site. Journal of neurochemistry 57: 1413–1421.
21. Jin R, Rummel A, Binz T, Brunger AT (2006) Botulinum neurotoxin B recognizes its protein receptor with high affinity and specificity. Nature 444: 1092–1095.
22. Swaminathan S, Eswaramoorthy S (2000) Structural analysis of the catalytic and binding sites of Clostridium botulinum neurotoxin B. Nature Structural & Molecular Biology 7: 693–699.
23. Montecucco C (1986) How do tetanus and botulinum toxins bind to neuronal membranes? Trends in Biochemical Sciences 11: 314–317.
24. Rummel A, Eichner T, Weil T, Karnath T, Gutcaits A, et al. (2007) Identification of the protein receptor binding site of botulinum neurotoxins B and G proves the double-receptor concept. Proceedings of the National Academy of Sciences 104: 359.
25. Muraro L, Tosatto S, Motterlini L, Rossetto O, Montecucco C (2009) The N-terminal half of the receptor domain of botulinum neurotoxin A binds to microdomains of the plasma membrane. Biochemical and biophysical research communications 380: 76–80.
26. Levy R, Forsyth C, LaPorte S, Geren I, Smith L, et al. (2007) Fine and domain-level epitope mapping of botulinum neurotoxin type A neutralizing antibodies by yeast surface display. Journal of molecular biology 365: 196–210.
27. Garcia-Rodriguez C, Geren I, Lou J, Conrad F, Forsyth C, et al. (2011) Neutralizing human monoclonal antibodies binding multiple serotypes of botulinum neurotoxin. Protein Engineering Design and Selection 24: 321–331.
28. Kumaran D, Eswaramoorthy S, Furey W, Navaza J, Sax M, et al. (2009) Domain organization in Clostridium botulinum neurotoxin type E is unique: its implication in faster translocation. Journal of molecular biology 386: 233–245.
29. Stenmark P, Dupuy J, Imamura A, Kiso M, Stevens RC (2008) Crystal Structure of Botulinum Neurotoxin Type A in Complex with the Cell Surface Co-

Receptor GT1b—Insight into the Toxin–Neuron Interaction. PLoS pathogens 4: e1000129.
30. Rummel A, Mahrhold S, Bigalke H, Binz T (2004) The HCC-domain of botulinum neurotoxins A and B exhibits a singular ganglioside binding site displaying serotype specific carbohydrate interaction. Molecular microbiology 51: 631–643.
31. Sharma S, Zhou Y, Singh BR (2006) Cloning, expression, and purification of C-terminal quarter of the heavy chain of botulinum neurotoxin type A. Protein expression and purification 45: 288–295.
32. Chapman ER, An S, Edwardson JM, Jahn R (1996) A novel function for the second C2 domain of synaptotagmin. Journal of Biological Chemistry 271: 5844.
33. Köhler G, Milstein C (1975) Continuous cultures of fused cells secreting antibody of predefined specificity. Nature 256: 495–497.
34. Jin C, Hantusch B, Hemmer W, Stadlmann J, Altmann F (2008) Affinity of IgE and IgG against cross-reactive carbohydrate determinants on plant and insect glycoproteins. Journal of Allergy and Clinical Immunology 121: 185–190. e182.
35. Kienle E, Buschmann H (1989) Specificity, cross-reactivity and competition profile of monoclonal antibodies to staphylococcal enterotoxins B and C 1 detected by indirect enzyme-linked immunosorbent assays. Medical Microbiology and Immunology 178: 127–133.
36. Perosa F, Luccarelli G, Prete M, Favoino E, Ferrone S, et al. (2003) β2-microglobulin-free HLA class I heavy chain epitope mimicry by monoclonal antibody HC-10-specific peptide. The Journal of Immunology 171: 1918.
37. Wang H, Li T, Shi J, Cai K, Hou X, et al. (2010) A new neutralizing antibody against botulinum neurotoxin B recognizes the protein receptor binding sites for synaptotagmins II. Microbes and Infection 12: 1012–1018.

10

Chemical Camouflage – A Frog's Strategy to Co-Exist with Aggressive Ants

Mark-Oliver Rödel[1,2]*, Christian Brede[2,5], Mareike Hirschfeld[1,2], Thomas Schmitt[2,6], Philippe Favreau[3], Reto Stöcklin[3], Cora Wunder[4], Dietrich Mebs[4]

1 Museum für Naturkunde Berlin, Leibniz Institute for Research on Evolution and Biodiversity, Berlin, Germany, **2** Theodor-Boveri-Institute (Biocenter of the University), Department of Animal Ecology and Tropical Biology (Zoology III), Würzburg, Germany, **3** Atheris Laboratories, Bernex, Switzerland, **4** Institute of Legal Medicine, University of Frankfurt, Frankfurt/M., Germany, **5** Medizinische Klinik und Poliklinik II, Zentrum für Experimentelle Molekulare Medizin, Würzburg, Germany, **6** Ecological Networks, Biology, Technische Universität Darmstadt, Darmstadt, Germany

Abstract

Whereas interspecific associations receive considerable attention in evolutionary, behavioural and ecological literature, the proximate bases for these associations are usually unknown. This in particular applies to associations between vertebrates with invertebrates. The West-African savanna frog *Phrynomantis microps* lives in the underground nest of ponerine ants (*Paltothyreus tarsatus*). The ants usually react highly aggressively when disturbed by fiercely stinging, but the frog is not attacked and lives unharmed among the ants. Herein we examined the proximate mechanisms for this unusual association. Experiments with termites and mealworms covered with the skin secretion of the frog revealed that specific chemical compounds seem to prevent the ants from stinging. By HPLC-fractionation of an aqueous solution of the frogs' skin secretion, two peptides of 1,029 and 1,143 Da were isolated and found to inhibit the aggressive behaviour of the ants. By *de novo* sequencing using tandem mass spectrometry, the amino acid sequence of both peptides consisting of a chain of 9 and 11 residues, respectively, was elucidated. Both peptides were synthesized and tested, and exhibited the same inhibitory properties as the original frog secretions. These novel peptides most likely act as an appeasement allomone and may serve as models for taming insect aggression.

Editor: Stephen C. Pratt, Arizona State University, United States of America

Funding: This study was part of the BIOLOG-program of the German Ministry of Education and Science (BMB+F, http://www.bmbf.de/; Project BIOTA-West III, amphibian projects, 01LC0617J). The funders had no role in study design, data collection and analysis, decision to publish, or preparation of the manuscript.

Competing Interests: This is to certify that Reto Stocklin is the co-founder, owner, president and CEO of Atheris Laboratories, a biotechnology company working in the field of discovery of bio-active molecules from animal venoms and other natural origins.

* E-mail: mo.roedel@mfn-berlin.de

Introduction

Interspecific associations ranging from mutually beneficial, neutral to parasitic are considered to be an important driver of evolution and are particularly well studied in host-parasite and plant-insect interactions [1,2,3]. Invertebrates of various families, orders or even classes are often living together such as ants and other insects [4]. These associations are mostly parasitic or neutral, mutualistic combinations, such as ant-aphid interactions being the exception [5]. Neutral or mutualistic associations between vertebrates and invertebrates are known from marine ecosystems, e.g. cleaner association of reef fishes and crustaceans [6], but are less known from terrestrial habitats; although the benefits of such associations are manifold and may comprise protection from predators or parasites [7], access to particular food sources [8] or a unique shelter [9].

Anuran amphibians are preyed upon by a variety of vertebrate and invertebrate species [10]. To avoid predation frogs developed a wide range of defence strategies such as concealing or aposematic colouration, special encounter behaviour, noxious or toxic skin secretions, but also seeking a safe shelter such as retreating to burrows or living underground [11]. The associations of frogs with invertebrates such as spiders (e.g. [12,13,14,15,16,17]) or ants [18,19,20], which are potential frog predators [11,21], might be a further very particular strategy to avoid predation. The proximate mechanisms facilitating these associations, however, are not yet understood.

The medium-sized (40–60 mm) microhylid frog *Phrynomantis microps* inhabits the savanna regions of West Africa [22,23] where it hides in burrows or empty termite mounds during the day and the dry season. However, the frog was also observed to occupy and live essentially unharmed in the nest of the highly aggressive ant *Paltothyreus tarsatus* ([24]; Appendix S1 and S2). This ponerine ant species has large workers, reaching 25 mm body length and builds huge underground nests. A colony may consist of several hundred to several thousand workers [25]. They have powerful mandibles and a sting connected with a venom gland [4]. As hunters and scavengers the ants predominantly prey on larger arthropods [26], but also on frogs and other small vertebrates (MOR, *unpubl. obs.*). The benefit for the frogs to live in the underground ant-nest is evident: (1) It provides a safe retreat and protection from (other) predators, since the ants effectively defend their nest against any intruder by fiercely stinging and injecting powerful venom, and (2) the nests produce a constantly humid atmosphere enabling the frogs to survive the long dry season.

Ant nests are known to be inhabited by many other, mostly arthropod, species, which need to evolve mechanisms that enable them to circumvent the colonies' defence strategies [4]. Some species can imitate a colony member due to the chemical resemblance of their cuticular hydrocarbon profiles (e.g. [27,28,29,30]). Other species exhibit only a very small amount of hydrocarbons and thus use their "invisibility" to be integrated into the host colony [31]. Furthermore, deterrents, propaganda and appeasement substances can be used by intruders to invade ant nests without being harmed [31,32,33,34].

Previously, it was demonstrated that the skin secretion of *P. microps* prevents attacks of the ants [24]. Other frog species, when presented to the ants, are immediately attacked and stung. Termites (*Macrothermes bellicosus*) and anurans which are otherwise killed right after contact with the ants are not stung when they had been wetted with the skin secretion of *P. microps* [24]. In the present study we identified the chemical nature of some compounds of the frog's secretion which inhibits the aggressive behavior of the ants. These compounds belong to a hitherto unknown class of peptides.

Materials and Methods

Ethics Statement

No Institutional Animal Care and Use Committee (IACUC) or ethics committee approved this study as this was not required by German law. According to the German Protection of Animal Act ("Tierschutzgesetz", latest adapted on 9 December 2010; http://www.gesetze-im-internet.de/tierschg/BJNR012770972.html; assessed on 12 June 2013) painless experiments and observations with vertebrates neither require permission nor disclosure (§ 1/§ 7 TierSchG). The vertebrates involved, *Phrynomantis microps*, experienced no pain, suffering, complaints or harm. The German Protection of Animal Act only applies to vertebrates, decapods and cephalopods. Experiments with other invertebrates, here: beetles (larvae of *Tenebrio molitor*), termites (*Macrothermes bellicosus*) and ants (*Paltothyreus tarsatus*), are not subject for approval. None of the vertebrates or invertebrates involved in our study are protected by any national or international law. All our work complied with the guidelines for the use of live amphibians and reptiles in field research compiled by the American Society of Ichthyologists and Herpetologists (ASIH), The Herpetologists' League (HL) and the Society for the Study of Amphibians and Reptiles (SSAR).

Research and collections permits in Africa were issued by the respective ministries as well as government bodies in Bénin: Faculté des Sciences Agronomiques, Département d'Aménagement et de Gestion de l'Environnement, Laboratoire d'Ecologie Appliquée, Université d'Abomey-Calavi on behalf of the Centre National de Gestion des Réserves de Faune and the Ministère de l'Environnement et de la Protection de la Nature; and Côte d'Ivoire: Ministère de l'Environnement et du Cadre de Vie, Direction de la Protection de la Nature; Ministre de l'Enseignement Supérieur et de la Recherche Scientifique, Direction de la Recherche; Ministère de la Construction et de l'Environnement, Direction de la Protection de la Nature; Ministère de l'Environnement et de la Forêt, Direction de la Protection de la Nature; Société de Développement des Forêts.

Animals

Frogs (*Phrynomantis microps*) and ants (*Paltothyreus tarsatus*) were collected in the Pendjari Biosphere Reserve (PBR), a savanna area in northern Bénin, West Africa (N 10°30′–11°30′, W 0°50′–2°00′). Thirteen adult frogs were taken alive to the University of Würzburg (UW), Germany, where they were kept in terraria (12/12 light cycle; 25±1°C), fed with crickets, *Achaeta domestica*, and *Drosophila* sp. flies. Mealworm larvae, *Tenebrio molitor*, were purchased from the pet trade. Termites, *Macrotermes bellicosus*, were excavated either in PBR or the Lamto Faunal Reserve (LFR), a savanna in central Côte d'Ivoire, and immediately used in field experiments. The ants used in the field experiments were excavated and housed in plastic containers (31.0×41.0×23.5 cm; filled with savanna soil) at least one week prior to the tests. Ants from different colonies were kept separately, supplied with water and fed with termites and locusts. Two excavated colonies with queens (227 and 258 workers, respectively) were taken to UW, kept in special containers for ant colonies (12/12 light cycle; 25±1°C; for method see [35]) and fed with *Achaeta domestica* and *Tenebrio molitor* larvae. One of the ant colonies successfully reproduced under laboratory conditions. These ants were used for the laboratory experiments.

As we had only termites available in Africa and only mealworms in Germany we first tested if mealworms are suitable for our approach by repeating the experiments of Rödel & Braun [24]. We therefore coated mealworms (UW) or termites (PBR) with either pure water (all controls), with lyophilized skin secretions of *P. microps* (mealworms), or rubbed on *P. microps* (termites). Subsequently, they were offered to the ants and time from first contact to stinging was recorded. After confirming the results from [24] with both insect species we aimed to identifying the compounds of the frog's skin secretions responsible for the lack of aggressive ant behaviour.

Collection and analysis of the skin secretion

Frog secretions were collected by placing *Phrynomantis microps* in a beaker (250 ml) containing 10 ml distilled water to stimulate skin secretion by shaking for two min. The frogs' secretions were frozen and lyophilized. For analysis, the sample was dissolved in 0.1% trifluoroacetic acid (TFA) and centrifuged for 10 min at 3,000 rpm. The supernatant was fractionated by reversed-phase HPLC (LiChroCART column, 125×4 mm, Merck, Darmstadt, Germany) which was eluted with a linear gradient of solvent A (0.1% (v/v) TFA in water) to 100% solvent B (60% (v/v) acetonitrile in 0.1% (v/v) TFA in water) over 60 min at a flow rate of 0.5 ml/min using the Agilent 1200 Series HPLC system. The eluant was monitored at 220 nm. The fractions with a visible absorbance were manually collected, lyophilized and subsequently applied to mealworms for testing in the laboratory.

Peptide analysis and synthesis

For analysis by liquid chromatography linked to time-of-flight mass spectrometry (LC/TOF-MS), the positively tested samples exhibiting no or significantly delayed stinging behaviour of ants (see below), were dissolved in 100 µl acetonitrile:water (50:50, v/v) containing 0.1% formic acid. 2 µl of the solution were analyzed using the Agilent 1100 Series HPLC system interfaced to an Agilent 1100 Series-TOF system (Waldbronn, Germany) operated in positive electrospray ionization mode (ESI). The amino acid sequence of the two active peptides was automatically analyzed by ESI-MS/MS using a Q-TOF micro mass spectrometer (Waters/Micromass, Miford, MA, USA) by nanospray in the positive ionisation mode. The native sample was dissolved in water/acetonitrile/formic acid (v/v/v 49.8/50/0.2), loaded in a PicoTip® emitter (New Objective, Woburn, USA) and analyzed by ESI-MS. Collision induced dissociation was manually adjusted for the best fragmentation pattern. The multiply-charged spectrum was transformed into a singly-charged axis using the MaxEnt3 option from Masslynx 4.0 (Waters, Milford) to allow *de novo* sequence analysis.

The peptides were synthesized using a multichannel peptide synthesizer adapted to Boc chemistry. Classical Boc protected amino acids were used during the assembly and deprotection. Leucine residues were chosen when isomeric amino acids leucine and isoleucine were identified by ESI-MS/MS. Cleavage from the resin was performed with HF. After purification by reverse-phase HPLC, the peptide purity and integrity were controlled by ESI-MS. The final product was found to be homogeneous upon standard HPLC and MS quality control analyses; their molecular weights were identical to those of the native peptides.

Bioassays in laboratory

The entire lyophilized skin secretions of *P. microps*, 32 HPLC fractions of the lyophilized skin secretions were separately dissolved in distilled water and applied to mealworms until they were coated with the solution. Coated mealworms were placed into a plastic arena containing 30 workers of *P. tarsatus* from the laboratory colony (surface of plastic containers: 41.0×31.0 cm). Time was recorded from the first antennal contact of an ant until the mealworm was stung by one of the ants. The experiment was terminated after 5 min when no stinging occurred. Mealworms and ants involved in interactions were removed from the arena before a new trial was started and not used again. Removed ants were replaced by naïve workers to keep the number of test ants constant. The size of the ant colony (app. 250–300 individuals) limited the number of possible replicates when testing the various HPLC fractions.

Field experiments

One mg of either of the two synthesized peptides (A & B) and a combination of both were kept in Eppendorf-tubes at 4°C. In LFR we excavated *P. tarsatus* workers and small *M. bellicosus* soldiers from colonies living in the wild. We used *M. bellicosus* termites as test organisms in the field as (1) under natural conditions they are killed by *P. tarsatus* within a few seconds after encounter, (2) were always about the same size and (3) were available in large numbers. After 2–3 days of acclimatisation 60 ants were placed in a plastic arena with a diameter of 45 cm and a central hiding place. The peptides were dissolved in 1 ml of mineral water. A termite soldier was completely immersed in one of the four solutions, i.e. mineral water (control group) or peptide solutions A, B or A+B, respectively. The termite was then placed into the arena. We stopped the time from first contact of an ant with the termite. In contrast to the above experiments we exclusively recorded the time period during which an ant and a termite were in direct contact (recording stopped when contact was interrupted). The experiment was terminated when a termite was either stung or survived 20 sec of contact. Ants and termites being in contact with each other were removed after each trial. Ants were added to the arena when their numbers were below 50, in order to keep the ant numbers almost constant. Statistical analyses of the data were performed using BiAS.8.2-2006 and R 2.14.1.

Results

In vivo experiments

The ants when encountering termites act with antennating followed by immediate biting and stinging (Fig. 1, inlets A & B). Under experimental conditions the ants stung untreated termites right after contact (median = 1.0 sec, n = 20). However, they stung significantly later when a termite was coated with the skin secretions of *P. microps* (termite rubbed over frog; median time

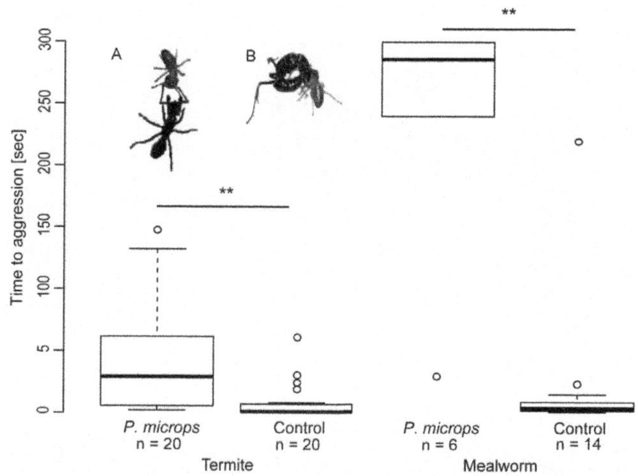

Figure 1. Time from first ant, *Paltothyreus tarsatus*, contact with termites (left; inlet A) or mealworms (right), coated with the skin secretion of *Phrynomantis microps*, until stinging (inlet B). Control groups are termites or mealworms coated with water. Boxplots show the median and the interquartiles of time from first ant contact with a termite or mealworm until stinging. Coated insects were stung significantly later than control insects.

from ant contact to stinging: termites$_{P.\ microps}$ = 29 sec; termites$_{control}$ = 1 sec; Wilcoxon-Test: W = 335.5, $p<0.001$, $n_{P\ microps} = 20$, $n_{control} = 20$; Fig. 1).

When the mealworms were coated with the lyophilized skin secretion of the frog, they were stung much later (median = 286 - sec, n = 6) compared to control insects wetted with distilled water only (median = 4 sec, n = 14; Wilcoxon-Test: W = 83, $p<.001$; Fig. 1).

In both prey species previous results [24] were confirmed as the skin secretion of the frogs significantly delayed the ants' stinging behaviour. Hence both insects were used in subsequent experiments.

Analysis of the skin secretion of *Phrynomantis microps*

The frogs' secretions did not depend on particular food items (Appendix S3). Testing the presence of hydrocarbons and alkaloids in the skin secretion by gas-chromatography/mass spectrometry (GC-MS), as found in the skin of dendrobatid frogs, or of terpenoids and steroids secreted by skin glands of toads [35] using liquid-chromatography/mass-spectrometry (LC-MS-TOF) techniques proved to be negative. By HPLC-fractionation of the secretion 32 fractions were obtained and lyophilized. After dissolving in distilled water these fractions were tested in a bioassay using mealworms presented to ants reared in the laboratory.

Mealworms coated with the entire lyophilized skin secretion of the frogs were stung significantly later than those coated with water only (see above, Fig. 1). Several fractions delayed the aggressive stinging reactions of the ants (results not shown); however, one fraction in particular inhibited the ants' aggressive behaviour (twice no aggressive behaviour at all; once stinging only after 206 sec). This fraction was further analysed (Fig. 2).

Liquid chromatography mass spectrometry analysis of this fraction allowed the identification of two compounds exhibiting molecular masses (*m/z*) of 1029.14 and 1143.25, respectively, suggesting a peptide structure. Due to their low amount and co-elution by liquid chromatography, their amino acid sequence was deduced by ESI-MS/MS suggesting the following, tentative

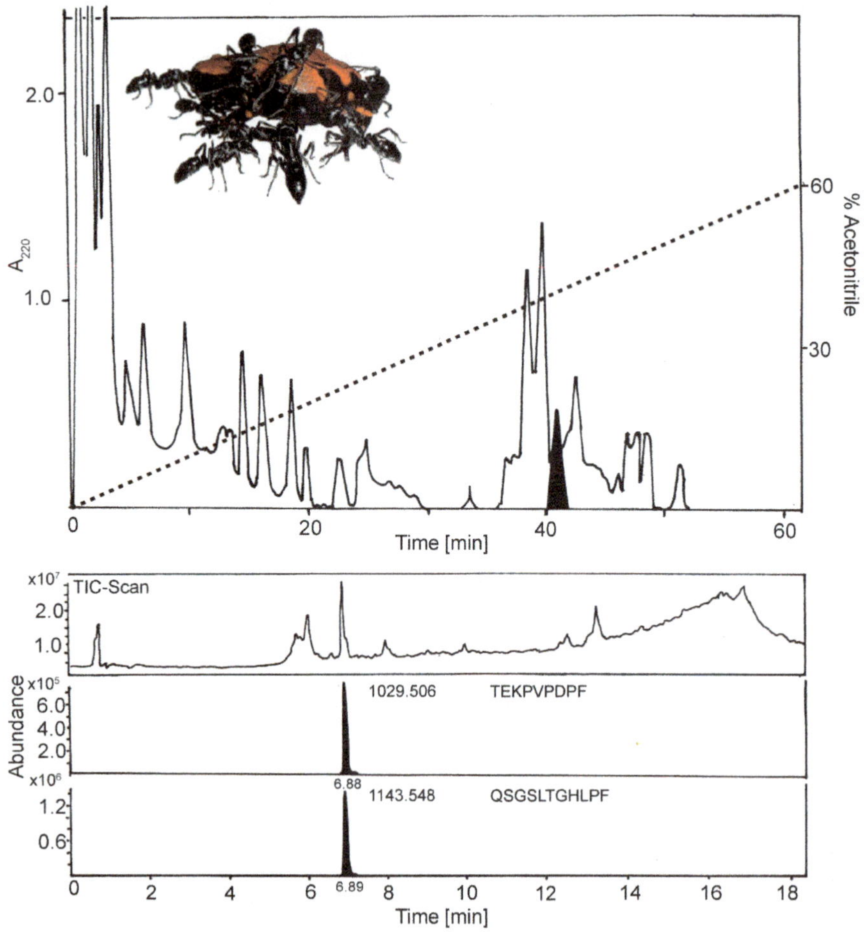

Figure 2. Fractionation of the skin secretion from *Phrynomantis microps* by reversed-phase HPLC on a LiChroCART column, 125×4 mm (Merck, Darmstadt) which was eluted with a linear gradient from 0 to 60% acetonitrile in 0.1% trifluoroacetic acid (dotted line) over 60 min at a flow rate of 0.5 ml/min. Absorbance was monitored at 220 nm. The active fraction is indicated in black (above). Below: Total ion chromatogram (TIC) of liquid chromatography-time-of-flight mass spectrometry (LC/TOF MS) analysis of the HPLC fraction containing two peptides: *m/z* 1029.506 and *m/z* 1143.548 (non-protonated). *De novo* sequencing of the peptides suggested the tentative sequences as indicated in the chromatograms. Inlet picture: adult *Phrynomantis microps* examined by *Paltothyreus tarsatus* workers.

primary structure: peptide A – TEKPVPDPF and peptide B – QSGSLTGHLPF. Both peptides consisting of 9 and 11 amino acids, respectively, are cysteine-free and share the common C-terminal sequence of proline-phenylalanine. No peptides exhibiting similar amino acid sequences were revealed by BLAST search of the protein database of the National Center for Biotechnology Information (NCBI, http//blast.ncbi.nlm.nih.gov) as per March 2013.

To confirm peptide identification, chemical synthesis of both peptides was performed by solid-phase technique and their activity was tested in the field (LFR) by presenting termites wetted with an aqueous peptide solution to *P. tarsatus*. Like the entire lyophilized skin secretion and the positive fraction from the frog skin, the synthetic peptides significantly delayed the ants' aggression (Fig. 3): Termites wetted with water only (control) were immediately stung by the ants (median = 2.9 sec; range: 0.4–20.0 sec; n = 50), whereas treatment of the termites with the peptides significantly delayed the stinging behaviour of the ants, 7.6 sec for peptide A (range: 0.9–20.0 sec), 9.7 sec for peptide B (1.1–20.0 sec), and 9.8 sec for a mixture of both peptides (range: 0.9–20.0 sec; n = 25 in each case; Kruskal-Wallis-test; $H = 29.6874$, $df = 3$, $p<0.0001$; DUNN-tests with ά corrected according to HOLM; treatments against control group: peptide A: $Z = 3.452$, $p<0.01$; peptide B: $Z = 4.069$, $p<0.0001$; peptides A+B: $Z = 4.486$, $p<0.0001$). The peptide treatments did not differ from each other.

Discussion

Skin secretions of amphibians contain a huge variety of substances [36,38] including amines, alkaloids, terpenes, steroids, peptides and proteins which are effectively used in defence against predators and microorganisms [36,39,40,41]. For example more than 800 alkaloids have been detected in skin extracts of the frog families Dendrobatidae, Mantellidae and Myobatrachidae as well as of toads of the genus *Melanophryniscus* [41]. It is well accepted that these compounds are of dietary origin and derive from the frog's food sources such as ants, mites, beetles and other arthropods (e.g. [42,43]). However, although *P. microps* feeds on ants and termites [22], our analyses did not reveal the presence of either alkaloids, terpenoids or steroids in the skin secretions.

In recent years it became evident that peptides are another important and active part of frog secretions. They have been identified in the skin secretion of various frog families [44,45,46,47,48], and show high activity against a wide array of

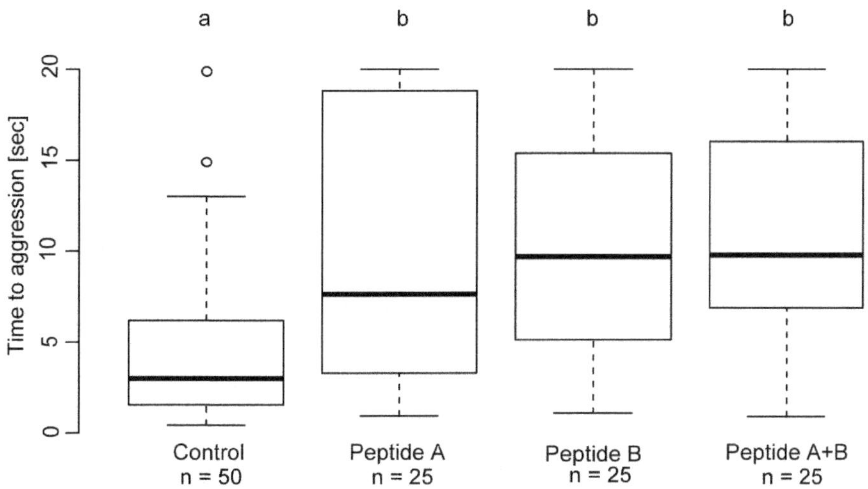

Figure 3. Effect of the two peptides from the skin secretion of *Phrynomantis microps* applied to termite, *Macrotermes bellicosus*, soldiers and delaying the aggressive behaviour and stinging of *Paltothyreus tarsatus* ants. Maximum observation time was 20 sec of full contact of an ant with a termite, after which an experiment trial was terminated. Boxplots show the median and the interquartiles of time from first ant contact with a termite until stinging. Compared to the control group, the peptides significantly delayed the stinging behaviour of the ants. The peptide treatments did not differ from each other.

microorganisms, including the amphibian specific fungal disease chytridiomycosis [49,50]. However, not all investigated frog species synthesize these peptides. It has been suggested that the peptides may play a role as effector molecules in the first line of defense against infections [44], but their precise biological function is still a matter of discussion [47,48]. These frog peptides usually comprise between 10 and 48 amino acid residues, possess a strong positive charge, contain at least 50% hydrophobic amino acids and adapt an alpha-helical conformation in solvents mimicking the environment of cell membranes [48].

The amino acid sequence of the peptides identified in the present study, which was obtained by tandem mass spectrometry, was considered as tentative. Some uncertainties such as the leucine/isoleucine differentiation could not be solved by Edman degradation due to the incomplete separation of the peptides and to their low amount in the skin secretion. However, the synthetic peptides exhibited molecular weights essentially identical to the fraction obtained from the crude skin secretion containing both peptides. In the bioassay, performed under realistic field conditions in the frog's habitat, the synthetic peptides proved to exert the same activity as the original skin secretions. The synthetic peptides are both structurally not related to antimicrobial peptides commonly present in amphibian skin secretions. Although they consist of about 50% hydrophobic amino acid residues (peptide A: 4 of 8, peptide B: 6 of 11) they are not particularly charged and represent a novel group of amphibian skin peptides. They seem to interact with the ants' antennal chemoreception in a way that the frog is either recognized as a nestmate or at least not as an intruder. When approaching the frog the ants are intensively sweeping the frog's skin with their antennae, but are not behaving aggressively, as it is always observed with other intruders. The observations that (i) *P. microps* fed with non-natural food items for prolonged periods continue producing active secretions and (ii) that even freshly metamorphosed *P. microps* are not attacked when placed in the ant colony, suggests that the presence of the peptides in the secretion is not induced, such as by contacts with ants, and not based on the uptake of particular food items, but rather an inherited property and a *de novo* synthesized product of the frog.

Although some studies have suggested that chemicals play a role in frog-invertebrate interaction [17,19,20,24], this study is the first demonstrating that specific peptides play a role in this interaction in particular and in the interactions of an insect with a vertebrate in general [51]. In associations of ants with other insects such as beetles or butterfly larvae, it has been shown that certain hydrocarbon profiles in the insect cuticula are concealing the intruder and provide protection from being killed or expelled from the colony by the ants (e.g. [27,28,29,30,31]). However, to the best of our knowledge, there is no report on hydrocarbons on frog skins and skin extracts of *P. microps* using hexane as solvent with subsequent GC/MS analysis did not reveal any hydrocarbons.

Some frogs are unpalatable to predatory insects [52], including ants [53], due to their toxicity. As another *Phrynomantis* species is known to be toxic [54,55], it could be argued that the herein observed behaviour is likewise simply a result of skin toxins. However, as social insects ants, at least in order to defend their colonies, should accept the death of some individuals. In contrast *P. microps* can freely move within *Paltothyreus tarsatus* colonies [24]. More importantly we could show that two isolated and synthesised peptides from the frogs' secretions showed the same effect on ants as observed in living frogs. This is strongly supporting the assumption that the substances are the proximate reasons of the absence of aggression in the ants when encountering *P. microps*. Most likely the peptides function as an appeasement allomone circumventing aggression of ants against the frog.

In our bioassays with termites and mealworms, a complete all-or-non-reaction was never achieved, but rather a significant reduction and/or delay in the aggressive behaviour of the ants. The lack of an all-or-non answer may be due to several reasons. Firstly, the insects can probably not be covered consistently with the peptides when wetted with a solution; in contrast to the frog where a multitude of glands provides the skin with a dense cover of secretion. Uncovered body parts of the insect, when in contact with the ants' antennae may then trigger stinging. Secondly, termite soldiers often reacted to the ants' contact with biting the latter. In such cases the information of the secretions is probably of secondary importance and as usual the ants respond with instant stinging. This is comparable to instances where *P. microps* steps on ants, which is also followed by an ant reaction of biting or even stinging [24]. In contrast mealworms, only covered in water, where also sometimes not stung. However, these beetle larvae are often very inactive, even when carried away by ants. Stinging to immobilize a potential prey is, therefore, not necessary. Lastly we cannot exclude the possibility that in addition to the two peptides identified herein, the frogs' secretion may contain other components which act solely or in combination with the peptides in preventing the ants from aggressions against the frogs. This was indicated by (i) the time differences from antennating to stinging between trials with the entire skin secretions and the peptides and (ii) several other HPLC-fractions showing similar effects in the bioassays. Unfortunately, the concentrations of the latter were far too low for further analysis. By rechromatography of these fractions their activity was mostly lost or poorly detectable. Therefore, we herein concentrated on the most active fraction in which we identified the two peptides. Nevertheless we could show that these peptides are protecting the frog from the ants' attack, particularly when they were tested under realistic field conditions in the frogs' and ants' natural habitat. The analyses and testing of potential further skin components has to be the focus of further research.

Conclusions

The two peptides identified in this study represent a new group of compounds which not only prevent the ants to attack, but may also inhibit their feeding behaviour and most likely that of other insects. In that respect the peptides could be potentially used as models for taming insect aggression.

Supporting Information

Appendix S1 Video showing adult *Phrynomantis microps* moving unharmed between *Paltothyreus tarsatus* ants.

Appendix S2 Video showing *Phrynomantis microps* metamorph moving unharmed between *Paltothyreus tarsatus* ants.

Appendix S3 Indication for *de novo* synthesis of the peptides by *Phrynomantis microps*.

Acknowledgments

We thank the "Centre National de Gestion des Réserves de Faune", in particular D.A. Tiomoko and A. Tehou, and B. Sinsin, University of Abomey-Calavi, as well as all other authorities in Bénin and Côte d'Ivoire for permits and logistic support. We acknowledge S. Konaté's permission to conduct the bio-assays at the Lamto Scientific Station, Ivory Coast, where N.A. Koné and M. Koné collected ants and termites for our experiments. K.E. Linsenmair and W. Rössler provided access to the laboratories in the Zoological Departments II and III at Würzburg University, respectively. A. Endler assured successful keeping and breeding of the ants under laboratory conditions. The critiques of C. Shaw and an anonymous reviewer helped to improve the manuscript.

Author Contributions

Conceived and designed the experiments: MOR TS DM. Performed the experiments: MOR MH CB DM TS PF RS CW. Analyzed the data: MOR DM CB TS. Contributed reagents/materials/analysis tools: DM PF RS. Wrote the paper: MOR DM. Field work: MOR MH CB.

References

1. Bronstein JL, Alarcon R, Geber M (2006) The evolution of plant-insect mutualisms. New Phytolo 172: 412–428.
2. Jones EI, Bronstein JL, Ferriere R (2012) The fundamental role of competition in the ecology and evolution of mutualisms. Ann NY Acad Scien 1256: 66–88. (doi:10.1111/j.1749-6632.2011.06552.x)
3. Poulin R, Forbes M (2012) Meta-analysis and research on host-parasite interactions: past and future. Evol Ecol 26: 1169–1185. (doi:10.1007/s10682-011-9544-0)
4. Hölldobler B, Willson EO (1990) The ants. Cambridge, Massachusetts, Harvard University Press.
5. Edwards DP (2009) The roles of tolerance in the evolution, maintenance and breakdown of mutualism. Naturwissenschaften 96: 1137–1145. (doi:10.1007/s00114-009-0559-0)
6. Ostlund-Nilsson S, Becker JHA, Nilsson GE (2005) Shrimps remove ectoparasites from fishes in temperate waters. *Biol Lett* 1, 454–456. (doi:10.1098/rsbl.2005.0363)
7. Erickson DM, Wood EA, Oliver KM, Billick I, Abbot P (2012) The effect of ants on the population dynamics of a protective symbiont of aphids, *Hamiltonella defensa*. Ann Entomol Soc Am 105: 447–453. (doi:10.1603/AN11176)
8. Becker JH, Grutter AS (2004) Cleaner shrimp do clean. Coral Reefs 23: 515–520. (doi:10.1007/s00338-004-0429-3)
9. Ricciardi F, Boyer M, Ollerton J (2010) Assemblage and interaction structure of the anemonefish-anemone mutualism across the Manado region of Sulawesi, Indonesia. Environ Biol Fishes 87: 333–347. (doi:10.1007/s10641-010-9606-0)
10. Toledo LF, Ribeiro RS, Haddad CFB (2007) Anurans as prey: an exploratory analysis and size relationships between predators and their prey. J Zool (London) 271: 170–177. (doi:10.1111/j.1469-7998.2006.00195.x)
11. Wells KD (2007) The ecology and behaviour of amphibians. Chicago & London, University of Chicago Press.
12. Blair WF (1936) Ecological notes on *Microhyla olivacea* (Hallowell). Copeia 1936: 115.
13. Hunt RH (1980) Toad sanctuary in a tarantula burrow. J Nat Hist 89: 48–53.
14. Cocroft RB, Humbler K (1989) Observations on a commensal relationship of the microhylid frog *Chiasmocleis ventrimaculata* and the burrowing theraphosid spider *Xenesthis immanis* in southeastern Peru. Biotropica 21: 2–8. (doi:10.2307/2388434)
15. Siliwal M, Ravichandran B (2008) Commensalism in microhylid frogs and mygalomorph spiders. Zoos' Print Mag 23: 13.
16. Karunarathna SDMS, Amarasinghe AAT (2009) Mutualism in *Ramanella nagaoi* Manamendra-Arachchi & Pethiyagoda, 2001 (Amphibia: Microhylidae) and *Poecilotheria* species (Arachnida: Thereposidae) from Sri Lanka. Taprobanica 1: 16–18.
17. Dundee HA, Shillington C, Yeary CM (2012) Interactions between tarantulas (*Aphonopelma hentzi*) and narrow-mouthed toads (*Gastrophryne olivacea*): support for a symbiotic relationship. Tul Stud Zool Bot 32: 31–38.
18. Dejean A, Amiet JL (1992) Un cas de myrmécophilie inattendu: la cohabitation de l'anoure *Kassina senegalensis* avec la fourmi *Megaponera foetens*. Alytes 10: 31–36.
19. Schüter A, Regös J (1996) *Lithodytes lineatus* (Schneider, 1799) (Amphibia: Leptodactylidae) as a dweller in nests of the leaf cutting ant *Atta cephalotes* (Linnaeus, 1758)(Hymenoptera: Attini). Amphibia-Reptilia 2: 117–121.
20. Schlüter A, Löttker P, Mebert K (2009) Use of an active nest of the leaf cutter ant *Atta cephalotes* (Hymenioptera: Formicidae) as a breeding site of *Lithodytes lineatus* (Anura: Leptodactylidae). Herpetol Notes 2: 101–105.
21. Toledo LF, Sazima I, Haddad CFB (2011) Behavioural defences of anurans: an overview. Ethol Ecol Evol 2011 23: 1–25. (doi:10.1080/03949370.2010.534321)
22. Rödel MO (2000) Herpetofauna of West Africa, Vol. I: Amphibians of the West African savanna. Frankfurt/M, Edition Chimaira.
23. Hirschfeld M, Rödel MO (2011) Variable reproductive strategies of an African savanna frog, *Phrynomantis microps* (Amphibia, Anura, Microhylidae). J Trop Ecol 27: 601–609. (doi:10.1017/S0266467411000320)
24. Rödel MO, Braun U (1999) Associations between anurans and ants in a West African savanna (Anura: Microhylidae, Hyperoliidae, and Hymenoptera: Formicidae). Biotropica 31: 178–183. (doi:10.2307/2663971)
25. Braun U, Peters C, Hölldobler B (1994) The giant nests of the African stink ant *Paltothyreus tarsatus* (Formicidae, Ponerinae). Biotropica 26: 308–311. (doi:10.2307/2388852)
26. Dejean A, Lachaud JP, Beugnon G (1993) Efficiency in the exploitation of patchy environments by the ponerine ant *Paltothyreus tarsatus*: An ecological consequence of the flexibility of prey capture behavior. J Ethol 11: 43–53. (doi:10.1007/BF02350005)
27. Vander Meer RK, Jouvenaz DP, Wojcik DP (1989) Chemical mimicry in a parasitoid (Hymenoptera: Eucharitidae) of fire ants (Hymenoptera: Formicidae). J Chem Ecol 15: 2247–2261. (doi:10.1007/BF01014113)
28. Howard RW (1993) Cuticular hydrocarbons and chemical communication. In Insect Lipids: Chemistry, Biochemistry and Biology, Edited by Stanley-

Samuelson DW, Nelson DR. Lincoln NE, University of Nebraska Press; 179–226.
29. Akino T (2002) Chemical camouflage by myrmecophilous beetles *Zyras comes* (Coleoptera: Staphylinidae) and *Diaritiger fossulatus* (Coleoptera: Pselaphidae) to be integrated into the nest of *Lasius fuliginosus* (Hymenoptera: Formicidae). Chemoecology 12: 83–89. (doi:10.1007/s00049-002-8330-4)
30. Bagnères AG, Lorenzi MC (2010) Chemical deception/mimicry using cuticular hydrocarbons. In Insect Hydrocarbons: Biology, Biochemistry, and Chemical Ecology, Edited by Blomquist GJ, Bagnères AG. Cambridge, Cambridge University Press; 282–324.
31. Lenoir A, D'Ettorre P, Errard C, Hefetz A (2001) Chemical ecology and social parasitism in ants. Ann Rev Entomol 46: 573–599.
32. Martin SJ, Jenner EA, Drijfhout FP (2007) Chemical deterrent enables a socially parasitic ant to invade multiple hosts. Proc Roy Soc B 274, 2717–2721. (doi:10.1098/rspb.2007.0795)
33. Nash DR, Boomsma JJ (2008) Communication between hosts and social parasites. In Sociobiology of Communication: an interdisciplinary perspective, Edited by D'Ettorre P, Hughes D. Oxford, Oxford University Press; 55–79.
34. Lhomme P, Ayasse M, Valterová I, Lecocq T, Rasmont P (2012) Born in an alien nest: How do social parasite male offspring escape from host aggression. PLoS ONE 7(9): e43053. (doi:10.1371/journal.pone.0043053)
35. Endler A, Liebig J, Hölldobler B (2006) Queen fertility, egg marking and colony size in the ant *Camponotus floridanus*. Behav Ecol Sociobiol 59: 490–499. (doi:10.1007/s00265-005-0073-0)
36. Daly JW, Myers CW, Whittaker N (1987) Further classification of skin alkaloids from Neotropical poison frogs (Dendrobatidae), with a general survey of toxic noxious substances in the amphibia. Toxicon 25: 1023–1095. (doi:10.1016/0041-0101(87)90265-0)
37. Barthalmus GT (1994) Biological roles of amphibian skin secretions. In Amphibian Biology, Vol. 1 The Integument, Edited by Heatwole H, Barthalmus GT. Chipping Norton, New South Wales, Surrey Beatty & Sons; 382–410.
38. Erspamer V (1994) Bioactive secretions of the amphibian integument. In Amphibian Biology, Vol. 1 The Integument, Edited by Heatwole H, Barthalmus GT. Chipping Norton, New South Wales, Surrey Beatty & Sons; 178–350
39. Bevins CL, Zasloff M (1990) Peptides from frog skin. Ann Rev Biochem 59: 395–414.
40. Clarke BT (1997) The natural history of amphibian skin secretions, their normal functioning and potential medical applications. Biol Rev Camb Philos Soc 72: 365–379. (doi:10.1017/S0006323197005045)
41. Saporito RA, Donnelly MA, Spande TF, Garraffo HM (2012) A review of chemical ecology in poison frogs. Chemoecology 22: 159–168. (doi:10.1007/s00049-011-0088-0)
42. Daly JW, Kaneko J, Wilham J, Garraffo HM, Spande TF, et al. (2002) Bioactive alkaloids from frog skin: Combinatorial bioprospecting reveals that pumiliotoxins have an arthropod source. PNAS 99: 13996–14001. (doi:10.1073/pnas.222551599)
43. Mebs D, Jansen M, Köhler G, Pogoda W, Kauert G (2010) Myrmecophagy and alkaloid sequestration in amphibians: a study on *Ameerga picta* (Dendrobatidae) and *Elachistocleis* sp. (Microhylidae) frogs. Salamandra 46: 11–15.
44. Simmaco M, Mignogna G, Barra D (1998) Antimicrobial peptides from amphibian skin: What do they tell us? Peptide Sci 47: 435–450. (doi: 10.1002/(SICI)1097-0282(1998)47:6<435::AID-BIP3>3.0.CO;2-8)
45. Rinaldi AC (2002) Antimicrobial peptides from amphibian skin: an expanding scenario. Curr Opin Chem Biol 6: 799–804. (doi:10.1016/S1367-5931(02)00401-5)
46. Nascimento AC, Fontes W, Sebben A, Castro MS (2003) Antimicrobial peptides from anurans skin secretions. Prot Pept Lett 10: 227–238. (doi:10.2174/0929866033478933)
47. Conlon JM, Iwamuro S, King JD (2009) Dermal cytolytic peptides and the system of innate immunity in anurans. Ann NY Acad Scien 1163: 75–82. (doi:10.1111/j.1749-6632.2008.03618.x)
48. Conlon JM (2011) Structural diversity and species distribution of host-defense peptides in frog skin secretions. Cell Mol Life Scien 68: 2303–2315. (doi:10.1007/s00018-011-0720-8)
49. Woodhams DC, Ardipradja K, Alford RA, Marantelli G, Reinert LK, et al. (2007) Resistance to chytridiomycosis varies among amphibian species and is correlated with skin peptide defenses. Anim Cons 10: 409–417. (doi:10.1111/j.1469-1795.2007.00130.x)
50. Rollins-Smith LA (2009) The role of amphibian antimicrobial peptides in protection of amphibians from pathogens linked to global amphibian declines. Biochem et Biophys Acta-Biomem 1788: 1593–1599. (doi:10.1016/j.bbamem.2009.03.008)
51. Weldon PJ, Carroll JF (2006) Vertebrate chemical defense: secreted and topically acquired deterrents of arthropods. In Insect Repellents. Principles, Methods, and Users, Edited by Debboun M, Frances SP, Strickman D. CRC Press Taylor & Francis Group; 47–74.
52. Szelistowski WA (1985) Unpalatability of the poison arrow frog *Dendrobates pumilio* to the ctenid spider *Cupiennius coccineus*. Biotropica 17: 345–346. (doi:10.2307/2388601)
53. Fritz G, Rand AS, de Pamphilis CW (1981) The aposematically colored frog, *Dendrobates pumilio*, is distasteful to the large ant, *Paraponera clavata*. Biotropica 13: 158–158. (doi:10.2307/2387719)
54. Jaeger RG (1971) Toxic reaction to skin secretions of the frog, *Phrynomantis bifasciatus*. Copeia 1971: 160–161.
55. Pickersgill M (1997) A case of mild contact poisoning by the rubber frog *Phrynomantis bifasciatus*. Herptile 22: 141.

What Can We Learn from the Evolution of Protein-Ligand Interactions to Aid the Design of New Therapeutics?

Alicia P. Higueruelo[1]*, Adrian Schreyer[1], G. Richard J. Bickerton[1¤], Tom L. Blundell[1], Will R. Pitt[1,2]

1 Department of Biochemistry, University of Cambridge, Cambridge, United Kingdom, 2 UCB Pharma, Slough, United Kingdom

Abstract

Efforts to increase affinity in the design of new therapeutic molecules have tended to lead to greater lipophilicity, a factor that is generally agreed to be contributing to the low success rate of new drug candidates. Our aim is to provide a structural perspective to the study of lipophilic efficiency and to compare molecular interactions created over evolutionary time with those designed by humans. We show that natural complexes typically engage in more polar contacts than synthetic molecules bound to proteins. The synthetic molecules also have a higher proportion of unmatched heteroatoms at the interface than the natural sets. These observations suggest that there are lessons to be learnt from Nature, which could help us to improve the characteristics of man-made molecules. In particular, it is possible to increase the density of polar contacts without increasing lipophilicity and this is best achieved early in discovery while molecules remain relatively small.

Editor: Narayanaswamy Srinivasan, Indian Institute of Science, India

Funding: Funding was provided by the BBSRC, Wellcome Trust, UCB Pharma, and the University of Cambridge. The funders had no role in study design, data collection and analysis, decision to publish, or preparation of the manuscript.

Competing Interests: WRP is an employee of UCB Pharma. However, the work was carried out while he was seconded to the University of Cambridge and he has no competing interests in the material in this manuscript.

* E-mail: alicia@cryst.bioc.cam.ac.uk

¤ Current address: Division of Biological Chemistry and Drug Discovery, College of Life Sciences, University of Dundee, Dundee, United Kingdom

Introduction

The decrease in productivity in drug discovery and development, as measured by the cost of bringing a new medical entity (NME) to the market, is a many fold problem [1]. One of the main reasons for this well-documented decline is the poor physical properties of drug candidates entering into clinical trials ([2–5] and the references therein). Current leads are far too lipophilic to have good chances of success as safe drugs, as logP (calculated octanol-water partition coefficient) correlates positively with *in vitro* compound promiscuity [2]. For oral drugs, lipophilicity is generally considered a requirement for their absorption by passive diffusion in membranes, although there is some debate regarding the contribution of carrier-mediated transport to cellular uptake [6,7]. Standard medicinal chemistry practices, driven by a "perceived need for potency" [8] and a sense of urgency to deliver leads into development, also contribute. In practical terms this has led to more lipophilic candidates, as the exploration of more risky molecules is discouraged [3]. This realization has led to the concept of lipophilic efficiency [2].

Nevertheless, trends in target selection are now providing further challenges. Although the focus on enzyme superfamilies such as protein kinases [9] and aspartic proteinases [10] has allowed knowledge and expertise to be transferred from one target to the next superfamily member, selectivity has proved more elusive than hoped. This has led to increasing interest in targeting protein-protein interactions [11], where the varied regulatory mechanisms within superfamilies promise a new avenue to selective drugs [12]. However, successful small molecule inhibitors of multi-protein complexes tend to be larger lipophilic molecules with few polar features [13], representing challenging starting points for the development of safe drugs. Here we ask whether this size and lipophilicity is a requirement that small molecules need to fulfill in order to bind to protein interfaces or merely a reflection on the nature of chemical matter that has been explored to date.

The aim of this analysis is to understand how Nature effects interactions and to migrate this knowledge to the design of small molecule modulators of biological targets. Although molecular recognition laws are far from simple, one can elucidate general trends in terms of atomic interactions from experimentally determined structures of natural protein complexes (not only multi-protein complexes but also endogenous small-molecule protein complexes) and compare them with trends from synthetic small molecule protein complexes. We therefore define interaction profiles in terms of polar and apolar contacts, with the aim of learning from natural patterns and incorporating them into the design of new therapeutics.

Methods

Compound set definition

Description and examples for each molecular set studied here can be found in File S1. PDB codes and redundancy flags can be downloaded from http://www-cryst.bioc.cam.ac.uk/members/alicia.

Contact definition

Software to calculate hydrogen bonds for all types of molecules (proteins, nucleic acids and small molecules), with the same level of

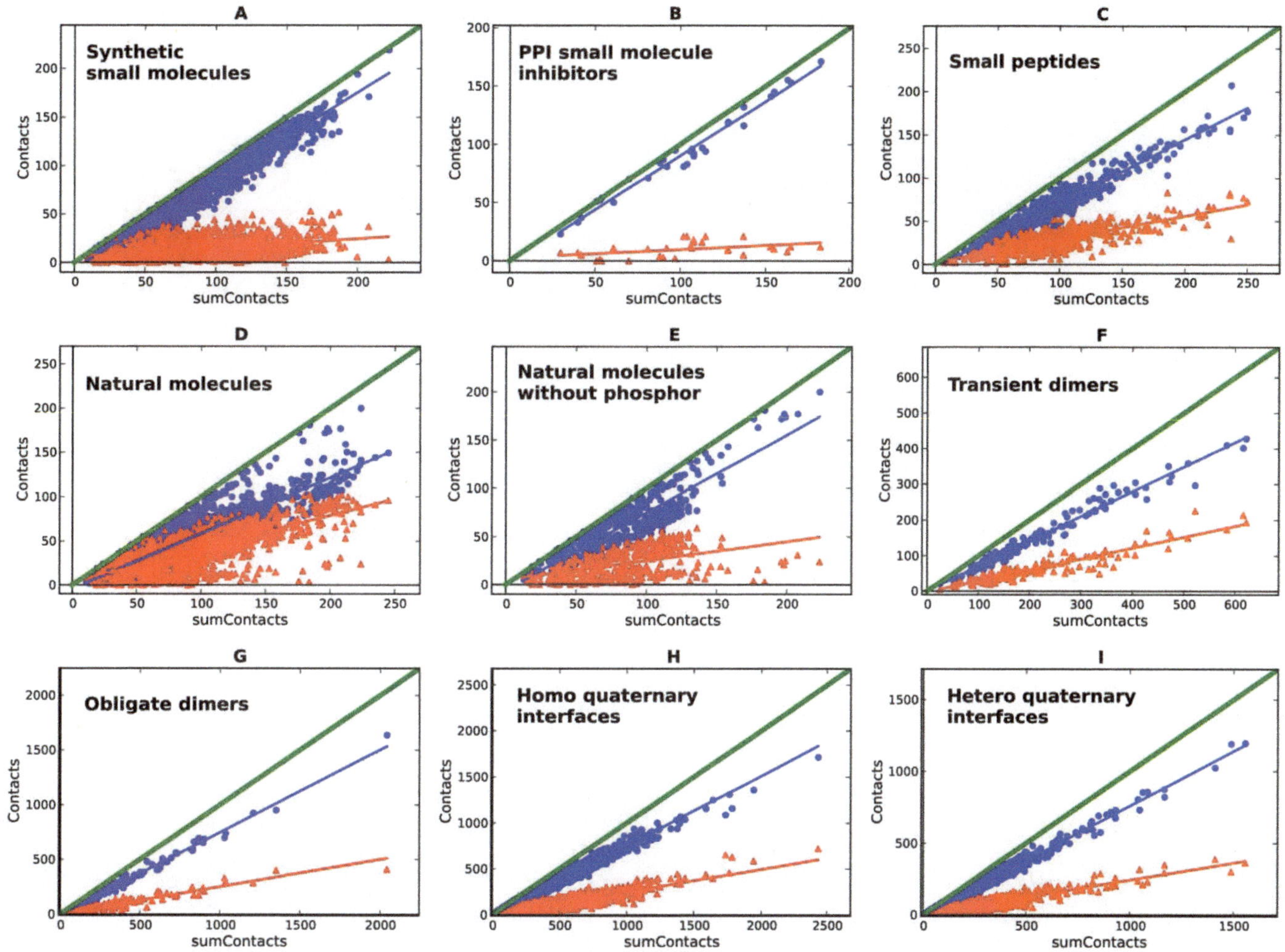

Figure 1. Scissors plots for the non-redundant-by-complex (Table S2) sets of protein complexes. a: synthetic small molecules bound to proteins; b: protein-protein interactions small molecule inhibitors bound to proteins; c: small peptides bound to proteins; d: natural small molecules bound to proteins; e: natural small molecules not containing phosphorus bound to proteins; f: transient protein-protein dimers; g: obligate protein-protein dimers; h: homo protein-protein interfaces from quaternary structures; i: hetero protein-protein interfaces from quaternary structures; polar (red) and apolar (blue) contacts are scattered against sum of contacts.

specificity, is not available at the moment. For this reason, simple polar and apolar contacts were defined using a simple distance cut-off of 4.5 Å. Contacts are labeled depending on the atom types of the proximal pair as follows:

Protein-protein complexes. Apolar contacts: C…C, C…S, S…S (not in Cys-Cys bridges).

Polar contacts: N…O, O…O, N…N, O…S, N…S (S from Cys).

Protein-small molecules complexes. Apolar contacts: C…C, C…S, C…X, S…X (X = Cl, Br, I).

Polar contacts: N…O, O…O, N…N, O…S, N…S, N…F, O…F, S…F (S from Cys).

This discrete count of atomic interactions correlates with buried surface area, which is used in other related studies. See File S2.

Calculation of molecular properties

Molecular properties have been calculated using Pipeline Pilot (http://accelrys.com/products/pipeline-pilot/). The partition coefficient used in this study is AlogP. Rotatable bonds are defined as single bonds between heavy atoms. This does not include ring bonds (regardless of ring size) or those bonds that connect two heavy atoms, one of which is attached to only hydrogens. Furthermore, amide bonds are not considered to be rotatable.

Statistical analysis

Distribution of polar/apolar contact ratio (normalized as polar/[polar+apolar]) between sets has been analyzed with non-parametrical tests (Kruskal-Wallis test for comparison of medians), as not all the sets have a normal distribution of the contact ratio.

Results

Data sets

A non-redundant set of protein-ligand complexes was extracted from the CREDO database [14] and classified into six sets by the type of small molecule involved: approved drugs, oral drugs, synthetic small molecules (for want of a better term, see File S1), protein-protein interactions inhibitors, natural small molecules and small peptides. These terms are used to describe the categories in the remaining sections of this paper. Five non-redundant sets of protein-protein complexes were extracted from the PICCOLO database [15] and classified as obligate dimers, transient dimers,

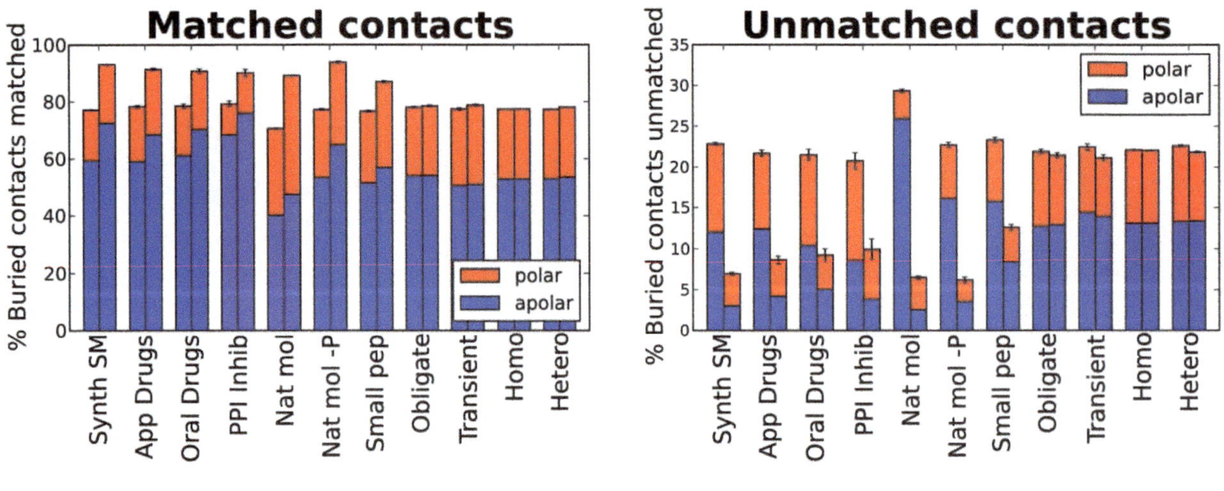

Figure 2. Matched and unmatched contacts. Mean of the percentage of buried atoms per molecule engaged in successful interactions (Matched contacts, left chart) and mean of the percentage of buried atoms without an appropriate partner in the other side of the interface (Unmatched contacts, right chart). The percentage is divided into polar (red) and apolar (blue) contribution. Each set has two bars, one on the left for the atoms in the protein and one on the right for the atoms in the ligand or smaller protein in the case of protein complexes. Error bars denote the standard error of the mean. Sets are ordered from left to right: Synthetic small molecules, approved drugs, oral drugs, protein-protein interaction small molecule inhibitors, natural molecules, natural molecules without phosphor, small peptides, obligate protein-protein dimers, transient protein-protein dimers, homo quaternary protein-protein interfaces and hetero quaternary protein-protein interfaces.

homo and hetero pairwise interfaces from quaternary assemblies, and protein complexes successfully inhibited by small molecules. See File S1. PDB [16] codes for each subset are available to download at: http://www-cryst.bioc.cam.ac.uk/members/alicia.

Polar and apolar contacts with protein molecules

Comparisons between different sets of molecules are based upon the number of polar and apolar atomic contacts. This discrete count of atomic interactions is analogous to the measurement of buried surface area used in other studies, providing a coarse description of the binding interfaces (see File S2).

For each set of interfaces, plotting the total number of contacts (as sum of contacts) versus the number of both polar and apolar contacts generates the 'scissors plot'. In such scissors plots, the

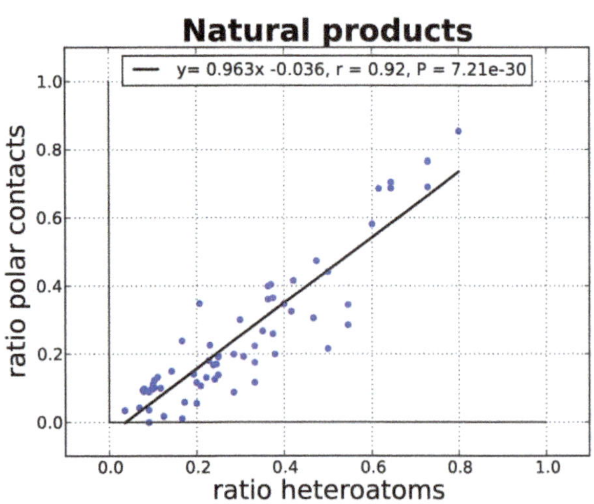

Figure 3. Linear regression of the proportion of heteroatoms to the polar ratio (as number of polar contacts divided by the total [polar + apolar] number of contacts) for the natural-product-like subset of molecules.

relationship between the number of contacts and the polar and apolar contributions can be shown to be linear, with the gradient varying depending of the interaction type. The angle between the trend lines reflects the ratio of polar versus apolar contributions. Figure 1 shows the scissors plots for synthetic small molecules and PPI inhibitors (scissors open), natural small molecules (scissors closed) and small peptides and protein complexes (scissors half way). For the synthetic small molecules, the larger the interface, the greater the degree to which molecular interactions are dominated by apolar contacts, whereas the polar contacts remain relatively constant. The same conclusion was reached by Olsson and co-workers in their analysis of the SCORPIO database [17]. This observation is more pronounced in small molecules inhibiting protein-protein interactions as shown previously [13]. Conversely, natural small molecules, small peptides and protein complexes follow a different trend, where the polar interactions proportionally make a greater contribution. Interestingly, the lower left quadrant of the graphs, where the smaller ligands (fragments) are mostly located, shows a more balanced ratio between polar and apolar contacts. See File S3 for a different representation of these data.

This result agrees with the Hann complexity model [18], which maintains that it is easier for a smaller molecule to match target features, and also it supports the strategy of fragment-based drug design where the initial fragments anchor in a site with specific interactions [19] and where the step-by-step "growing" of the molecules, often guided by structure, tends to deliver less lipophilic hits [3],[5]. Natural molecules have a bimodal distribution as shown in Figure 1 (d and e) and Figure S1 (a); this is due to the presence of a subpopulation of steroid-like molecules, which have an apolar profile, amongst the population as a whole which tend to have more polar interactions. To evaluate whether this relatively high proportion of polar contacts is because many of the natural molecules have phosphate groups, Figure 1e shows the scissors plot for the set of natural molecules, which do not contain phosphorus. Despite the smaller sample size the same trend is observed, and the bimodal distribution is maintained. See File S4 for statistical significance of these plots.

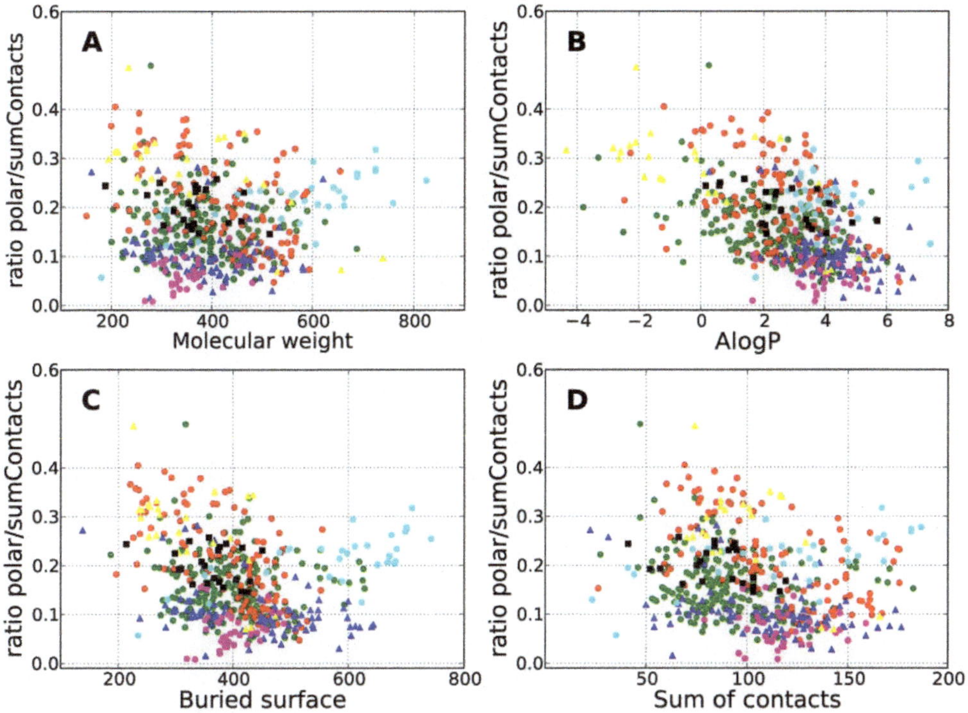

Figure 4. Calculated properties versus interaction profile. The polar ratio (as number of polar contacts divided by the total [polar + apolar] number of contacts) versus molecular weight (A), AlogP (B), buried area upon binding (C) and sum of contacts (D) for protein complexes with synthetic small molecules. Different colors denote SCOP families: Protein kinase catalytic subunit (green), nuclear receptor ligand-binding domain (blue), eukaryotic proteases (red), retroviral proteases – retropepsin (cyan), reverse transcriptase (magenta), Higher-molecular weight phosphotyrosine protein phosphatases (yellow), HSP90 N-terminal domain (black). For clarity, only SCOP families binding to more than 20 different ligands are shown.

Comparison of Figure 1a with Figure 1d and Figure 1e illustrates that natural molecules have a higher proportion of polar contacts than their synthetic counterparts. Likewise on average small peptides also engage in more polar contacts with their targets than synthetic molecules (Figure 1c versus 1a). In order to understand the determinants of these observed differences, the chemical structures of the different ligand sets were analyzed in terms of heteroatom content, rotatable bonds and the proportion of matched and unmatched atoms at the interfaces. The results of these analyses are shown below.

Atomic composition and molecular flexibility

The heteroatom content of each molecule was measured by calculating the ratio of the number of non-carbon atoms to the total number of heavy (i.e. non hydrogen) atoms. The flexibility of each molecule was estimated by dividing the number of rotatable bonds by the number of heavy atoms. The average values for these metrics were then calculated for each of our sets, see File S5. The more polar interaction profile presented by natural molecules compared to synthetic small molecules coincides with a higher content of heteroatoms (26% on average in synthetic small molecules raises to 45% on average in natural molecules). Whereas peptides have a lower content of heteroatoms compared to natural molecules (36% on average). Peptides are also more flexible, the proportion of rotatable bonds by heavy atom is on average 18%, 22% and 42% for synthetic, natural molecules and small peptides respectively. This flexibility should make them more able to match the more directionally constrained polar interactions. In contrast to the natural molecules and peptide sets, the predominately apolar nature of the interactions for synthetic small molecules coincides with a lower heteroatom content and greater rigidity. However, examples exist of natural molecules that are rigid and lipophilic, including steroids such as testosterone. In fact, natural molecules cover a broad range of polarity, but the overall character is predominantly polar, especially when compared with synthetic molecules.

Matched and unmatched atoms at the binding interfaces

As we have shown, the higher content of heteroatoms in non-peptidic natural molecules often leads to more polar interactions with the protein. It is also clear from our analyses that this is not the case for small peptides but the lower heteroatom content may be compensated for by their greater flexibility, allowing them to form more polar interactions. However, it is important to understand what proportion of heteroatoms in each class of molecule is making polar contacts or is unmatched (i.e. heteroatoms within the 4.5 Å threshold that do not satisfy any polar criteria). In other words, what are the polar contact efficiencies?

Figure 2 shows the means of the proportions of matched and unmatched buried atoms. Small molecule ligands, which are represented on the left hand side of the figure, are more contact efficient than the protein to which they are bound. On average around 90% of the ligand atoms are matched in all sets, whereas only 70–80% of the protein atoms are matched. Natural molecules without phosphorus are the most contact efficient, 70–80% of the protein atoms are matched. The small molecule atoms are more exposed and able to contact the protein, whereas the atoms in the protein can be less accessible probably due to the concave nature of ligand binding pockets. Furthermore, studies of ligand-binding

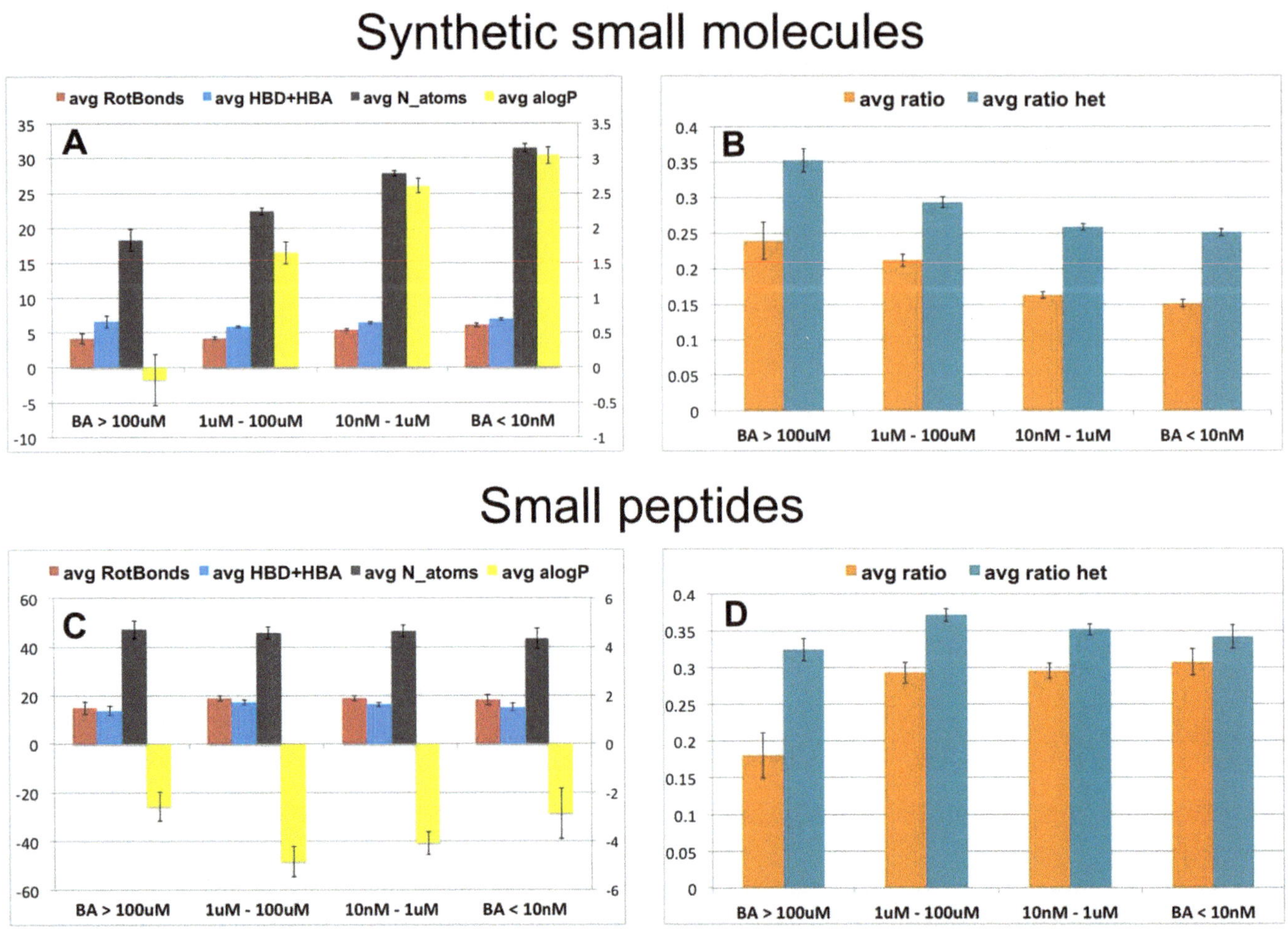

Figure 5. Binding affinity versus interaction profile. Binned binding affinity (BA) data for synthetic small molecules (A and B) and for small peptides (C and D). Bars in (A) and (C) denote the average of molecular properties for each affinity bin: rotatable bonds (red), sum of hydrogen bond donors and acceptors (blue), number of atoms (black) and AlogP (yellow, in the secondary right axis scale for clarity). Bars in (B) and (D) denote the average of polar ratio (orange) and the average heteroatom content (cyan). Error bars are the standard error of the mean.

pockets have shown that they are on average one and a half to three times bigger than the ligand they encapsulate [20,21] (the exact ratio varies depending on the dataset studied and the algorithm used). Therefore, in proportion more atoms in the protein will be at the periphery of the ligand (our cut-off here was 4.5 Å) without making useful interactions. Another interesting result from this analysis is that synthetic molecules have a higher proportion of unmatched polar atoms (on both ligand and protein

Table 1. Examples of polar ratio in protein complexes.

Target	PDB p-p	Ratio p-p	Affinity	PDB p-sm	Ratio p-sm	Affinity	refs
IL-2	1Z92 (A:B)	**0.35**	10nM (Kd)	1PY2 (A)	**0.21**	60nM (IC50)	[27],[28]
Bcl-XL	2BZW (A:B)	**0.19**	6nM (Kd)	2YXJ (B)	**0.08**	0.6nM	[29]
MDM2	1YCR (A:B)	**0.14**	600nM (Kd)	1T4E (A)	**0.03**	67nM (Kd)	[30],[31]
XIAP	1G3F (A:B)	**0.22**		1TFT (A)	**0.12**		[32,33]
ZipA	1F47 (B:A)	**0.10**	21.6uM (Kd)	1Y2F (A)	**0.00**	12uM (Kd)	[34],[35]
TNF	1TNF (AB:C)	**0.30**	-	2AZ5 (C&D)	**0.12**	13uM	[36,37]
S100B	1DT7 (A:X)	**0.34**	-	3GK1 (A)	**0.12**		[38],[39]

Polar ratio is the number of polar contacts divided by the total [polar + apolar] number of contacts. These protein complexes are for proteins that bind to both protein partners (ratio p-p, left) and synthetic small molecules (ratio p-sm, right). The PDB code includes the interacting chains, for example 1TNF(AB:C) denotes chain A and B interacting with chain C of the TNF trimer, whereas 2AZ5 (C&D) denotes chains C and D interacting with the small molecule. When available, affinity measure and units is specified in table.

side) than the natural ones. Therefore, if one wants to increase the polar contacts synthetic molecules make, then there is opportunity for improvement. Oral drugs need to have a limited polar surface area in order to be absorbed by the body [22]. However, Figure 2 shows that approved and oral drugs are not making the most of their polar composition. Nevertheless, improving enthalpic contacts is not a trivial task, not only because of the difficulty of designing chemical geometries that match polar constraints, but also because of the increased desolvation penalty and the reduction of conformational entropy [23] that can result.

The question remains, does Nature make the most of the polar composition of its ligands? Figure 3 shows a linear correlation between heteroatom content (the number heteroatoms divided by the total number of atoms) and the ratio of polar interactions (the number polar contacts divided by the sum of contacts, henceforth referred to as the polar ratio), for the natural-product-like subset in natural molecules. For these small molecules, the increase in polar features translates into more polar interactions with the protein.

Synthetic small molecule complexes. Calculated properties versus interaction profile

An analysis of the distribution of the polar ratio plotted against molecular weight, AlogP, surface area buried upon binding and sum of contacts has been carried out for the synthetic small molecules. Figure 4 shows these distributions color-coded by SCOP [24] family.

Synthetic molecules bound to protein kinases (green dots in Figure 4) tend to have a high AlogP and a low polar ratio (usually below 30%). Even those molecules with a negative AlogP do not appear to make many polar contacts. However, there are a few molecules with 50% of polar contacts, for example HET-ID 3C3 in PDB entry 2CGW. In the case of nuclear receptor ligand-binding domain (NR-LBD, blue triangles in Figure 4), all the molecules have AlogP>1 and most have less than 15% of polar contacts, but as with the kinase ligands there are examples that buck the trend, for example HET-ID 444 in PDB entry 1UHL with a polar ratio of 34%). Eukaryotic proteases (red dots in Figure 4) bind to a wide range of molecules with molecular weights ranging from 200 to 700 Da and with AlogPs between −2 and 6. They also have a wide range of polar ratios. Bigger and more lipophilic molecules are found to bind to retroviral proteases (cyan dots in Figure 4) with similar polar binding patterns as found for eukaryotic enzymes, although polar fragments are also found. The proteins belonging to the reverse transcriptase SCOP family have similar characteristics to nuclear receptor ligand-binding domain, and bind to molecules with AlogP>1, all of which have less than 15% polar contacts. Most of the molecules bound to phosphotyrosine protein phosphatases (PTPP, yellow triangles in Figure 4) have around 30% of polar contacts with a low AlogP range (−2 to 2), although there are also three apolar binders (HET-ID 892 in PDB entry 1T49, HET-ID BB3 in PDB entry 1T48 and HET-ID FRJ in PDB entry 1T4J). However, these apolar molecules are inhibitors binding to an allosteric site. Finally, HSP90 domains (black squares in Figure 4) bind to molecules with a wide range of AlogP (0–6) with between 15–25% of polar contacts.

The characteristics of synthetic small molecules depend heavily upon their targets. However, there is no correlation between calculated lipophilicity or molecular weight and the polar ratio. Thus, it is possible to achieve more polar contacts without compromising the physical properties of the drug leads. See Table S1 for a list of approved and oral drugs with more than 40% of polar contacts. The low lipophilicity of these molecules is noteworthy. In a recent analysis of patented small molecules from 18 pharmaceutical companies, Leeson and St-Gallay [5] show that here is indeed a target dependency of the lead lipophilicity. However, the outcomes of different companies working on the same target also highlight that the major determinant of the lipophilicity of the compounds is the discovery practices of each organization and not target requirements.

Synthetic small molecule complexes. Affinity versus interaction profile

Affinity data from the implementation of PDBBind [25] in CREDO [14] are available for almost 700 of the synthetic small molecule set and more than one hundred of the small peptides. For these molecules there is no correlation between the binding energy and the proportion of polar contacts they make with their protein partners (File S6). Small, weakly binding synthetic molecules have high polar ratios, whereas small peptides can engage more polar contacts across a wide range of affinities.

Indeed, Figure 5a shows that the most potent synthetic molecules have the highest average number of atoms and AlogP while the average count of hydrogen bond acceptors and donors remains constant across the whole potency range. This result is in agreement with the much discussed general tendency in drug discovery of gaining potency by adding lipophilic groups to the small molecules; see for example [2]. In the set studied here, this translates (Fig. 5b) into lower heteroatom content and a lower polar ratio.

Figure 5 (c and d) shows that small peptides have on average the same property profile regardless of their potency. This result suggests that small peptides do not achieve tight binding through increase of lipophilicity. Furthermore the proportion of polar contacts and the heteroatom content is maintained across the whole range of affinities with the exception of weak binders where the polar ratio is lower (although there are only seven complexes in this category). The important point to highlight here is that peptides, unlike synthetic small molecules, can bind specifically to their receptors and have a high number of polar contacts. We argue that this is in part due to their inherent flexibility and low lipophilicity. Note that the average number of rotatable bonds in small peptides is four-fold higher than in synthetic molecules. In fact, the small peptides studied here are often bigger than the synthetic small molecules, as no size limit was applied to select the small peptide set, whereas synthetic molecules larger than 900 Da were omitted. However, there is no correlation between the number of atoms and the free energy of binding for small peptides, see File S7, in agreement with the conclusions previously reached by Kuntz [26]. Nevertheless, they are less efficient than small synthetic molecules as they use more atoms to achieve the same affinity.

Small molecule inhibitors of protein-protein interactions

We now return to the question of whether small molecules inhibiting protein-protein interactions need to be large and lipophilic. Although these molecules tend to be lipophilic with few polar features, Figure 2 shows that they have polar atoms unmatched in the binding site. Using the TIMBAL database [13], we have extracted the seven cases where there is structural information for both the small molecule-protein and the protein-protein complexes. In all cases studied, there are unutilized potential polar contacts on the protein interface (see Table 1 and File S8 for a graphical representation of these data). These cases illustrate a common pattern for synthetic molecules: they typically comprise few anchor points (which can be considered as more constrained polar contacts) and more potency-boosting hydrophobic interactions. Small molecule inhibitors of protein-protein interactions do not take advantage of all the available opportunities for polar contacts in the interfaces as only a few of them are engaged. For these seven cases, comparison of the interacting residues in the target protein highlights that small molecules tend to use more aromatic and fewer charged residues than the protein partner, see Figure S2.

Discussion

In this study we have analyzed the atomic contacts between different sets of molecules, some natural and some synthetic. The results presented here show that natural complexes typically engage in more polar contacts than synthetic molecules bound to proteins. This latter set, also have a higher proportion of unmatched heteroatoms than the natural sets and, probably for this reason, show no correlation between lipophilicity and proportion of polar contacts. These differences may be due to the fact that the dynamics of evolutionary processes are more likely to provide a more highly optimized configuration than is possible by medicinal chemistry efforts. Importantly it must be borne in mind that endogenous molecules have not been constrained by the requirements for oral absorption and distribution across membranes into cellular compartments.

For synthetic small molecules in general, but in particular for the inhibitors of protein-protein interactions, we conclude that efforts should be invested to maximize polar contacts to better resemble the interaction patterns that natural molecules present. As the ratio of polar versus apolar contacts is greater when the size of the synthetic molecules is smaller, this will be easiest to achieve when ligands remain small. However, it will almost always be harder to match polar contacts than to gain affinity through the addition of hydrophobic substituents. Nevertheless, the recent analysis of Leeson and St-Gallay [5] shows that for the same target, a range of compound lipophilicity can be achieved using different strategies, thus encouraging efforts that may lead to better molecules. If we are to minimize ligand promiscuity and ultimately lower the costs of delivering safe and effective drugs to the market, it must be worthwhile to "go back to Nature" and see what can be learnt from the exquisite specificity – albeit low ligand efficiency – of protein-ligand interactions that has evolved to be selectively advantageous in living organisms.

Supporting Information

Figure S1 Normalized distributions of the ratio of polar contacts (represented by polar/[polar+apolar]), each chart compares synthetic small molecules against the other sets.

Figure S2 Comparisons of residue propensities at the binding sites for small molecule protein-protein inhibitors versus protein-protein complexes inhibited by them.

File S1 Detailed description of each data set.

File S2 Buried surface area versus atomic contacts.

File S3 More polar contacts for synthetic fragments.

File S4 Comparison of interaction profiles.

File S5 Heteroatom content and rotatable bond count.

File S6 Free energy of ligand binding versus the polar ratio of contacts [polar/(polar+apolar)] for the synthetic small molecules set and the small peptide set.

File S7 Free energy of ligand binding versus the number of atoms of the ligand for the synthetic small molecules set and the small peptide set.

File S8 Protein-protein complexes inhibited by small molecules.

Table S1 Approved and oral small molecule drugs that engage more than 40% polar contacts with their bound protein.

Table S2 Number of entries in each set of molecules.

Acknowledgments

APH would like to thank Nikolay Todorov and Colin Groom for the many helpful discussions.

Author Contributions

Conceived and designed the experiments: APH TLB WRP. Performed the experiments: APH. Analyzed the data: APH. Contributed reagents/materials/analysis tools: AS GRJB. Wrote the paper: APH TLB WRP. Reviewed and edited the final version of the manuscript: APH AS GRJB TLB WRP.

References

1. Garnier JP (2008) Rebuilding the R&D engine in big pharma. Harv Bus Rev 86: 68–70.
2. Leeson P, Springthorpe B (2007) The influence of drug-like concepts on decision-making in medicinal chemistry. Nat Rev Drug Discov 6: 881–890.
3. Keserü GM, Makara GM (2009) The influence of lead discovery strategies on the properties of drug candidates. Nat Rev Drug Discov 8: 203–212.
4. Gleeson MP, Hersey A, Montanari D, Overington J (2011) Probing the links between in vitro potency, ADMET and physicochemical parameters. Nature Reviews Drug Discovery 10: 197–208.
5. Leeson PD, St-Gallay SA (2011) The influence of the 'organizational factor' on compound quality in drug discovery. Nat Rev Drug Discov 10: 749–765.
6. Dobson PD, Kell DB (2008) Carrier-mediated cellular uptake of pharmaceutical drugs: an exception or the rule? Nat Rev Drug Discov 7: 205–220.
7. Sugano K, Kansy M, Artursson P, Avdeef A, Bendels S, et al. (2010) Coexistence of passive and carrier-mediated processes in drug transport. Nat Rev Drug Discov 9: 597–614.
8. Hann MM (2011) Molecular obesity, potency and other addictions in drug discovery. MedChemComm 2: 349–355.
9. Davis MI, Hunt JP, Herrgard S, Ciceri P, Wodicka LM, et al. (2011) Comprehensive analysis of kinase inhibitor selectivity. Nat Biotechnol 29: 1046–1051.
10. Deu E, Verdoes M, Bogyo M (2012) New approaches for dissecting protease functions to improve probe development and drug discovery. Nat Struct Mol Biol 19: 9–16.
11. Wells J, McClendon C (2007) Reaching for high-hanging fruit in drug discovery at protein-protein interfaces. Nature 450: 1001–1009.
12. Cochran AG (2000) Antagonists of protein-protein interactions. Chem Biol 7: R85–R94.
13. Higueruelo AP, Schreyer A, Bickerton GRJ, Pitt WR, Groom CR, et al. (2009) Atomic Interactions and Profile of Small Molecules Disrupting Protein-Protein Interfaces: the TIMBAL Database. Chem Biol Drug Des 74: 457–467.
14. Schreyer A, Blundell T (2009) CREDO: A Protein-Ligand Interaction Database for Drug Discovery. Chem Biol Drug Des 73: 157–167.
15. Bickerton GR, Higueruelo AP, Blundell TL (2011) Comprehensive, atomic-level characterization of structurally characterized protein-protein interactions: the PICCOLO database. BMC Bioinformatics 12: 313.
16. Berman HM, Westbrook J, Feng Z, Gilliland G, Bhat TN, et al. (2000) The Protein Data Bank. Nucleic Acids Res 28: 235–242.

17. Olsson TSG, Williams MA, Pitt WR, Ladbury JE (2008) The Thermodynamics of Protein-Ligand Interaction and Solvation: Insights for Ligand Design. J Mol Biol 384: 1002–1017.
18. Hann MM, Leach AR, Harper G (2001) Molecular Complexity and Its Impact on the Probability of Finding Leads for Drug Discovery. J Chem Inf Comput Sci 41: 856–864.
19. Congreve M, Chessari G, Tisi D, Woodhead AJ (2008) Recent Developments in Fragment-Based Drug Discovery. J Med Chem 51: 3661–3680.
20. An J, Totrov M, Abagyan R (2005) Pocketome via Comprehensive Identification and Classification of Ligand Binding Envelopes. Molecular & Cellular Proteomics 4: 752–761.
21. Kahraman A, Morris RJ, Laskowski RA, Thornton JM (2007) Shape variation in protein binding pockets and their ligands. J Mol Biol 368: 283–301.
22. Veber DF, Johnson SR, Cheng HY, Smith BR, Ward KW, et al. (2002) Molecular properties that influence the oral bioavailability of drug candidates. J Med Chem 45: 2615–2623.
23. Ferenczy GG, Keserű GM (2010) Thermodynamics guided lead discovery and optimization. Drug Discov Today 15: 919–932.
24. Murzin A, Brenner S, Hubbard T, Chothia C (1995) SCOP: a structural classification of proteins database for the investigation of sequences and structures. J Mol Biol 247: 536–540.
25. Wang R, Fang X, Lu Y, Wang S (2004) The PDBbind database: collection of binding affinities for protein-ligand complexes with known three-dimensional structures. J Med Chem 47: 2977–2980.
26. Kuntz ID, Chen K, Sharp KA, Kollman PA (1999) The maximal affinity of ligands. Proc Natl Acad Sci U S A 96: 9997–10002.
27. Rickert M, Wang X, Boulanger MJ, Goriatcheva N, Garcia KC (2005) The structure of interleukin-2 complexed with its alpha receptor. Science 308: 1477–1480.
28. Thanos CD, Randal M, Wells JA (2003) Potent Small-Molecule Binding to a Dynamic Hot Spot on IL-2. J Am Chem Soc 125: 15280–15281.
29. Lee EF, Czabotar PE, Smith BJ, Deshayes K, Zobel K, et al. (2007) Crystal structure of ABT-737 complexed with Bcl-xL: implications for selectivity of antagonists of the Bcl-2 family. Cell Death Differ 14: 1711–1713.
30. Kussie PH, Gorina S, Marechal V, Elenbaas B, Moreau J, et al. (1996) Structure of the MDM2 oncoprotein bound to the p53 tumor suppressor transactivation domain. Science 274: 948–953.
31. Grasberger BL, Lu T, Schubert C, Parks DJ, Carver TE, et al. (2005) Discovery and cocrystal structure of benzodiazepinedione HDM2 antagonists that activate p53 in cells. J Med Chem 48: 909–912.
32. Liu Z, Sun C, Olejniczak ET, Meadows RP, Betz SF, et al. (2000) Structural basis for binding of Smac/DIABLO to the XIAP BIR3 domain. Nature 408: 1004–1008.
33. Oost TK, Sun C, Armstrong RC, Al-Assaad AS, Betz SF, et al. (2004) Discovery of potent antagonists of the antiapoptotic protein XIAP for the treatment of cancer. J Med Chem 47: 4417–4426.
34. Mosyak L, Zhang Y, Glasfeld E, Haney S, Stahl M, et al. (2000) The bacterial cell-division protein ZipA and its interaction with an FtsZ fragment revealed by X-ray crystallography. EMBO J 19: 3179–3191.
35. Rush TS, Grant JA, Mosyak L, Nicholls A (2005) A shape-based 3-D scaffold hopping method and its application to a bacterial protein-protein interaction. J Med Chem 48: 1489–1495.
36. Eck MJ, Sprang SR (1989) The structure of tumor necrosis factor-alpha at 2.6 A resolution. Implications for receptor binding. J Biol Chem 264: 17595–17605.
37. He M, Smith A, Oslob J, Flanagan W, Braisted A, et al. (2005) Small-Molecule Inhibition of TNF-alpha. Science 310: 1022–1025.
38. Rustandi RR, Baldisseri DM, Weber DJ (2000) Structure of the negative regulatory domain of p53 bound to S100B(betabeta). Nat Struct Biol 7: 570–574.
39. Charpentier T, Wilder P, Liriano M, Varney K, Zhong S, et al. (2009) Small Molecules Bound to Unique Sites in the Target Protein Binding Cleft of Calcium-Bound S100B As Characterized by Nuclear Magnetic Resonance and X-ray Crystallography. Biochemistry 48: 6202–6212.

Phage Display and Synthetic Peptides as Promising Biotechnological Tools for the Serological Diagnosis of Leprosy

Silvana Maria Alban[1], Juliana Ferreira de Moura[2], Vanete Thomaz-Soccol[1]*, Samira Bührer Sékula[3], Larissa Magalhães Alvarenga[3], Marcelo Távora Mira[4], Carlos Chávez Olortegui[5], João Carlos Minozzo[6]

1 Department of Biotechnology and Bioprocess Engineering, Federal University of Parana, Curitiba, Parana, Brazil, **2** Basic Pathology Department, Federal University of Parana, Curitiba, Parana, Brazil, **3** Immunology Department, Tropical Pathology and Public Health Institute, Federal University of Goias, Goiania, Goias, Brazil, **4** Health Sciences Postgraduate Program, School of Medicine, Pontifical Catholic University of Parana, Curitiba, Parana, Brazil, **5** Department of Biochemistry and Immunology, Federal University of Minas Gerais, Belo Horizonte, Minas Gerais, Brazil, **6** Center for Production and Research of Immunobiological Products, Parana State Department of Health, Piraquara, Parana, Brazil

Abstract

Background: The diagnosis of leprosy is primarily based on clinical manifestations, and there is no widely available laboratory test for the early detection of this disease, which is caused by *Mycobacterium leprae*. In fact, early detection and treatment are the key elements to the successful control of leprosy.

Methodology/Principal Findings: Peptide ligands for antibodies from leprosy patients were selected from phage-displayed peptide libraries. Three peptide sequences expressed by reactive phage clones were chemically synthesized. Serological assays that used synthetic peptides were evaluated using serum samples from leprosy patients, household contacts (HC) of leprosy patients, tuberculosis patients and endemic controls (EC). A pool of three peptides identified 73.9% (17/23) of multibacillary (MB) leprosy patients using an enzyme-linked immunosorbent assay (ELISA). These peptides also showed some seroreactivities to the HC and EC individuals. The peptides were not reactive to rabbit polyclonal antisera against the different environmental mycobacteria. The same peptides that were conjugated to the carrier protein bovine serum albumin (BSA) induced the production of antibodies in the mice. The anti-peptide antibodies that were used in the Western blotting analysis of *M. leprae* crude extracts revealed a single band of approximately 30 kDa in one-dimensional electrophoresis and four 30 kDa isoforms in the two-dimensional gel. The Western blotting data indicated that the three peptides are derived from the same bacterial protein.

Conclusions/Significance: These new antigens may be useful in the diagnosis of MB leprosy patients. Their potentials as diagnostic reagents must be more extensively evaluated in future studies using a large panel of positive and negative sera. Furthermore, other test approaches using peptides should be assessed to increase their sensitivity and specificity in detecting leprosy patients. We have revealed evidence in support of phage-displayed peptides as promising biotechnological tools for the design of leprosy diagnostic serological assays.

Editor: Mauricio Martins Rodrigues, Federal University of São Paulo, Brazil

Funding: The study was financially supported by Conselho Nacional de Desenvolvimento Científico e Tecnológico (www.cnpq.br), Ministério da Ciência, Tecnologia e Inovação (MCTI) under grant number 473061/2007-4, grants from the Fundação Araucária (www.fundacaoaraucaria.org.br), Secretaria de Estado da Ciência, Tecnologia e Ensino Superior (SETI) and grant from the Coordenação de Aperfeiçoamento de Pessoal de Nível Superior (www.capes.gov.br) to S.M.A. The funders had no role in study design, data collection and analysis, decision to publish, or preparation of the manuscript.

Competing Interests: The authors have declared that no competing interests exist.

* Email: vanetesoccol@gmail.com

Introduction

Leprosy, which is caused by *Mycobacterium leprae* [1], remains a currently relevant disease. As of 2012, cases were reported in 115 countries with the vast majority being concentrated in India, Brazil, and Indonesia [2]. Despite the reduction in the global prevalence rate from 5.4 million in 1985 to 181,941 at the beginning of 2012, the number of new cases that are detected has remained stable over recent years [2], indicating the continuity of its transmission [3].

The clinical manifestations are determined by the immune response of the patient to *M. leprae*. Individuals with lepromatous leprosy present with high bacterial loads and exacerbated humoral immune responses. In contrast, those with tuberculoid leprosy show few bacilli in their lesions and intense cellular immune responses that can be evaluated by the lepromin test [1,4]. Most patients present borderline leprosy clinical forms [1]. The

Table 1. Amino acid sequences of the peptides obtained by phage display.

Peptide	Amino acid sequence	Length
5A	APDDPAWQNIFNLRR	15
6A	ICPRDPAWSYCN	12
1B	NNIHAHRYWGTDLNA	15

diagnosis of this disease is essentially clinical and is occasionally accompanied by bacteriological or histological examinations [5]. With regard to treatments, the World Health Organization (WHO) has proposed clinical classifications, including the numbers of skin lesions and nerves that are involved, for grouping the leprosy patients into MB and paucibacillary (PB) categories [6]. Patients with up to five lesions are classified as PB, and those with more than five lesions are classified as MB. However, classifications that are based solely on the numbers of lesions impair the proper diagnosis of this disease. Many MB populations with few skin lesions are incorrectly classified as PB; therefore, they are inadequately treated and run the risk of relapse [7].

Regarding immunological diagnostic assays, the presence of antibodies against phenolic glycolipid I (PGL-I) has been extensively studied. While anti-PGL-I serology can detect the majority of MB patients, it has limited value in identifying PB patients. In addition, false positive results in the areas of endemicity are relatively high (>10%) [8,9]. Consequently, it has been recommended that PGL-I-based tests be used in support of the clinical examinations to direct the clinicians towards appropriate treatment and none of these PGL-I-based tests have been widely implemented in field situations [10]. Thus, tools that permit the correct and early diagnosis of *M. leprae* infection in patients who are displaying symptoms are a priority in leprosy research. The search for antigens for immunological diagnoses was initially based on research using total extracts and subcellular fractions of *M. leprae* followed by advances that were achieved using recombinant DNA technology and, more recently, on studies involving comparative genomic analyses and bioinformatics. The main difficulty that is encountered involves obtaining reagents that are more sensitive and specific or that distinguish *M. leprae* exposure from infection. The low specificity of the antigens is a result of cross-reactivity with other mycobacteria, which becomes even more problematic in countries with high incidence rates of tuberculosis and routine *M. bovis* bacillus Calmette-Guerin (BCG) vaccinations [11].

Accordingly, this study proposes the use of the phage-display technique as a tool to identify new reagents that may be effectively used in immunological assays. We have extended our previous observations by evaluating the potentials of peptide mimotopes of *M. leprae* antigens selected by the screening of phage-displayed random peptide libraries as potential serological test reagents for leprosy diagnosis. The mimotopes were tested on leprosy patients, HC, EC, tuberculosis patients and with immune sera that were raised against several mycobacteria in animals. The results indicate that peptide mimotopes may be useful for the diagnosis of leprosy.

Methods

Ethics statement

All of the animal care and experimental procedures and human blood sample collections were performed in accordance with the institutional guidelines. All of the individuals provided written informed consent prior to venipuncture. The animal protocols were conducted in compliance with the Brazilian rules of animals used for experimental purposes. These protocols are based on national and international guidelines, such as those that have been disclosed by the following organizations: the International Guiding Principles for Biomedical Research Involving Animals (CIOMS), the American Association for Laboratory Animal Science (AALAS), and the Brazilian College of Animal Experimentation (COBEA).

Both animal procedure and experiment involving human subjects were approved by the Research Ethics Committee of the Federal University of Parana (UFPR), Curitiba, Brazil (protocol number CEP/SD 428.108.07.10).

Human sera

The leprosy patients and HC were recruited from the Clinical Dermatology Hospital of Parana (Piraquara, Brazil), the Barão Regional Specialties Center (Curitiba, Brazil), the Pro-Hansen Foundation (Curitiba, Brazil), and the Humanitas Philanthropic Society (São Jerônimo da Serra, Brazil). Sera from 10 PB and 23 MB leprosy patients were included, of whom 14 were newly diagnosed and 19 had received up to four months of treatment. In addition, sera were included from 26 HC, 30 patients with pulmonary tuberculosis (TB) who had been treated for up to three

Table 2. ELISA results using peptides in the groups studied.

Group	Number of serum samples tested	Number (%) of positive serum samples by ELISA with peptides
PB	10	3 (30)
MB	23	17 (73.9)
TB	30	16 (53.3)
HC	26	12 (46.2)
EC	30	5 (16.7)

MB: MB leprosy patients. PB: PB leprosy patients. TB: TB patients. HC: healthy household contacts of MB leprosy patients. EC: endemic control.

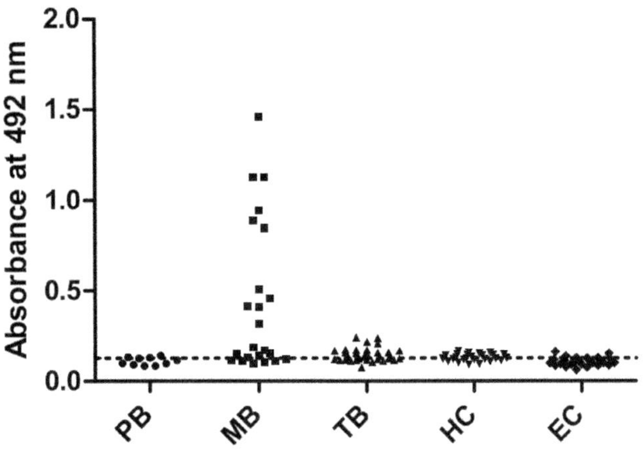

Figure 1. ELISA reactivities of synthetic peptides with sera from leprosy patients and controls. Microtiter plates were coated with 1.5 μg/mL of the peptide pool (5A, 6A, and 1B) and incubated with serum that was diluted 1:50. The detection of the reaction was performed with the anti-human IgG (Fc-specific)-biotin antibody and streptavidin-peroxidase. MB, MB leprosy patients (n = 23); PB, PB leprosy patients (n = 10); TB, TB patients (n = 30); HC, healthy household contacts of MB leprosy patients (n = 26); EC, endemic controls (n = 30). The dotted line represents the cut-off value (as determined using the ROC curve with serum samples from the EC). Each symbol represents the absorbance obtained with a single serum.

months in specialized centers, and 30 healthy EC with no known previous exposure to *M. leprae* or *M. tuberculosis*.

Synthetic phage library derived-peptides evaluated by ELISA assay

Procedures for the biopanning of phage libraries with the anti-*M. leprae* antibodies of MB patients were performed as previously described [12,13]. The peptide sequences in the reactive phage clones were synthesized using a 9-fluorenylmethoxycarbonyl (Fmoc)-based solid-phase synthesis technique.

The ELISA test was optimized with regard to antigen concentration and serum and conjugate dilutions. Microtiter plates (Corning) were coated with 100 μL of the peptide pool at 1.5 μg/mL in 0.05 M carbonate buffer (pH 9.6) overnight at 4°C. After washing with a solution containing 0.9% NaCl and 0.05% Tween 20, the plates were blocked with Protein-Free Blocking Buffer (Thermo Fisher Scientific) for 1 h at 37°C. Then, the plates were washed and incubated for 1 h at 37°C with sera in duplicate dilutions of 1:50 in a phosphate-buffered saline (PBS) solution at pH 7.4, which also contained 0.1% BSA. The plates were washed and incubated with anti-human IgG (Fc-specific)-biotin antibodies (Sigma-Aldrich) that were diluted 1:2 500 in PBS with 0.1% BSA for 1 h at 37°C. The plates were then washed and incubated with streptavidin-peroxidase (Sigma-Aldrich) that was diluted 1:2 500 in PBS with 0.1% BSA for 1 h at 37°C and detected using *o*-phenylenediamine (OPD) dihydrochloride (Sigma-Aldrich). The reactivities of the rabbit anti-mycobacterial sera with the peptides were also analyzed (data not shown). In this analysis, the antibodies that were bound to the peptides were detected by anti-rabbit IgG-peroxidase antibodies that were diluted to 1:4 000.

Immunoassay with cellulose membrane-bound peptides

The peptides and their alanine analogs were prepared by Spot Synthesis [14]. Peptide synthesis was performed using Fmoc-protection chemistry on cellulose membranes that had been derivatized with polyethylene glycol spacers (Intavis Bioanalytical Instruments AG) and MultiPep RS (Intavis Bioanalytical Instruments AG). Following overnight incubation with a blocking solution (0.05% phosphate buffered saline tween-20 (PBST) and 3% BSA), the membrane was probed with the sera of the MB patients at 1:250 dilutions in PBST for 1 h at room temperature. The reaction was developed with the anti-human IgG (Fc-specific)-peroxidase antibody that was diluted to 1:120 000 in PBST for 1 h in addition to ECL Plus (GE Healthcare) and Hyperfilm ECL (GE Healthcare). The spot intensities were determined using the ImageJ software (version 1.43).

Research of native proteins in *M. leprae* using anti-peptide antibodies

The synthetic peptides were conjugated to BSA using glutaraldehyde as a cross-linker [15]. Conjugates (20 μg) that were emulsified in incomplete Freund's adjuvant (Sigma-Aldrich) were administered via subcutaneous (first and second doses) and intraperitoneal (third to fifth doses) injections to eight-week-old Swiss female mice (n = 5 per group). The injections were performed on days 1, 14, 28, 42, and 56. Non-immune serum was used as the negative control, and immune blood was collected one week after the final dose. To assess the production of anti-peptide antibodies, microtiter plates (Corning) were coated with 100 μL of the peptide at final concentrations of 1.5 μg/mL in 0.05 M carbonate buffer (pH 9.6) overnight at 4°C. After washing in a solution containing 0.9% NaCl and 0.05% Tween 20, the plate was blocked (2% casein in PBS, pH 7.4) for 1 h at 37°C. The plate was then washed and incubated for 1 h at 37°C with sera that had been diluted in incubation buffer (0.25% casein in 0.05% PBST). After washing, the detection of the reaction was performed using the anti-mouse IgG (γ-specific)-peroxidase antibody (Sigma-Aldrich) at a 1:2 500 dilution and OPD dihydrochloride as a chromogen.

The *M. leprae* whole cells were provided by Dr. J. S. Spencer from Colorado State University, Fort Collins, USA through the National Institute of Allergy and Infectious Diseases/National Institutes of Health under the contract N01-AI-25469 for the production of soluble antigens for use in electrophoresis. The cells were resuspended in 0.9% NaCl, mixed with 100 μg/mL phenylmethanesulfonyl fluoride (PMSF) and 2 mM ethylenediaminetetraacetic acid (EDTA) and sonicated (Sonopuls HD 2200) for four cycles of 15 min each [16]. The protein concentration of the soluble fraction that was obtained by centrifugation at 10 000 *g* for 20 min at 4°C was determined using the Quant-iT

Table 3. Diagnostic performance of ELISA using peptides for the diagnosis of leprosy.

Serological test	Sensitivity in PB+MB groups % (95% CI)	Specificity in PB+MB groups % (95% CI)	AUC	Youden index J
ELISA-Peptides	60.6 (42.1–77.1)	90.0 (73.5–97.9)	0.779	0.5061

Samples from leprosy patients (n = 33, PB+MB leprosy patients) and endemic control (n = 30) were tested. CI: confidence interval. AUC: area under the ROC curve.

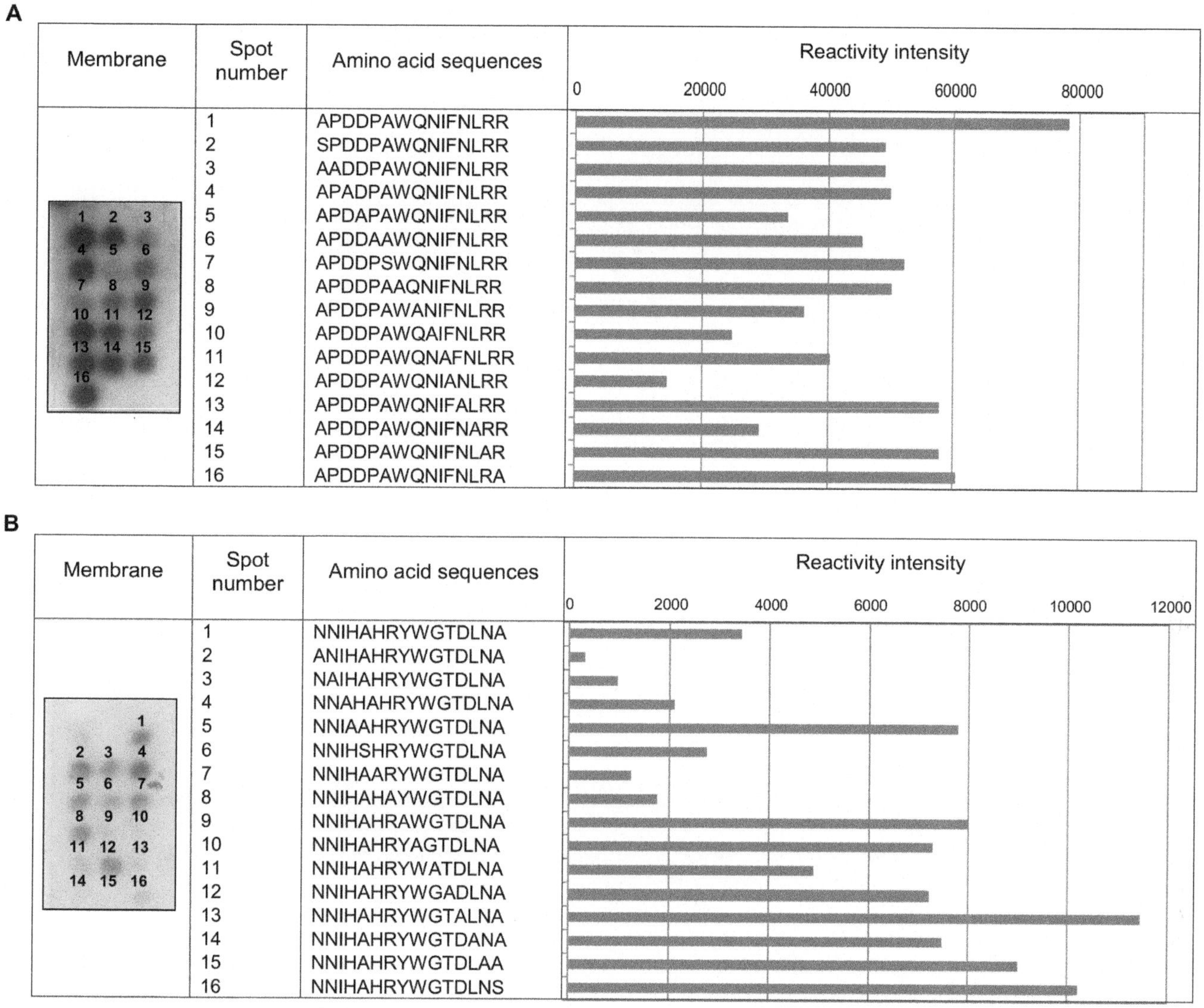

Figure 2. Identification by alanine scanning of key residues in peptides. Reactivities of antibodies from leprosy patients (1:250) with membrane-bound peptides detected using the anti-human (Fc-specific)-peroxidase antibody (1:120000). (A) The reference peptide 5A is in position 1, and the other spots correspond with the alanine analogs of peptide 5A. (B) The reference peptide 1B is in position 1 followed by its alanine analogues. Each bar represents the intensity of the antibody binding to the reference peptide or an analog peptide. When alanine was present in the generation position of the analog, the alanine residue was substituted by serine.

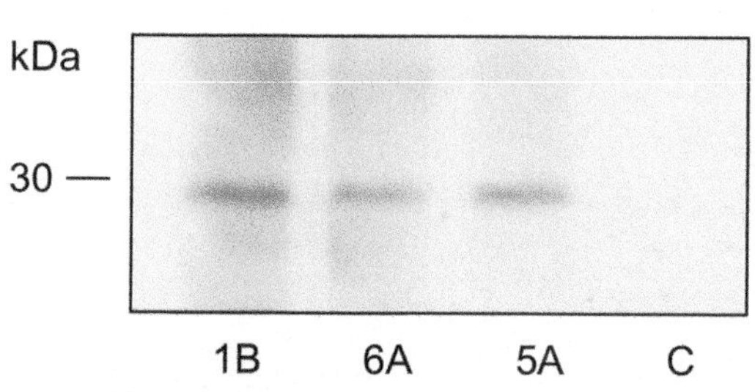

Figure 3. Reactivity of anti-peptide antibodies to *M. leprae* by Western blotting. Forty micrograms of total *M. leprae* extract were separated by 15% SDS-PAGE and, following Western blotting, were reacted with anti-peptide serum 1B (1:12000), anti-peptide serum 6A (1:9000), anti-peptide serum 5A (1:4000), and non-immune serum (C) (1:4000). The detection of the reaction was performed with the anti-mouse IgG (γ-specific)-peroxidase antibody and chemiluminescence.

Protein Assay Kit (Invitrogen). *M. leprae* protein (40 µg) were then electrophoresed on a 15% sodium dodecyl sulfate-polyacrylamide gel electrophoresis (SDS-PAGE) and transferred to a PVDF membrane that was blocked with 0.3% PBST, washed with 0.05% PBST and incubated with anti-peptide serum in 0.05% PBST. The membrane was washed, incubated with anti-mouse IgG (γ-specific)-peroxidase antibodies that were diluted to 1:5 000 in 0.05% PBST, and processed using ECL Plus.

For the identification of the proteins in *M. leprae*, 2-dimensional gel electrophoresis (2-DE) followed by Western blotting were performed. Two hundred µg of *M. leprae* protein were purified using a 2D Clean-Up Kit (GE Healthcare) and solubilized in rehydration buffer (8 M urea, 2% CHAPS, 20 mM DTT, 0.5% IPG buffer, and 0.002% bromophenol blue). Samples were in gel rehydrated and run on linear 13 cm IPG strips at pH levels ranging from 4–7 (GE Healthcare) with the Ettan IPGphor II (GE Healthcare) for 21400 Vh. The focused strips were equilibrated in a solution that contained 6 M urea, 50 mM Tris (pH 8.8), 30%

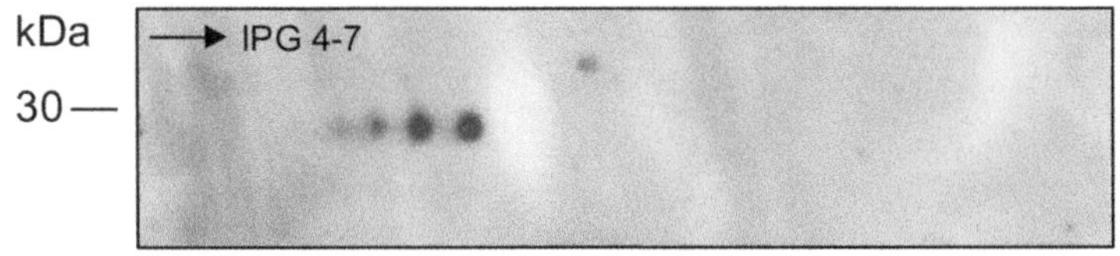

Figure 4. Western blotting analysis of total *M. leprae* extract 2D gel using mouse anti-peptide antibodies. Two hundred micrograms of protein were separated in the first dimension by isoelectric focusing on a 13 cm IPG strip at pH 4-7 and in the second dimension by SDS-PAGE on a 15% gel. Western blotting with anti-peptide serum 1B at 1:4000 is shown. The other two anti-peptide sera showed the same reactivity profiles as that of anti-peptide serum 1B (data not shown). The detection of the reaction was performed with the anti-mouse IgG (γ-specific)-peroxidase antibody and chemiluminescence.

glycerol, 2% SDS, 0.002% bromophenol blue and 0.065 M DTT for 15 min, followed by incubation in the same solution that was prepared with 0.14 M iodoacetamide instead of DTT for an additional 15 min. The second-dimensional separation was performed using 15% SDS-PAGE by the SE 600 Ruby (GE Healthcare) electrophoresis unit. Western blotting was performed as previously described.

Statistical analyses

A cut-off point for optimal sensitivity and specificity for the ELISA tests was determined using the Receiver Operating Characteristic (ROC) curve analysis [17,18]. The accuracy of each test was evaluated according to the area under the curve (AUC) and the Youden index J [19]. The statistical analyses were performed using the MedCalc statistical software (version 13.2.0).

The differences in seroreactivities between the groups were analyzed with the two-tailed Mann-Whitney U test for nonparametric distribution using the GraphPad Prism software (version 5.03). The *P* values were corrected for multiple comparisons. The *P* value indicating statistical significance was set to <0.05.

Results

Peptide sequences

Table 1 shows the peptide sequences that were selected by phage display using the affinity purified anti-*M. leprae* MB patients' antibodies as previously reported [12]. The consensus sequence DPAW was observed between peptides 5A and 6A.

ELISA using synthetic peptides

The peptides 5A, 6A, and 1B were chemically synthesized and evaluated by ELISA against the sera from the leprosy patients. The peptides were capable of detecting 30% (3/10) of the PB and 73.9% (17/23) of the MB patients (Table 2). Compared with the PB leprosy patients, MB patients, in general, showed strong positive responses. Of the 30 serum samples from the tuberculosis patients, 16 (53.3%) were reactive to the peptides (Table 2). Weak positive responses were observed in 12 out of 26 of the HC sera and in 5 out of the 30 EC sera that were tested in this experiment (Table 2 and Fig. 1). These findings indicate the existence of significant differences between the leprosy (PB and MB groups) and EC groups and the TB and EC groups. No significant differences were observed between HC and EC groups. The sensitivity, specificity, AUC and Youden index J values were calculated for ELISA assay (Table 3). The peptides were not reactive against the rabbit antimycobacterial sera using the ELISA (data not shown).

Identification of key residues for antibody binding in peptides

For peptide 5A, the proline, aspartate and alanine residues were important for antibody recognition. The substitution of these amino acids by alanine or serine reduced the reactivity of the peptide compared with that of the unmodified peptide. No reactivity was observed for peptide 6A. For peptide 1B, the residues of greatest importance to the antibody interactions were tyrosine, tryptophan, glycine, aspartate, leucine and asparagine (Fig. 2).

Identification of natural antigens in *M. leprae*

The peptides that were conjugated to BSA induced mouse anti-peptide antibody production and these antibodies reacted with antigens of *M. leprae* after electrophoresis, which was followed by Western blotting (Fig. 3). The Western blotting of the *M. leprae* proteins that was resolved by 2 DE and probed with the anti-peptide antibody revealed four spots of approximately 30 kDa in size (Fig. 4). The three anti-peptide antisera showed the same reactivity profiles with *M. leprae* (data not shown). The estimated isoelectric points (pI) for the four spots were 4.56, 4.64, 4.75 and 4.9.

In silico analyses of peptides

The three peptide sequences that were identified demonstrated significant alignments with the antigen 85B (fbpB) of *M. leprae* using BLASTp. Antigen 85B is a member of the antigen 85 complex. It has a molecular mass of 31 kDa and a theoretical isoelectric point of 4.9.

Discussion

In this study we evaluated the serological responses to peptides that were selected from random peptide phage-display libraries in an attempt to identify antigens that could be useful for the serological diagnoses of leprosy patients. The MB leprosy patients were easier to identify and the ELISA showed positive results in 73.9% of the samples that were tested. For the PB leprosy patients, 30% of the serum samples were reactive against the peptides as shown by the peptide-based ELISA and weaker responses were observed in this group compared with the MB group.

Protein antigens require binding to human leukocyte antigen (HLA) molecules to be presented by T cells. Because responses to mycobacterial antigens are HLA-restricted, not all individuals respond to the same antigens [20]. Reasons other than HLA restriction exist that explain the seronegativities between the patients that have been diagnosed with MB leprosy by the peptide-based ELISA. Recent research suggests that for tuberculosis, heterogeneous antigen recognition may result from the production of different antigens at different stages of the disease [21]. Another reason is related to differences in the strains of *M. leprae* [8]. However, research has shown that different strains have highly conserved genomes, which indicates that antigenic variations are likely to be negligible [22]. Because of the differences in the immune responses that occur among individuals, it has been suggested that immunological tests should be based on a combination of several different antigens to permit greater coverage in individuals that have been infected with *M. leprae* [3,23].

The detection of *M. leprae* infection in PB patients through serological assays is difficult because once the immune response develops in these individuals, it is predominantly cellular. Analyses of T cell responses as interferon-gamma (IFN-γ) release assays are required for the diagnosis of PB [9]. Thus, the immunological diagnosis of leprosy may be based on both antibody and T cell assays [24]. The combination of these assays would permit the correct classification of patients and consequently, the initiation of effective treatments that may improve the prognoses for each case.

Seroreactivities against peptides were observed in the tuberculosis and HC and EC individuals. The recognition of peptides by tuberculosis patient sera was expected because *M. leprae* shares antigens with *M. tuberculosis*. Moreover, the diagnosis of leprosy or tuberculosis does not rely on a single test but must be confirmed by a combination of several tests and further associated with a clinical examination.

The capacity of the peptide-based ELISA test to predict disease development in the HC patients was not possible in this work because it requires that they be monitored over a prolonged period of time. Seroreactivity with some endemic control samples suggests that peptides not are capable of discriminate infection of exposure to *M. leprae*. Further investigation needs to be conducted testing the peptides with non-endemic controls to clarify this issue.

The understanding of the molecular basis of peptide recognition by patient antibodies may assist in the design of novel peptides with improvements in specificity and sensitivity for diagnostic purposes. The contribution of each amino acid residue in the peptide sequence to antibody recognition was deduced through alanine scanning. In this assay, for peptide 5A, the residues that were critical for antibody recognition were proline, alanine, and aspartate. Peptide 6A did not indicate reactivity. One potential explanation for these findings is that the conformational flexibility of the peptides on the membranes may be restricted, thus reducing the affinities of the peptides for the antibody [25]. Peptide 1B showed low levels of reactivity. However, sufficient levels were present for the identification of important residues that interacted with the antibody. These results indicate that the residues of the C-terminal region were responsible for binding to the antibody. Because this portion of the sequence is immobilized in the membrane, the accessibility of the antibody to the residues is more difficult and most likely explains the low reactivities of the peptides that were derived from 1B in the membrane assay.

The peptides that were evaluated in this study were shown to be antigenic and immunogenic. This latter property allowed for the characterization of these peptides as mimotopes of *M. leprae*.

The *M. leprae* database analysis suggests that the epitope protein that is mimicked by the peptides is antigen 85B, which is a 31 kDa protein with a pI of 4.9 that possesses properties that are consistent with the information that was experimentally obtained (Fig. 4).

Antigen 85B is a component of the fibronectin-binding protein complex that presents with very interesting antigenic and immunogenic properties because of its capacity to induce strong humoral and cellular responses, including the *in vitro* proliferation of T cells [26]. Its immunological features and predominant role as a *M. leprae* antigen explain why only mimotopes of this antigen have been bioselected. These epitope-mimicking peptides contain amino acid sequences that are part of two sequences (AA 206–230 and 291–325) from three regions of antigen 85B, which have been recognized as important antigenic domains for the diagnosis of patients with lepromatous leprosy [27].

The search for antigens may not only improve the sensitivities and specificities of the diagnostic tests but also contribute to a better understanding of the immune response that is developed against the antigens in affected individuals. The results of these observations may be used to improve antigenic recognition and identify novel peptides. In addition, the peptides are very feasible for widespread use in serodiagnostic tests because they can be easily prepared in sufficient purities and large quantities [27].

This work demonstrates the usefulness of this biotechnological tool in overcoming diagnostic constraints by selecting, synthesizing, and confirming certain peptide sequences *in vitro* for their uses as diagnostic laboratory tests. The peptides that were identified were able to recognize antibodies that are primarily found in MB leprosy patients. MB cases are the principal sources of leprosy infection in the population, and affected individuals are at greater risks of developing disabilities compared with those with PB leprosy [23]. Therefore, the early detection and treatment of MB cases will contribute to a reduction in the transmission of the bacterium. Ideally, a serological test should identify all leprosy patients (MB and PB). In fact, it may be possible to diagnose all of the patients by combining peptides and assays that are based on humoral and cellular immunity. Hence, it is important to identify new peptides using the approach that is described in this work while also evaluating different diagnostic approaches using these peptides to increase the specificity and sensitivity of the detection of MB and particularly PB leprosy patients.

Acknowledgments

The authors would like to thank the representatives and members of the Clinical Dermatology Hospital of Parana, the Barão Regional Specialties Center, the Humanitas Philanthropic Society, São Sebastião da Lapa Regional Hospital, and the Pro-Hansen Foundation for allowing us access to the premises to collect patient samples. The authors are also grateful to the Center for the Production and Research of Immunobiological Products for their collaboration.

Author Contributions

Conceived and designed the experiments: SMA JFM VTS. Performed the experiments: SMA SBS JCM. Analyzed the data: SMA JFM VTS. Contributed reagents/materials/analysis tools: SBS MTM CCO JFM VTS. Contributed to the writing of the manuscript: SMA VTS JFM SBS LMA MTM.

References

1. Britton WJ, Lockwood DNJ (2004) Leprosy. Lancet 363: 1209–1219.
2. World Health Organization (2012) Global leprosy situation, 2012. Available: http://www.who.int/wer/2012/wer8734/en/index.html. Accessed 21 September 2012.
3. Geluk A, van der Ploeg J, Teles ROB, Franken KLMC, Prins C, et al. (2008) Rational combination of peptides derived from different *Mycobacterium leprae* proteins improves sensitivity for immunodiagnosis of *M. leprae* infection. Clin Vaccine Immunol 15: 522–533.
4. Alban SM, Sella SR, Miranda RN, Mira MT, Thomaz-Soccol V (2009) PCR-restriction fragment length polymorphism analysis as a tool for *Mycobacterium* species identification in lepromas for lepromin production. Lepr Rev 80: 129–142.

5. Lockwood DNJ (2005) Leprosy. Medicine 33: 26–29.
6. World Health Organization (1997) WHO Seventh Expert Committee Report, June 1997. Available: http://www.who.int/lep/resources/Expert.pdf. Accessed 1 July 2010.
7. Bührer-Sékula S, Visschedijk J, Grossi MA, Dhakal KP, Namadi AU, et al. (2007) The ML flow test as a point of care test for leprosy control programmes: potential effects on classification of leprosy patients. Lepr Rev 78: 70–79.
8. Duthie MS, Goto W, Ireton GC, Reece ST, Cardoso LPV, et al. (2007) Use of protein antigens for early serological diagnosis of leprosy. Clin Vaccine Immunol 14: 1400–1408.
9. Stefani MM (2008) Challenges in the post genomic era for the development of tests for leprosy diagnosis. Rev Soc Bras Med Trop (Suppl 2): 89–94.
10. Duthie MS, Ireton GC, Kanaujia GV, Goto W, Liang H, et al. (2008) Selection of antigens and development of prototype tests for point-of-care leprosy diagnosis. Clin Vaccine Immunol 15: 1590–1597.
11. Geluk A, Spencer JS, Bobosha K, Pessolani MC, Pereira GM, et al. (2009) From genome-based in silico predictions to ex vivo verification of leprosy diagnosis. Clin Vaccine Immunol 16: 352–359.
12. Alban SM, de Moura JF, Minozzo JC, Mira MT, Thomaz Soccol V (2013) Identification of mimotopes of *Mycobacterium leprae* as potential diagnostic reagents. BMC Infect Dis 13: 42.
13. Thomaz Soccol V, Alban SM, de Moura JF (2011) Uso de peptídeos miméticos de *Mycobacterium leprae* para diagnóstico e vacinas. Brazilian Patent PI0151110000997.
14. Frank R (1992) Spot-synthesis: an easy technique for the positionally addressable, parallel chemical synthesis on a membrane support. Tetrahedron 48: 9217–9232.
15. Harlow E, Lane D (1988)Antibodies: A laboratory manual. Cold Spring Harbor Laboratory Press. 726 p.
16. Pessolani MC, Brennan PJ (1992) *Mycobacterium leprae* produces extracellular homologs of the antigen 85 complex. Infect Immun 60: 4452–4459.
17. Metz CE (1978) Basic principles of ROC analysis. Semin Nucl Med 8: 283–289.
18. Zweig MH, Campbell G (1993) Receiver-operating characteristic (ROC) plots: a fundamental evaluation tool in clinical medicine. Clin Chem 39: 561–577.
19. Youden WJ (1950) Index for rating diagnostic tests. Cancer 3: 32–35.
20. Meyer CG, May J, Stark K (1998) Human leukocyte antigens in tuberculosis and leprosy. Trends Microbiol 6: 148–154.
21. Lyashchenko K, Colangeli R, Houde M, Al Jahdali H, Menzies D, et al. (1998) Heterogeneous antibody responses in tuberculosis. Infect Immun 66: 3936–3940.
22. Monot M, Honoré N, Garnier T, Zidane N, Sherafi D, et al. (2009) Comparative genomic and phylogeographic analysis of *Mycobacterium leprae*. Nat Genet 41: 1282–1289.
23. Aráoz R, Honoré N, Banu S, Demangel C, Cissoko Y, et al. (2006) Towards an immunodiagnostic test for leprosy. Microbes Infect 8: 2270–2276.
24. Aseffa A, Brennan P, Dockrell H, Gillis T, Hussain R, et al. (2005) Report on the first meeting of the IDEAL (Initiative for Diagnostic and Epidemiological Assays for Leprosy) consortium held at Armauer Hansen Research Institute, ALERT, Addis Ababa, Ethiopia on 24–27 October 2004. Lepr Rev 76: 147–159.
25. Van Regenmortel MH (2001) Antigenicity and immunogenicity of synthetic peptides. Biologicals 29: 209–213.
26. Thole JE, Schöningh R, Janson AA, Garbe T, Cornelisse YE, et al. (1992) Molecular and immunological analysis of a fibronectin-binding protein antigen secreted by *Mycobacterium leprae*. Mol Microbiol 6: 153–163.
27. Filley E, Thole JE, Rook GA, Nagai S, Waters M, et al. (1994) Identification of an antigenic domain on *Mycobacterium leprae* protein antigen 85B, which is specifically recognized by antibodies from patients with leprosy. J Infect Dis 169: 162–169.

13

PKA Compartmentalization via AKAP220 and AKAP12 Contributes to Endothelial Barrier Regulation

Mariya Y. Radeva, Daniela Kugelmann, Volker Spindler, Jens Waschke*

Institute of Anatomy and Cell Biology, Ludwig-Maximilians-University Munich, Munich, Germany

Abstract

cAMP-mediated PKA signaling is the main known pathway involved in maintenance of the endothelial barrier. Tight regulation of PKA function can be achieved by discrete compartmentalization of the enzyme via physical interaction with A-kinase anchoring proteins (AKAPs). Here, we investigated the role of AKAPs 220 and 12 in endothelial barrier regulation. Analysis of human and mouse microvascular endothelial cells as well as isolated rat mesenteric microvessels was performed using TAT-Ahx-AKAPis peptide, designed to competitively inhibit PKA-AKAP interaction. *In vivo* microvessel hydraulic conductivity and *in vitro* transendothelial electrical resistance measurements showed that this peptide destabilized endothelial barrier properties, and dampened the cAMP-mediated endothelial barrier stabilization induced by forskolin and rolipram. Immunofluorescence analysis revealed that TAT-Ahx-AKAPis led to both adherens junctions and actin cytoskeleton reorganization. Those effects were paralleled by redistribution of PKA and Rac1 from endothelial junctions and by Rac1 inactivation. Similarly, membrane localization of AKAP220 was also reduced. In addition, depletion of either AKAP12 or AKAP220 significantly impaired endothelial barrier function and AKAP12 was also shown to interfere with cAMP-mediated barrier enhancement. Furthermore, immunoprecipitation analysis demonstrated that AKAP220 interacts not only with PKA but also with VE-cadherin and ß-catenin. Taken together, these results indicate that AKAP-mediated PKA subcellular compartmentalization is involved in endothelial barrier regulation. More specifically, AKAP220 and AKAP12 contribute to endothelial barrier function and AKAP12 is required for cAMP-mediated barrier stabilization.

Editor: Yulia Komarova, University of Illinois at Chicago, United States of America

Funding: This work was supported by a grant from the Deutsche Forschungsgemeinschaft (SFB 688, TP A4) (http://www.sfb688.de/). JW received the funding. The funders had no role in study design, data collection and analysis, decision to publish, or preparation of the manuscript.

Competing Interests: The authors have declared that no competing interests exist.

* Email: jens.waschke@med.uni-muenchen.de

Introduction

The vascular endothelium lining the intima of blood vessels precisely regulates the passage of solutes, macromolecules, and leukocytes between the blood and the underlying tissue. Under inflammatory conditions, mainly in post-capillary venules, loss of this primary function leads to formation of intercellular gaps and increased vascular permeability. The latter is a hallmark of several pathological processes (i.e. inflammation, allergy, arteriosclerosis, edema, tumor growth and sepsis) and contributes to multi-organ failure and death [1,2]. Therefore, understanding of the mechanisms maintaining endothelial barrier functions under resting conditions, as well as the signaling pathways leading to barrier impairment or recovery are of great biological and clinical importance.

Paracellular permeability is tightly regulated by coordinate opening and closing of mainly two types of endothelial cell-cell junctions, namely tight junctions (TJs) and adherens junctions (AJs). While TJs seal the intercellular cleft between cells, the AJs are providing mechanical strength. However, the junctional composition of intracellular clefts varies across the vascular tree [3]. Both junctional types are composed of transmembrane proteins, i.e. the tight junction protein claudin-5 and the adherens junction protein VE-cadherin. These junctional markers are associated with the cortical actin cytoskeleton via several adaptor molecules such as zonula occludens (ZO) proteins and catenins, respectively [4]. Numerous studies showed that modulation of endothelial barrier functions via actin cytoskeleton remodeling and cell junction integrity can be controlled by members of the Rho family of small GTPases, i.e. RhoA, Rac1 and Cdc42 as well as by the Ras family GTPase Rap1. Although it is suggested that fine balance between activation and/or inactivation of these small GTPases is required for barrier maintenance, it is generally assumed that activation of RhoA impairs barrier function, while Rac1 and Cdc42 are considered to primarily stabilize barrier integrity [1,5,6,7]. It is now widely recognized that a number of barrier-stabilizating mediators activate Rac1 either directly or indirectly via an increase in the concentration of the cellular second messenger cAMP. cAMP- dependent Rac1 activation can be achieved by both, exchange protein activated by cAMP1 (Epac1)/Ras-related protein 1 (Rap1), and cAMP-dependent protein kinase A (PKA) signaling pathways. The latter is generally believed to be the predominant cAMP mechanism that exerts significant protection against the increase in endothelial paracellular permeability [8]. Furthermore, it is assumed that precise spatiotemporally regulated activation is essential for the response specificity of the PKA pathways. Thus, it was found that a key role in tight regulation and compartmentalization of PKA-dependent signaling is played by A kinase-anchoring proteins (AKAPs) [9]. AKAPs are a large structurally diverse family of functionally related proteins that contain a conserved amphipathic helix PKA binding motif and function to localize PKA-AKAP complexes at discrete compartments within the cell such as plasma membrane, endoplasmic reticulum, mitochondria or Golgi complex [10]. By anchoring the inactive PKA to defined cellular sites, AKAPs allow specific placement of the holoenzyme at regions of cAMP production and thus to propagate confined phosphorylation of only a subset of potential substrates located in close proximity. AKAPs are also scaffolding proteins tethering not only PKA, but also other molecules involved in cAMP signaling such as adenylyl cyclases (ACs), phosphodiesterases (PDEs), Epac1, which is guanine nucleotide exchange factor (GEF) of Rap1 and protein phosphatases [11,12,13,14,15,16]. Thus, AKAP complexes assemble PKA with a determined set of signal transduction (i.e. kinases) and termination (i.e. phosphatases, phosphodiesterases) molecules as well as with a variety of other members of different signaling pathways. Therefore, AKAPs organize crosstalk across diverse paths in the cell's signaling networks [9,17].

Although the protective effects of cAMP/PKA signaling for endothelial barrier regulation are well recognized, it is not yet clear by which mechanisms PKA is located close to cell junctions. Based on our previous investigations, we speculated that compartmentalized cAMP-signaling by AKAPs contribute to endothelial barrier integrity. Thus, we investigated the importance of AKAP function for maintenance of the cAMP/PKA-dependent endothelial barrier *in vivo* and *in vitro*.

In order to modulate AKAP function, we used a modified analog of a cell-permeable synthetic peptide designed to competitively inhibit PKA-AKAP interaction [18]. This peptide, named TAT-Ahx-AKAPis, is comprised of two functional peptides, TAT and AKAPis, connected via an aminohexanoic (Ahx) linker. AKAPis is a precisely designed sequence with high-affinity binding and specificity for the PKA regulatory subunit (RII) which enables a higher dissociation effect on the PKA-AKAP anchoring than the widely used Ht31 synthetic peptides [18]. The second functional unit, commonly denoted as TAT, is a cell-penetrating peptide derived from the TAT protein of human immunodeficiency virus (HIV-1). The TAT peptide possesses a high ability to mediate the import of membrane-impermeable molecules such as DNA, RNA, peptides and even whole proteins into the cell [19].

Although approximately 50 AKAPs have been identified in different cell types, little is known about the AKAP expression profile and function in endothelial cells. In the current investigation, besides AKAP12, which has already been found in endothelium and its involvement in regulation of endothelial integrity has been reported [20,21,22,23], we focused on AKAP220. The latter was recently shown to contribute to the integrity of the cortical actin cytoskeleton [24], but was also suggested to link cAMP signaling to cell adhesion [25]. Both AKAP220 and AKAP12 are expressed in endothelial cells according to microarray data published in GeneCards database (http://www.genecards.org/).

In this study, by using *in vivo* and *in vitro* techniques, we provide evidence that AKAP-mediated PKA subcellular compartmentalization contributes to endothelial barrier integrity. Our data furthermore suggest AKAP220 and AKAP12 to be involved in these processes.

Materials and Methods

Cell culture

Human Dermal Microvascular Endothelial Cells (HDMECs) were obtained from PromoCell (Heidelberg, Germany). The cells were grown in Endothelial Cell Growth Medium MV containing supplement mix provided by the same company. Passage of the cells was performed by using Detach kit (PromoCell). All experimental procedures were carried out with HDMEC from passage 2 to 5 [26]. Additionally, an immortalized microvascular endothelial cells (MyEnd), previously isolated from murine myocardial tissue were used for the experiments. Cell generation and characterization have been described elsewhere [27,28]. Cells were cultured in DMEM (Life Technologies, Karlsruhe, Germany), supplemented with Penicillin G/Streptomycin (50 U/ml) (Sigma-Aldrich, Munich, Germany) and 10% Fetal Calf Serum (FCS) (Biochrom, Berlin, Germany).

Both cell types were cultured at 37°C in a humidified atmosphere of 5% CO_2.

Antibodies and test reagents

The polyclonal rabbit anti-PKA RII alpha (sc-908) and the goat anti-VE-cadherin (sc-6458) antibodies were purchased from Santa Cruz Biotechnology (Santa Cruz, U.S.A.). Detection of VE-cadherin in MyEnd cells was performed by using rat anti-VE-cadherin mAb (clone 11D4.1; undiluted hybridoma). The mouse monoclonal anti-PKA RII beta (Cat.# 610625) and anti-ß-catenin (Cat.# 610154) antibodies were obtained from BD Biosciences (Franklin Lakes, U.S.A.). The mouse monoclonal anti-AKAP12 antibody (ab49849) was acquired from Abcam (United Kingdom). The rabbit polyclonal anti-AKAP220 antibody was kindly provided by John Scott (Howard Hughes Medical Institute, Department of Pharmacology, University of Washington School of Medicine).

To increase cAMP levels, Forskolin (F), (Cat.# F6886) and Rolipram (R), (Cat.#R6520) purchased from Sigma-Aldrich (St. Louis, U.S.A.) were used in combination for 1 hour at concentrations of 5 and 10 µM, respectively. In addition, cell-permeable synthetic peptide TAT-Ahx-AKAPis (GRKKRRQRRRXQIEYLAKQIVDDNAI) was utilized to competitively inhibit the interaction between the PKA regulatory subunit II (RII) and AKAPs. By using the ECIS system, preliminary concentration- effect experiments determined the effectiveness of the peptide on endothelial barrier stability (data not shown). The analysis revealed that 30 µM inhibitory peptide, dissolved in sterile distilled water with 10% DMSO, is the most effective peptide concentration for modification of endothelial barrier integrity. In parallel, experiments were conducted with TAT-Ahx-mhK77 scrambled synthetic peptide (GRKKRRQRRRXQFSSQSAFSSRSRRARS). The latter is similar to the inhibitory peptide regarding molecular weight, isoelectric point and amino acid composition. Both peptides were synthesized by Peptide Specialty Laboratories GmbH (Heidelberg, Germany). Simultaneously, a control condition (vehicle) was run. This internal control is composed of medium containing DMSO in a concentration corresponding to the one used for dissolving the peptides.

Rac1 activation assay

To quantify the levels of Rac1 activity in cells, a Rac G-Lisa Activation Assay Biochem Kit (Cytoskeleton, Denver, CO; Cat.#BK126) was used. Briefly, confluent monolayers of HDMEC and/or MyEnd cells were incubated either in the absence or the presence of TAT-Ahx-AKAPis. Parallel administration of scrambled TAT-Ahx-mhK77 peptide was carried out. Besides F/R application, vehicle was applied as an additional control. Cells were lysed and the lysates were processed according to the manufacturer's instructions. The absorption was measured at 490 nm using a TECAN, Infinite 200 PRO microplate reader (Tecan, Austria).

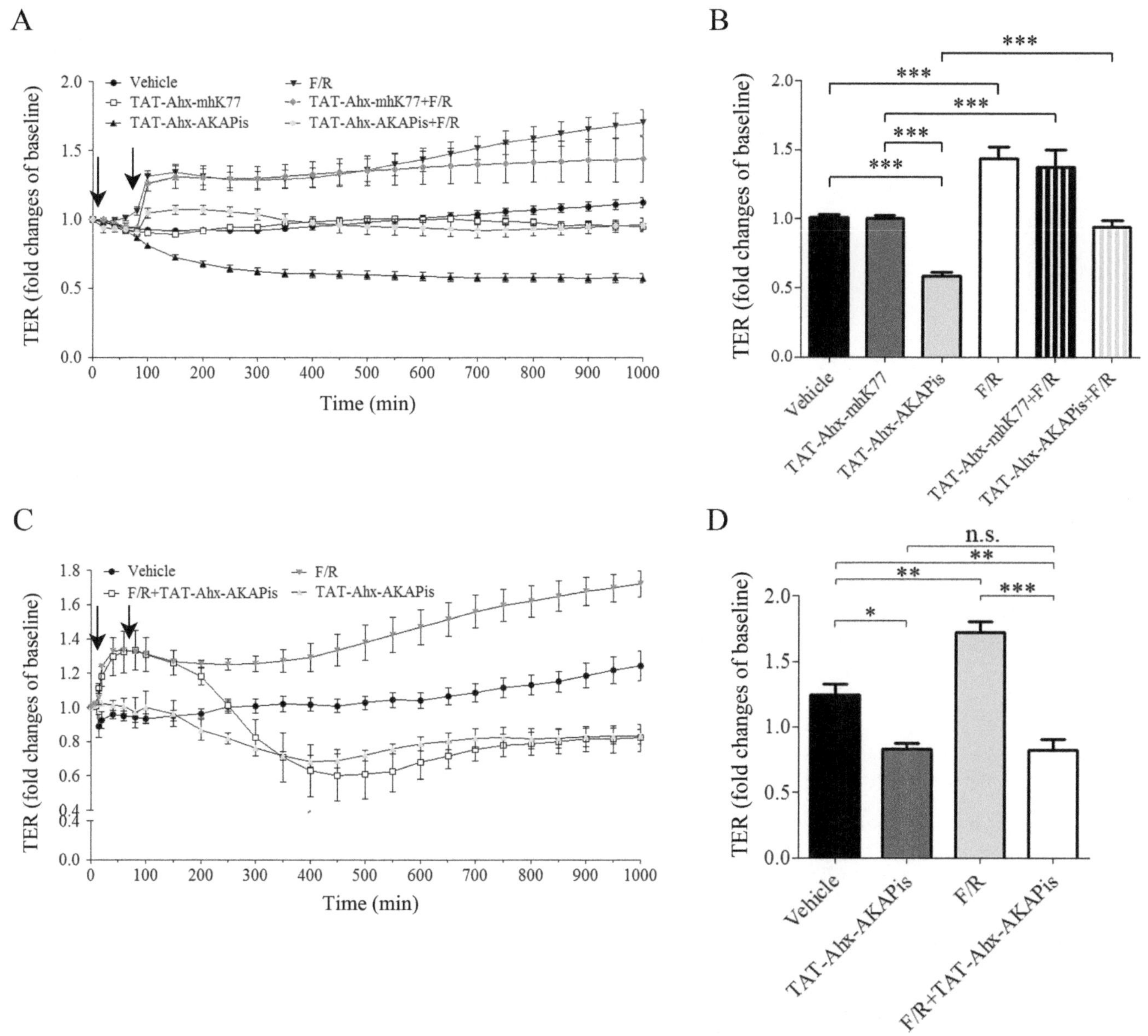

Figure 1. TAT-Ahx-AKAPis significantly reduced microvascular endothelial barrier function and reverted F/R- mediated barrier stabilization. TER measurements were carried out to monitor endothelial barrier alterations in response to different mediators and synthetic peptides. (A) displays the time course of TER measurements under various experimental conditions for HDMEC. The first arrow indicated the time point of TAT-Ahx-AKAPis and/or TAT-Ahx-mhK77 addition. 1 hour after the first application, F/R was added (second arrow). (B) summarizes the results after 600 min, the time point at which the monitored effects reached their peaks. TAT-Ahx-AKAPis significantly decreased TER compared to control/ vehicle condition and treatment with scrambled peptide (TAT-Ahx-mhK77) starting at 80 min after application for HDMEC. F/R addition resulted in pronounced and continuous increase of TER after 1 hour. A similar effect was detected after pre-incubation with TAT-Ahx-mhK77 scrambled peptide. In contrast, 1 hour pre-treatment with TAT-Ahx-AKAPis initially reduced, but subsequently did not abolished the effect of F/R. (C) To further test the effect of TAT-Ahx-AKAPis on F/R- mediated enhancement of endothelial barrier function, HDMEC cells were exposed to F/R for 1 hour (first arrow) and post- incubated with TAT-Ahx-AKAPis inhibitory peptide (second arrow). (D) graphically represents the statistical outcome of the data presented in panel (C). Based on ANOVA multiple analysis, the most significant peaks of the monitored effects were determined at 1000 min. At that time point (1000 min), the similarly responding TAT-Ahx-AKAPis- and F/R+TAT-Ahx-AKAPis- cell monolayers displayed TER significantly lower than the one in control monolayers. In contrast, F/R- mediated enhancement in TER remained constant over time. Data were collected from more than three independent experiments ($N \geq 3$, $n \geq 4$–10). *** $p \leq 0.001$, ** $p \leq 0.01$, * $p \leq 0.05$, indicate statistically significant difference between examined groups. n.s. – not significant.

Measurement of transendothelial resistance (TER)

An ECIS Z Theta system (Applied Biophysics Inc, Troy, U.S.A.) was used to assess the endothelial barrier integrity of confluent and subconfluent cell monolayers as previously described [29]. In short, the cells were grown to confluency on gold microelectrodes 8W10E+ (Ibidi, Germany, Cat.# 70040). MyEnd were seeded on gelatin-coated gold electrodes, HDMEC were grown on uncoated arrays. HDMEC cells reached confluency in between 8 to 10 days, while MyEnd formed a confluent monolayer within 3 to 4 days. Directly before the experiment, the medium was exchanged and the arrays were mounted onto the holders of the ECIS system, placed in an incubator (37°C, 5% CO_2). For both cell types, the optimal frequency to analyze the changes in the transendothelial resistance was identified as 4000 Hz. After short equilibration for approximately 15 to 20 min, the baseline resistance was recorded for another 15 min, followed by application of the test reagent to each well yielding a final well volume of 400 µl.

For some experiments, the effect of AKAPs on endothelial barrier formation was tested by TER measurements initiated in subconfluent cell monolayers transiently transfected with specific siRNA. Cell confluency was determined by a light microscopy, subsequently the basal transendothelial electrical resistance was measured by ECIS system. Subconfluent MyEnd cells had a basal resistance of 550 to 800 ohm. The resistance monitored in confluent cell monolayers was above 800 ohm and reached values of approximately 1400 ohm.

Immunoprecipitation (IP) and Western blot (WB) analysis

MyEnd cells were grown to confluence. After quick washing with ice-cold PBS, the cells were incubated for 10 min with pre-chilled precipitation buffer (0.5% Nonidet-40, 10 mM HEPES (pH 7.9), 1.5 mM $MgCl_2$, 10 mM KCl, 5 mM EDTA), to which cOmplete protease inhibitor cocktail (Cat.# 11697498001, Roche) was freshly added. Cells were scraped and the homogenized suspension was clarified by centrifugation at 6000 g for 5 min, at 4°C. The supernatant was transferred to a fresh reaction vial and the protein concentration was estimated using the Bradford assay [30]. For immunoprecipitation with either rabbit anti-PKA or goat anti-VE-cadherin antibodies, cell lysates with 400–600 µg of total protein amount were pre-cleaned (for 1 hour at 4°C) by incubation with Protein G Plus-Agarose (Calbiochem Merck, Darmstadt, Germany Cat.#IP08). The resulting supernatant was incubated for 3 hours at 4°C with 1, 5 µg of defined antibody. To each sample, 25 µl of beads were added and overnight incubation at 4°C on a rocker platform was carried out. The beads with attached immunoprecipitated complexes were collected by centrifugation at 6000 g for 2 min, washed 3 times with cold washing buffer (0,05% Nonidet-40, 10 mM HEPES (pH 7.9), 1.5 mM $MgCl_2$, 10 mM KCl, 5 mM EDTA) and resuspended in Laemmli buffer. Boiling at 95°C for 5 min serves to denature proteins and detach them from the beads. The immunoprecipitated complexes were separated on a SDS-PAGE gel, transferred to a nitrocellulose membrane and analyzed by Western blot (WB) assay [29]. The AKAP220 antibody was used in 1:500 dilution, detection of AKAP12 was done with antibody diluted 1:5000. All other antibodies were used in 1:1000 dilution. Secondary antibodies were diluted 1:3000.

ImageJ software was used to quantitatively assess the Western blot data. Same size rectangular areas were drawn around each band of interest and the signal intensity within the area was measured. Similarly, the fluorescence intensity determined at identically sized rectangles surrounding regions under or above the bands served as background. For the immunoprecipitation experiments, subtraction of the intensity determined within the control IP (IP: Beads, no Ab) was performed as well. Then the signal intensity was normalized either to the fluorescence intensity of housekeeping proteins (loading control) or to the intensity of the protein used to pull down desired protein complex. In order to combine all experiments, the data were presented as percent of control.

Immunofluorescence (IF) analysis

Detailed description of the assay was previously published [29]. In short, HDMEC were cultured to confluency on uncoated glass cover slips. Control cells or cells treated with different test reagents as specified below were washed with PBS, fixed with 2% formaldehyde for 10 min at room temperature (RT) and, after washing with PBS, were permeabilized with 0,1% Triton X-100 in PBS. The adherent cell monolayer was blocked for an hour with 10% normal goat or donkey serum in 1% BSA/PBS followed by overnight incubation at 4°C with primary antibodies. After washing with PBS, the respective secondary antibodies conjugated with Cy3 (1 hour, RT) were applied. In addition, for visualization of filamentous actin, 1:100 diluted ALEXA488-conjugated–phalloidin was added to the secondary antibody. Images were obtained by using a confocal microscopy setup (Leica SP5, Mannheim and Wetzlar, Germany) equipped with a HCX PL APO Lambda blue 63x 1.4.oil objective (Leica), with the same settings for all conditions tested.

Quantification of protein distribution was carried out using ImageJ computer software. For intensity determination, a rectangle was drawn over more than 15 randomly chosen areas of HDMEC plasma membrane in at least three independent experiments. The mean intensity distribution of proteins examined was recorded for each depicted rectangle. A region next to the cell without specific fluorescence was selected as the respective background. In order to combine and compare all immunofluorescence experiments, the data was presented as percent of control (vehicle).

Transfection with small interfering RNA (siRNA)

Down-regulation of mouse specific AKAP12- and AKAP220 mRNA was obtained by using ON-Target SMARTpool siRNA (Dharmacon Thermo Fisher Scientific, Waltham, U.S.A.; Cat.# L-046094-00-0005371,00 and Cat.#L-048015-00-0005, respectively). As a negative control, ON-TARGET plus Non-Targeting siRNAs (Dharmacon, Cat.# D-001810-10-20) was applied. The siRNA was delivered into the cells by applying TurboFect *in vitro* transfection reagent (Fermentas, St. Leonrot, Germany, Cat.# R0531). The transient transfection was carried out according to the manufacturer's protocol. Briefly, after 20 min pre-incubation at RT, the transfection solution composed of TurboFect, siRNA and serum- free DMEM was diluted into serum-containing medium and added dropwise to MyEnd cells with 60–70% confluence. 48 hours after transfection, AKAPs-depletion was confirmed by Western blot.

For ECIS-based measurements, MyEnd were transfected with specific siRNA at 70% confluency. 24 hours after transfection medium was exchanged and TER was monitored. Basal and cAMP- stimulated Rac1 activities were examined 48 hours after siRNA transfection in control cells or cells treated with F/R, respectively.

Animal preparation and measurement of hydraulic conductivity (L_p) of the microvessel wall

A detailed description of the animal preparation and the microvessel L_p measurements was reported elsewhere [27,31]. All experimental protocols and procedures were consistent with the requirements of the National Institute of Health "Guide for the Care and the use of Laboratory Animals" and approved by Government of Lower Franconia (Az 55.2-2531.01-24/09). Wistar rats (Charles River Laboratories, Sulzfeld, Germany), with body weight ranging from 250 to 450 g, were anesthetized by subcutaneous injection of pentobarbital sodium at a dose of 65 mg/kg. The anesthetic substance and its way of application were chosen not to interfere with blood vessel permeability. In addition, depth of anaesthesia was checked regularly by animal's reaction to foot pad pinching. Supplemental anaesthetic was given only if the above mentioned reaction was positive. The experiments were carried out using straight, non-branched segments of mesenteric venular microvessels (25 to 35 μm in diameter). As descried earlier, the L_p measurements of the microvessel wall are based on the modified Landis technique, which measures the volume flux of fluid per unit surface area of the vessel (J_v/S), which was canulated with a glass micropipette and occluded in advance [32,33]. During measurements, the hydraulic pressure of usually 50 cm H_2O was constant with the assumption that the net effective pressure determining fluid flow (Peff) was equal to the applied hydraulic pressure minus 3 cm H_2O (the approximate oncotic pressure contributed by BSA in all perfusates (10 mg/ml)) [27]. For each occlusion, L_p was estimated as (J_v/S)/Peff. All perfusates were mammalian Ringer's solution containing 10% BSA (Sigma-Aldrich) with or without TAT-Ahx-AKAPis inhibitory synthetic peptide. Measurements were performed every 10 min for up to 120 min in the presence and absence of the synthetic peptide. Each animal was used for a single experiment only. After experimental completion, the unconscious rats were euthanized by intracardial injection of saturated KCl solution.

Statistical analysis

The data were processed using Prism Software version 5 (Graph Pad, San Diego, U.S.A.). All values are expressed as means ± SEM. While 2-tailed paired Student's t-test was used to compare two groups, One-way Analysis of Variance (ANOVA) followed by a Tukey's or Bonferoni's Multiple comparison tests were applied to determine the differences among multiple groups. The results were considered statistically significant at $p \leq 0.05$.

Results

TAT-Ahx-AKAPis peptide impaired endothelial barrier properties *in vitro*

To study the role of AKAP-mediated PKA subcellular anchoring for maintenance of endothelial barrier properties, we used a cell-permeable synthetic peptide (TAT-Ahx-AKAPis) designed to disrupt the endogenous PKA-AKAP complex [18]. Confluent microvascular endothelial cells were treated with TAT-Ahx-AKAPis (30 μM) and subjected to continuous TER measurements. In HDMEC, application of TAT-Ahx-AKAPis significantly decreased TER after 80 min incubation (80.4±2.04% of baseline values) compared to control conditions (vehicle: 92±1.2% of baseline ($p \leq 0,01$) and TAT-Ahx-mhK77:90±1.93% of baseline ($p \leq 0,05$) (Figure 1, A). Microvascular murine MyEnd cells responded similarly but slower to the inhibitory peptide (Figure S1). Nonetheless, in both cell types, TAT-Ahx-AKAPis induced a slowly progressive decrease in TER. In HDMEC, after 600 min of application, values reached 58±3%, whereas the values in control cells treated with vehicle remained stable (101±2.4%) (Figure 1, A - B). In contrast, treatment with TAT-Ahx-mhK77 did not affect TER in HDMEC (100±2.0%) (Figure 1, A - B).

The AC activator forskolin and the PDE IV inhibitor rolipram (F/R) increase cAMP levels and subsequently activate Rac1, the latter was found to be at least in part required for maintenance of endothelial barrier properties [7,29,34,35]. Thus, in line with previous studies [29,36], treatment with F/R (5 μM/10 μM) increased TER within 1 hour in both cell lines, with the effect being more prominent in HDMEC. Within the first hour, F/R rapidly increased TER in HDMEC (139±5.6%) and remained constantly elevated while in controls no significant changes were detected (92±1.2%). After 600 min, F/R reached its maximal effect compared to control conditions with values being 143±8.4% (Figure 1, A–B). Interestingly, in contrast to the pretreatment with scrambled peptide (TAT-Ahx-mhK77), where no significant alterations in the F/R effect (detected in control conditions) were observed, pretreatment with the TAT-Ahx-AKAPis inhibitory peptide diminished the F/R-mediated increase in TER relative to control levels. This effect of TAT-Ahx-AKAPis was also clearly detectable in MyEnd cells but again was evident later (Figure S1). To further evaluate the effect of TAT-Ahx-AKAPis on F/R- mediated endothelial barrier stabilization, HDMEC monolayers treated for 1 hour with F/R were subsequently subjected to TAT-Ahx-AKAPis peptide (Figure 1, C). The analysis revealed that 1 hour after TAT-Ahx-AKAPis application TER was similar in cells treated with F/R+TAT-Ahx-AKAPis or F/R alone (F/R+TAT-Ahx-AKAPis: 126±7%; FR: 127±1%). However, starting 100 min after peptide addition TER decline and after 230 min was comparable to experiments in which cells were treated with TAT-Ahx-AKAPis alone (at 300 min: F/R+TAT-Ahx-AKAPis: 82.4±10.7%; TAT-Ahx-AKAPis: 76±1.4%, Figure 1, C). Thus, the most significant peaks of the monitored effects were determined at 1000 min, the time point where similarly responding TAT-Ahx-AKAPis and F/R+TAT-Ahx-AKAPis monolayers displayed TER significantly lower than that of the control cells. In contrast, treatment of HDMEC with F/R significantly increased TER compared to controls as TER remained continuously elevated during the entire experiment (Figure 1, D).

Taken together, these results indicate that TAT-Ahx-AKAPis was sufficient to disrupt microvascular endothelial barrier properties, presumably via preventing AKAP-PKA complexation. Additionally, post-treatment with TAT-Ahx-AKAPis reverted F/R- mediated barrier stabilization whereas the pretreatment with the synthetic peptide was uneffective to completely abolish the barrier enhancing effect of F/R.

TAT-Ahx-AKAPis-induced endothelial barrier disruption was paralleled by VE-cadherin reorganization and actin cytoskeleton remodeling

Endothelial barrier functions are dependent on the organization of junctional complex and the actin cytoskeleton [35]. Therefore, possible alterations of these structures accompanying the TAT-Ahx-AKAPis-induced decrease in TER were investigated by immunofluorescence studies in HDMEC. Subsequently, measurements of the fluorescence intensity along cell borders served to quantitatively assess changes in the distribution of membrane associated proteins. Beside the adherens junction protein VE-cadherin, filamenteous (F-) actin, and PKA, stainings for AKAP12 and 220 were also performed. The AKAP 12 and 220 expression profiles were initially evaluated by Western blot in human and mouse microvascular endothelial cells (Figure 2, A and B). The

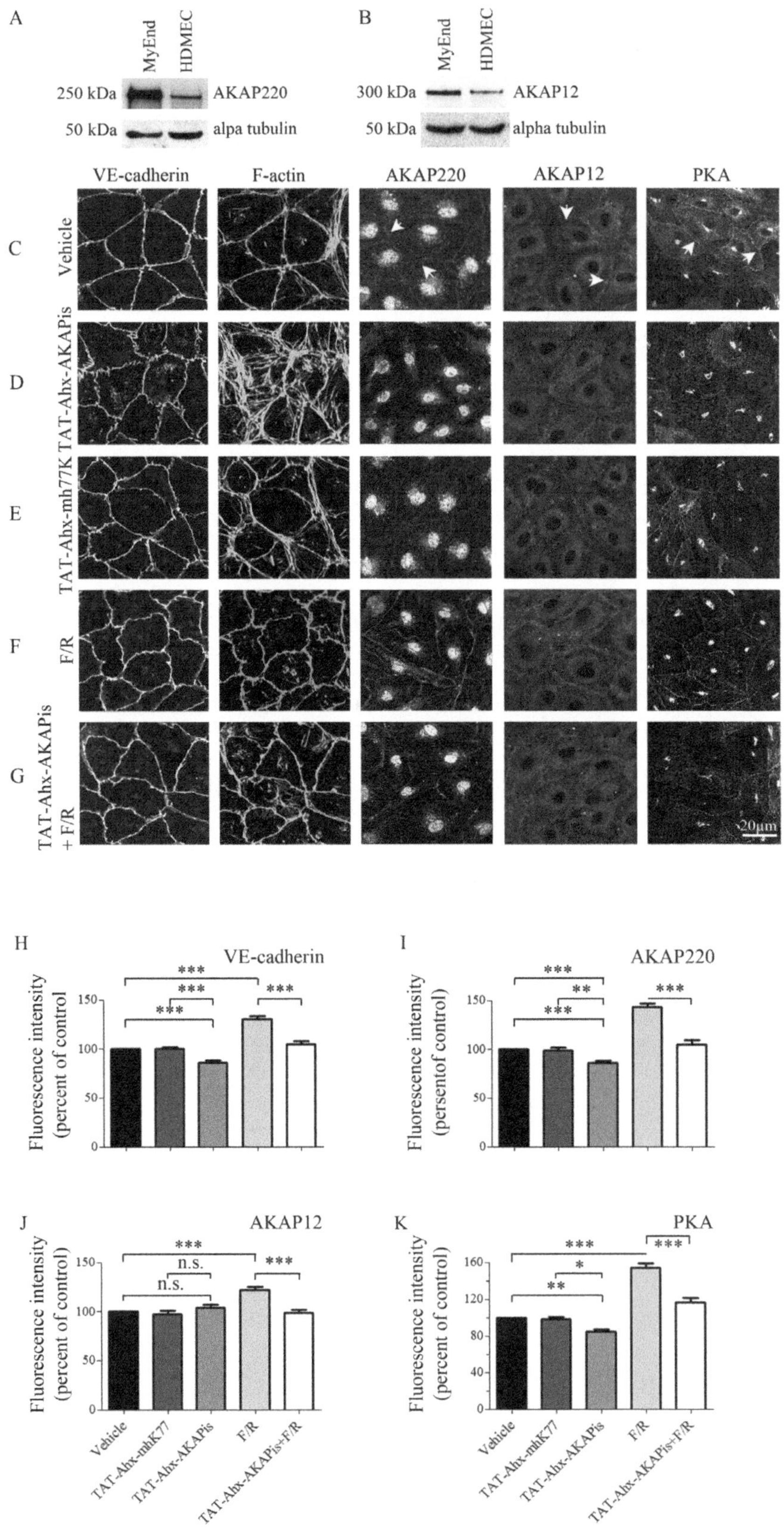

Figure 2. Effects of PKA compartmentalization on adherens junctions, actin cytoskeleton and AKAP organization. To determine the AKAPs protein expression profile, equal amounts of MyEnd and HDMEC cell lysates (20 µg per line) were subjected to WB analysis. The expression patterns of AKAP220 (A) and AKAP12 (B) were analyzed. Loading was controlled by alpha tubulin expression. Data are representative of three or more independent experiments (N≥3). In addition, HDMEC monolayers were treated for 1 hour either with vehicle, TAT-Ahx-AKAPis, TAT-Ahx-mhK77, F/R or with a combination of 1 hour TAT-Ahx-AKAPis pretreatment followed by 1 hour F/R application. The distributions of VE-cadherin, PKA, AKAP220

and AKAP12 were assayed by immunofluorescence. Additionally, ALEXA-488-conjugated phalloidin was used for visualization of F-actin. (C) Under control condition, VE-cadherin displayed slightly interdigitated but continuous staining along cell borders, and the actin cytoskeleton was preferentially organized cortically. The staining of AKAP220, AKAP12 and PKA was detectable at cell borders (arrows). (D) In clear contrast, exposure to TAT-Ahx-AKAPis increased interdigitations and significantly reduced the intensity of VE-cadherin staining. Profound reorganization of the actin cytoskeleton was paralleled by substantial reduction of AKAP220 and PKA, but not of AKAP12 membrane staining. (E) However, cell monolayers incubated with TAT-Ahx-mhK77 showed immunofluorescence staining similar to control for all proteins under investigation. (F) Not surprisingly, F/R treatment resulted in pronounced and linearized VE-cadherin appearance, intensified cortical actin cytoskeleton, and pronounced membrane staining for AKAP220, AKAP12 and PKA compared to control conditions. (G) 1 hour pre-incubation with TAT-Ahx-AKAPis followed by 1 hour treatment with F/R resulted in monolayers largely similar to controls, but not to F/R incubation alone. Images are representative of three or more independent experiments (N $\geq$3). Scale bar = 20 μm. The above presented data were confirmed by quantification of signal intensity distribution at cell borders. (H–K) demonstrates the mean intensity peak observed at cell borders (n$\geq$15). *** p$\leq$0.001, ** p$\leq$0.01, * p$\leq$0.05, indicate statistically significant difference between examined groups; n.s., not significant.

analysis revealed more prominent expression of AKAPs in MyEnd cells.

HDMEC monolayers were treated for 1 hour either with vehicle solution, TAT-Ahx-AKAPis, corresponding scrambled peptide or F/R. In addition, a regimen of 1 hour TAT-Ahx-AKAPis pretreatment followed by 1 hour F/R application was carried out. The cell monolayers incubated with vehicle solution displayed slightly interdigitated but continuous VE-cadherin staining along cell borders as well as few stress fibers localized predominantly in cortical regions (Figure 2, C). Peripheral membrane localization was partially visible for AKAP220, AKAP12 and PKA (Figure 2, C, arrows). However, when HDMEC were treated with TAT-Ahx-AKAPis, pronounced reorganization of the actin cytoskeleton accompanied by enhanced interdigitations and decreased staining intensity of VE-cadherin were detectable (Figure 2, D, H). This was paralleled by considerable reduction of PKA and AKAP220 but not AKAP12 membrane staining (Figure 2, D, I–K) indicating that at least in the case of AKAP220 the peptide was effective in disrupting PKA anchorage at sites of cell contacts. In contrast, the proteins under investigation showed distributions similar to controls when monolayers were treated with scrambled synthetic peptide (Figure 2, E, H–K). Compared to controls, as reported previously [26,29], F/R treatment resulted in more intense and linearized VE-cadherin staining (Figure 2, F, H). Moreover, membrane staining for AKAP12, AKAP220 and PKA was also more pronounced (Figure 2, F, I–K). This was accompanied by intensified cortical actin staining (Figure 2, F). In good agreement with the TER data pre-incubation with the inhibitory peptide interfered with the initial effect of F/R. HDMEC monolayers appeared more similar to controls (Figure 2, G–K). In summary, the above presented data showed that TAT-Ahx-AKAPis induced reorganization of both endothelial adherens junctions and the actin cytoskeleton as well as caused AKAP220 and PKA relocation from the membrane.

In endothelial adherens junctions, VE-cadherin in addition to a variety of structural proteins (i.e. ß-catenin, p120-catenin and plakoglobin) associates with several molecules participating in cAMP signaling such as PKA, PDE IV and Epac1 [37]. On the other hand, it is well known that PKA is tethered by AKAP220 [38] and the latter was suggested to be connected to cytoskeletal structures [39]. Therefore, we speculated that PKA via AKAP220 interacts with junctional complexes which may be required for stabilization of the endothelial barrier. To test this hypothesis, MyEnd lysates were subjected to immunoprecipitation. The analysis confirmed a complex consisting of AKAP220, PKA, ß-catenin and VE-cadherin. Both, pulling down VE-cadherin or PKA, respectively, yielded the same results (Figure 3, A and B). In addition, to monitor the changes in the complex composition as a result of TAT-Ahx-AKAPis and/or F/R treatment, PKA pull-down in lysates derived from cells treated either with synthetic inhibitory peptide or with F/R was carried out. PKA pull-down in cells subjected to scrambled peptide was used as respective control. Compared to TAT-Ahx-mhK77 treatment, application of TAT-Ahx-AKAPis reduced the band intensities for AKAP220 as well as for VE-cadherin and ß-catenin indicating decreased association with PKA. In contrast, F/R enhanced ß-catenin-, VE-cadherin- and AKAP220- band intensities (Figure 3, C–D).

AKAP12 and AKAP220 are involved in regulation of endothelial barrier function

To further investigate the role of AKAPs, the effect of AKAP220- and AKAP12- specific depletion on endothelial barrier function was determined and compared to treatment with TAT-Ahx-AKAPis. Subconfluent MyEnd cells were transiently transfected either with AKAP220- or AKAP12- specific siRNA or with n.t siRNA, respectively. 24 hours after siRNA application, TER measurements were initiated (Figure 4, A). The beginning of the TER measurements was also the initial point of TAT-Ahx-AKAPis peptide application. The experiments were continued for additional 46 hours. The time window was estimated by Western blot analysis validating the efficiency of the gene silencing in MyEnd treated with AKAP-specific siRNAs (Figure 4 C, D).

Control cells transfected with n.t. siRNA increased TER over time to values of 128.6±3.95% of baseline (Figure 4, A). In contrast, siRNA-mediated AKAP12 and AKAP220 knockdown initially decreased TER and subsequently abolished barrier stabilization (104±4.3% and 108±3.6%, respectively). Similar, but more significant was the effect upon TAT-Ahx-AKAPis inhibitory treatment (98±7.6%). Thus, these data indicate that besides AKAP12 and AKAP220 possibly other AKAPs are involved in the regulation of endothelial barrier function.

In order to estimate the effect on cAMP-mediated endothelial barrier function, F/R was applied to cells either transiently depleted of specific AKAPs or treated with n.t. siRNA (Figure 4, B). The results indicate that depletion of AKAP12, but not of AKAP220 significantly decreases the effect of cAMP-mediated endothelial barrier stabilization. These data suggest that both AKAPs alter endothelial barrier function but only AKAP12 modifies the subsequent cAMP-mediated endothelial barrier enhancement.

Disruption of the PKA-AKAP endogenous complex reduced Rac1 activity

Our data demonstrate that TAT-Ahx-AKAPis-mediated disruption of the endogenous PKA–AKAP complex attenuated endothelial barrier functions under resting conditions. Since cumulative evidence shows that cAMP governs microvascular barrier properties, at least in part, in a Rac1-dependent manner [35,40,41], we investigated the effect of TAT-Ahx-AKAPis on Rac1 localization and activity. Immunofluorescence analysis in HDMEC revealed that, under control conditions, Rac1 staining

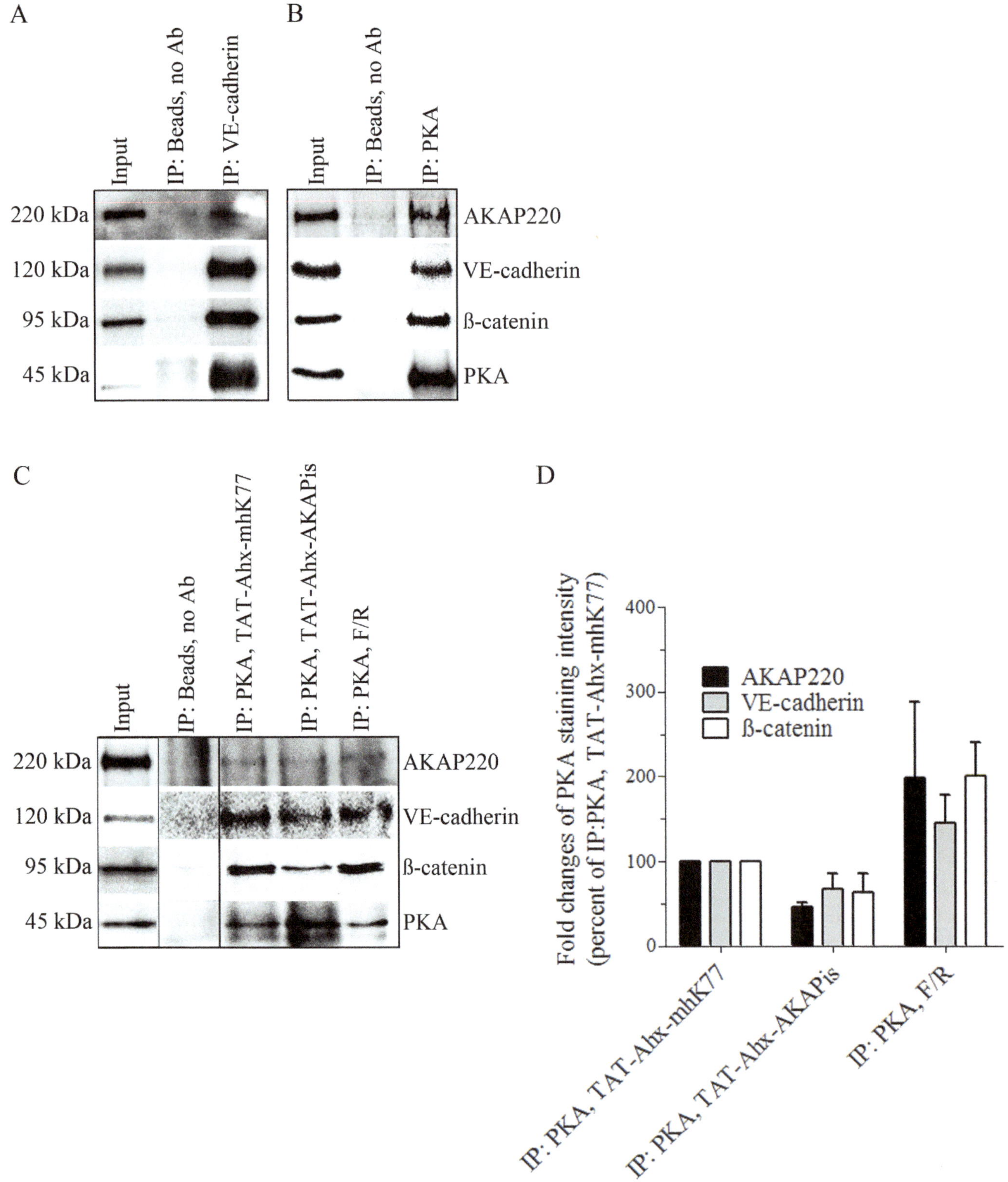

Figure 3. PKA and AKAP220 form a complex with junctional-associated proteins VE-cadherin and ß-catenin. A complex consisting of PKA, ß-catenin, VE-cadherin and AKAP220 was detected in MyEnd either by pulling down VE-cadherin (A) or PKA (B). Immunoprecipitated proteins and their binding partners were detected in the lysates used for immunoprecipitation. The lysates were loaded at the first lane and denoted as an "Input". Immunoprecipitation without antibody (IP: Beads, no Ab) was used as a negative control. Every other lane represents proteins in complex under different experimental conditions. (C) To test the role of PKA compartmentalization on the stability of the determined complex, PKA pull down was performed in cells treated either with inhibitory or with the respective scrambled peptide. Additionally, the effect of F/R on complex stability was analyzed. (D) summarizes data collected from three independent immunoprecipitation experiments. Firstly, intensity readings for each band were background corrected. Secondly, data was modified by substraction of "non-specific background intensities" detected in IP without Ab (negative

control). Since PKA was used to pull-down the complex, signal intensity of the immunoprecipitated proteins was normalized to the band intensity of PKA (represented as fold changes of PKA staining intensity). As a final, the data was presented as percent of IP: PKA, TAT-Ahx-mhK77. The analysis revealed decreased complex association of ß-catenin, VE cadherin and AKAP220 after TAT-Ahx-AKAPis treatment. In contrast, F/R application led to stabilization of the complex indicated by more prominent intensities for bands representing ß-catenin, VE-cadherin and AKAP220. Noncontiguous bands run on the same gel are separated by a black line. Images are representative of three independent experiments (N = 3).

was in part detectable along cell borders, (Figure 5, A, arrows). Such membrane localization of Rac1 was previously correlated with an increase in its activity. In this respect, our previous study showed that constitutively active Rac1 localized to cell- cell borders in endothelial cells whereas this effect was not observed in cells transfected with dominant negative Rac1 [42]. However, strong reduction of Rac1 membrane staining and relocation to the cytoplasm were detected after TAT-Ahx-AKAPis application (Figure 5, C). Further densitometric assessment of the immunofluorescent data confirmed these observations (Figure 5, E). Consistently, Rac1 rearrangement was paralleled by altered GTPase activity in HDMEC and MyEnd cells as measured by G-LISA Rac activation assay (Figure 5 F–G). However, treatment with TAT-Ahx-mhK77 neither showed changes in Rac1 localization nor in Rac1 activity when compared to control condition (Figure 5, B, E–G). In contrast, application of F/R dramatically

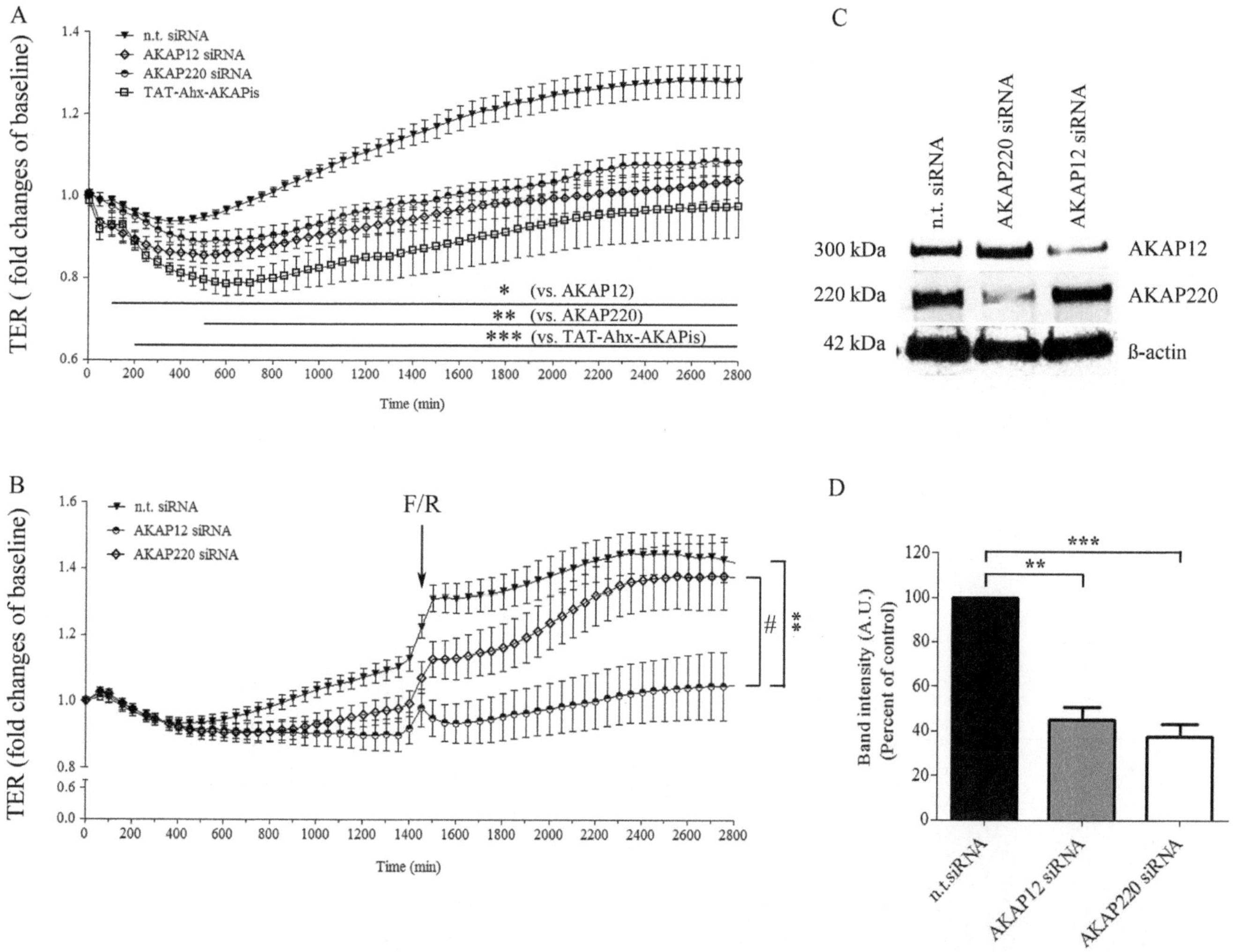

Figure 4. siRNA-mediated AKAP12 and AKAP220 knockdown significantly impaired endothelial barrier function. (A) Barrier function of subconfluent MyEnd cells was monitored by TER measurements following transfection with siRNA specific for AKAP12 and AKAP220. Non-targeting siRNA was used as a control. The results were compared to cells treated with TAT-Ahx-AKAPis inhibitory peptide. AKAP12 and AKAP220 depletion lead to a significant decrease in TER compared to monolayers transfected with n.t. siRNA. Similar, but more prominent, was the effect obtained after TAT-Ahx-AKAPis application. Data were collected from three or more independent experiments (N ≥3, n ≥12). (B) To test the role of specific AKAPs on cAMP-mediated endothelial barrier formation, TER measurements in cells treated with F/R, 48 hours after successful AKAP-depletion were performed; (n≥10). (C) AKAP12 and AKAP220 knockdown was confirmed by Western blot, 48 hours after transfection. ß-actin was used as an internal gel-loading control. (D) Western blot data were analyzed densitometrically. Normalized intensities for AKAP signals are presented as a bar graph. Data were collected from more than three independent experiments (N ≥3). * p≤0.05, **p≤0.01 and ***p≤0.001 indicate statistically significant difference between the examined group and n.t. siRNA; # p≤0.05 shows statistically significant difference between AKAP12 siRNA and AKAP220 siRNA upon treatment with F/R. The difference was significant starting at 1500 min.

enriched the staining of Rac1 at the membrane (Figure 5, D–E). Consistent with the immunofluorescence analysis, F/R caused a significant increase of Rac1 activity in both cell types (for HDMEC, F/R: 148±5% and for MyEnd, F/R: 139±14%; Figure 5 F–G). In HDMEC, the latter was approximately 48% more than the activity determined in controls or scrambled-treated cells (vehicle: 100%; TAT-Ahx-mhK77:99±3%). The effect in MyEnd cells was similar, but slightly smaller (more than 33%), (vehicle: 100%; scrambled: 105±3% (n.s.)). ELISA-based Rac1 activity measurements also demonstrated that peptide-application significantly reduced Rac1 activity to 83±2% of control conditions in HDMECs and 71±6% in MyEnd cells (Figure 5 F–G).

To further evaluate the effect of specific AKAPs on Rac1 activity, we silenced AKAP12 or AKAP220 by siRNA and assessed Rac1 activity 48 hours after knockdown in MyEnd cells. Neither down-regulation of AKAP12 and/or AKAP220 mRNA alone nor parallel silencing of both AKAPs altered basal Rac1 activity (Figure 6, A). Nevertheless, cAMP-mediated Rac1 activation was significantly reduced in cells simultaneously depleted for AKAP12 and AKAP220 but not in cells in which only one of the two AKAPs was silenced (Figure 6, A). Effective mRNA down-regulation was validated by Western blot (Figure 6, B).

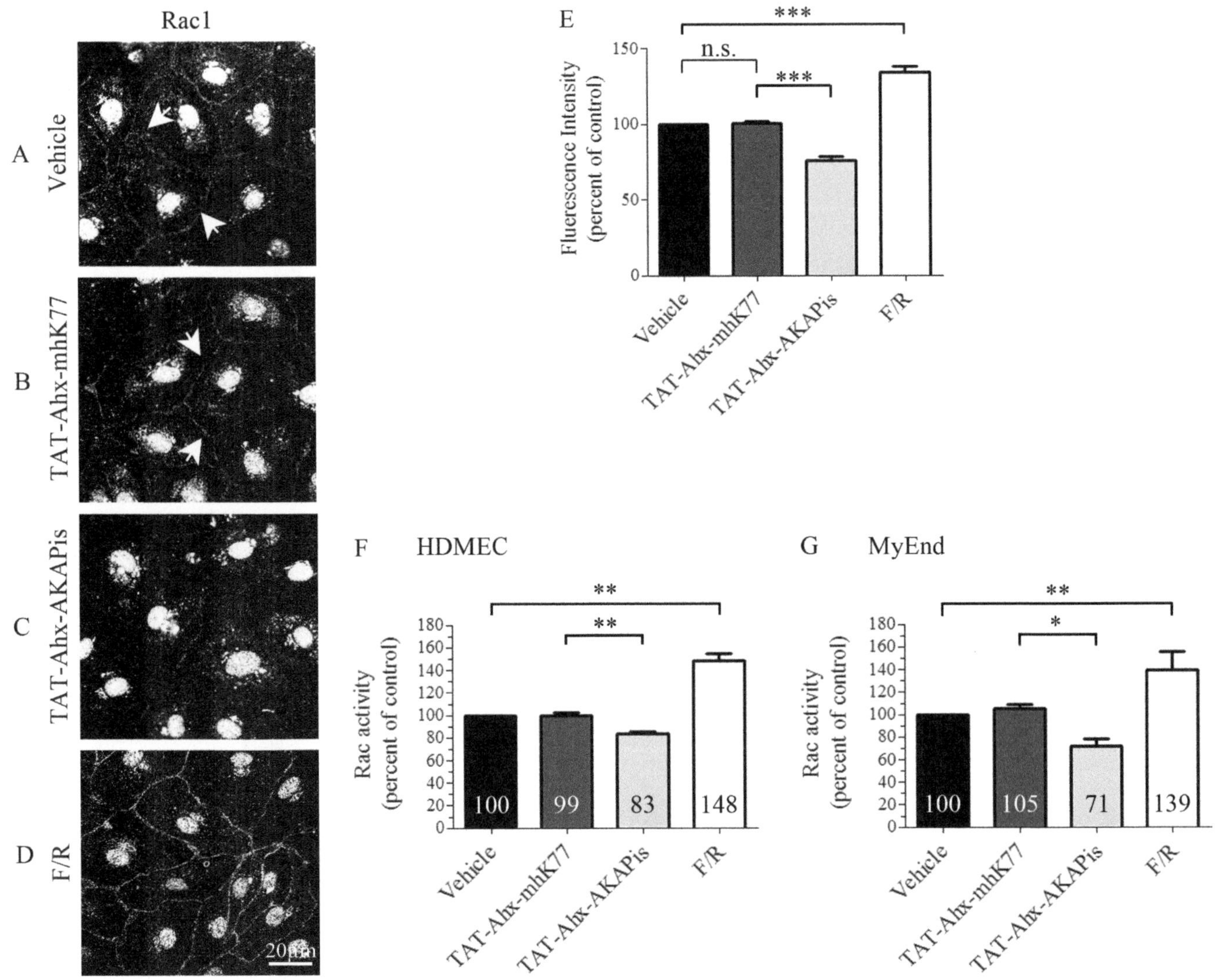

Figure 5. TAT-Ahx-AKAPis reduced Rac1 activity. HDMEC monolayers grown on glass coverslips were subjected to immunostaining with anti-Rac1 antibody. (A) In controls, Rac1 was partially present along cell borders (arrows). (B) Similar Rac1 staining was observed after application of TAT-Ahx-mhK77. (C) In contrast, treatment with TAT-Ahx-AKAPis strongly reduced Rac1 membrane staining and induced re-localization of the molecule to the cytoplasm. (D) However, the cAMP enhancers F/R led to pronounced and intensified Rac1 membrane staining. Images are representative of three or more experiments (N≥3). Scale bar = 20 μm. (E) Quantification of the signal intensity distributed along cell-cell junctions supported the above mentioned observations. In order to summarize all experiments, data were presented as percent of control (N≥3n≥10). (F) In HDMEC, ELISA-based measurements revealed a significant decrease of Rac1 activity in response to 1 hour treatment with TAT-Ahx-AKAPis. Application of F/R also significantly increased Rac1 activity when compared to control. As expected, TAT-Ahx-mhK77 administration had no effect on Rac1 activity. (G) The effects observed in MyEnd cells were similar. In agreement with TER, the effect of TAT-Ahx-AKAPis inhibitory peptide was evident later. Therefore, the MyEnd samples were collected 6 hours after peptide application. The results are representative of three independent experiments (N = 3, n ≥6). * defines statistically significant difference; * p≤0.05, ** p≤0.05, *** p≤0.001. n.s. not significant.

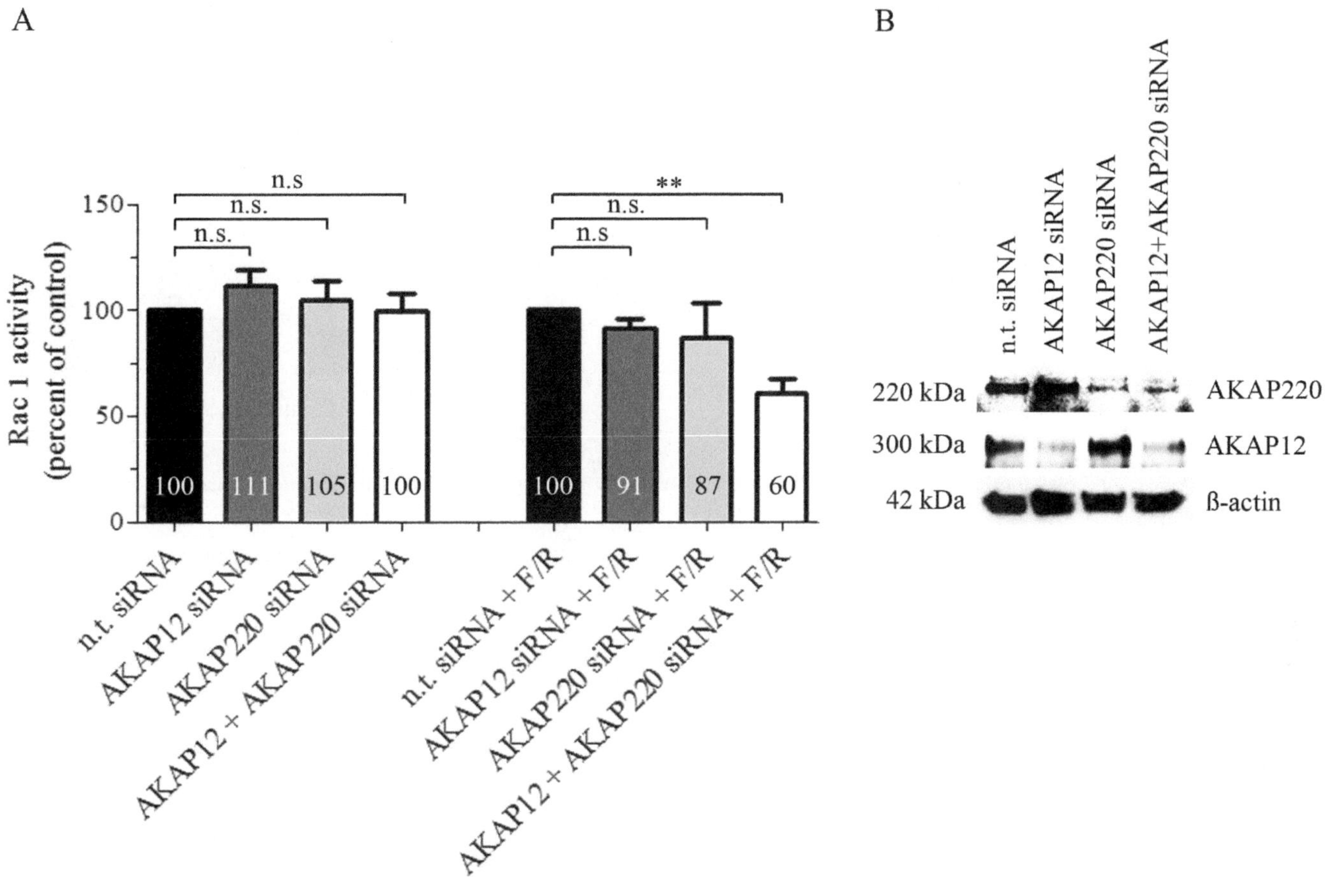

Figure 6. Simultaneous knockdown of AKAP12 and AKAP220 mRNA impaired cAMP-mediated Rac1 activation. (A) Subconfluent monolayers of MyEnd cells were transfected either with single AKAP siRNA or with a pool of AKAP12 and AKAP220 specific siRNAs. Basal Rac1 activity was tested 48 hours after initial siRNA depletion. For evaluation of cAMP-mediated Rac1 activation, transfected cells were subjected to F/R for 1 hour. Non-target siRNA (n.t. siRNA) alone or in combination with F/R (1 hour) was used as a respective control. Bars represent mean values ± SEM. Basal activity of Rac1 was unaffected after single or combined silencing of AKAPs. In contrast, cAMP-mediated Rac1 activation was significantly impaired following simultaneous AKAP12 and AKAP220 mRNA knockdown. (B) Successful knockdown was verified by Western blot 48 hours after initial siRNA transfection; ß-actin was used as an internal gel-loading control. ** indicates p≤0.01 versus cells transfected with n.t. siRNA treated for an hour with F/R.

TAT-Ahx-AKAPis treatment increased hydraulic conductivity (L_p) of single perfused rat postcapillary venules *in vivo*

To investigate whether the application of TAT-Ahx-AKAPis also affects microvascular barrier properties *in vivo*, we applied the single microvessel perfusion technique [27] in rat mesentery postcapillary venules. Intact microvessels were constantly perfused either with TAT-Ahx-AKAPis or with vehicle solution and hydraulic conductivity was measured every 10 min. No significant L_p differences of TAT-Ahx-AKAPis-treated and control vessels were monitored within the first 60 min (Figure 7, A and B). However, L_p increased starting 60 min after TAT-Ahx-AKAPis addition. After 120 min of treatment, L_p of TAT-Ahx-AKAPis-perfused vessels reached a mean value of $4.57 \pm 0.91 \times 10^{-7}$ [cm/(s×cm H_2O)]) which was significantly higher compared to controls ($2.53 \pm 0.47 \times 10^{-7}$ [cm/(s×cm H_2O)]), (Figure 7, B). Thus, TAT-Ahx-AKAPis impaired microvascular endothelial barrier properties both *in vivo* and *in vitro*.

Discussion

The goal of the current report was to study in more details the mechanisms underlying cAMP-mediated regulation of endothelial barrier properties. Specifically, we examined the functional relevance of AKAPs which on one hand are well established to tightly regulate PKA function by facilitating its discrete cellular compartmentalization, and on the other hand are known to form multivalent complexes to integrate cAMP signaling with various other pathways. We provided evidence that AKAPs are required for maintenance of microvascular endothelial barrier integrity both *in vivo* and *in vitro* under resting conditions. We also showed that under resting conditions AKAPs are relevant for correct signal transduction because Rac1 activity was decreased following inhibition of PKA-AKAP anchoring. Particularly, we also found that AKAP12 and AKAP220 are involved in regulation of endothelial barrier function. Further examinations revealed that simultaneous down-regulation of AKAP12 and AKAP220 expression impaired F/R-mediated Rac1 activation. We also demonstrated that AKAP12, but not AKAP220, is important for subsequent cAMP-mediated endothelial barrier improvement. Finally, we showed that AKAP220 forms a complex with PKA,

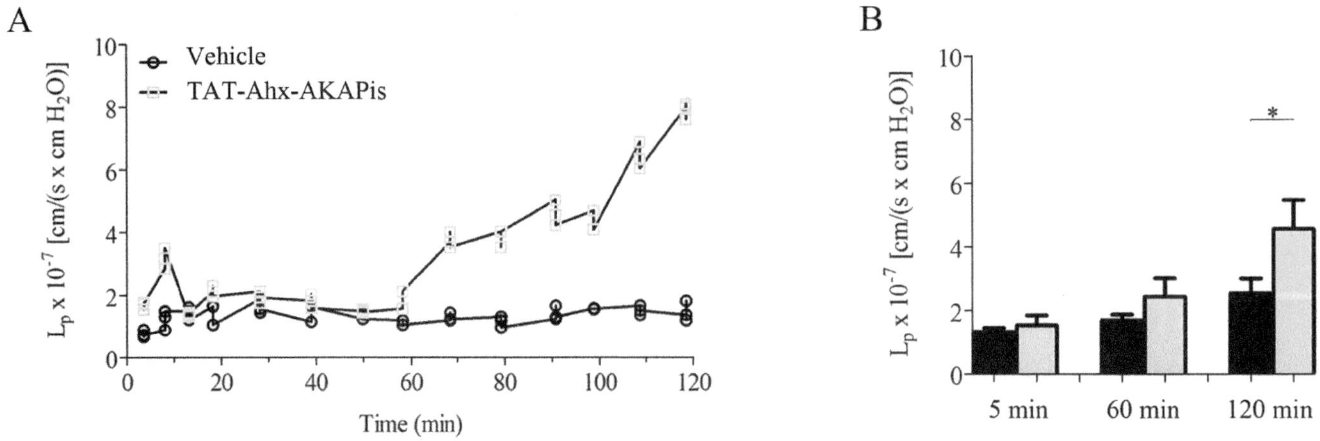

Figure 7. TAT-Ahx-AKAPis application increased hydraulic conductivity (L_p) in rat mesenteric microvessels *in vivo.* Hydraulic conductivity (L_p) of isolated post-capillary microvessels from rat mesentery was monitored over time. (A) displays data from a representative vessel constantly perfused either with vehicle or with TAT-Ahx-AKAPis. (B) Bar diagram summarizes the mean values of all five independent experiments. 120 min after TAT-Ahx-AKAPis perfusion, L_p was significantly increased to $4.57 \pm 0.91 \times 10^{-7}$ [cm/(s×cm H_2O)] compared to the corresponding vehicle condition ($2.53 \pm 0.47 \times 10^{-7}$ [cm/(s×cm H_2O)]). *indicates $p \leq 0.05$ versus control condition; N = 5.

ß-catenin and VE-cadherin, and therefore can spatially link cAMP signaling to endothelial adherens junctions.

cAMP signaling needs to be compartmentalized to exert barrier protective effects

cAMP is well established to be important for barrier formation and maintenance [7,33]. Furthermore, a variety of barrier destabilizing mediators decrease cAMP levels leading to increased permeability [26,31,43]. This decrease in cAMP can be initiated by a rapid Ca^{2+}-influx [26,33,44] which subsequently may inhibit AC activity [45]. Downstream of cAMP, Rac1 is an important signaling molecule for barrier stabilization [7,35,46]. Two distinct signaling pathways involving on one hand PKA and on the other hand Epac1/Rap1 have been shown to merge on and activate Rac1 [47]. Rac1 via a plethora of mechanisms [35,40,46] induces cortical actin remodeling and reorganization of intercellular junctions which ultimately affects paracellular permeability [1,34]. Not surprisingly, it is important where in the cell cAMP is generated in order to provide barrier protective effects [48]. cAMP generated by membrane-attached ACs is protective, whereas cAMP increased by a soluble AC is not. The latter in contrast increases permeability [49]. Thus, to provide distinct effects, cAMP signaling needs to be tightly regulated in space and time. Beside ACs, other key players involved in this regulation are PDEs, which locally hydrolyze cAMP [50]. Similarly, AKAPs facilitate compartmentalization of PKA signaling downstream of cAMP. Our data provide a mechanism, by which the function of PKA can be directed to cell junctions.

AKAPs are essential for maintenance and stabilization of endothelial barrier properties

Under resting conditions, TAT-Ahx-AKAPis destabilized barrier functions both *in vitro* and *in vivo*. This effect was qualitatively similar in two microvascular cell types and postcapillary venules, indicating that AKAP function is an important factor for endothelial barrier maintenance. Similar to our observation, a recent study demonstrated that low expression of AKAP12 may lead to blood-retinal barrier dysfunction [21]. Further investigations in this direction reported the role of AKAP12 in maintenance of the vascular integrity by modulation of the actin cytoskeleton dynamic through PAK2 and AF6 [20]. Another member of the AKAP-family, i.e. AKAP9 was also discovered to be required for microtubule growth, integrin adhesion at cell-cell borders and endothelial barrier function via Epac1-dependent pathway [51]. Thus, besides PKA, AKAPs can also be associated with Epac1 [51,52]. Therefore, AKAPs may serve as coordinators not only of PKA- but also of Epac1- induced regulation of endothelial barrier properties [52].

Furthermore, we found that inhibition of AKAP function via TAT-Ahx-AKAPis also interfered with barrier stabilization in response to increased cAMP. In HDMEC, this approach was effective to revert F/R-induced barrier stabilization. In line with that, earlier we reported that incubation with a cell permeable PKA inhibitor blocked the F/R-mediated increase in TER [36]. Herein, we also showed that depletion of AKAP12 but not of AKAP220 significantly decreased cAMP-mediated endothelial barrier integrity as examined by TER (Figure 4). Moreover, simultaneous depletion of AKAP12 and AKAP220 but not of a single AKAP impaired cAMP-mediated Rac1 activation (Figure 6) which is indicative for a redundant function of those AKAPs in the regulation of Rac1 activity. Taken together, these results also demonstrate that AKAP12 may interfere with cAMP-mediated endothelial barrier stabilization in a manner which at least in part is independent of Rac1 (because AKAP12 down- regulation did not alter Rac1 activation). In agreement with this presumption is our recent study revealing that F/R- induced Rac1 activation and barrier augmentation were not affected by the Rac1 inhibitor NSC-23766. Therefore, we argue that GTPases other than Rac1 may also account for the F/R- induced enhancement of endothelial barrier properties [34]. Additionally, one can speculate that besides Rac1, AKAP12 may take part in different cAMP-induced signaling pathways involved in endothelial barrier stabilization. In this respect, a recent study determined AKAP12 molecule as a dynamic platform for signal transduction complexing a number of signaling molecules such as PKA, PKC, calmodulin, F- actin and ß-adrenergic receptors [53].

Similar to AKAP12, we also showed that depletion of AKAP220 impaired the function of the endothelial barrier in MyEnd cells. However, the effect of silencing specific AKAPs was less prominent than the one observed upon TAT-Ahx-AKAPis application (Figure 4). This supports the idea that several AKAPs including AKAP220 and AKAP12 are involved in modulation of endothelial barrier function.

AKAP220 contributed to endothelial barrier integrity by forming a multivalent complex that links cAMP signaling to adherens junctions

Besides PKA anchoring, several AKAPs were found to act as scaffolding proteins thereby participating in various signal transduction processes. Formation of multivalent complexes provides a high level of specificity and temporal regulation to cAMP/PKA signaling [17]. As mentioned above, we examined the role of AKAP220 which was already reported to organize multivalent complexes. In this respect, AKAP220 was shown to form a complex with IQGAP1 and E-cadherin in MCF-7 cells and to link cAMP signaling to cell adhesion [25]. Furthermore, recent investigations provided evidence that AKAP220 forms a complex with IQGAP2 that favors PKA-dependent recruitment of Rac1 to strengthen cortical actin [24]. Thus, AKAP220 not only provides substrate specificity by tight subcellular localization of PKA, but also regulates and restricts the activity of several effectors which are part of this complex [24,54,55,56]. Similar to AKAP79/150, which was found to localize on the cell membrane and to assemble a ternary complex with E-cadherin and ß-catenin in epithelial cells [57], we detected AKAP220 to co-immunoprecipitate with VE-cadherin and ß-catenin as well as to localize at cell borders similar to VE-cadherin, PKA and Rac1 in microvascular endothelial cells (Figure 2 and 3). Moreover, we demonstrated that F/R- mediated endothelial barrier stabilization was paralleled by increased membrane localization and association of PKA with AKAP220 and VE-cadherin in a complex. The latter observations are consistent with the idea that cAMP via PKA may allow compartmentalized Rac1 activation close to adherens junctions and the cortical actin cytoskeleton. This may be physiologically relevant because TAT-Ahx-AKAPis induced prominent cytoskeletal rearrangement and VE-cadherin interdigitation under conditions of a destabilized endothelial barrier. These effects were associated with decreased PKA, AKAP220, and Rac1 membrane staining, as well as reduced Rac1 activity (Figure 2 and 5). Additionally, TAT-Ahx-AKAPis decreased the association of AKAP220, VE-cadherin and ß-catenin with PKA demonstrating that AKAPs are required to localize PKA to endothelial adherens junctions (Figure 3). Consistent with our assumptions is a study demonstrating that PKA, Epac1, PDE4D and AKAP79 are recruited to VE-cadherin-based complexes in response to cell-cell-contact formation [37].

In conclusion, we showed that AKAPs, and specifically AKAP12 and AKAP220, contribute to regulation of microvascular endothelial barrier function in Rac1- dependent and – independent manner. Our data also indicate that AKAP220 forms a multivalent protein complex linking cAMP signaling to adherens junctions.

Supporting Information

Figure S1 TAT-Ahx-AKAPis interfered with endothelial barrier function in MyEnd cells. (A) shows TER time courses for different conditions applied in MyEnd cell monolayers. In comparison to control and TAT-Ahx-mhK77- treated endothelial cells, monolayers subjected to TAT-Ahx-AKAPis significantly decreased TER over time. The effect was slower than in HDMEC cells but prominent approximately 600 min after initial application. Moreover, in vehicle and TAT-Ahx-mhk77- treated monolayers F/R application led to a progressive TER increase over time. In contrast, TER in TAT-Ahx-AKAPis- pretreated monolayers transiently increased after addition of F/R but declined afterwards to finally reach values similar to control conditions after 450 min. (B) outlines the results after 800 min, the time point at which the monitored effects reached their peaks. Data were collected from more than three independent experiments ($N \geq 3$). * $p \leq 0.05$ and *** $p \leq 0.001$ indicate statistically significant difference between examined groups; n.s., not significant.

Acknowledgments

We are grateful to John Scott for providing an AKAP220 antibody. We thank Nadja Niedermeier, Andrea Wehmeyer, Tetjana Frantzeskakis and Veronica Heimbach for their skilful technical assistance; Angela Wölfel for her help in manuscript editing.

Author Contributions

Conceived and designed the experiments: MYR JW. Performed the experiments: MYR DK VS. Analyzed the data: MYR. Contributed to the writing of the manuscript: MYR JW. Carried out the statistical analysis: MYR. Contributed to confocal imaging: MYR DK VS. Performed all experiments involving siRNA-mediated gene silencing: DK. Read and approved the final version of the manuscript: MYR DK VS JW.

References

1. Mehta D, Malik AB (2006) Signaling mechanisms regulating endothelial permeability. Physiological reviews 86: 279–367.
2. Cinel I, Dellinger RP (2007) Advances in pathogenesis and management of sepsis. Current opinion in infectious diseases 20: 345–352.
3. Aird WC (2007) Phenotypic heterogeneity of the endothelium: I. Structure, function, and mechanisms. Circulation research 100: 158–173.
4. Drenckhahn D, Ness W (1997) The endothelial contractile cytoskeleton In: Vascular Endothelium: Physiology, Pathology, and, Therapeutic, Opportunities, editors. Stuttgart, Germany: Schattauer. 1–25.
5. Wojciak-Stothard B, Ridley AJ (2002) Rho GTPases and the regulation of endothelial permeability. Vascular pharmacology 39: 187–199.
6. Hoshi S, Ohyama C, Hagisawa S, Ono K, Satoh M, et al. (2003) Complete regression of bone metastases on super bone scan, by low-dose cisplatin, UFT, diethylstilbestrol diphosphate, and dexamethasone in a patient with hormone-refractory prostate cancer. International journal of clinical oncology 8: 118–120.
7. Schlegel N, Waschke J (2013) cAMP with other signaling cues converges on Rac1 to stabilize the endothelial barrier- a signaling pathway compromised in inflammation. Cell and tissue research.
8. Qiao J, Huang F, Lum H (2003) PKA inhibits RhoA activation: a protection mechanism against endothelial barrier dysfunction. American journal of physiology Lung cellular and molecular physiology 284: L972–980.
9. Wong W, Scott JD (2004) AKAP signalling complexes: focal points in space and time. Nature reviews Molecular cell biology 5: 959–970.
10. Bauman AL, Goehring AS, Scott JD (2004) Orchestration of synaptic plasticity through AKAP signaling complexes. Neuropharmacology 46: 299–310.
11. Dodge KL, Khouangsathiene S, Kapiloff MS, Mouton R, Hill EV, et al. (2001) mAKAP assembles a protein kinase A/PDE4 phosphodiesterase cAMP signaling module. The EMBO journal 20: 1921–1930.
12. Tasken KA, Collas P, Kemmner WA, Witczak O, Conti M, et al. (2001) Phosphodiesterase 4D and protein kinase a type II constitute a signaling unit in the centrosomal area. The Journal of biological chemistry 276: 21999–22002.
13. Dodge-Kafka KL, Soughayer J, Pare GC, Carlisle Michel JJ, Langeberg LK, et al. (2005) The protein kinase A anchoring protein mAKAP coordinates two integrated cAMP effector pathways. Nature 437: 574–578.
14. Bauman AL, Soughayer J, Nguyen BT, Willoughby D, Carnegie GK, et al. (2006) Dynamic regulation of cAMP synthesis through anchored PKA-adenylyl cyclase V/VI complexes. Molecular cell 23: 925–931.
15. Dessauer CW (2009) Adenylyl cyclase–A-kinase anchoring protein complexes: the next dimension in cAMP signaling. Molecular pharmacology 76: 935–941.
16. Kapiloff MS, Piggott LA, Sadana R, Li J, Heredia LA, et al. (2009) An adenylyl cyclase-mAKAPbeta signaling complex regulates cAMP levels in cardiac myocytes. The Journal of biological chemistry 284: 23540–23546.
17. Tasken K, Aandahl EM (2004) Localized effects of cAMP mediated by distinct routes of protein kinase A. Physiological reviews 84: 137–167.
18. Faruque OM, Le-Nguyen D, Lajoix AD, Vives E, Petit P, et al. (2009) Cell-permeable peptide-based disruption of endogenous PKA-AKAP complexes: a

tool for studying the molecular roles of AKAP-mediated PKA subcellular anchoring. American journal of physiology Cell physiology 296: C306–316.
19. Duchardt F, Fotin-Mleczek M, Schwarz H, Fischer R, Brock R (2007) A comprehensive model for the cellular uptake of cationic cell-penetrating peptides. Traffic 8: 848–866.
20. Kwon HB, Choi YK, Lim JJ, Kwon SH, Her S, et al. (2012) AKAP12 regulates vascular integrity in zebrafish. Experimental & molecular medicine 44: 225–235.
21. Choi YK, Kim JH, Kim WJ, Lee HY, Park JA, et al. (2007) AKAP12 regulates human blood-retinal barrier formation by downregulation of hypoxia-inducible factor-1alpha. The Journal of neuroscience : the official journal of the Society for Neuroscience 27: 4472–4481.
22. Barth N LA, Fisslthaler B., Fleming I (2009) Interfering with the association of A Kinase Anchoring Proteins (AKAPs) to PKA prevents the fluid shear stress induced phosphorylation of ENOs in endothelial cells. Acta Physiologica 2009, Volume 195, Supplement 669, The 88th Annual Meeting of the German Physiological Society, Giessen, Germany.
23. Jarnaess E, Tasken K (2007) Spatiotemporal control of cAMP signalling processes by anchored signalling complexes. Biochemical Society transactions 35: 931–937.
24. Logue JS, Whiting JL, Tunquist B, Langeberg LK, Scott JD (2011) Anchored protein kinase A recruitment of active Rac GTPase. The Journal of biological chemistry 286: 22113–22121.
25. Logue JS, Scott JD (2009) AKAP220 links the cAMP signaling pathway to cell adhesion. FASEB J, April 2009, 23, Meeting Abstract, 530 10.
26. Baumer Y, Spindler V, Werthmann RC, Bunemann M, Waschke J (2009) Role of Rac 1 and cAMP in endothelial barrier stabilization and thrombin-induced barrier breakdown. Journal of cellular physiology 220: 716–726.
27. Waschke J, Baumgartner W, Adamson RH, Zeng M, Aktories K, et al. (2004) Requirement of Rac activity for maintenance of capillary endothelial barrier properties. American journal of physiology Heart and circulatory physiology 286: H394–401.
28. Adamson RH, Curry FE, Adamson G, Liu B, Jiang Y, et al. (2002) Rho and rho kinase modulation of barrier properties: cultured endothelial cells and intact microvessels of rats and mice. The Journal of physiology 539: 295–308.
29. Baumer Y, Drenckhahn D, Waschke J (2008) cAMP induced Rac 1-mediated cytoskeletal reorganization in microvascular endothelium. Histochemistry and cell biology 129: 765–778.
30. Bradford MM (1976) A rapid and sensitive method for the quantitation of microgram quantities of protein utilizing the principle of protein-dye binding. Analytical biochemistry 72: 248–254.
31. Schlegel N, Baumer Y, Drenckhahn D, Waschke J (2009) Lipopolysaccharide-induced endothelial barrier breakdown is cyclic adenosine monophosphate dependent in vivo and in vitro. Critical care medicine 37: 1735–1743.
32. Michel CC, Mason JC, Curry FE, Tooke JE, Hunter PJ (1974) A development of the Landis technique for measuring the filtration coefficient of individual capillaries in the frog mesentery. Quarterly journal of experimental physiology and cognate medical sciences 59: 283–309.
33. Michel CC, Curry FE (1999) Microvascular permeability. Physiological reviews 79: 703–761.
34. Spindler V, Peter D, Harms GS, Asan E, Waschke J (2011) Ultrastructural Analysis Reveals cAMP-Dependent Enhancement of Microvascular Endothelial Barrier Functions via Rac1-Mediated Reorganization of Intercellular Junctions. The American journal of pathology 178: 2424–2436.
35. Spindler V, Schlegel N, Waschke J (2010) Role of GTPases in control of microvascular permeability. Cardiovascular research 87: 243–253.
36. Schlegel N, Waschke J (2009) VASP is involved in cAMP-mediated Rac 1 activation in microvascular endothelial cells. American journal of physiology Cell physiology 296: C453–462.
37. Rampersad SN, Ovens JD, Huston E, Umana MB, Wilson LS, et al. (2010) Cyclic AMP phosphodiesterase 4D (PDE4D) Tethers EPAC1 in a vascular endothelial cadherin (VE-Cad)-based signaling complex and controls cAMP-mediated vascular permeability. The Journal of biological chemistry 285: 33614–33622.
38. Michel JJ, Scott JD (2002) AKAP mediated signal transduction. Annual review of pharmacology and toxicology 42: 235–257.
39. Reinton N, Collas P, Haugen TB, Skalhegg BS, Hansson V, et al. (2000) Localization of a novel human A-kinase-anchoring protein, hAKAP220, during spermatogenesis. Developmental biology 223: 194–204.
40. Schlegel N, Waschke J (2010) Vasodilator-stimulated phosphoprotein: crucial for activation of Rac1 in endothelial barrier maintenance. Cardiovascular research 87: 1–3.
41. Waschke J, Drenckhahn D, Adamson RH, Curry FE (2004) Role of adhesion and contraction in Rac 1-regulated endothelial barrier function in vivo and in vitro. American journal of physiology Heart and circulatory physiology 287: H704–711.
42. Schlegel N, Burger S, Golenhofen N, Walter U, Drenckhahn D, et al. (2008) The role of VASP in regulation of cAMP- and Rac 1-mediated endothelial barrier stabilization. American journal of physiology Cell physiology 294: C178–188.
43. Schlegel N, Waschke J (2009) Impaired integrin-mediated adhesion contributes to reduced barrier properties in VASP-deficient microvascular endothelium. Journal of cellular physiology 220: 357–366.
44. Werthmann RC, von Hayn K, Nikolaev VO, Lohse MJ, Bunemann M (2009) Real-time monitoring of cAMP levels in living endothelial cells: thrombin transiently inhibits adenylyl cyclase 6. The Journal of physiology 587: 4091–4104.
45. Cioffi DL, Moore TM, Schaack J, Creighton JR, Cooper DM, et al. (2002) Dominant regulation of interendothelial cell gap formation by calcium-inhibited type 6 adenylyl cyclase. The Journal of cell biology 157: 1267–1278.
46. Beckers CM, van Hinsbergh VW, van Nieuw Amerongen GP (2010) Driving Rho GTPase activity in endothelial cells regulates barrier integrity. Thrombosis and haemostasis 103: 40–55.
47. Birukova AA, Burdette D, Moldobaeva N, Xing J, Fu P, et al. (2010) Rac GTPase is a hub for protein kinase A and Epac signaling in endothelial barrier protection by cAMP. Microvascular research 79: 128–138.
48. Fischmeister R (2006) Is cAMP good or bad? Depends on where it's made. Circulation research 98: 582–584.
49. Sayner SL, Alexeyev M, Dessauer CW, Stevens T (2006) Soluble adenylyl cyclase reveals the significance of cAMP compartmentation on pulmonary microvascular endothelial cell barrier. Circulation research 98: 675–681.
50. Rich TC, Fagan KA, Tse TE, Schaack J, Cooper DM, et al. (2001) A uniform extracellular stimulus triggers distinct cAMP signals in different compartments of a simple cell. Proceedings of the National Academy of Sciences of the United States of America 98: 13049–13054.
51. Sehrawat S, Ernandez T, Cullere X, Takahashi M, Ono Y, et al. (2011) AKAP9 regulation of microtubule dynamics promotes Epac1-induced endothelial barrier properties. Blood 117: 708–718.
52. Oldenburger A, Maarsingh H, Schmidt M (2012) Multiple facets of cAMP signalling and physiological impact: cAMP compartmentalization in the lung. Pharmaceuticals 5: 1291–1331.
53. Choi YK, Kim KW (2008) AKAP12 in astrocytes induces barrier functions in human endothelial cells through protein kinase Czeta. The FEBS journal 275: 2338–2353.
54. Schillace RV, Scott JD (1999) Association of the type 1 protein phosphatase PP1 with the A-kinase anchoring protein AKAP220. Current biology : CB 9: 321–324.
55. Schillace RV, Voltz JW, Sim AT, Shenolikar S, Scott JD (2001) Multiple interactions within the AKAP220 signaling complex contribute to protein phosphatase 1 regulation. The Journal of biological chemistry 276: 12128–12134.
56. Tanji C, Yamamoto H, Yorioka N, Kohno N, Kikuchi K, et al. (2002) A-kinase anchoring protein AKAP220 binds to glycogen synthase kinase-3beta (GSK-3beta) and mediates protein kinase A-dependent inhibition of GSK-3beta. The Journal of biological chemistry 277: 36955–36961.
57. Gorski JA, Gomez LL, Scott JD, Dell'Acqua ML (2005) Association of an A-kinase-anchoring protein signaling scaffold with cadherin adhesion molecules in neurons and epithelial cells. Molecular biology of the cell 16: 3574–3590.

14

Cateslytin, a Chromogranin A Derived Peptide Is Active against *Staphylococcus aureus* and Resistant to Degradation by Its Proteases

Rizwan Aslam[1], Céline Marban[1], Christian Corazzol[2], François Jehl[2], François Delalande[3], Alain Van Dorsselaer[3], Gilles Prévost[2], Youssef Haïkel[1,4], Corinne Taddei[4], Francis Schneider[5], Marie-Hélène Metz-Boutigue[1]*

1 Inserm UMR-1121, Université de Strasbourg, Strasbourg, France, **2** EA-7290, Virulence bactérienne précoce, Fédération de Médecine Translationnelle de Strasbourg, Institut de Bactériologie, Université de Strasbourg – CHRU Strasbourg, Strasbourg, France, **3** CNRS UMR-7178, Institut Pluridisciplinaire Hubert Curien, Université de Strasbourg, Strasbourg, France, **4** Faculté de chirurgie dentaire, Université de Strasbourg, Strasbourg, France, **5** Service de Réanimation Médicale, Hôpital de Hautepierre, Université de Strasbourg, Strasbourg, France

Abstract

Innate immunity involving antimicrobial peptides represents an integrated and highly effective system of molecular and cellular mechanisms that protects host against infections. One of the most frequent hospital-acquired pathogens, *Staphylococcus aureus*, capable of producing proteolytic enzymes, which can degrade the host defence agents and tissue components. Numerous antimicrobial peptides derived from chromogranins, are secreted by nervous, endocrine and immune cells during stress conditions. These kill microorganisms by their lytic effect at micromolar range, using a pore-forming mechanism against Gram-positive bacteria, filamentous fungi and yeasts. In this study, we tested antimicrobial activity of chromogranin A-derived peptides (catestatin and cateslytin) against *S. aureus* and analysed *S. aureus*-mediated proteolysis of these peptides using HPLC, sequencing and MALDI-TOF mass spectrometry. Interestingly, this study is the first to demonstrate that cateslytin, the active domain of catestatin, is active against *S. aureus* and is interestingly resistant to degradation by *S. aureus* proteases.

Editor: Stefan Bereswill, Charité-University Medicine Berlin, Germany

Funding: The authors thank the University of Strasbourg, the Inserm for financial support and the Faculty of Odontology and Higher Education Commission (HEC) of Pakistan for the PhD grant of RA. The funders had no role in study design, data collection and analysis, decision to publish, or preparation of the manuscript.

Competing Interests: The authors have declared that no competing interests exist.

* E-mail: marie-helene.metz@inserm.fr

Introduction

Chromogranins (Cgs) constitute the predominant family of proteins enclosed in secretory vesicles of chromaffin cells. [1] They are naturally processed to produce numerous peptides with various biological activities. [2–4] During the past decade, our group characterized several new antimicrobial peptides (AMPs) derived from chromogranin A (CgA) [5–8] and chromogranin B (CgB). [3,9] These peptides are released by stimulated chromaffin cells of adrenal medulla and also by activated polymorphonuclear neutrophils (PMNs). [5,7] Sequences of these peptides are highly conserved during evolution, suggesting that they are well-integrated in innate immune system. [10] Among these AMPs: chromofungin (CHR, CgA_{47-66}) and catestatin (CAT, $CgA_{344-364}$), derived from bovine CgA, activate PMNs and induce a calcium influx into immune cells. [11].

Staphylococcus aureus is the most frequently isolated pathogen in Gram-positive sepsis, often involved in blood clotting disorders and destruction of endocardial tissue. [12] *S. aureus* has developed several mechanisms to avoid immune response including resistance to AMPs, [13] impairment of phagocyte recruitment, [14] escape from neutrophil extracellular traps, [15] interference with complement, [16] neutrophil lysis, resistance to oxidative burst [17] and non-specific binding and degradation of immunoglobulins. [18] The AMPs evasion mechanisms deployed by *S. aureus* include proteolytic degradation by extracellular proteases of three major catalytic classes, namely metallo-, serine- and papain-like cysteine proteases. [19] The expression of proteolytic enzymes is controlled directly by global regulators of virulence factors such as *agr*, *sar* [20,21] and indirectly by RsbU that controls Sigma(B) activity.[22] Moreover, *SarA* is also a regulator of methicillin resistance factor (fmtA).[23] It has been previously reported that *S. aureus* metallo-protease aureolysin can cleave and inactivate human cathelicidin LL-37, thereby contributing to bacterial escape from the innate immune system. [13].

As staphylococci easily colonize skin and epithelia, regardless of the expression of antimicrobial Cgs-derived peptides, [24] we aimed to investigate the antimicrobial effects of CAT and its shorter fragment cateslytin (CTL, $CgA_{344-358}$) against *S. aureus*. Taking into account the difference in the activity of these peptides, using HPLC and proteomic analysis (sequencing and MALDI-TOF mass spectrometry), we examined the degradation of these CgA-derived peptides by staphylococcal proteases released into bacterial supernatants. Here, we report the relationship between

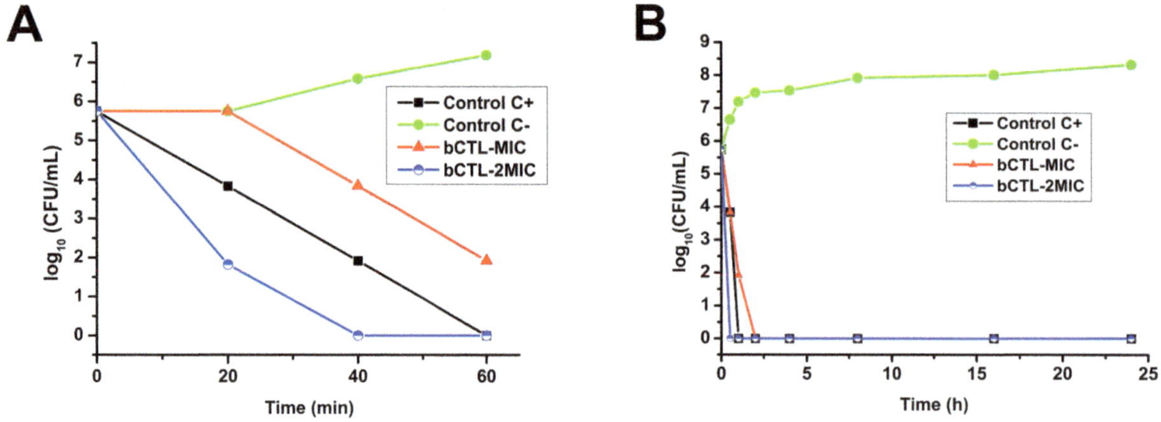

Figure 1. The bacterial killing kinetics of the bCTL against the *S. aureus* ATCC 25923. Different concentrations (MIC 40 µg/mL and 2× MIC 80 µg/mL) of bCTL were used. C+, represents antibiotic control (Cefotaxime 0.1 µg/mL + Tetracycline 10 µg/mL) and C-, represents phosphate buffer saline control. (A): *S. aureus* killing kinetics at time zero to 60 min. (B): *S. aureus* killing kinetics over 24 h time period.

peptide sequences and their sensitivity to bacterial proteases, as well as the possibility to use them as new antimicrobial agents in combination with antibiotics.

Materials and Methods

Preparation and analysis of synthetic antimicrobial peptides

Synthetic peptides were prepared on an Applied Biosystems 433A peptide synthesizer (Foster City, USA), using the stepwise solid-phase approach with 9-fluorenylmethoxycarbonyl (Fmoc) chemistry. Sequences of the synthetic peptides are as follows: bCAT (bCgA$_{344-364}$: RSMRLSFRARGYGFRGPGLQL), hCAT (hCgA$_{352-372}$: SSMKLSFRARGYGFRGPGPQL), bCTL (bCgA344–358: RSMRLSFRARGYGFR) and LL-37 (hCAP$_{134-170}$ LLGDFFRKSKEKIGKEFKRIVQRIKDFLRNLVPRTES). Then, the synthetic peptides were purified by a Dionex HPLC system (Ultimate 3000; Sunnyvale, CA USA) on a Macherey Nagel Nucleosil RP 300–5C18 column (10×250 mm; particle size 5 µm and pore size 100 nm). Synthetic peptides were analysed by an automated Edman sequencing on an Applied Sequencing System Procise (Applied Biosystems, Foster City, USA). [9] Finally, MALDI-TOF mass measurements were carried out on an Ultraflex^{TM} TOF/TOF (BrukerDaltonics, USA), as previously described. [25].

Isolation and characterization of *Staphylococcus aureus* strains

Different *S. aureus* strains were used to demonstrate the peptide antimicrobial activity and subsequently, the peptide degradation: strains ATCC 25923, ATCC49775, S1 and S2, were provided by the Institute of Bacteriology, Strasbourg, France. S1 was isolated from the blood of an 83 y. o. patient and S2 was isolated from the *sputum* of a 12 days old neonate. After isolation and identification, *S. aureus* strains were assessed for their susceptibility to various antibiotics, using the agar disc diffusion method. [26] S1 was found Methicillin resistant (MRSA) and is also resistant to Amoxicillin, Oxacillin, Amikacin, Tobramycin, Fluoroquinolones, Erythromycin and Clindamycin. However, it was susceptible to Gentamicin, Synercid, Co-trimoxazole, Rifampicin, Fusidic acid, Vancomycin, Teicoplanin and Linezolid. In contrast, S2 was found to be Methicillin susceptible (MSSA) and is sensitive to all the antibiotics tested.

Antibacterial activity against *S. aureus*

Different *S. aureus* strains described above were first pre-cultured aerobically at 37°C for 20 h in a Mueller-Hinton-Broth (MHB) medium, pH 7.3 (Difco Laboratories, Detroit, MI). Bacteria were suspended at absorbance of 0.001 at 620 nm in the MHB medium. Antibacterial activity was tested for 24 h incubation at 37°C with shaking by measuring the inhibition of bacterial growth. Ten µl final volumes (10–200 µg/mL) of synthetic peptides (LL-

Table 1. Antibacterial assays of catestatin (bovine, bCAT and human, hCAT) and cateslytin (bovine, bCTL) against different *S. aureus* strains, compared to the cathelicidin antimicrobial peptide-18 (LL-37).

Antimicrobial peptides	MIC (µg/mL)			
	S. aureus ATCC49775	*S. aureus* ATCC25923	*S. aureus* S1 (MRSA)	*S. aureus* S2 (MSSA)
bCAT	100 aA	100 aA	100 aA	115 bA
hCAT	125 aB	135 aB	130 aB	150 bB
bCTL	37 aC	40 aC	37 aC	45 bC
LL-37	30 aD	30 aD	30 aC	35 bC

Results are presented as MIC (µg/mL) of each peptide against four *S. aureus* strains. Values represent the means of the triplicate (n = 3) wells. Means with same letters are not significantly different ($p<0.05$). Small letters (a, b, c) represents significance between different *S. aureus* strains, while capital letters (A, B, C) represent significantly difference between peptides.

37, bCAT, hCAT and bCTL) were incubated in microtitration plates with 90 µl of a mid-logarithmic phase culture of bacteria, with a starting absorbance of 0.001 at 620 nm. In the initial inoculi, bacteria were quantified by the agar plate spreading method which was 5×10^5 colony forming units (CFU)/mL. [27] Tetracycline (10 mg/L) and Cefotaxime (0.1 mg/L) were used as positive controls. Microbial growth was assessed by the increase of absorbance after 24 h incubation at 37°C. [28,29] The $A_{620\ nm}$ value of control cultures growing in the absence of peptide and antibiotics was defined as 100% growth. $A_{620\ nm}$ zero with the antibiotics (Tetracycline and Cefotaxime) was taken as 100% inhibition. Absence of bacterial growth was verified by agar plate spreading. Each assay was performed in triplicates.

Killing kinetics

Bacterial strains were first grown in MHB medium as described above. Bacteria killing kinetic activity was measured according to the previously described method, after making few modifications. [30] Initial inoculum was prepared at a concentration of 5×10^5 CFU/mL with absorbance of 0.001 at 620 nm, which was calculated by agar plate spreading method. [27] Bacteria were incubated with different concentrations of peptides (MIC and 2× MIC), determined by MIC assay *via* microdilution method. [27] Viable bacterial count was then assessed at different time intervals up to 24 h. Briefly, aliquots were taken at different time intervals and were diluted in phosphate buffer saline (pH 7.4). After appropriate shaking 100 µL of the dilutions were plated on MH agar plates. Plates were incubated at 37°C and colony count was performed to determine CFU/mL after 24 h.

Preparation of *S. aureus* supernatants

S. aureus strains were precultured and plated on the agar plates and cultivated for 24 h at 37°C. After incubation, one colony *per* isolate was transferred to 10 mL of the MHB medium and incubated at 37°C for about 30 h to late stationary growth phase. After incubation, cultures were centrifuged at 10,000×*g* for 15 min and supernatants were filtered using a 0.22 µM Millex®-GV (Millipore, Carrigtwohill, Ireland) to eliminate bacteria. The supernatants were stored at −20°C until further use. In order to check sterility, 1 mL of each supernatant was incubated at 37°C for 48 h. Absence of a colony was interpreted as lack of viable bacteria.

Degradation analysis of the synthetic antimicrobial peptides by *S. aureus* supernatants

Synthetic peptides catestatin (bovine and human) and cateslytin (bovine) were incubated at 37°C for 24 h, with the different strain supernatants, at a concentration corresponding to the MIC values. Triplicate wells were treated for each concentration of peptide. In parallel, several controls were performed: with and without peptides in MHB medium.

In order to prevent degradation, the above experiment was repeated. For which, protease inhibitor cocktail (complete protease inhibitors except metalloproteases) (Roche Diagnostic GmbH, Mannheim, Germany) supplemented with 100 µM of 1–10 phenanthroline (metalloprotease inhibitor) (Sigma Aldrich GmbH, Steinheim, Germany) was added to bacterial supernatants before the addition of synthetic peptides. The whole suspension was incubated at 37°C for 24 h prior to being analyzed by RP-HPLC.

RP-HPLC purification of CgA-derived peptides after incubation with *S. aureus* supernatants

The synthetic peptides treated with bacterial supernatants and different controls including culture medium, bacterial supernatants, and synthetic peptides were separated using a Dionex HPLC system (Ultimate 3000; Sunnyvale, CA USA) on a Nucleosil reverse-phase 300–5C18-column (4×250 mm; particle size, 5 µm; porosity, 300 Å) (Macherey Nagel Hoerdt, France). Absorbance was monitored at 214 nm, and the solvent system consisted of 0.1% (v/v) TFA in water (solvent A) and 0.09% (v/v) TFA in 70% (v/v) acetonitrile-water (solvent B). Elution was performed at a flow rate of 700 µL/min, with the gradient as indicated on chromatograms. Fractions were collected with one min time interval. They were subsequently concentrated by evaporation by using a speed-vac, however not allowing them to dry completely.

Automatic Edman sequencing of CgA-derived peptides

The N-terminal sequence of purified peptides was determined by automatic Edman degradation analysis on a Procise micro-sequencer (Applied Biosystems, Courtaboeuf, France). Samples purified by HPLC were loaded on polybrene-treated glass-fiber filters. Phenylthiohydantoin-amino acids (Pth-Xaa) were identified by chromatography on a C_{18} column (PTH C-18, 2.1 mm×200 mm). [2].

Proteomic analysis

Mass determination was carried out on a Brucker BIFLEX™ Matrix-Assisted Laser Desorption Ionization – Time-Of-Flight mass spectrometer (MALDI-TOF) (equipped with the high resolution optics (SCOUT™) with X–Y multi-sample probe, a gridless reflector and with the HIMAS™ linear detector). With a maximum accelerating potential of 30 kV, the system can be operated either in linear or the reflective mode. Ionization was carried out with a 337-nm beam from a nitrogen laser with a repetition rate of 3 Hz. The output signal from the detector was digitized at a sampling rate of 250 MHz in linear mode and 500 MHz in reflector mode using a 1 GHz digital oscilloscope (Lecroy model). Brucker supplied software hosted in Sun sparcworkstation was used for data processing and the instrumental control. [31] These studies were realized using the matrix α-cyano-4-hydroxycinnamic acid, obtained from Sigma, and prepared as a saturated solution in acetone. A total of 1–2 µl of the sample matrix aliquot solution was deposited on the probe and dried by ambient air. A thin layer matrix crystal was obtained after rapid spreading and evaporation. [32] A micromolar analyte solution was applied to the matrix and allowed to dry under moderate vacuum. The whole preparation was washed by 1 µl of trifluoroacetic acid (0.5%) aqueous solution. This cleaning procedure helps remove remaining alkaline cations and often leads to an increase in sensitivity.

Statistical analysis

The MIC values (µg/mL) are reported as means of three independent experiments. To determine significance, between different peptides (rows) and bacterial strains (columns), one way ANOVA was used. Overall significance and correlation was evaluated by 4×4 factorial ANOVA for independent samples. Significance was accepted at $p\leq0.05$.

Results

Antibacterial activity of CgA-derived peptides against *S. aureus*

The aim of present study was to examine antibacterial activity of bovine and human CAT ($bCgA_{344-364}$ and $hCgA_{352-372}$) and bovine CTL ($bCgA_{344-358}$), against *S. aureus*. In comparison with the other peptides tested, CTL was found to be most active against all *S. aureus* strains tested (ATCC49775, ATCC25923, S1, and S2).

Activity of these CgA derived peptides was compared to the well-known C-terminal peptide (LL-37) of hCAP-18 (Table 1). Activity of bCTL was comparable to that of LL-37 ($p>0.05$) against S1 and S2, but significantly different ($p<0.05$) for ATCC49775 and ATCC25923. However, MIC (presenting 100% growth inhibition) values for bCAT and hCAT were highly significantly different ($p<0.01$) that were about 2–4 fold higher than short fragment CTL (Table 1). MIC value for CTL varies between different *MRSA* and *MSSA* strains ($p<0.05$) and can effectively kill bacteria at 45 µg/mL. While both human and bovine CAT have more than 100 µg/mL MIC values. In order to detail CTL antimicrobial activity, we further determined the killing rate of CTL. Time-killing kinetic assays were performed against *S. aureus* (ATCC 25923). As shown in Figure 1, CTL acts very rapidly against *S. aureus*. AT concentration 2×MIC CTL present 4 logs bacterial killing at within 20 min of treatment and 100% within 40 min (Figure 1A). Moreover, at MIC, CTL reaches 100% bacterial killing within 60 min of treatment. This killing assay was continued till 24 h (Figure 1B) and bacteria were unable to grow further during 24 h. As CTL is the most active, we hypothesized this fragment is more resistant to proteases produced by *S. aureus*. In order to demonstrate the stability of CTL compared to human and bovine CAT, we incubated the three peptides with the supernatants of four different *S. aureus* strains.

Analysis of the proteolytic cleavage of catestatin (h/bCAT) and cateslytin (CTL) in the presence of *S. aureus* strain supernatants

S. aureus agr and *sar* regulate expression of proteases, which often are modulated by *SarA* regulator during biofilm formation. [33] These proteases are mostly expressed in the late growth phase and secreted in the extracellular environment to facilitate the bacterial spread. [34] As described above, *S. aureus* supernatants were prepared by the extended incubation period and are rich in proteases. Four *S. aureus* strains were tested for these experiments: two of them are referenced strains (ATCC 49775 and ATCC 25923), while two strains (S1 and S2) were isolated from the patients of Strasbourg hospital. S1 is a *MRSA* (methicillin resistant *S. aureus*) and S2 is a *MSSA* (methicillin sensitive *S. aureus*). The synthetic peptides were incubated with *S. aureus* supernatants and were separated by RP-HPLC (Figure 2 A, C, E). The HPLC profiles of CAT and CTL are compared with HPLC profiles of MHB medium and *S. aureus* supernatants. Peaks resulting from the bacterial degradation were analyzed by automatic Edman sequencing and MALDI-TOF. In contrast to bovine and human CAT (Figure 2A, 2C), CTL is not degraded (Figure 2E).

HPLC of bCAT (Figure 2A) indicates 2 major peaks, eluted at 38.7 and 40 min, corresponding to full-length peptide according to sequencing and MALDI-TOF mass spectrometry (2426 Da) (Figure 3). Presence of these 2 isoforms may be related to beta-turn conformation induced by proline residue in position 360 and its isomeric state (cis/trans). [35] In MHB medium, bCAT was not processed. Whereas, in presence of culture supernatants (S49775, S25923 and S1), bCAT was processed to generate fragments eluted at 38.3 and 38.6 min (Figure 2A). Sequencing and MALDI-TOF analysis indicate that these 2 fragments correspond to the isoforms of $bCgA_{349-364}$ (1782 Da) (Figure 3). In presence of culture supernatant of S2, bCAT was largely cleaved to generate fragments eluted at 28.0, 36.6 and 37.5 min (Figure 2A). Sequencing and MALDI-TOF analysis indicate that these fragments correspond to $bCgA_{350-356}$ (826 Da) and the 2 isoforms of $bCgA_{357-364}$ (887 Da) (Figure 3).

Moreover, the HPLC profile of hCAT (Figure 2C) showed one major peak at 38 min that corresponds to the full-length peptide according to sequencing and MALDI-TOF mass spectrometry (2327 Da) (Figure 3). In presence of MHB medium, hCAT was not cleaved. However, after incubation with S49775, S25923 and S1, hCAT was partially processed to generate a fragment eluted at 36 min. Sequencing and MALDI-TOF analysis indicated that this fragment corresponds to $hCgA_{357-372}$ (1780 Da) (Figure 3). In presence of supernatant S2, hCAT was completely cleaved to generate fragments eluted at 28.5 and 33 min (Figure 2C). Sequencing and MALDI-TOF analysis indicated that these fragments correspond to $hCgA_{358-364}$ (840 Da) and $hCgA_{365-372}$ (871 Da) (Figure 3).

Finally, the shortest peptide bCTL (bCTL, $CgA_{344-358}$), is eluted as a single peak (Figure 2E) and is not degraded by any of the staphylococcal strains tested.

Interestingly, incubation of synthetic peptides with bacterial supernatants supplemented with protease inhibitors, prevents the peptide degradation (Figure 2B, D, F). HPLC profiles of bCAT (Figure 2B: chromatogram 2–5) indicate that it is not cleaved by any of the four strain supernatants tested in presence of protease inhibitors. Similar results were obtained for hCAT (Figure 2D: chromatogram 2–5). In addition, CTL remains unaffected by *S. aureus* supernatants with or without protease inhibitors (Figure 2E, F).

The proteolytic cleavage sites of bCAT and hCAT obtained after incubation with S49775, S25923 and S1 were identical (Figure 3). They corresponded to the peptide bonds $L_{348}-S_{349}$ and $L_{356}-S_{357}$ for bCAT and hCAT, respectively. With S2, they correspond to the peptide bonds $S_{349}-F_{350}+G_{356}-F_{357}$ and $S_{357}-F_{358}+G_{364}-F_{365}$ for bCAT and hCAT, respectively. It is important to point out that the short peptide bCTL ($bCgA_{344-358}$) derived from CAT was not cleaved by *S. aureus* proteases (Figure 2E, F) demonstrating that CTL resists to the proteolytic cocktail produced by *S. aureus*.

Discussion

Due to continuous resistance development by *S. aureus*, the interest for natural AMPs is increasing because some of them display potent antibacterial activity. However, secondary reactions such as hemolytic activity, [36] limit their use as systemic anti-infective drugs. Until recently, no such reaction was reported for the antimicrobial Cgs-derived peptides in the literature. Several CgA-derived peptides were previously characterized to display antimicrobial properties, [9] however, data concerning killing of *S. aureus* is not reported. CAT was previously reported, to be active against different bacterial strains at micromolar range. [24] CTL is well characterized for antimicrobial activity against *Micrococcus luteus*, few yeasts and fungal strains. Moreover, it is not hemolytic for human erythrocytes. [5]. In the present study, for the first time, we brought to light the antibacterial activity of CTL against different strains of *S. aureus* with a MIC value of 37–45 µg/mL (Table 1).

In vivo, a cross talk is established *via* dynamics of the interactions between host and bacteria. Many bacterial strains express a variety of proteases, ranging from non-specific and powerful enzymes that degrade many proteins involved in innate immunity, to proteases that are extremely specific in their mode of action. [37] Here, we questioned whether *S. aureus* proteases can cleave the three CgA-derived peptides studied. This proteolytic degradation could explain the 3–4 fold difference observed in antibacterial activity. To address this issue, we hypothesized that the human and bovine CAT peptides are degraded by the staphylococcal proteases and that CTL resists to the degradation and maintains its activity. Therefore, we tested the effect of four staphylococcal supernatants and demonstrated that bovine and human CAT are similarly

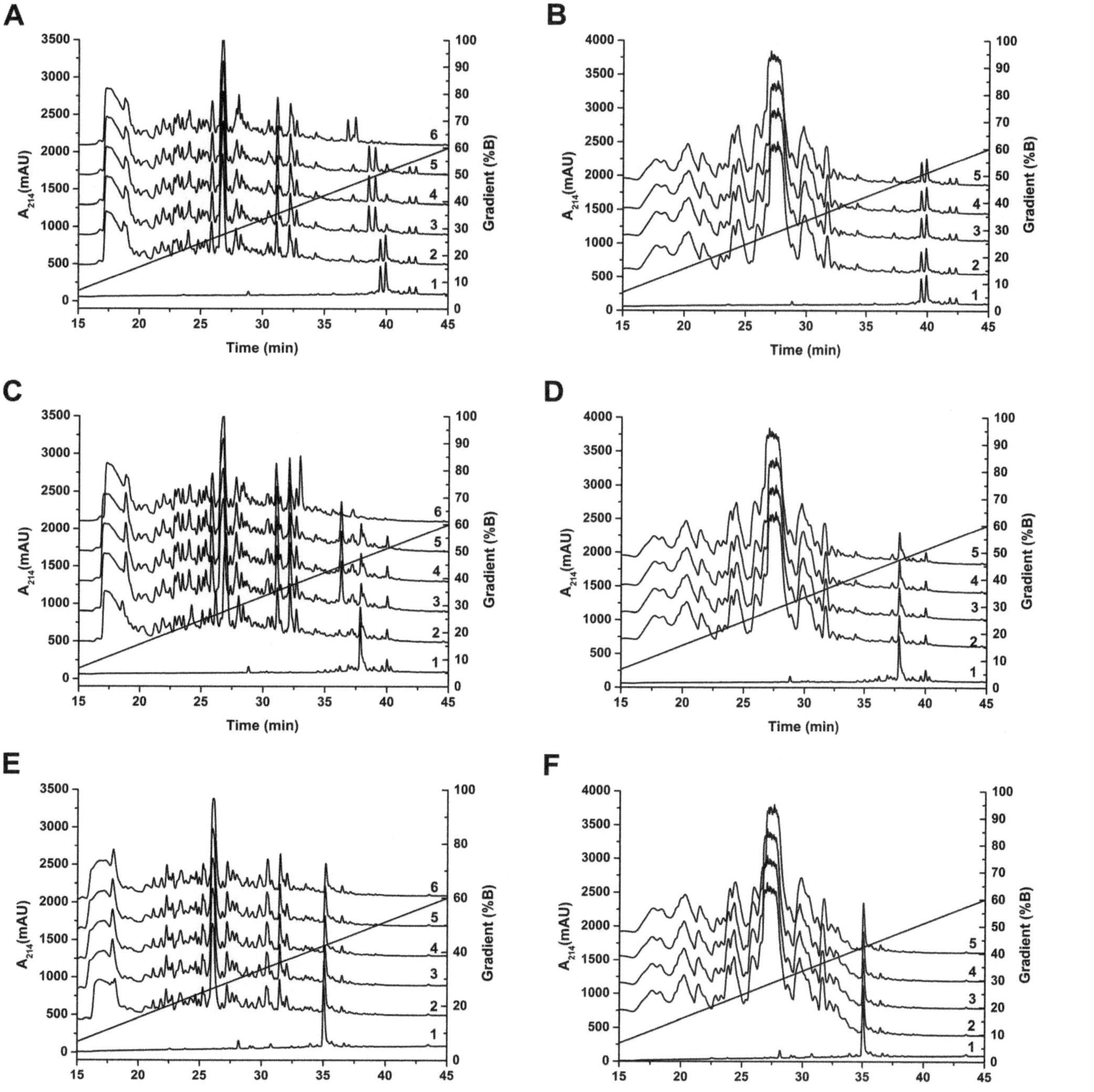

Figure 2. HPLC chromatograms of bCAT, hCAT and CTL alone or with different bacterial strain supernatants, with or without protease inhibitors. (A): Alignment of the HPLC chromatograms corresponding to: (1) bCAT, (2) bCAT+MHB, (3) bCAT+S49775, (4) bCAT+S25923 (5) bCAT+S1 (6) bCAT+S2. (B): Alignment of the HPLC chromatograms corresponding to: (1) bCAT, (2) bCAT+Pi+S49775, (3) bCAT+Pi+S25923 (4) bCAT+Pi+S1 (5) bCAT+Pi+S2. (C): Alignment of the HPLC chromatograms corresponding to: (1) hCAT, (2) hCAT+MHB, (3) hCAT+S49775, (4) hCAT+S25923 (5) hCAT+S1 (6) hCAT+S2. (D): Alignment of the HPLC chromatograms corresponding to: (1) hCAT, (2) hCAT+Pi+S49775, (3) hCAT+Pi+S25923 (4) hCAT+Pi+S1 (5) hCAT+Pi+S2. (E): Alignment of the HPLC chromatograms corresponding to: (1) bCTL, (2) bCTL+MHB, (3) bCTL+S49775, (4) bCTL+S25923 (5) bCTL+S1 (6) bCTL+S2. (F): Alignment of the HPLC chromatograms corresponding to: (1) bCTL, (2) bCTL+Pi+S49775, (3) bCTL+Pi+S25923 (4) bCTL+Pi+S1 (5) bCTL+Pi+S2.

degraded to generate inactive fragments, while CTL is not degraded. For bovine and human CAT, the cleavage sites induced by the supernatants are different. For S2 these correspond to the cleavage of S-F/G-F. Whereas, for ATCC49775, ATCC25923, and S1, it corresponds to cleavage of L-S (Figure 3). Upon cleavage by *S. aureus* proteases, the CAT peptides are no longer active against *S. aureus*. Previous data showed that bovine $CgA_{344-351}$ and $CgA_{348-358}$ display an antimicrobial activity against *Micrococcus luteus* at 20 μM, suggesting that L348/356 is important for its antibacterial activity. [5] Furthermore, in the protease inhibition assay, we demonstrated that degradation of human and bovine CAT can be prevented by the use of protease inhibitors. CTL maintains its position either with or without protease inhibitors indicating that proteases have no effect on the CTL.

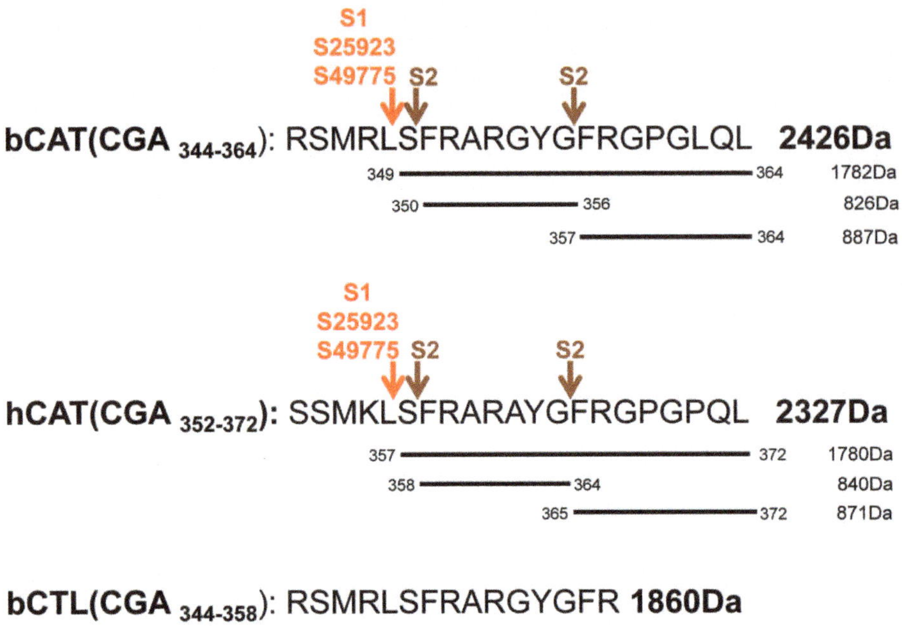

Figure 3. The proteolytic cleavage sites induced after treatment of bCAT, hCAT and CTL with different *S. aureus* strains (ATCC25923, ATCC49775, S1 and S2). The experimental molecular masses obtained after MALDI-TOF analysis, are indicated.

The biosynthesis of CTL results by the action of prohormone convertases PC1/2 present in the granular matrix of chromaffin cells. [2,38] The prohormone thiol protease (PTP) was also reported to be essential for CAT synthesis (bCgA$_{344-364}$) by cleaving D-R and L-R. [38] In addition, in chromaffin secretory vesicles, the cysteine protease cathepsin L (CTSL) [39] generates CTL by the additional cleavage R-G of CAT. CAT fragment is known to activate the neutrophils by inducing extra-cellular calcium influx in human neutrophils *via* calmodulin-regulated calcium independent phospholipase A2. [11] Here, we demonstrate that CTL is resistant to the degradation by staphylococcal proteases, which strengthens the involvement of this CgA-derived domain in innate immunity. [10,40].

The primary structures of CAT and CTL are highly conserved during evolution (Figure 4). Arginine ratio of these peptide sequences is important, since it modulates the interaction with negatively charges of the microorganism's membranes. In addition, it was previously shown that arginine residues have a high tendency to interact with the lipids as suggested for other peptides, such as HIV-1 transcriptional activator Tat protein. [41,42] For hCAT, bCAT, and bCTL the arginine ratios are 15%, 23%, and 33% respectively. High arginine ratio of bCTL supports a strong interaction with the negatively charged lipid bilayer as compared to the both CAT. Structure activity relationship of CTL with bacterial membrane is also demonstrated by recent experiments concerning the CTL derived peptide with a hydrophobic N-terminal end FLE-CTL (FLE-RSMRLSFRARGYGFR) (5R). With the help of HPLC and antimicrobial assays, we have shown that this synthetic peptide is inactive against *S. aureus* at 400 µg/mL and interacts strongly with bacterial membrane. In contrast, the shorter peptide that is lacking the C-terminal end (YGFR) of CTL: FLE-RSMRLSFRARG, displays antibacterial activities at 200 µg/mL (data not shown). Thus, CTL sequence possesses the cationic amphipathic features to exhibit potent antimicrobial activities.

Concerning its secondary structure, an aggregated antiparallel beta sheet structure was previously reported for CTL [43] by CD (Circular Dichroism) and ATR (Attenuated total reflectance) experiments, whereas hCAT and bCAT form short helical structures (residue 7 to residue 11) in the presence of high concentrations of DPC, as confirmed by CD and NMR (Nuclear magnetic resonance) experiments. [44] CTL adopts a major β-sheet character only on negatively charged membranes, whereas it is essentially unstructured in water. These β-sheet formations give a much more stability to peptide than the helical formation reported for CAT. [44] In addition, the (HR-MAS) ^{1}H NMR analysis of CTL indicates that the arginine and hydrophobic residues are in close proximity, thus creating a deep penetration of charged residues into the membrane. [42] Therefore, the electrostatic interaction between positively charged arginine residues and negatively charged lipids appears to be responsible for binding of CTL to the lipid bilayer and the aromatic residues stabilize the lipid-peptide interaction.

In our group, we recently evaluated the synergistic effect of three CgA-derived peptides (CAT, CTL and amidated CTL) with antibiotics and demonstrated that this co-treatment induce a reduction of the antibiotics concentration used and can potentiate their activities. Antimicrobial assays were carried out by combining the AMPs at concentration below the MIC. The comparison was made with antibiotic or peptide separately at the same doses. For all these experiments, we evaluated the Fractional Inhibitory Concentration (FIC) of CgA-derived peptides combined with Minocyclin or Voriconazole. FIC corresponds to a synergistic effect in the range of ≤ 0.5, an additive effect when it is between >0.5 and <2, and an antagonistic effect when it is >2. [45] For the combination of amidated CTL and Minocyclin, we obtained a FIC of 0.37 against *S. aureus,* and for CTL and Voriconazole, we obtained a FIC of 0.25, and 0.5 against *Candida albicans* and *Candida tropicalis* respectively. [10] Regarding these data, one could imagine a mechanism in which, the peptides could favor destabilization of bacterial membrane, thus allowing the antibiotics to rapidly penetrate inside bacterial cells to reaching their site of action. These *in vitro* results are likely to occur *in vivo*, especially during systemic inflammation. Indeed, sepsis triggers numerous

Organism	Code	Sequence	%
		5 10 15 20	
Bovine	P05059	RSMRL SFRAR GYGFR GPGLQ L	100
Human	P10645	SSMKL SFRGR GYGFR GPGPQ L	85
Human G364S	P10645	SSMKL SFRGR GYSFR GPGPQ L	80
Human P370L	P10645	SSMKL SFRGR GYGFR GPGLQ L	80
Macacafascicularis	Q4R4V1	RSMKL SFRAR AYGFR GPGPQ L	90
Macacamulata	12CTA3	RSMKL SFRAR AYGFR GPGPQ L	90
Pan troglodytes	H2Q8T7	SSMKL SFRAR AYGFR GPGPQ L	85
Gorillagorillagorilla	G3SJY7	SSMKL SFRAR AYGFR GPGPQ L	85
Nomascusleucogenys	G1S5U4	SSMKL SFRAR AYGFR GPGPQ L	85
Pongo abelii	H2NM30	WSMKL PFRAR AYGFR GPGPQ L	75
Otolemurgarnettii	HOWYW2	RSMKL SLQTR AYDFR GPGPQ L	76
Horse	Q9XS63	RSMKL SFRAR AYGFR GPGLQ L	95
Sus scrofa	P04404	RSMKL SFRAP AYGFR GPGLQ L	90
Caviaporcellus	H0VR23	RSMKL SFRAQ AYGFP GPEPQ L	80
Ailuropodamelanoleuca	GIM4X7	RSMKL SFRAR AYDFR GPGLQ L	90
Canislupus	C5NM83	RSMKL SFRAR AYDFR GPGLP L	85
Rattusnorvegicus	P10354	RSMKL SFRAR AYGFR DPGPQ L	85
Heterocephalusglaber	G5B2Y1	RSMKL SFQAP AYDFR GSGPQ L	76
Mus musculus	P26339	RSMKL SFRTR AYGFR DPGPQ L	80
Oryctolaguscuniculus	g1TSN0	RSMKL SFQPR AYGFR GPGPQ L	85
Cricetulusgriseus	G3HU91	RSMKL SFRAR AYGFR DPRPQ L	85
Callithrixjacchus	F3T376	HSMKL SFQAR AYDFR GPGPQ L	80
Loxodontaafricana	G3TBQ8	RSMKL SFRAQ AYGFP GPEPQ L	80
Consensus		+SM+L SF+a+ aYaF+ aPaPQ L	

Figure 4. Sequence alignment of bovine catestatin ($CgA_{344-364}$) with corresponding fragments from several species. For each position predominant identical residues are indicated in bold letters. Homology sequence is indicated (%).The data base used is UniProtKB. (+, basic residue; a, for A/G/T/P).

changes in proteases and inhibitors activity. [46] These regulations have been strongly related with sepsis severity. [47] Under such influences, CAT and CTL could even be difficult to release, owing to structural modifications of CgA, [48] due to sepsis-related oxidative stress. These modifications include oxidation of methionine, aromatic residues, glycanes, phosphorylation and also aggregation of the complete protein that might prevent the processing of CgA to produce active antimicrobial peptides. The structure-function relationship of AMPs is also very important to unravel the regular pattern of antibacterial activity. It has been previously demonstrated, that selective end tagging or short hydrophobic stretches can be, added to increase AMPs activity. [49,50] Recently, we studied the properties of a modified CTL by inserting cysteine at C-terminal. This modified CTL was used to develop antimicrobial polymer coatings *via* alternative deposition of polyelectrolytes conjugated to CTL-C. [51] Activity of CTL is not altered, even with this modification or when inserted into PEM (Polyelectrolyte multilayer) coatings. Moreover, it is non-toxic for human gingival fibroblast. [51].

To conclude, for the first time we present the involvement of staphylococcal proteases on the cleavage of antimicrobial CgA-derived peptides. As the CTL domain is resistant to staphylococcal proteases, it constitutes a promising natural antibacterial peptide for further studies.

Acknowledgments

We are grateful for the generous gift of bacterial strains from the bacteriology Institute of Civil Hospital of Strasbourg. Authors thank Cosette Betcha and Ouria Tahar for their excellent technical assistance. We also acknowledge Azhar Ayyaz Pirzado and Adnan Khan Niazi for their help in writing.

Author Contributions

Conceived and designed the experiments: RA CC FD FS MHM. Performed the experiments: RA CM CC FD. Analyzed the data: RA CM FJ FD AVD GP MHM. Contributed reagents/materials/analysis tools: AVD GP YH CT MHM. Wrote the paper: RA GP FS MHM.

References

1. Helle KB (2010) Regulatory peptides from chromogranin A and secretogranin II: putative modulators of cells and tissues involved in inflammatory conditions. Regul Pept 165: 45–51.
2. Metz-Boutigue MH, Garcia-Sablone P, Hogue-Angeletti R, Aunis D (1993) Intracellular and extracellular processing of chromogranin A. Determination of cleavage sites. Eur J Biochem 217: 247–257.
3. Strub JM, Garcia-Sablone P, Lonning K, Taupenot L, Hubert P, et al. (1995) Processing of chromogranin B in bovine adrenal medulla. Identification of secretolytin, the endogenous C-terminal fragment of residues 614-626 with antibacterial activity. Eur J Biochem 229: 356–368.
4. Montero-Hadjadje M, Vaingankar S, Elias S, Tostivint H, Mahata SK, et al. (2008) Chromogranins A and B and secretogranin II: evolutionary and functional aspects. Acta Physiol (Oxf) 192: 309–324.
5. Briolat J, Wu SD, Mahata SK, Gonthier B, Bagnard D, et al. (2005) New antimicrobial activity for the catecholamine release-inhibitory peptide from chromogranin A. Cell Mol Life Sci 62: 377–385.
6. Lugardon K, Chasserot-Golaz S, Kieffer AE, Maget-Dana R, Nullans G, et al. (2001) Structural and biological characterization of chromofungin, the antifungal chromogranin A-(47–66)-derived peptide. J Biol Chem 276: 35875–35882.

7. Lugardon K, Raffner R, Goumon Y, Corti A, Delmas A, et al. (2000) Antibacterial and antifungal activities of vasostatin-1, the N-terminal fragment of chromogranin A. J Biol Chem 275: 10745–10753.
8. Strub JM, Goumon Y, Lugardon K, Capon C, Lopez M, et al. (1996) Antibacterial activity of glycosylated and phosphorylated chromogranin A-derived peptide 173–194 from bovine adrenal medullary chromaffin granules. J Biol Chem 271: 28533–28540.
9. Metz-Boutigue MH, Goumon Y, Lugardon K, Strub JM, Aunis D (1998) Antibacterial peptides are present in chromaffin cell secretory granules. Cell Mol Neurobiol 18: 249–266.
10. Aslam R, Atindehou M, Lavaux T, Haikel Y, Schneider F, et al. (2012) Chromogranin A-derived peptides are involved in innate immunity. Curr Med Chem 19: 4115–4123.
11. Zhang D, Shooshtarizadeh P, Laventie BJ, Colin DA, Chich JF, et al. (2009) Two chromogranin a-derived peptides induce calcium entry in human neutrophils by calmodulin-regulated calcium independent phospholipase A2. PLoS One 4: e4501.
12. Ohbayashi T, Irie A, Murakami Y, Nowak M, Potempa J, et al. (2011) Degradation of fibrinogen and collagen by staphopains, cysteine proteases released from Staphylococcus aureus. Microbiology 157: 786–792.
13. Sieprawska-Lupa M, Mydel P, Krawczyk K, Wojcik K, Puklo M, et al. (2004) Degradation of human antimicrobial peptide LL-37 by Staphylococcus aureus-derived proteinases. Antimicrob Agents Chemother 48: 4673–4679.
14. Postma B, Poppelier MJ, van Galen JC, Prossnitz ER, van Strijp JA, et al. (2004) Chemotaxis inhibitory protein of Staphylococcus aureus binds specifically to the C5a and formylated peptide receptor. J Immunol 172: 6994–7001.
15. Sumby P, Barbian KD, Gardner DJ, Whitney AR, Welty DM, et al. (2005) Extracellular deoxyribonuclease made by group A Streptococcus assists pathogenesis by enhancing evasion of the innate immune response. Proc Natl Acad Sci U S A 102: 1679–1684.
16. Rooijakkers SH, Ruyken M, Roos A, Daha MR, Presanis JS, et al. (2005) Immune evasion by a staphylococcal complement inhibitor that acts on C3 convertases. Nat Immunol 6: 920–927.
17. Karavolos MH, Horsburgh MJ, Ingham E, Foster SJ (2003) Role and regulation of the superoxide dismutases of Staphylococcus aureus. Microbiology 149: 2749–2758.
18. Prokesova L, Potuznikova B, Potempa J, Zikan J, Radl J, et al. (1995) Cleavage of human immunoglobulins by proteinase from Staphylococcus aureus. Adv Exp Med Biol 371A: 613–616.
19. Dubin G (2002) Extracellular proteases of Staphylococcus spp. Biol Chem 383: 1075–1086.
20. Abdelnour A, Arvidson S, Bremell T, Ryden C, Tarkowski A (1993) The accessory gene regulator (agr) controls Staphylococcus aureus virulence in a murine arthritis model. Infect Immun 61: 3879–3885.
21. Chien Y, Manna AC, Projan SJ, Cheung AL (1999) SarA, a global regulator of virulence determinants in Staphylococcus aureus, binds to a conserved motif essential for sar-dependent gene regulation. J Biol Chem 274: 37169–37176.
22. Palma M, Cheung AL (2001) sigma(B) activity in Staphylococcus aureus is controlled by RsbU and an additional factor(s) during bacterial growth. Infect Immun 69: 7858–7865.
23. Zhao Y, Verma V, Belcheva A, Singh A, Fridman M, et al. (2012) Staphylococcus aureus methicillin-resistance factor fmtA is regulated by the global regulator SarA. PLoS One 7: e43998.
24. Radek KA, Lopez-Garcia B, Hupe M, Niesman IR, Elias PM, et al. (2008) The neuroendocrine peptide catestatin is a cutaneous antimicrobial and induced in the skin after injury. J Invest Dermatol 128: 1525–1534.
25. Sizova D, Charbaut E, Delalande F, Poirier F, High AA, et al. (2007) Proteomic analysis of brain tissue from an Alzheimer's disease mouse model by two-dimensional difference gel electrophoresis. Neurobiol Aging 28: 357–370.
26. Davis MF, Peterson AE, Julian KG, Greene WH, Price LB, et al. (2013) Household Risk Factors for Colonization with Multidrug-Resistant Staphylococcus aureus Isolates. PLoS One 8: e54733.
27. National Committee for Clinical Laboratory Standards (2003) Methods for Dilution Antimicrobial Susceptibility Tests for Bacteria That Grow Aerobically–*Eighth Edition: Approved Standard M07-A8.*: NCCLS, Wayne, PA, USA.
28. Bulet P, Dimarcq JL, Hetru C, Lagueux M, Charlet M, et al. (1993) A novel inducible antibacterial peptide of Drosophila carries an O-glycosylated substitution. J Biol Chem 268: 14893–14897.
29. Wu M, Hancock RE (1999) Interaction of the cyclic antimicrobial cationic peptide bactenecin with the outer and cytoplasmic membrane. J Biol Chem 274: 29–35.
30. Pag U, Oedenkoven M, Papo N, Oren Z, Shai Y, et al. (2004) In vitro activity and mode of action of diastereomeric antimicrobial peptides against bacterial clinical isolates. J Antimicrob Chemother 53: 230–239.
31. Pons AM, Zorn N, Vignon D, Delalande F, Van Dorsselaer A, et al. (2002) Microcin E492 is an unmodified peptide related in structure to colicin V. Antimicrob Agents Chemother 46: 229–230.
32. Kussmann M, Lassing U, Sturmer CA, Przybylski M, Roepstorff P (1997) Matrix-assisted laser desorption/ionization mass spectrometric peptide mapping of the neural cell adhesion protein neurolin purified by sodium dodecyl sulfate polyacrylamide gel electrophoresis or acidic precipitation. J Mass Spectrom 32: 483–493.
33. Mrak LN, Zielinska AK, Beenken KE, Mrak IN, Atwood DN, et al. (2012) saeRS and sarA act synergistically to repress protease production and promote biofilm formation in Staphylococcus aureus. PLoS One 7: e38453.
34. Dunman PM, Murphy E, Haney S, Palacios D, Tucker-Kellogg G, et al. (2001) Transcription profiling-based identification of Staphylococcus aureus genes regulated by the agr and/or sarA loci. J Bacteriol 183: 7341–7353.
35. Vanhoof G, Goossens F, De Meester I, Hendriks D, Scharpe S (1995) Proline motifs in peptides and their biological processing. Faseb J 9: 736–744.
36. Conlon JM, Mechkarska M, Prajeep M, Sonnevend A, Coquet L, et al. (2012) Host-defense peptides in skin secretions of the tetraploid frog Silurana epitropicalis with potent activity against methicillin-resistant Staphylococcus aureus (MRSA). Peptides 37: 113–119.
37. Potempa J, Pike RN (2009) Corruption of innate immunity by bacterial proteases. J Innate Immun 1: 70–87.
38. Lee JC, Taylor CV, Gaucher SP, Toneff T, Taupenot L, et al. (2003) Primary sequence characterization of catestatin intermediates and peptides defines proteolytic cleavage sites utilized for converting chromogranin a into active catestatin secreted from neuroendocrine chromaffin cells. Biochemistry 42: 6938–6946.
39. Biswas N, Rodriguez-Flores JL, Courel M, Gayen JR, Vaingankar SM, et al. (2009) Cathepsin L colocalizes with chromogranin a in chromaffin vesicles to generate active peptides. Endocrinology 150: 3547–3557.
40. Sternberg EM (2006) Neural regulation of innate immunity: a coordinated nonspecific host response to pathogens. Nat Rev Immunol 6: 318–328.
41. Henriques ST, Melo MN, Castanho MA (2006) Cell-penetrating peptides and antimicrobial peptides: how different are they? Biochem J 399: 1–7.
42. Ziegler A (2008) Thermodynamic studies and binding mechanisms of cell-penetrating peptides with lipids and glycosaminoglycans. Adv Drug Deliv Rev 60: 580–597.
43. Jean-Francois F, Castano S, Desbat B, Odaert B, Roux M, et al. (2008) Aggregation of cateslytin beta-sheets on negatively charged lipids promotes rigid membrane domains. A new mode of action for antimicrobial peptides? Biochemistry 47: 6394–6402.
44. Sugawara M, Resende JM, Moraes CM, Marquette A, Chich JF, et al. (2010) Membrane structure and interactions of human catestatin by multidimensional solution and solid-state NMR spectroscopy. Faseb J 24: 1737–1746.
45. Serra P, Brandimarte C, Martino P, Carlone S, Giunchi G (1977) Synergistic treatment of enterococcal endocarditis: in vitro and in vivo studies. Arch Intern Med 137: 1562–1567.
46. Kwiecinski J, Josefsson E, Mitchell J, Higgins J, Magnusson M, et al. (2010) Activation of plasminogen by staphylokinase reduces the severity of Staphylococcus aureus systemic infection. J Infect Dis 202: 1041–1049.
47. Michel A, Agerer F, Hauck CR, Herrmann M, Ullrich J, et al. (2006) Global regulatory impact of ClpP protease of Staphylococcus aureus on regulons involved in virulence, oxidative stress response, autolysis, and DNA repair. J Bacteriol 188: 5783–5796.
48. Strub JM, Sorokine O, Van Dorsselaer A, Aunis D, Metz-Boutigue MH (1997) Phosphorylation and O-glycosylation sites of bovine chromogranin A from adrenal medullary chromaffin granules and their relationship with biological activities. J Biol Chem 272: 11928–11936.
49. Pasupuleti M, Schmidtchen A, Chalupka A, Ringstad L, Malmsten M (2009) End-tagging of ultra-short antimicrobial peptides by W/F stretches to facilitate bacterial killing. PLoS One 4: e5285.
50. Malmsten M, Kasetty G, Pasupuleti M, Alenfall J, Schmidtchen A (2011) Highly selective end-tagged antimicrobial peptides derived from PRELP. PLoS One 6: e16400.
51. Cado G, Aslam R, Séon L, Garnier T, Fabre R, et al. (2013) Self-defensive biomaterial coating against bacteria and yeasts: polysaccharide multilayer film with embedded antimicrobial peptide. Adv functional materials [DOI: 10.1002/adfm.201300416].

15

A Common Minimal Motif for the Ligands of HLA-B*27 Class I Molecules

Alejandro Barriga[1], Elena Lorente[1], Carolina Johnstone[1], Carmen Mir[1], Margarita del Val[2], Daniel López[1]*

1 Centro Nacional de Microbiología, Instituto de Salud Carlos III, Majadahonda, Madrid, Spain, **2** Centro de Biología Molecular Severo Ochoa, CSIC/Universidad Autónoma de Madrid, Madrid, Spain

Abstract

$CD8^+$ T cells identify and kill infected cells through the specific recognition of short viral antigens bound to human major histocompatibility complex (HLA) class I molecules. The colossal number of polymorphisms in HLA molecules makes it essential to characterize the antigen-presenting properties common to large HLA families or supertypes. In this context, the HLA-B*27 family comprising at least 100 different alleles, some of them widely distributed in the human population, is involved in the cellular immune response against pathogens and also associated to autoimmune spondyloarthritis being thus a relevant target of study. To this end, HLA binding assays performed using nine HLA-B*2705-restricted ligands endogenously processed and presented in virus-infected cells revealed a common minimal peptide motif for efficient binding to the HLA-B*27 family. The motif was independently confirmed using four unrelated peptides. This experimental approach, which could be easily transferred to other HLA class I families and supertypes, has implications for the validation of new bioinformatics tools in the functional clustering of HLA molecules, for the identification of antiviral cytotoxic T lymphocyte responses, and for future vaccine development.

Editor: Anna Carla Goldberg, Albert Einstein Institute for Research and Education, Brazil

Funding: This work was supported by grants to D. L. from the Spanish "Ministerio de Economía y Competitividad". The funders had no role in study design, data collection and analysis, decision to publish, or preparation of the manuscript.

Competing Interests: The authors have declared that no competing interests exist.

* Email: dlopez@isciii.es

Introduction

The proteolytic degradation of newly synthesized pathogen proteins in the cytosol through the combined actions of proteasomes and different peptidases continuously generates peptides of typically 8 to 11 residues long, and these fragments were translocated to the endoplasmic reticulum (ER) lumen through a transporter associated with antigen-processing (TAP) molecules. These short peptides are subsequently assembled with nascent human leukocyte antigen (HLA) class I heavy chain and β_2-microglobulin molecules [1]. This assembly likely occurs through the interaction of the lateral chains of anchor residues at position 2 (P2) and the C-terminus (PΩ) of the antigenic peptide [2,3], and these chains are inserted deeply into specific pockets of the antigen recognition groove of the HLA class I molecule [4,5]. The stable HLA/peptide complexes are eventually exported to the cell membrane and presented for cytotoxic T lymphocyte (CTL) recognition [6]. The detection of pathogen peptides by specific T cell receptors results in the killing of pathogen-infected cells.

HLA class I is the largest polymorphic biological system described. More than 7,000 HLA class I alleles have been identified to date (Immuno Polymorphism Database, http://www.ebi.ac.uk/ipd), and classic HLA serologies have been largely divided into complex HLA gene families with increasing numbers of expressed protein subtypes. For example, to date, HLA-B*27 (a well-studied HLA class I family) comprises at least 100 different alleles. Although the presence of Arg at P2 is necessary for HLA-B*27 ligands (SYFPEITHI Database [3]), only a partial overlapping of the peptide repertoire has been observed in different HLA-B*27 subtypes [7]. Individual HLA-B*27 subtypes could present or not specific ArgP2-containing peptides, or the same ligand could bind to different HLA-B*27 subtypes with a broad range of affinity values [8]. Thus, the existence of HLA-B*27 ligands with additional binding motifs for presentation by all or most of the different HLA-B*27 subtypes remains unknown. To address this question, the binding affinity of a homogeneous set of nine naturally processed viral HLA-B*2705 ligands with different sequences, identified using mass spectrometry analysis of complex HLA-bound peptide pools isolated from large amounts of Human respiratory syncytial virus (HRSV)-infected cells [9], was examined using seven phylogenetically and functionally different major HLA-B*27 subtypes [10,11]. This analysis revealed a common minimal peptide motif for efficient binding to different HLA-B*27 subtypes.

Materials and Methods

HLA-B*27 cell lines and antibodies

RMA-S is a TAP-deficient murine cell line that expresses the mouse H-2^b haplotype [12]. Transfected RMA-S cell lines

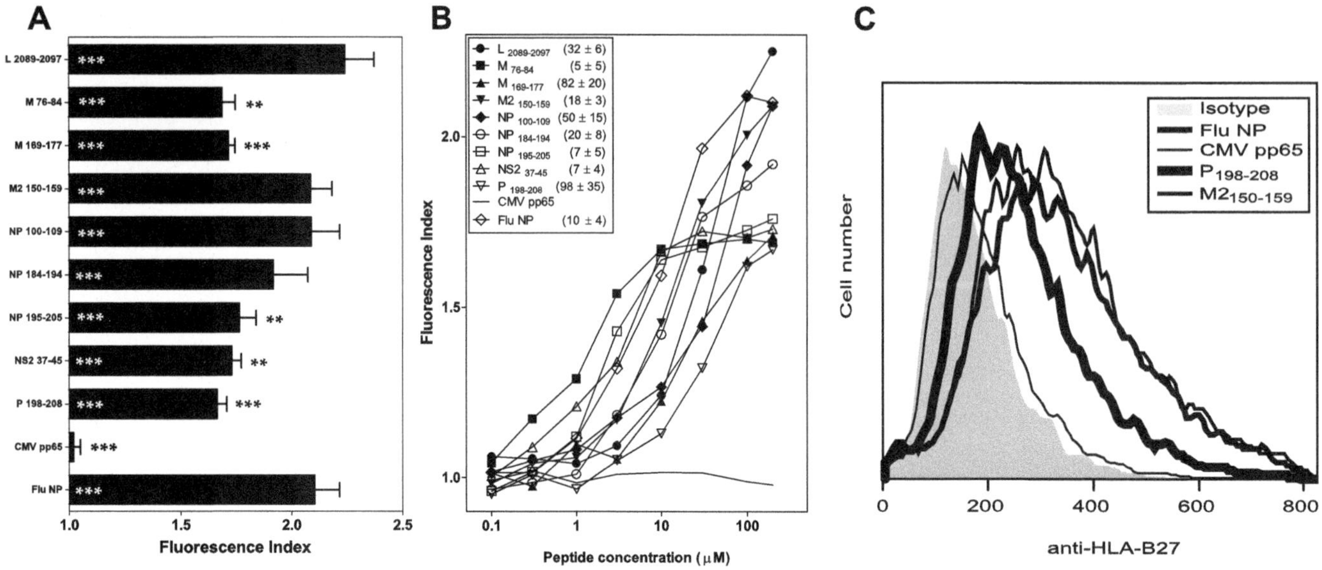

Figure 1. HLA-B*2705 stabilization with synthetic HRSV ligands. The stability of HLA-B*2705/peptide complexes on the cell surface of transfected RMA-S cells was measured through flow cytometry. Panel A: The indicated peptides were used at 200 μM. The CMV pp65 and Flu NP peptides were used as negative and positive controls, respectively. The mAb ME1 was used for staining. The results, calculated as fluorescence indexes, show the means ± SD of 4-5 independent experiments. Significant P values (***, p<0.001; **, p<0.01; *, p<0.05) versus negative or positive peptide controls are represented as white or black asterisks, respectively. Panel B: The titration curves for the indicated synthetic peptides with HLA-B*2705 are depicted. The results show the mean values obtained from three or four independent experiments. The calculated EC_{50} values (μM) (means ± SD) are shown in the legend of panel B. A representative experiment is depicted in panel C.

expressing HLA-B*2701 [13], -B*2702 [13], -B*2703 [14], -B*2704 [8], -B*2705 [15], -B*2706 [8], or -B*2709 [16] have been previously described (summarized in Figure S1). All cell lines were cultured in RPMI 1640 medium supplemented with 10% heat-inactivated fetal bovine serum and 50 μM β-mercaptoethanol. ME1, a monoclonal antibody (mAb) specific for HLA-B27, -B7, and -Bw22 [17] and goat anti-mouse IgG-FITC (AbD Serotec, Kidlington, UK) were used in this study.

Synthetic peptides

The peptides were synthesized in a peptide synthesizer (model 433A; Applied Biosystems, Foster City, CA) and subsequently purified through reversed-phase HPLC. The molecular mass of the peptides was established using MALDI-TOF MS, and the peptide composition was determined through μLC-MS/MS.

HLA/Peptide Stability Assays

The synthetic peptide CMV $pp65_{294\text{-}302}$ (VAFTSHEHF, HLA-C*012-restricted)[18] was used as a negative control in complex stability assays. In addition, for some HLA-B*27 subtypes, the Flu NP peptide (SRYWAIRTR, HLA-B27-restricted) [19] was used as a positive control. The transfected RMA-S B*27 cell lines were incubated at 26°C for 16 h to promote the expression of empty HLA class I molecules (without antigenic peptide) at the cell membrane, as these molecules are stable at 26°C but not at 37°C. The cells were washed and incubated for 2 h at 26°C with various concentrations of peptide in medium without fetal bovine serum. The cells were maintained at 37°C for an additional 2 h to facilitate the internalization of empty MHC class I molecules. Subsequently, the cells were collected for flow cytometry to discriminate between bound or unbound peptides. MHC expression was measured using 100 μl of hybridoma culture supernatant containing the ME1 (anti-HLA-B27) mAb and the secondary antibody as previously described [20]. The data were acquired using a FACSCanto flow cytometer (BD Biosciences, San Jose, CA, USA) and analyzed using BD FACSDiva software version 6 (BD Bioscience). The cells alone exhibited peak fluorescence intensities similar to the background staining observed with secondary Ab alone. The fluorescence index was calculated as the ratio of the mean peak channel fluorescence of the sample to that of the control incubated without peptide. Peptide binding was expressed as the EC_{50}, the molar concentration of the peptide at 50% of the maximum fluorescence obtained at a concentration range of 0.01–200 μM.

Statistical analysis

Unpaired Student's *t* test was used to analyze statistical significance. *P* values<0.05 were considered significant.

Results

HRSV specific ligands bind to the B*2705 molecule with a broad range of affinities

Nine HLA-B*2705-restricted ligands endogenously processed and presented in the HRSV-infected cells were previously identified (Table S1) [9]. To confirm that HLA-B*2705 is the Major Histocompatibility Complex (MHC) class I molecule responsible for the presentation of these ligands, MHC/peptide complex stability assays were performed using TAP-deficient RMA-S cells transfected with the HLA-B*2705 molecule. Four HRSV synthetic peptides (L $_{2089\text{-}2097}$, M2 $_{150\text{-}159}$, NP $_{100\text{-}109}$ and NP $_{184\text{-}194}$) showed stable numbers of HLA-peptide surface complexes similar to those of the well known HLA-B*2705 epitope from influenza virus (Figure 1A and Table 1). For the

Table 1. Summary of HLA stabilization assays using synthetic ligands.

Peptide	Sequence	B*2705	B*2703	B*2704	B*2706	B*2701	B*2702	B*2709
L $_{2089\text{-}2097}$	GRNEVFSNK	2.2±0.1[a]	1.8±0.1	1.5±0.2	1.7±0.1	1.0±0.1	1.2±0.1	1.6±0.1
M $_{76\text{-}84}$	SRSALLAQM	1.7±0.1	1.8±0.2	1.8±0.1	2.6±0.4	1.6±0.1	1.3±0.1	1.7±0.1
M $_{169\text{-}177}$	VRNKDLNTL	1.7±0.1	1.5±0.1	1.8±0.2	3.0±0.2	1.2±0.2	1.4±0.1	1.6±0.1
M2 $_{150\text{-}159}$	KRLPADVLKK	2.1±0.1	1.7±0.1	1.8±0.1	1.5±0.1	1.0±0.1	1.4±0.1	1.6±0.1
NP $_{100\text{-}109}$	HRQDINGKEM	2.1±0.1	1.8±0.1	1.7±0.2	2.3±0.1	1.2±0.1	1.3±0.1	1.5±0.1
NP $_{184\text{-}194}$	RRANNVLKNEM	1.9±0.2	1.8±0.1	1.9±0.2	3.0±0.3	1.6±0.1	1.5±0.1	1.8±0.1
NP $_{195\text{-}205}$	KRYKGLLPKDI	1.8±0.1	1.7±0.1	2.1±0.2	2.6±0.4	1.7±0.1	1.7±0.1	1.6±0.1
NS2 $_{37\text{-}45}$	HRFIYLINH	1.7±0.1	1.9±0.1	1.5±0.1	2.0±0.1	1.1±0.2	1.5±0.1	1.5±0.1
P $_{198\text{-}208}$	LRNEESEKMAK	1.7±0.1	1.3±0.1	1.1±0.1	1.2±0.1	1.0±0.1	1.0±0.1	1.5±0.1
CMV pp65	VAFTSHEHF	1.0±0.1	1.0±0.1	1.0±0.1	1.0±0.1	1.0±0.1	1.0±0.1	1.1±0.1
Flu NP	SRYWAIRTR	2.1±0.1	1.7±0.2	2.0±0.2	2.6±0.2	1.0±0.1	1.1±0.1	1.5±0.1

[a]Data are expressed as fluorescence index when peptides were used at 200 μM ± S.D. The results show the mean of three to five independent experiments. All data show significant P values versus the negative control CMV pp65 peptide, except the seven values underlined (see figures).

Table 2. Affinity values of HRSV ligands for different HLA-B27 subtypes.

Peptide	Sequence	B*2705	B*2703	B*2704	B*2706	B*2701	B*2702	B*2709
L $_{2089\text{-}2097}$	GRNEVFSNK	**32±6** [a]	71±28	169±57	>200	-	>200	68±24
M $_{76\text{-}84}$	SRSALLAQM	**5±5**	**6±3**	**15±6**	**1±1**	**15±11**	>200	**6±2**
M $_{169\text{-}177}$	VRNKDLNTL	82±20	>200	**56±16**	**3±1**	>200	94±25	**19±7**
M2 $_{150\text{-}159}$	KRLPADVLKK	**18±3**	78±10	73±9	>200	-	136±63	**22±9**
NP $_{100\text{-}109}$	HRQDINGKEM	**50±15**	96±43	67±27	**3±2**	>200	132±53	**24±7**
NP $_{184\text{-}194}$	RRANNVLKNEM	**20±8**	**60±29**	**19±10**	**2±1**	**39±9**	**43±27**	**7±1**
NP $_{195\text{-}205}$	KRYKGLLPKDI	**7±5**	**20±11**	**16±3**	**1±1**	**8±1**	**9±1**	**7±2**
NS2 $_{37\text{-}45}$	HRFIYLINH	**7±4**	**10±5**	78±33	**45±19**	-	77±14	**33±9**
P $_{198\text{-}208}$	LRNEESEKMAK	98±35	>200	-	>200	-	-	86±12
HIV gp160 $_{500\text{-}508}$	KRAVGIGAL	**12±3**	**14±7**	**24±8**	**1±1**	**17±3**	**16±6**	**8±1**
MV F $_{438\text{-}466}$	RRYPDAVYL	**3±2**	**4±3**	**18±4**	**2±1**	**5±2**	**11±4**	**4±1**
Flu PB1 $_{238\text{-}246}$	RRAIATPGM	**3±2**	**2±1**	**9±1**	**2±1**	**5±1**	**12±4**	**2±1**
RRA$_7$I	RRAAAAAAAI	**4±3**	**7±5**	**7±2**	**2±1**	**7±2**	**28±16**	**2±1**
Flu NP	SRYWAIRTR	**10±4**	**23±3**	**19±9**	**18±4**	-	>200	**26±3**

[a]Data are expressed as EC_{50} (μM) ± S.D and are the mean of three to five independent experiments. Affinity values indicating intermediate affinity peptides ($20<EC_{50}\leq60$ μM) are marked in bold and values indicating high affinity peptides ($EC_{50}\leq20$ μM) are also underlined. – indicates no statistical difference in fluorescence index compared to the negative control.

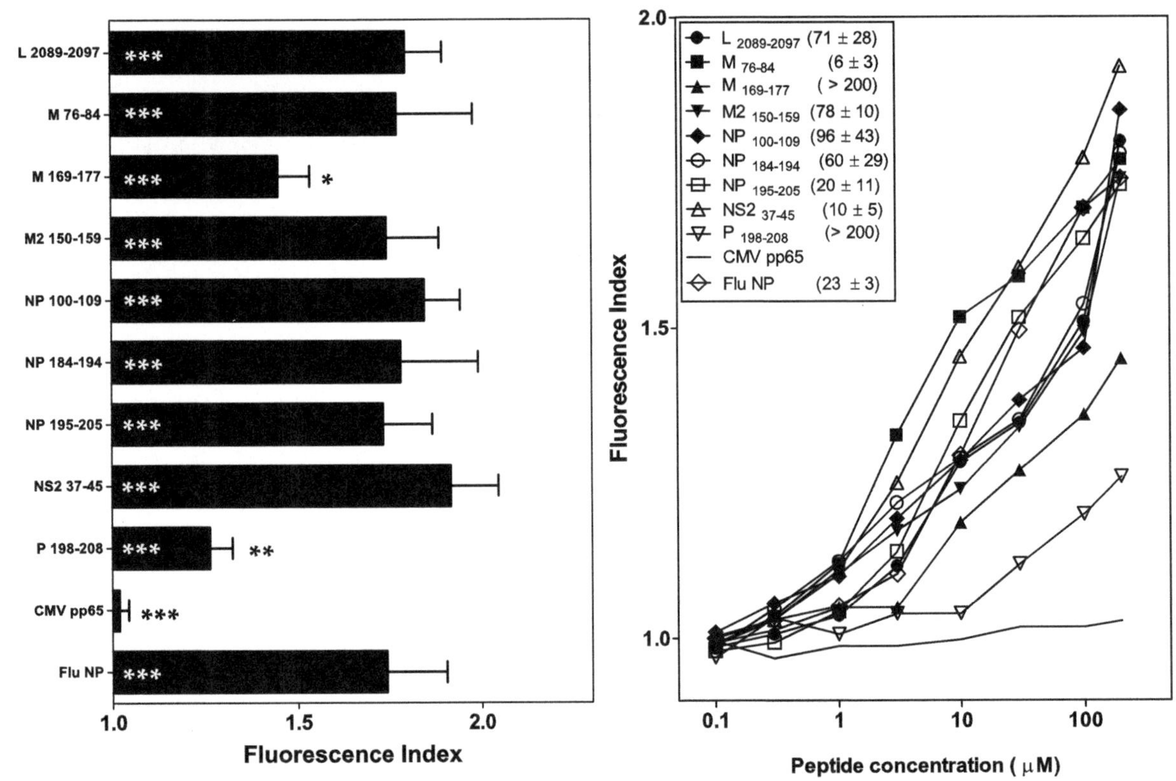

Figure 2. HLA-B*2703 stabilization with synthetic HRSV ligands. The stability of HLA-B*2703/peptide complexes on the cell surface of transfected RMA-S cells was measured using flow cytometry. The results shown in left and right panels are depicted as in Fig. 1A and 2B, respectively, showing the mean values obtained from three or four independent experiments.

other five HRSV synthetic peptides (M $_{76-84}$, M $_{169-177}$, NP $_{195-205}$, NS2 $_{37-45}$ and P $_{198-208}$), the induction of HLA-peptide surface complexes was somewhat lower than that of the positive control (Figure 1A and Table 1). In addition, the relative MHC class I affinity was determined for all HRSV peptides, according to the EC_{50} value. The peptides bound to HLA-B*2705 class I molecules at a broad range of EC_{50} values commonly observed among natural ligands: high (M $_{76-84}$, M2 $_{150-159}$, NP $_{184-194}$, NP $_{195-205}$ and NS2 $_{37-45}$), medium (L $_{2089-2097}$, and NP $_{100-109}$) or low affinity (M $_{169-177}$ and P $_{198-208}$) (Figure 1B and Table 2). A representative experiment is shown in Figure 1C. These data confirm that all ligands detected in HRSV-infected cells were endogenously presented in association with the B*2705 molecule.

The His59Tyr change in the A pocket induces a moderate decrease in the binding of HRSV-specific ligands to HLA-B*27

The B*2703 molecule differs from the prototypical subtype HLA-B*2705, reflecting a single amino acid change at heavy-chain residue 59, which is responsible for anchoring the N-terminus of the peptide within the A pocket of the class I molecule (Figure S1). The analysis of MHC/peptide complex stability assays using RMA-S cells transfected with the HLA-B*2703 molecule showed either no effect (M $_{76-84}$, NP $_{195-205}$ and NS2 $_{37-45}$) for three HRSV synthetic peptides or a moderate decrease in the EC_{50} value (L $_{2089-2097}$, M $_{169-177}$, M2 $_{150-159}$, NP $_{100-109}$, NP $_{184-194}$ and P $_{198-208}$) for six HRSV synthetic peptides in response to the His59Tyr change (Figure 2 and Tables 1 and 2).

Basic, but not nonpolar, C-terminal residues reduce binding to the B*2704 molecule of viral B*2705-restricted ligands

In the binding groove, the B*2704 subtype differs from the B*2705 molecule, reflecting two amino acid changes at residues 77 (Asp to Ser) and 152 (Val to Glu) located in the F and E pockets, respectively (Figure S1). The analysis of peptide binding to HLA-B*2704 in transfected RMA-S cells showed almost no effect with the five HRSV synthetic peptides harboring nonpolar amino acids at the PΩ residue (M $_{76-84}$, M $_{169-177}$, NP $_{100-109}$, NP $_{184-194}$, and NP $_{195-205}$; Tables 1 and 2, and Figure 3) compared with HLA-B*2705 affinity. In contrast, according to the B*2704-EC_{50}, three HRSV ligands with basic PΩ residues showed a moderately decreased affinity to B*2704 compared with B*2705 (L $_{2089-2097}$, M2 $_{150-159}$, and NS2 $_{37-45}$). Moreover, P $_{198-208}$, the low affinity ligand for the B*2705 molecule, did not bind to the B*2704 subtype (Figure 3 and Tables 1 and 2).

HRSV B*2705-restricted ligands with nonpolar PΩ residues exhibit high-affinity binding to B*2706

B*2704 and B*2706 differ at two amino acids, His114Asp and Asp116Tyr, located in the same strand of the β-pleated sheet floor of the peptide binding site of HLA-B*27 (Figure S1). These changes highly stabilized (in the range of μM) the binding of the five HRSV ligands harboring nonpolar PΩ residues (M $_{76-84}$, M $_{169-177}$, NP $_{100-109}$, NP $_{184-194}$, and NP $_{195-205}$) to B*2706 (Figure 4 and Tables 1 and 2). In contrast, a moderate increase (NS2 $_{37-45}$ and P $_{198-208}$) or decrease (L $_{2089-2097}$, and M2 $_{150-159}$) in affinity to

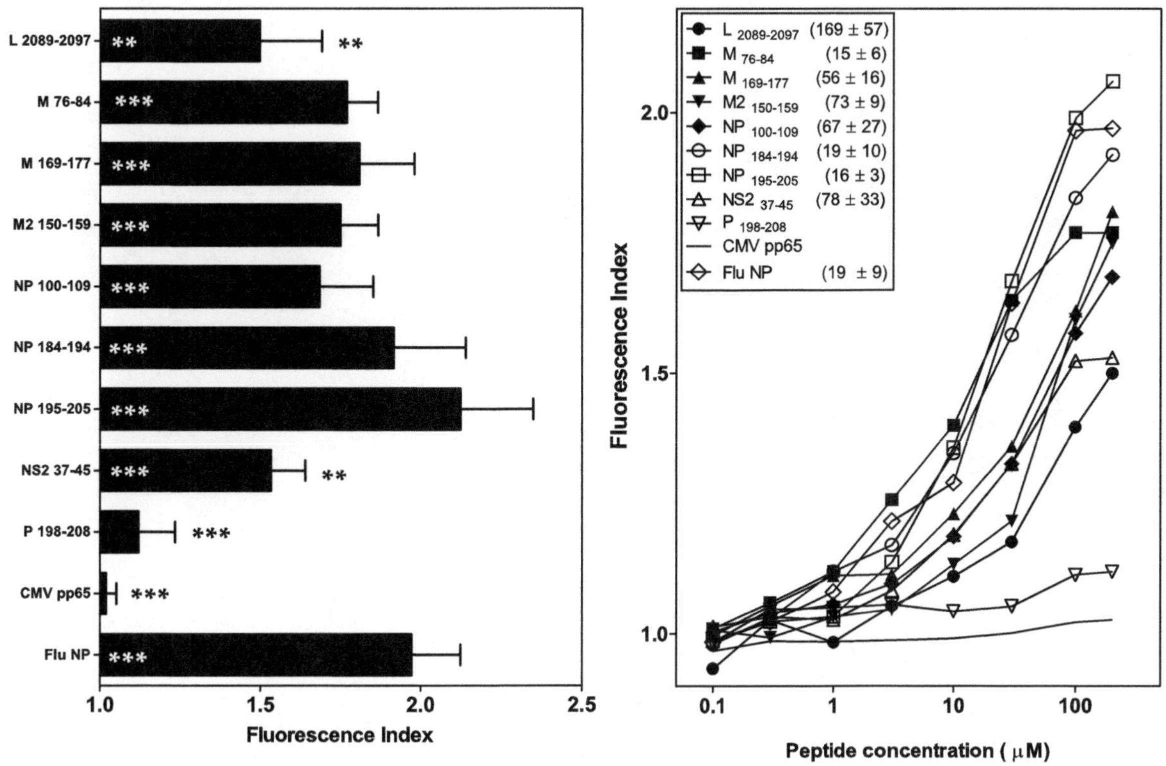

Figure 3. HLA-B*2704 stabilization with synthetic HRSV ligands. The stability of HLA-B*2704/peptide complexes on the cell surface of transfected RMA-S cells was measured through flow cytometry. The results shown in left and right panels are depicted as in Fig. 1A and 2B, respectively, showing the mean values obtained from three or four independent experiments.

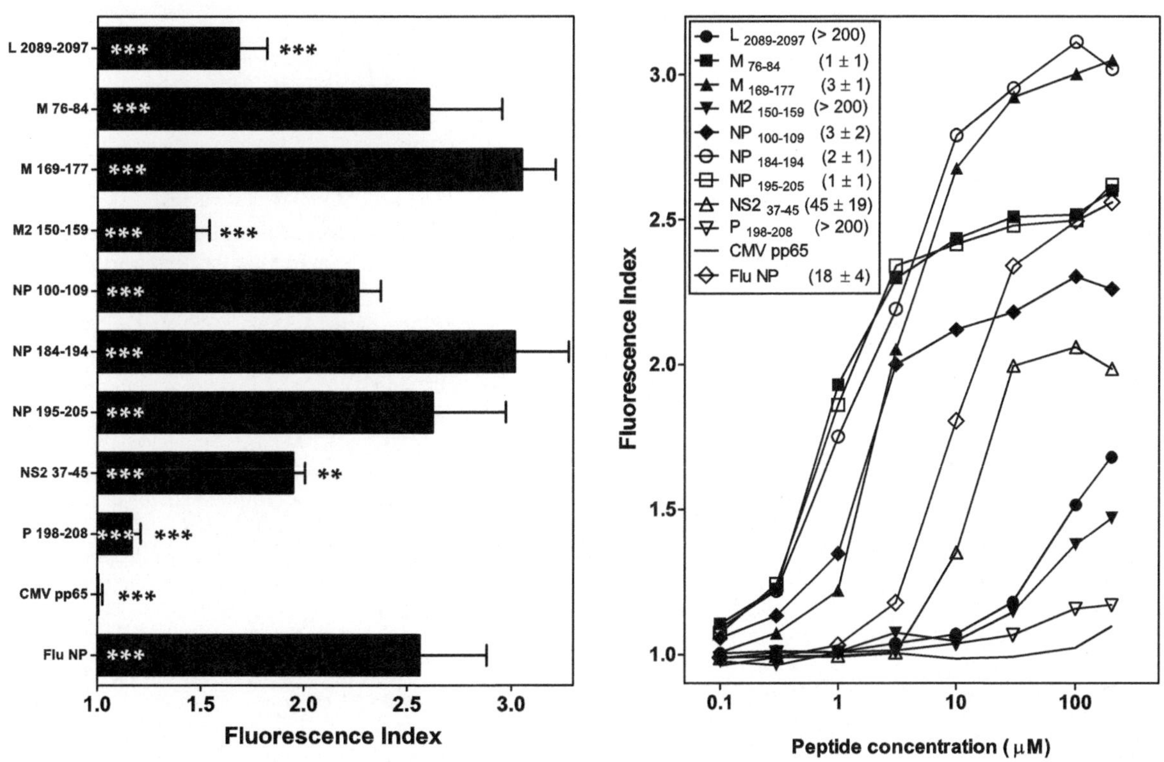

Figure 4. HLA-B*2706 stabilization with synthetic HRSV ligands. The stability of HLA-B*2706/peptide complexes on the cell surface of transfected RMA-S cells was measured through flow cytometry. The results shown in left and right panels are depicted as in Fig. 1A and 2B, respectively, showing the mean values obtained from three or four independent experiments.

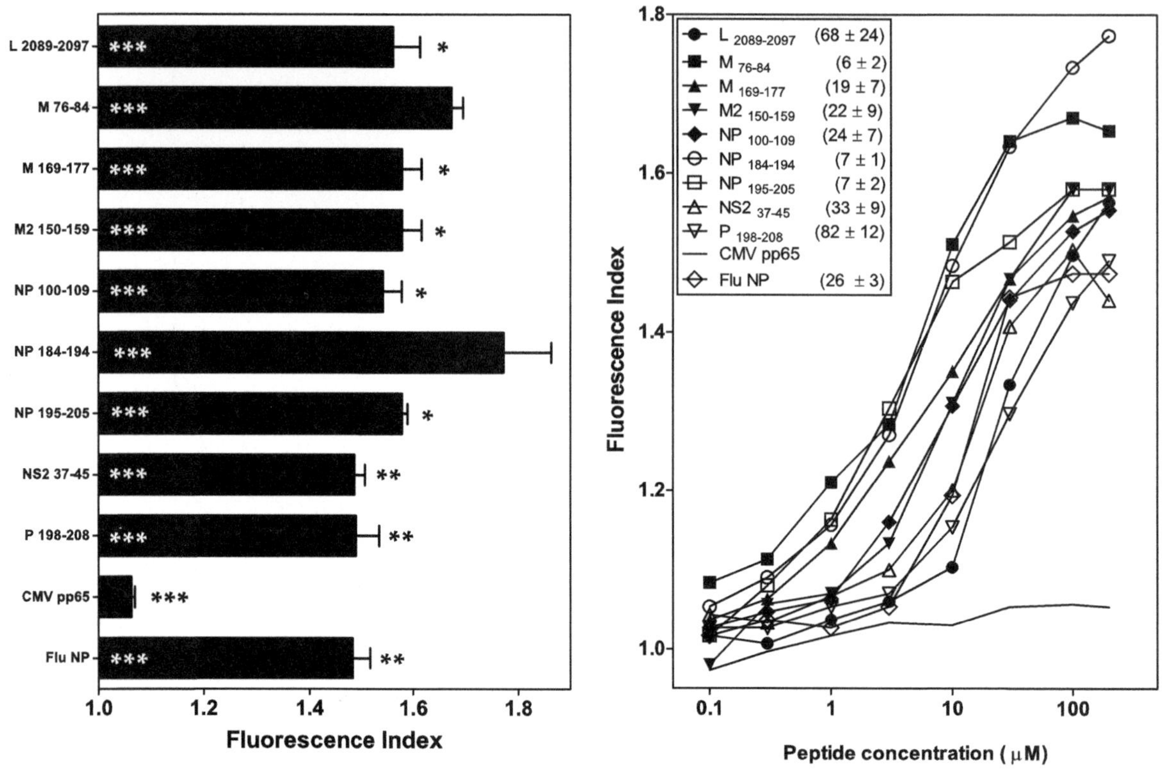

Figure 5. HLA-B*2709 stabilization with synthetic HRSV ligands. The stability of HLA-B*2709/peptide complexes on the cell surface of transfected RMA-S cells was measured through flow cytometry. The results shown in left and right panels are depicted as in Fig. 1A and 2B, respectively, showing the mean values obtained from three or four independent experiments.

B*2706 was observed for HRSV peptides containing basic amino acids in the PΩ residue compared with binding to B*2704. Thus, nonpolar PΩ residues could be used as additional strong auxiliary anchor motifs in B*2706 ligands.

High-affinity binding to B*2709 of viral ligands with nonpolar PΩ residues

B*2709 differs from B*2705 by a single amino acid change, Asp116His, located in a strand of the β-pleated sheet floor of the peptide binding site of HLA-B*27 (Figure S1). This change stabilized (M $_{76\text{-}84}$, NP $_{184\text{-}194}$, and NP $_{195\text{-}205}$) or increased (M $_{169\text{-}177}$, and NP $_{100\text{-}109}$) the binding of the five HRSV ligands harboring nonpolar PΩ residues to B*2709 compared with binding to the prototypical subtype B*2705 (Figure 5 and Tables 1 and 2). In contrast, a moderate decrease in affinity was observed for two HRSV peptides containing basic amino acids in the PΩ residue (L $_{2089\text{-}2097}$, and NS2 $_{37\text{-}45}$) compared with binding to B*2705, except with M2 $_{150\text{-}159}$ ligand where a basic amino terminal residue could compensate the loss of interaction in the F pocket. Thus, nonpolar PΩ residues are sufficient for efficient binding to the B*2709 class I molecule.

Nonpolar C-terminal residues in high affinity B*2705-restricted viral ligands preserve the binding affinity to the B*2701 molecule

B*2701 differs from B*2705, reflecting three amino acid changes, Asp74Tyr, Asp77Asn, and Leu81Ala, located in the C/F cavity of the HLA-B*27 peptide binding site (Figure S1). These changes destabilized the binding of all HRSV peptides except those ligands with nonpolar amino acids in the PΩ residue, which exhibited high affinity for the B*2705 molecule (M $_{76\text{-}84}$, NP $_{184\text{-}194}$, and NP $_{195\text{-}205}$) (Figure 6 and Tables 1 and 2). Thus, nonpolar PΩ residues are necessary, but not sufficient, for efficient binding to the B*2701 class I molecule.

Hydrophobic C-terminal residues, such as Leu or Ile, preserve the binding to the B*2702 molecule of HRSV B*2705-restricted ligands

The B*2702 subtype differs from the B*2705 molecule, reflecting three amino acid changes at positions 77 (Asp to Asn), 80 (Thr to Ile) and 81 (Asp to Tyr), located in the F pocket (Figure S1). Only the two HRSV peptides with Leu/Ile PΩ residues, M $_{169\text{-}177}$ and NP $_{195\text{-}205}$, similarly stabilized the binding to both B*2705 and B*2702 molecules (Figure 7 and Tables 1 and 2). In contrast, the binding affinity to B*2702 was moderately decreased for the NP$_{184\text{-}194}$ peptide, and strongly reduced for the other six viral ligands (Tables 1 and 2). Thus, hydrophobic PΩ residues, such as Leu/Ile, serve as additional auxiliary anchor motifs for B*2702-binding.

A common minimal motif for ligands of the seven different HLA-B*27 subtypes

Table 3 summarizes the HLA-B*27-binding patterns of the different HRSV B*2705-restricted ligands used in this study (Table S1). Most of the viral ligands bound to various HLA-B*27 subtypes, despite amino acid differences in relevant residues that contribute to the antigen binding site; however, only two of these ligands (NP $_{184\text{-}194}$, and NP $_{195\text{-}205}$) efficiently bound to all seven

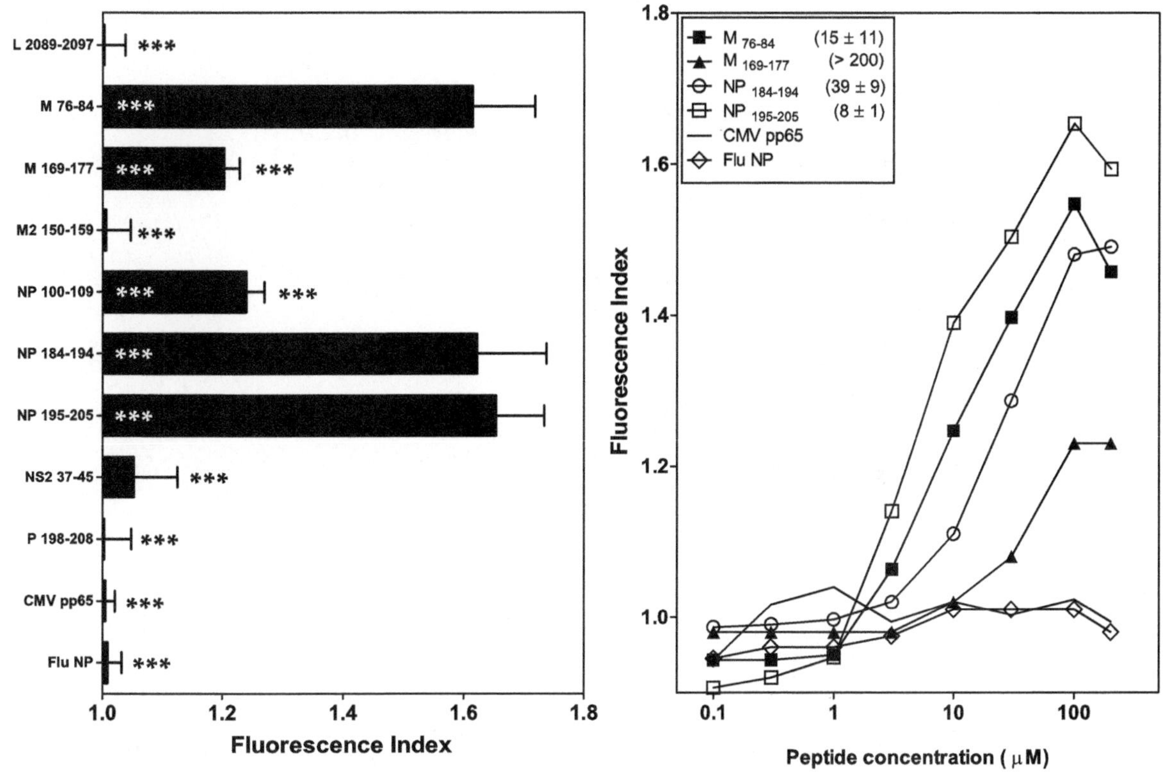

Figure 6. HLA-B*2701 stabilization with synthetic HRSV ligands. The stability of HLA-B*2701/peptide complexes on the cell surface of transfected RMA-S cells was measured through flow cytometry. The results shown in left and right panels are depicted as in Fig. 1A and 2B, respectively, showing the mean values obtained from three or four independent experiments.

HLA-B*27 subtypes studied. Both ligands contain basic amino acids in the N-terminus and a large and hydrophobic lateral chain in the C-terminal residues. This structure establishes the Arg/Lys-Arg-X_n-Ile/Met sequence as the common minimal peptide motif for efficient binding to seven different HLA-B*27 subtypes. To test this prediction three different peptides, HIV gp160 $_{500-508}$ [21], MV F $_{438-446}$ [22], and Influenza virus PB1 $_{238-246}$ [23], harboring the identified minimal peptide motif (basic residue at P1 and large hydrophobic aliphatic residue at PΩ), were studied. Interestingly, MHC/peptide complex stability assays showed efficient binding of these synthetic peptides to all seven HLA-B*27 subtypes tested (Tables 2 and 3). In addition, a synthetic peptide with only the common minimal peptide motif (RRAAAAAAI) efficiently bound to all seven HLA-B*27 subtypes analyzed (Tables 2 and 3). Conversely, the exchange for Ala of the Met residue at CΩ position in both NP $_{184-194}$, and NP $_{195-205}$ peptides abolished the interaction with B*2701 subtype (Table 4). Also, this monosubstitution significantly decreased the binding of A11-NP $_{195-205}$ peptide to B*2702 subtype (Table 4). Moreover the exchange for Ala of the Arg residue at P1 position in NP $_{184-194}$ peptide considerably decreased the binding to B*2703 subtype (Table 4).

Compared with the affinity of the nine HRSV ligands for B*2705, 7 ligands exhibited decreased affinity for B*2701 and B*2702, 6 ligands showed decreased affinity for B*2703, 4 ligands showed reduced affinity for B*2704 and B*2706, and only 3 ligands exhibited decreased affinity for B*2709 (summarized in Table 3). These data do not correlate with either the amino acid differences between subtypes or the individual interactions in the respective pockets, indicating compensatory effects of changes in some residues. These effects were more evident with the B*2706 subtype, showing increased affinity for some HRSV ligands with nonpolar PΩ residues compared with the B*2705 subtype. Thus, based on the affinity patterns with HLA-B*27 subtypes, a functional relationship between B*2705 and the other subtypes could be established as B*2705>B*2709>B*2706>B*2704>B*2703>B*2701 = B*2702.

Discussion

Using large scale mass spectrometry analysis, an extensive knowledge of HLA-B*2705 ligandome (with approximately 2,000 peptides identified) has been reported [7,24]. In contrast, few endogenous natural ligands have been identified in other HLA-B*27 subtypes: 32, 49, 38 and 50 from B*2703, B*2704, B*2706 and B*2709, respectively, and only 8 and 15 from B*2701 and B*2701, respectively (summarized in [7] and SYFPEITHI database [3]). Except for the ArgP2 residue, no additional anchor or auxiliary anchor motifs were identified in these studies. Thus, in the present study the Arg/Lys-Arg-X_n-Ile/Leu/Met sequence was identified as the minimal common peptide motif for efficient binding to the seven major, phylogenetically (Figure S2) and functionally different HLA-B*27 subtypes [10,11], thereby establishing the criteria to analyze the influence of the HLA-B*27 polymorphism on pathogen peptide presentation and T cell epitope predictions for the rational design of vaccines to treat large HLA families.

Extensive analyses of alloreactive T cell responses from HLA-B*27-negative individuals against B*2702 [25], B*2703 [10],

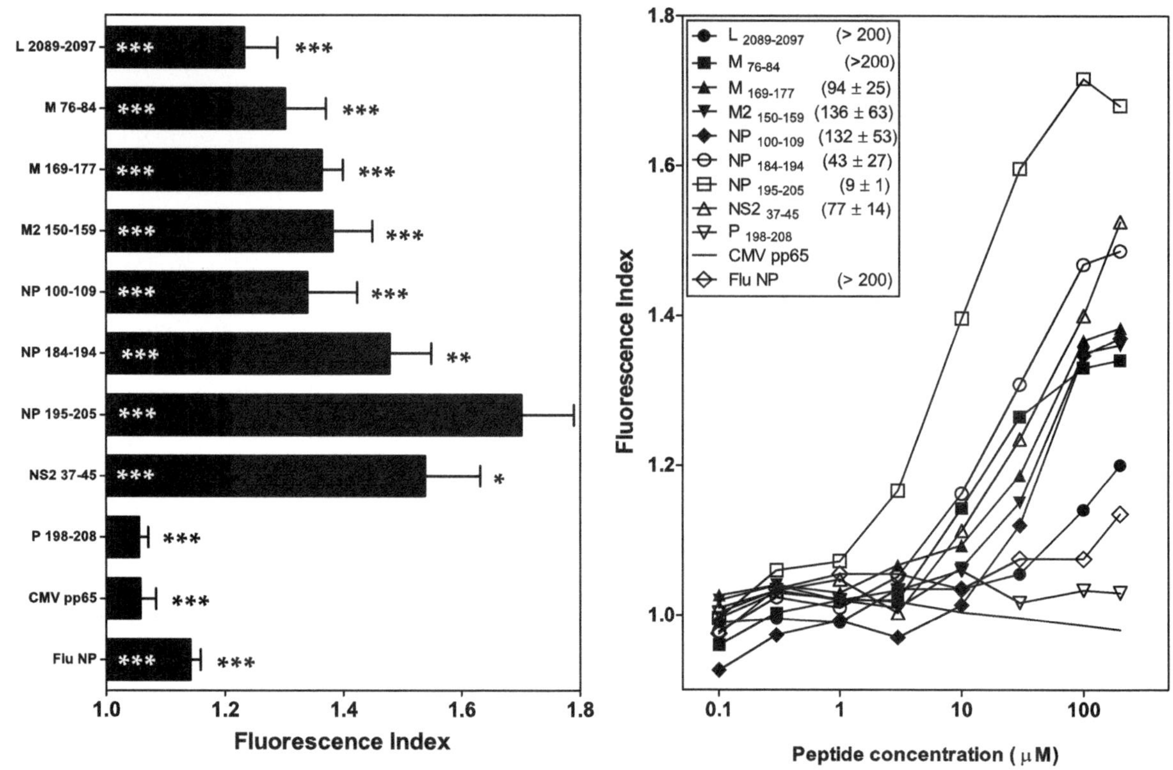

Figure 7. HLA-B*2702 stabilization with synthetic HRSV ligands. The stability of HLA-B*2702/peptide complexes on the cell surface of transfected RMA-S cells was measured through flow cytometry. The results shown in left and right panels are depicted as in Fig. 1A and 2B, respectively, showing the mean values obtained from three or four independent experiments.

B*2704 [26], B*2705 [10], and B*2709 [27] subtypes at the clonal level to assess T cell epitope sharing among HLA-B*27 subtypes have been previously described. These studies defined the functional relationship as B*2705>B*2709>B*2703>B*2702> B*2701>B*2704>B*2706, reflecting amino acid differences between subtypes. However, in contrast to self-restricted pathogen recognition in normal cellular immune responses, the allogenic anti-B*27 CTL clones recognized both polymorphic allo-MHC residues and self-derived peptides [14]. Thus, the direct contribution of these polymorphic HLA-B*27 residues to the CTL epitope could disguise both cross-presenting antigen peptide properties and the degree of functional relationship between the different HLA-B*27 subtypes. As the antigen processing and presentation machinery is similar, if not identical, in all cell lines expressing the different HLA-B*27 subtypes, the nine HLA-B*2705-restricted ligands endogenously processed and presented in the HRSV-infected cells might also be presented by the diverse HLA-B*27 subtypes according to affinity. Thus, the functional relationship between subtypes established with the HRSV ligands through affinity assays in the current report could be a better approximation to determine the antigen peptide-presenting properties of HLA-B*27 class I molecules, and this information could be applicable to the rational design of vaccines.

Few studies have analyzed antigen binding and/or presentation of the same viral ligands to different HLA-B27 subtypes. Four EBV ligands (EBNA3C $_{258-266}$, RRIYDLIEL; EBNA3B $_{243-253}$, RRARSLSAERY; LMP2 $_{236-244}$, RRRWRRLTV; EBNA3C $_{343-351}$, FRKAQIQGL) in B*2702, B*2704, and B*2705 subtypes were tested, and only the EBNA3C $_{258-266}$ ligand tended to be immunodominant and was recognized in the context of all three B27 subtypes studied, whereas the LMP $_{236-244}$ ligand was only recognized associated to B*2704 [28]. In contrast, in another study, the LMP $_{236-244}$ ligand was recognized by one of four CTL clones in the context of five HLA-B*27 (B*2702, B*2704, B*2705, B*2706 and B*2709) subtypes analyzed [29]. In addition two ligands, HIV gag $_{265-279}$ (KRWIILGNKIVRMYC) and Flu NP $_{380-393}$ (ELRSRYWAIRTRSG), were presented by both B*2702 and B*2705 subtypes [30]. Only one of these six viral epitopes harbors the minimal peptide motif for the efficient binding described in the current study, and this peptide was restricted by the three HLA-B27 subtypes analyzed (B*2702, B*2704 and B*2705). However, only an endogenous ligand derived from human histone H3.3 has been previously described for binding to B*2701 [13], B*2702 [31], B*2703 [32], B*2704 [33], B*2705 [34], B*2706 [33] and B*2709 [16] subtypes. This ligand, containing the RRYQKSTEL sequence, also harbors a basic amino acid in the N-terminal residue and a large and hydrophobic lateral chain in C-terminal residue, consistent with the motif defined in the present study.

Thus, studies examining different HLA class I families and supertypes are needed to determine the conserved anchor or auxiliary motifs common to these HLA clusters and validate new bioinformatics tools for the functional clustering of MHC molecules [35]. These data are also relevant for the identification of antiviral cytotoxic T lymphocyte responses and vaccine development.

Table 3. Summary of the relative affinity of HRSV ligands for different HLA-B27 subtypes.

Peptide	Sequence[a]	B*2705	B*2703	B*2704	B*2706	B*2701	B*2702	B*2709
L $_{2089-2097}$	GRNEVFSNK	+++ [b]	++	++	+	-	+	++
M $_{76-84}$	SRSALLAQM	++++	++++	++++	++++	++++	+	++++
M $_{169-177}$	VRNKDLNTL	++	+	+++	++++	+	++	++++
M2 $_{150-159}$	KRLPADVLKK	++++	++	++	+	-	++	+++
NP $_{100-109}$	HRQDINGKEM	+++	++	++	++++	+	++	+++
NP $_{184-194}$	RRANNVLKNEM	++++	+++	++++	++++	+++	+++	++++
NP $_{195-205}$	KRYKGLLPKDI	++++	++++	++++	++++	++++	++++	++++
NS2 $_{37-45}$	HRFIYLINH	++++	++++	++	+++	-	++	+++
P $_{198-208}$	LRNEESEKMAK	++	+	-	+	-	-	++
HIV gp160 $_{500-508}$	KRAVGIGAL	++++	++++	+++	++++	++++	++++	++++
MV F $_{438-466}$	RRYPDAVYL	++++	++++	++++	++++	++++	++++	++++
Flu PB1 $_{238-246}$	RRAIATPGM	++++	++++	++++	++++	++++	++++	++++
RRA$_7$I	RRAAAAAAAI	++++	++++	++++	++++	++++	+++	++++
Flu NP	SRYWAIRTR	++++	+++	++++	++++	-	+	+++

[a]The new HLA-B27 anchor motifs for efficient HLA binding are underlined.
[b]+, ++, +++ and ++++ indicate EC_{50} values>200 μM, 200-61 μM, 60-20 μM, and <20 μM, respectively. – indicates no statistical difference in fluorescence index compared to the negative control. All positive EC_{50} data show significant P values (P<0.01) versus the negative control.

Table 4. HLA stabilization assay with monosubstituted Ala analogs of HRSV NP $_{184-194}$ and NP $_{195-205}$ synthetic ligands.

Peptide	Sequence	B*2705	B*2703	B*2704	B*2706	B*2701	B*2702	B*2709
NP $_{184-194}$	RRANNVLKNEM	1.9±0.2 [a]	1.8±0.2	1.9±0.2	3.0±0.3	1.6±0.2	1.5±0.1	1.8±0.1
A1-NP $_{184-194}$	ARANNVLKNEM	2.4±0.3	**1.4±0.1**	2.2±0.2	3.0±0.3	2.0±0.1	2.0±0.3	1.8±0.1
A11-NP $_{184-194}$	RRANNVLKNEA	2.1±0.1	1.7±0.2	2.0±0.2	2.7±0.2	1.1±0.1	1.7±0.1	1.8±0.1
NP $_{195-205}$	KRYKGLLPKDI	1.8±0.1	1.7±0.1	2.1±0.2	2.6±0.3	1.7±0.1	1.7±0.1	1.6±0.1
A1-NP $_{195-205}$	ARYKGLLPKDI	2.4±0.3	2.1±0.2	2.2±0.3	2.7±0.3	1.9±0.1	1.8±0.1	1.8±0.2
A11-NP $_{195-205}$	KRYKGLLPKDA	2.2±0.2	1.9±0.1	1.9±0.3	2.5±0.2	1.0±0.1	**1.4±0.1**	1.6±0.2
CMV pp65	VAFTSHEHF	1.0±0.1	1.0±0.1	1.0±0.1	1.0±0.1	1.0±0.1	1.0±0.1	1.1±0.1

[a]Data are expressed as fluorescence index when peptides were used at 200 μM ± S.D. The results show the mean of three to five independent experiments. All data show significant P values (P<0.01) versus the negative control CMV pp65, except the two values underlined. In addition, the fluorescence index of A1-NP $_{184-194}$ with B*2703 subtype or A11-NP $_{195-205}$ peptide with B*2702 subtype (marked in bold) also shows significant P values (P<0.01) versus either the negative control or the NP $_{184-194}$ and NP $_{195-205}$ peptides, respectively.

Acknowledgments

The authors would like to thank Dr. J. A. López de Castro (Centro de Biología Molecular Severo Ochoa, Madrid, Spain) and Dr. David Yu (University of California, Los Angeles, CA) for providing the cell lines, Dr. M. Ramos for helpful discussions and Dr. J. Martínez-Laso for their help with the phylogenetic and molecular evolutionary analyses.

Author Contributions

Conceived and designed the experiments: DL. Performed the experiments: AB EL CJ CM. Analyzed the data: AB MV DL. Wrote the paper: DL.

References

1. Shastri N, Schwab S, Serwold T (2002) Producing nature's gene-chips: the generation of peptides for display by MHC class I molecules. Annu Rev Immunol 20: 463–493.
2. Parker KC, Bednarek MA, Coligan JE (1994) Scheme for ranking potential HLA-A2 binding peptides based on independent binding of individual peptide side-chains. J Immunol 152: 163–175.
3. Rammensee HG, Bachmann J, Emmerich NPN, Bachor OA, Stevanovic S (1999) SYFPEITHI: database for MHC ligands and peptide motifs. Immunogenetics 50: 213–219.
4. Bjorkman PJ, Saper MA, Samraoui B, Bennett WS, Strominger JL, et al. (1987) Structure of the human class I histocompatibility antigen, HLA-A2. Nature 329: 506–512.
5. Bjorkman PJ, Saper MA, Samraoui B, Bennett WS, Strominger JL, et al. (1987) The foreign antigen binding site and T cell recognition regions of class I histocompatibility antigens. Nature 329: 512–518.
6. York IA, Goldberg AL, Mo XY, Rock KL (1999) Proteolysis and class I major histocompatibility complex antigen presentation. Immunol Rev 172: 49–66.
7. Lopez de Castro JA, Alvarez I, Marcilla M, Paradela A, Ramos M, et al. (2004) HLA-B27: a registry of constitutive peptide ligands. Tissue Antigens 63: 424–445.
8. Galocha B, Lamas JR, Villadangos JA, Albar JP, Lopez de Castro JA (1996) Binding of peptides naturally presented by HLA-B27 to the differentially disease-associated B*2704 and B*2706 subtypes, and to mutants mimicking their polymorphism. Tissue Antigens 48: 509–518.
9. Infantes S, Lorente E, Barnea E, Beer I, Cragnolini JJ, et al. (2010) Multiple, non-conserved, internal viral ligands naturally presented by HLA-B27 in human respiratory syncytial virus-infected cells. Mol Cell Proteomics 9: 1533–1539.
10. López D, Garcia Hoyo R, Lopez de Castro JA (1994) Clonal analysis of alloreactive T cell responses against the closely related B*2705 and B*2703 subtypes. Implications for HLA- B27 association to spondyloarthropathy. J Immunol 152: 5557–5571.
11. López D, Rojo S, Calvo V, Lopez de Castro JA (1992) Peptide-presenting similarities among functionally distant HLA-B27 subtypes revealed by alloreactive T lymphocytes of unusual specificity. J Immunol 148: 996–1002.
12. Ljunggren HG, Karre K (1985) Host resistance directed selectively against H-2-deficient lymphoma variants. Analysis of the mechanism. J Exp Med 162: 1745–1759.
13. Garcia F, Galocha B, Villadangos JA, Lamas JR, Albar JP, et al. (1997) HLA-B27 (B*2701) specificity for peptides lacking Arg2 is determined by polymorphism outside the B pocket. Tissue Antigens 49: 580–587.
14. Paradela A, Garcia-Peydro M, Vazquez J, Rognan D, Lopez de Castro JA (1998) The same natural ligand is involved in allorecognition of multiple HLA-B27 subtypes by a single T cell clone: role of peptide and the MHC molecule in alloreactivity. J Immunol 161: 5481–5490.
15. Villadangos JA, Galocha B, Lopez de Castro JA (1994) Unusual topology of an HLA-B27 allospecific T cell epitope lacking peptide specificity. J Immunol 152: 2317–2323.
16. Vazquez MN, Lopez de Castro JA (2005) Similar cell surface expression of beta2-microglobulin-free heavy chains by HLA-B27 subtypes differentially associated with ankylosing spondylitis. Arthritis Rheum 52: 3290–3299.
17. Ellis SA, Taylor C, McMichael A (1982) Recognition of HLA-B27 and related antigen by a monoclonal antibody. Hum Immunol 5: 49–59.
18. Kondo E, Akatsuka Y, Kuzushima K, Tsujimura K, Asakura S, et al. (2004) Identification of novel CTL epitopes of CMV-pp65 presented by a variety of HLA alleles. Blood 103: 630–638.
19. Wang M, Lamberth K, Harndahl M, Roder G, Stryhn A, et al. (2007) CTL epitopes for influenza A including the H5N1 bird flu; genome-, pathogen-, and HLA-wide screening. Vaccine 25: 2823–2831.
20. Samino Y, López D, Guil S, Saveanu L, van Endert PM, et al. (2006) A long N-terminal-extended nested set of abundant and antigenic major histocompatibility complex class I natural ligands from HIV envelope protein. J Biol Chem 281: 6358–6365.
21. Roberts CG, Meister GE, Jesdale BM, Lieberman J, Berzofsky JA, et al. (1996) Prediction of HIV peptide epitopes by a novel algorithm. AIDS Res Hum Retroviruses 12: 593–610.
22. Herberts CA, Stittelaar KJ, van der Heeft E, van Gaans-Van den Brink, Poelen MC, et al. (2001) A measles virus glycoprotein-derived human CTL epitope is abundantly presented via the proteasomal-dependent MHC class I processing pathway. J Gen Virol 82: 2131–2142.
23. Terajima M, Cruz J, Leporati AM, Orphin L, Babon JA, et al. (2008) Influenza A virus matrix protein 1-specific human CD8+ T-cell response induced in trivalent inactivated vaccine recipients. J Virol 82: 9283–9287.
24. Ben Dror L, Barnea E, Beer I, Mann M, Admon A (2010) The HLA-B*2705 peptidome. Arthritis Rheum 62: 420–429.
25. Marti M, Alvarez I, Montserrat V, Lopez de Castro JA (2001) Large sharing of T-cell epitopes and natural ligands between HLA-B27 subtypes (B*2702 and B*2705) associated with spondyloarthritis. Tissue Antigens 58: 351–362.
26. Montserrat V, Marti M, Lopez de Castro JA (2003) Allospecific T cell epitope sharing reveals extensive conservation of the antigenic features of peptide ligands among HLA-B27 subtypes differentially associated with spondyloarthritis. J Immunol 170: 5778–5785.
27. Garcia-Peydro M, Marti M, Lopez de Castro JA (1999) High T cell epitope sharing between two HLA-B27 subtypes (B*2705 and B*2709) differentially associated to ankylosing spondylitis. J Immunol 163: 2299–2305.
28. Crotzer VL, Christian RE, Brooks JM, Shabanowitz J, Settlage RE, et al. (2000) Immunodominance among EBV-derived epitopes restricted by HLA-B27 does not correlate with epitope abundance in EBV-transformed B-lymphoblastoid cell lines. J Immunol 164: 6120–6129.
29. Fiorillo MT, Greco G, Maragno M, Potolicchio I, Monizio A, et al. (1998) The naturally occurring polymorphism ASP(116)-HIS(116), differentiating the ankylosing spondylitis-associated HLA-B-2705 from the non-associated HLA-B-2709 subtype, influences peptide-specific CD8 T cell recognition. European Journal of Immunology 28: 2508–2516.
30. Pazmany L, Rowland-Jones S, Huet S, Hill A, Sutton J, et al. (1992) Genetic modulation of antigen presentation by HLA-B27 molecules. J Exp Med 175: 361–369.
31. Raghavan M, Lebron JA, Johnson JL, Bjorkman PJ (1996) Extended repertoire of permissible peptide ligands for HLA-B*2702. Protein Sci 5: 2080–2088. 10.1002/pro.5560051014 [doi].
32. Boisgerault F, Tieng V, Stolzenberg MC, Dulphy N, Khalil I, et al. (1996) Differences in endogenous peptides presented by HLA-B-2705 and B-2703 allelic variants - Implications for susceptibility to spondyloarthropathies. Journal of Clinical Investigation 98: 2764–2770.
33. Garcia F, Marina A, Lopez de Castro JA (1997) Lack of carboxyl-terminal tyrosine distinguishes the B*2706-bound peptide repertoire from those of B*2704 and other HLA-B27 subtypes associated with ankylosing spondylitis. Tissue Antigens 49: 215–221.
34. Jardetzky TS, Lane WS, Robinson RA, Madden DR, Wiley DC (1991) Identification of self peptides bound to purified HLA-B27. Nature 353: 326–329.
35. Thomsen M, Lundegaard C, Buus S, Lund O, Nielsen M (2013) MHCcluster, a method for functional clustering of MHC molecules. Immunogenetics 65: 655–665.
36. Robinson J, Halliwell JA, McWilliam H, Lopez R, Parham P, et al. (2013) The IMGT/HLA database. Nucleic Acids Res 41: D1222–D1227.
37. Tamura K, Peterson D, Peterson N, Stecher G, Nei M, et al. (2011) MEGA5: molecular evolutionary genetics analysis using maximum likelihood, evolutionary distance, and maximum parsimony methods. Mol Biol Evol 28: 2731–2739.

Impairment of IFN-Gamma Response to Synthetic Peptides of *Mycobacterium tuberculosis* in a 7-Day Whole Blood Assay

Hannah Priyadarshini Gideon[1☯¤], Melissa Shea Hamilton[2☯], Kathryn Wood[1], Dominique Pepper[1], Tolu Oni[1], Ronnett Seldon[1], Claire Banwell[1], Paul R. Langford[2], Robert J. Wilkinson[1,2,3], Katalin A. Wilkinson[1,3]*

1 Clinical Infectious Disease Research Initiative, Institute of Infectious Diseases and Molecular Medicine, University of Cape Town, Cape Town, South Africa, **2** Division of Medicine, Imperial College London, Norfolk Place, London, United Kingdom, **3** MRC National Institute for Medical Research, London, United Kingdom

Abstract

Studies on *Mycobacterium tuberculosis* (MTB) antigens are of interest in order to improve vaccine efficacy and to define biomarkers for diagnosis and treatment monitoring. The methodologies used for these investigations differ greatly between laboratories and discordant results are common. The IFN-gamma response to two well characterized MTB antigens ESAT-6 and CFP-10, in the form of recombinant proteins and synthetic peptides, was evaluated in HIV-1 uninfected persons in both long-term (7 day) and 24 hour, commercially available QuantiFERON TB Gold in Tube (QFT-GIT), whole blood assays. Our findings showed differences in the IFN-gamma response between 24 hour and 7 day cultures, with recombinant proteins inducing a significantly higher response than the peptide pools in 7 day whole blood assays. The activity of peptides and recombinant proteins did not differ in 24 hour whole blood or peripheral blood mononuclear cell (PBMC) based assays, nor in the ELISpot assay. Further analysis by SELDI-TOF mass spectrometry showed that the peptides are degraded over the course of 7 days of incubation in whole blood whilst the recombinant proteins remain intact. This study therefore demonstrates that screening antigenic candidates as synthetic peptides in long-term whole blood assays may underestimate immunogenicity.

Editor: Pere-Joan Cardona, Fundació Institut d'Investigació en Ciències de la Salut Germans Trias i Pujol. Universitat Autònoma de Barcelona. CIBERES, Spain

Funding: This study was funded with grants from the European Union SANTE/2005/105-061-102, The Wellcome Trust 084323, 088316, and the Medical research Council of the UK (U1175.02.002.00014.01). ESAT-6 (HYB 76-8CS) and CFP-10 (K8493) antibodies were kindly provided by the Statens Serum Institute, Copenhagen, Denmark. The funders had no role in study design, data collection and analysis, decision to publish, or preparation of the manuscript.

Competing Interests: The authors have declared that no competing interests exist.

* E-mail: Katalin.Wilkinson@uct.ac.za

¤ Current address: Department of Microbiology and Molecular Genetics, University of Pittsburgh School of Medicine, Pittsburgh, Pennsylvania, United States of America

☯ These authors contributed equally to this work.

Introduction

Assay of the T cell response to antigens of *Mycobacterium tuberculosis* (MTB) is a research priority and critical in the evaluation of new antigenic candidates to improve immunodiagnosis and vaccines. The immunogenicity of candidate antigens is important to decide which should be selected for further studies, and the ability to compare results is important. Various laboratories use different techniques and readout methods to compare the performance of new candidates. Moreover, these potential comparisons are confounded by variations in individual laboratory approaches that use different readout methodologies to define the immunogenicity of such candidates [1].

The measurement of MTB antigen specific T cell Interferon (IFN) -gamma production is used widely for determination of immune recognition in antigen discovery. Early IFN-gamma release assays used purified protein derivative of MTB, which has also been used in the Tuberculin Skin Test (TST) for over one hundred years [2,3]. Therefore these early assays had compromised specificity due to cross reactivity of the immune response to homologous proteins in tuberculous, non-tuberculous and environmental mycobacteria. Newer IFN-gamma release assays use MTB specific ESAT-6 and CFP-10, two co-expressed immunodominant antigens encoded by the region of difference (RD)-1 of MTB. These antigens were first evaluated in a 6-day lymphocyte stimulation test (LST) and found to be highly specific to infer tuberculosis infection [4,5]. IFN-gamma release assays were subsequently developed that differed from the classical LST in terms of cell type to be stimulated (whole blood or peripheral blood mononuclear cells), the formulation of the antigenic candidate (overlapping synthetic peptides or recombinant proteins), duration of incubation (from <24 hours to 7 days), or the methods used to read out the results (proliferation, ELISA, ELISpot, flow cytometry) [1,6–30]. Proliferation and IFN-gamma production using PBMCs or MTB specific cell lines has been shown to correlate irrespective of the antigenic formulation (recombinant or synthetic overlapping peptides) of ESAT-6 and CFP-10 [5]. IFN-gamma results from the 24 hour ELISpot assay (using PBMCs) and a 72 hour diluted whole blood assay (WBA) were also found to correlate, when recombinant ESAT-6 and CFP-10 were used [9]. Discordance between reported results have been attributed mainly to the variation in the duration of incubation and have been suggested to be a result of the measurement of various components of the T cell response where short-term assays detect circulating effector cells while long-term assays detect the secondary effector cells that differentiate from central memory cells during culture [28,31–34].

In the present study, we compared the commercially available 24 hour QuantiFERON TB Gold in Tube (QFT-GIT) with a widely used 7 day WBA [1,34], as well as the ELISpot assay, using ESAT-6 and CFP-10 to induce an IFN-gamma response. The performance of these two antigens in the form of synthetic overlapping peptide pools and recombinant proteins in 7 day WBAs was also evaluated. We report differences between the efficacy of the peptide pools and recombinant proteins to induce IFN-gamma in the short-term, compared to the long-term assays, using whole blood. Moreover, we used SELDI-TOF mass spectrometry to show that the peptides are degraded over the course of 7 days of incubation in whole blood whilst the recombinant proteins remain intact.

Materials and Methods

Ethics Statement

The University of Cape Town Faculty of Health Sciences Human Research Ethics Committee (REC336/2004; REC 296/2007; REC 245/2009) approved this study. All participants provided written informed consent.

Study Subjects

Twenty-nine healthy laboratory and clinical staff (volunteers) with probable occupational exposure to tuberculosis were recruited. TST or HIV-1 testing was not performed. None of the volunteers had symptoms of active TB. Ten were male, 19 female, with median age of 27 years and 25 were BCG vaccinated.

QuantiFERON-TB Gold In Tube Assay

Commercially available QuantiFERON-TB Gold In Tube (QFT-GIT) tests were performed as per the manufacturer's instructions, with the optical density values at 450 nm used to calculate results using QuantiFERON TB Gold analysis software v2.50 (Cellestis, Australia). A positive result is defined by >0.35 IU/ml of IFN-γ measured in the TB antigen minus nil tube, and negative when this was <0.35 IU/ml.

Seven Day Whole Blood Assay

A diluted whole blood assay of 7-days' incubation was performed as previously described [13,16,34]. Briefly, whole blood was diluted 1:10 in RPMI 1640 medium containing 1% (w/v) L-Glutamine. A 48 well (flat bottom) plate was set up with the antigens and controls to a final volume of 1 ml. ESAT-6 and CFP-10 were used as stimulants, either as peptide pools (3 µg/ml of each peptide) or recombinant proteins (5 µg/ml). Phytohemagglutinin (PHA, 5 µg/ml, Sigma) was included as a positive control and RPMI 1640 medium with 1% L-Glutamine as a negative control. The antigens were added at 100 µl per well, followed by 900 µl of diluted blood. The plate was sealed with micro-pore tape to avoid evaporation during incubation at 37°C with 5% CO_2. After 7 days of incubation, the supernatants were harvested and stored at −20°C until assayed for IFN-gamma by ELISA.

IFN-gamma ELISA

Detection of IFN-gamma by ELISA was performed as previously described [35,36] using the mouse anti-human IFN-gamma monoclonal capture antibody (BD Pharmingen 551221) and the biotinylated mouse anti-human IFN-gamma detection antibody (BD Pharmingen 554550). This assay has a dynamic range of approximately 10–3000 pg/ml. The lowest sensitivity of

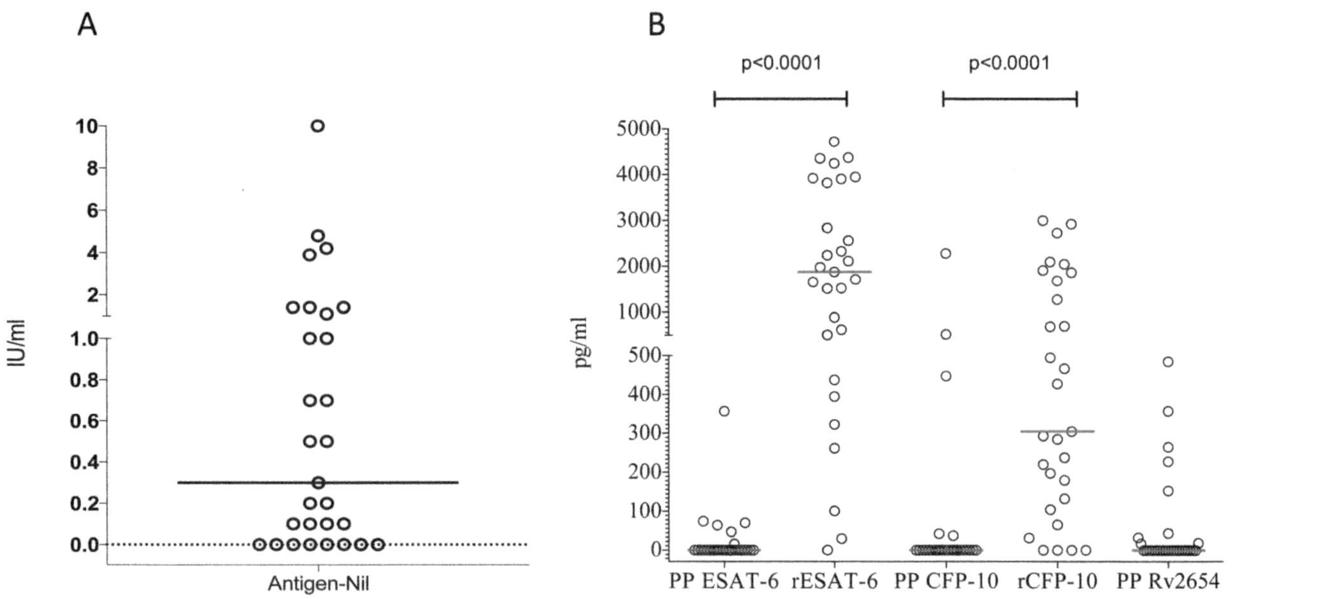

Figure 1. QFT-IT results, representing the Antigen-Nil IU/ml INF-gamma values (n = 29, Panel A). The solid line represents the median. **IFN-gamma response to peptide pools (covering ESAT-6, CFP-10, and Rv2654) and recombinant ESAT-6 and CFP-10 in the 7 day WBA (n = 29, Panel B).** "PP" indicates peptide pool and "r" indicates the recombinant proteins. Recombinant proteins induced significantly higher responses than the peptide pools (p<0.0001 for both ESAT-6 and CFP-10). IFN-gamma concentrations are shown as pg/ml. Lines indicate median response.

Table 1. Correlation between QFT-GIT and IFN-gamma response in the 7-day whole blood assay.

Parameter	ESAT-6 peptide pool	CFP-10 peptide pool	Rv2654 peptide pool	rESAT-6	rCFP-10
n*	28	28	28	28	28
Spearman r	0.45	0.15	0.29	0.33	0.59
95% Confidence interval	0.08 to 0.71	−0.25 to 0.50	−0.10 to 0.60	−0.06 to 0.63	0.26 to 0.79
p value (two-tailed)	**0.017**	0.45	0.13	0.086	**0.001**

*1 person had QFT-GIT IFN-gamma response >10 IU/ml and was therefore excluded from this analysis.

the assay was 14 pg/ml and calculated values below this were reassigned 'zero'. Of note, 14 pg/ml IFN-gamma is equivalent to 0.35 IU/ml (www.cellestis.com). A positive response for ELISA was defined as any response >14 pg/ml for peptide pool and >100 pg/ml for the recombinant stimulation. Antigen specific responses were corrected for the background response by subtraction.

Cell Culture

Peripheral blood mononuclear cells (PBMC) were separated over Ficoll as previously described [37]. Cells were frozen and stored in liquid nitrogen until analyzed, when they were set up at 250000/200 µl RPMI/10% FCS (R10) per well in 96U plates in the presence of various stimuli. Incubation periods ranged from overnight to 3, 5, and 7 days at 37°C in 5% CO_2. At the end of the incubation period, plates were centrifuged to pellet the cells, cell free supernatants were collected and stored at −20°C until the ELISA was performed in batches.

IFN-gamma ELISPOT Assay

The IFN-gamma ELISpot assay was performed as previously described [37], using mAb 1-D1K Pre-coated One-step 96-well plates (Mabtech; 3420-2ATP-10) washed with sterile PBS and blocked with R10 for 30 min at room temperature. PBMC at 2.5×10^5 were added in 100 µl of R10 per well, followed by the respective antigenic stimuli. Control stimuli included anti-CD3 mAb at 100 ng/ml final concentration and unstimulated wells. After incubation for 16–18 h at 37°C with 5% CO_2, plates were washed with PBS, and 100 µl of secondary antibody, mAb 7-B6-1-ALP conjugate, at 0.5 µg/ml final concentration was added. After 2 h incubation at room temperature, 100 µl substrate solution (BCIP/NBT-plus) was added until spots emerged, when the wells were washed with tap water and allowed to dry. Spot forming cells (SFC) were enumerated using the immunospot counter (CTL, Cellular Technology Ltd) and confirmed by microscope (X4). Results are quoted as spot forming cells (SFC) per 10^6 PBMC. All ELISpot responses presented are corrected for the background.

Antigens

Tuberculin Purified Protein derivative (PPD) was obtained from Evans Vaccines (Liverpool, UK). 10,000 U/ml stocks in PBS were stored at −80°C and used at 1000 U/ml final concentration. Recombinant rESAT-6 and rCFP-10 were obtained from Apronex, Czech Republic. The endotoxin content of these proteins was 1723 IU (EU)/mg for ESAT-6 and 260 IU (EU)/mg for CFP-10. Recombinant proteins were reconstituted in PBS/2% BSA and were used at a final concentration of 5 µg/ml. Overlapping peptides covering ESAT-6, CFP-10 and Rv2654 were 15-mers overlapping by 10 residues obtained from Peptide Protein Research Ltd, Oxford, UK. The individual peptides were solubilized in 100 µl DMSO and diluted to 10 mg/ml using PBS/2%BSA (as stock solutions). Mixtures of 17 peptides for ESAT-6 (one peptide pool), 18 peptides for CFP-10 (one peptide pool) and

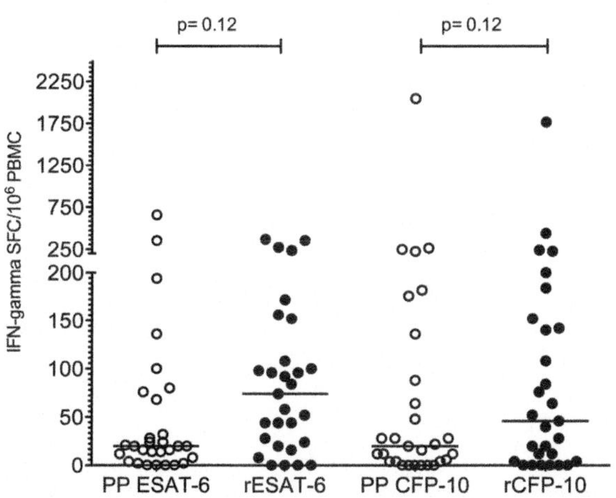

Figure 2. IFN-gamma response of PBMC to peptide pools and recombinant proteins in the overnight ELISpot assay (n = 29). "PP" indicates the peptide pools and "r" indicates recombinant proteins of ESAT-6 and CFP-10. No significant difference was found between the peptide pool and recombinant protein for either ESAT-6 or CFP-10 (Wilcoxon matched pairs test). Lines indicate median response.

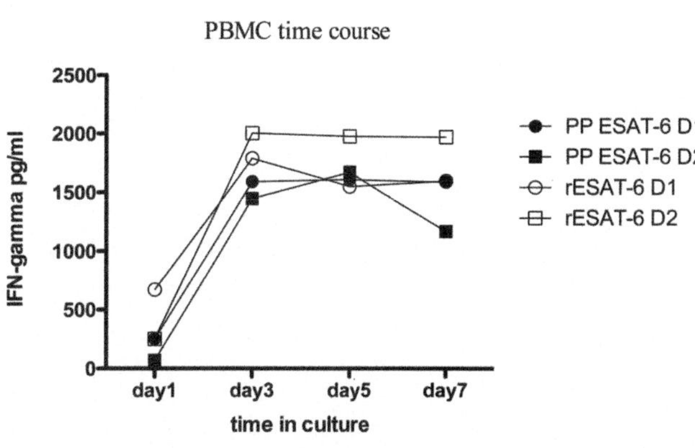

Figure 3. Peptides and recombinant protein have similar activity in prolonged PBMC culture. PBMC of two donors were set up with either peptide pool or recombinant ESAT-6 and cultured for 1,3,5,7 days. IFN-gamma was measured in the culture supernatant by ELISA. Filled symbols represent peptide stimulation, open symbols represent stimulation with recombinant protein. Donor 1 is represented by circles, and Donor 2 by squares.

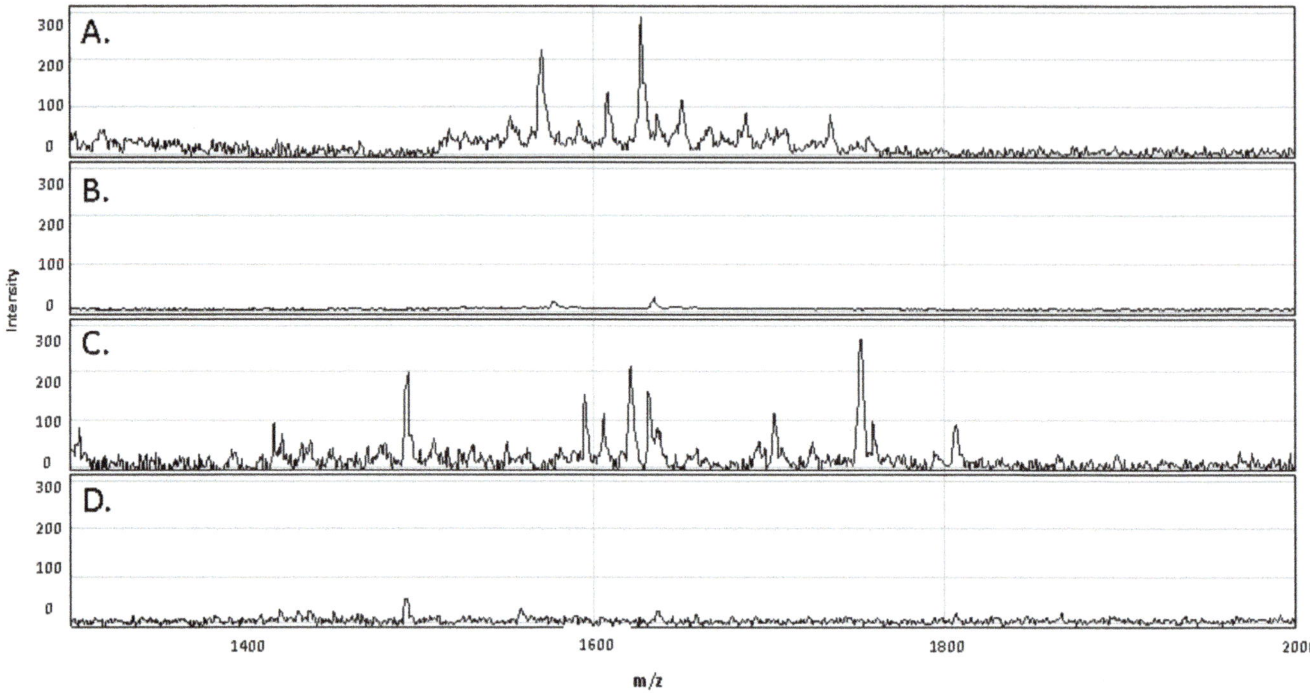

Figure 4. SELDI-TOF mass spectrometry analysis of whole blood supernatant stimulated with ESAT-6 and CFP-10 peptide pools. (A) ESAT-6 peptide pool stimulated whole blood at Day 0 with expected masses from 1419 Da to 1738 Da. (B) ESAT-6 peptide pool stimulated whole blood harvested after 7 days of incubation. (C) CFP-10 peptide pool stimulated whole blood at Day 0 with expected masses from 1415 Da to 1804 Da. (D) CFP-10 peptide pool stimulated whole blood harvested after 7 days of incubation.

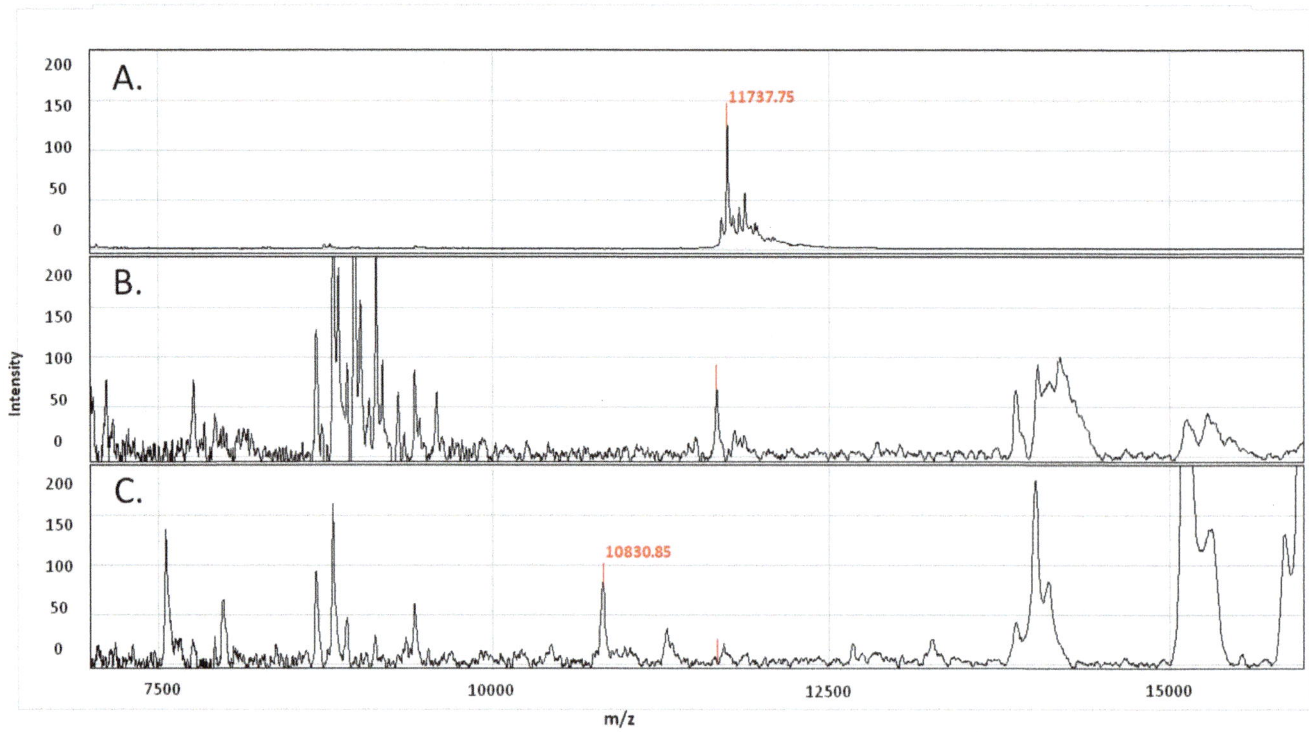

Figure 5. Recombinant ESAT-6 protein profiles on SELDI-TOF mass spectrometry. (A) Recombinant rESAT-6 protein. (B) Recombinant ESAT-6 stimulated whole blood at Day 0 showing a peak at the predicted molecular weight of 11.7 kDa. (C) Recombinant ESAT-6 stimulated whole blood after 7 days of incubation showing that rESAT-6 cannot be detected, while a new peak appears at 10.8 kDa.

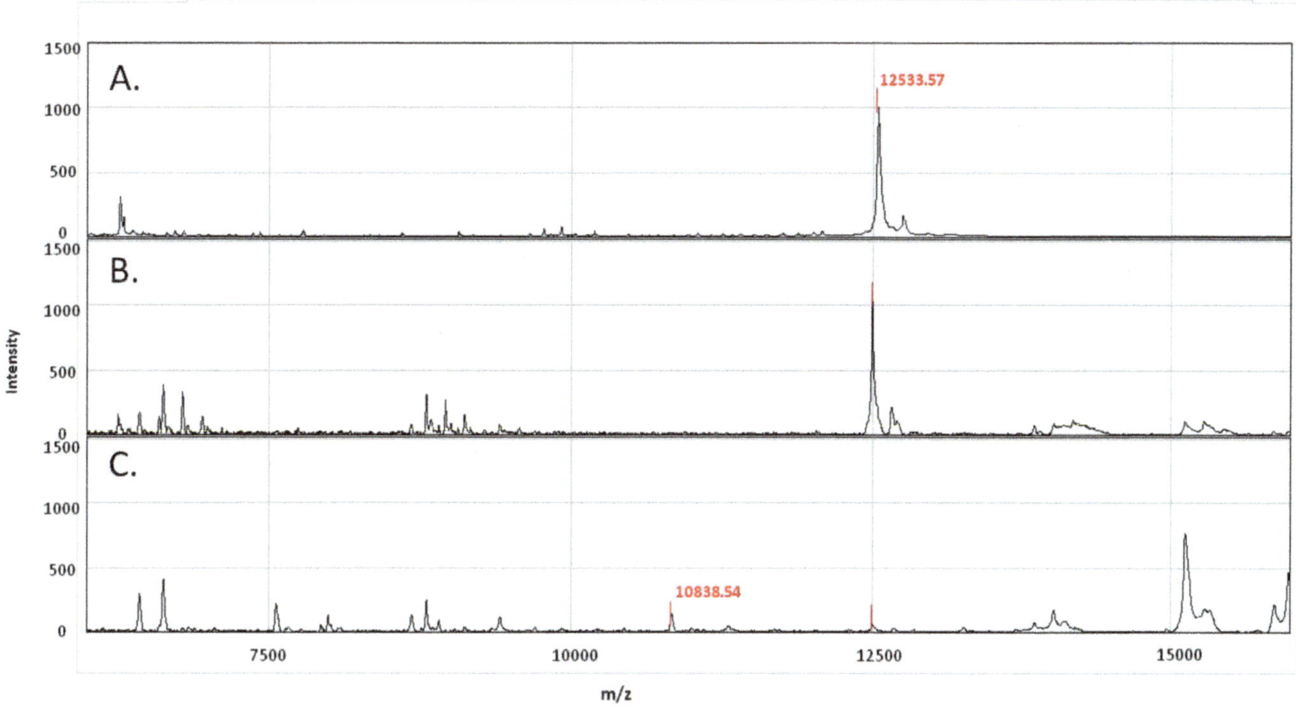

Figure 6. Recombinant CFP-10 protein profiles on SELDI-TOF mass spectrometry. (A) Recombinant rCFP-10 protein; (B) recombinant CFP-10 stimulated whole blood at Day 0 showing a peak at the predicted molecular weight of 12.5 kDa; (C) recombinant CFP-10 stimulated whole blood after 7 days of incubation showing that rCFP-10 cannot be detected, while a new peak appears at 10.8 kDa.

14 peptides for Rv2654 (one peptide pool) were prepared in PBS/ 2% BSA and the final dilution of each peptide was 3 µg/ml in the assays.

SELDI-TOF Mass Spectrometry

NP20 ProteinChip®arrays (BioRad) were primed with 5 µl of HPLC grade water (Sigma). Five microliters of neat supernantant were applied to the array surface 3 times, allowing the sample to air dry between each application. Five microliters of recombinant ESAT-6 or CFP-10 (1 mg/ml solution) were applied twice, allowing the sample to air dry between each application. The arrays were then rinsed 3 times with HPLC water and allowed to air dry for 10 min. Saturated sinapinic acid (SPA) or α-cyano-4-hydroxycinnamic acid (CHCA) were applied twice (2×0.7 µl) to each spot on the arrays, allowing the matrix to air dry between each application.

Time-of-flight spectra were generated using a PCS-4000 mass spectrometer (BioRad). Spectra from the peptide pools were obtained at a laser energy of 2000 nJ (one warming shot of 2200 nJ), a focus mass of 1500 and the matrix attenuated to 1000. Spectra from the recombinant proteins were obtained at a laser energy of 2500 nJ (one warming shot of 2750 nJ), a focus mass of 12,000 and the matrix attenuated to 5000. Ten shots were obtained per position. Mass accuracy was calibrated externally using All-in-One Peptide or Protein molecular mass standards (BioRad). Molecular weight shirts were predicted using PAWS software (Genomic Solutions).

1-D PAGE and Western Immunoblotting

1-D gels were performed using an Invitrogen XCell SureLock electrophoresis system. Twenty microliters of protein supernatant were separated on a 10% Bis-Tris gel using 2-(*N*-morpholino) ethanesulfonic acid (MES) buffer at 100 V for 70–80 min and compared against SeeBlue Plus 2 pre-stained protein standard (all Invitrogen, Carlsbad, CA). Protein gels were transferred to Hybond-C Extra nitrocellulose membranes (GE Healthcare, Pittsburgh, PA) in an XCell II blot electrophoretic cell (Life Technologies, Grand Island, NY) at 20 V for 60–90 min. Each

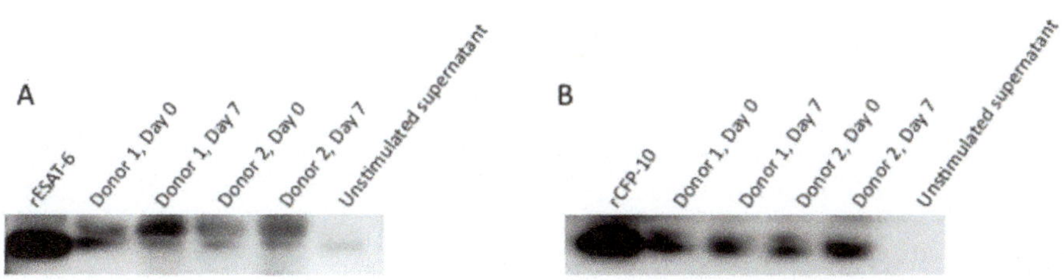

Figure 7. Detection of recombinant proteins by Western blot: Supernatant of whole blood stimulated with (A) rESAT-6 and (B) rCFP-10 at Day 0 and after 7 days of incubation, in 2 donors, showing the presence of both recombinant proteins after 7 days of incubation, including positive (recombinant proteins alone) and negative (supernatant of unstimulated whole blood) controls.

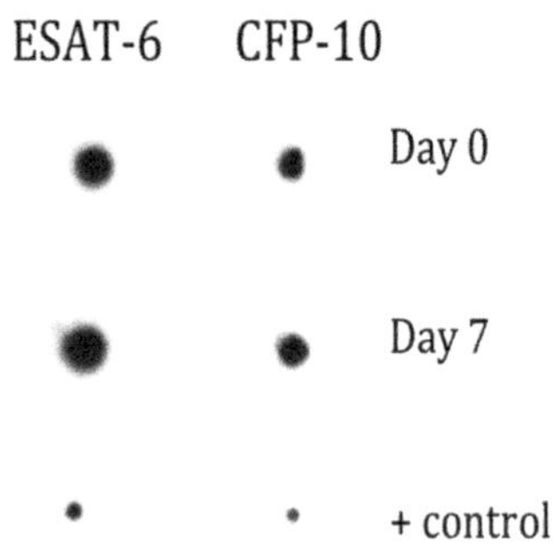

Figure 8. Detection of His-Tagged recombinant proteins by Western blot: in the supernatant of whole blood stimulated with rESAT-6 and rCFP-10 at Day 0 and after 7 days of incubation compared to a positive control, showing that the His-tag is detectable at both Day 0 and Day 7 of culture.

membrane was blocked in PBS/3% BSA/0.1% Tween 20 for 1 hr at RT and probed with monoclonal anti-ESAT-6 (HYB 76-8CS), polyclonal anti-CFP-10 (K8493) antibodies (both from Statens Serum Institute, Copenhagen, Denmark), or Calgranulin A (FL-83,Santa Cruz Biotechnology) overnight at 4°C. Dot blots to confirm his-tagged recombinant ESAT-6 and CFP-10 were performed by adding 5 µl of supernatant onto Hybond-C Extra nitrocellulose membranes (Amersham, Amersham, UK) and a his-tagged DsbA control [38]. After blocking, the membrane was probed with Penta Anti-His Antibody Selector Kit (Qiagen) shaking at RT for 30 min.

Statistical Analysis

The normality of data was assessed by the D'Agostino and Pearson omnibus test using Graphpad Prism 5.0 software (www.graphpad.com). Parametric continuous variables were assessed by student's paired and unpaired t-tests, and non-parametric variables by Wilcoxon matched pairs, Kruskal Wallis test with Dunn's post test correction or Mann Whitney U tests. Contingency analysis was by Fisher's exact test of probability. The kappa statistic was used to test the agreement between two tests. Correlation was assessed by non-parametric Spearman correlation coefficient. Differences were considered significant when $p<0.05$.

Results

The T Cell Response to Peptide Pools is Significantly Lower than the Recombinant Proteins in the 7 Day Whole Blood Assay

We first compared the commercially available overnight QuantiFERON TB Gold in Tube (QFT-GIT) with the widely used 7 day diluted WBA. Of the 29 tested volunteers, 14 (48%) were found positive by the QFT-GIT test (median IFN-gamma 0.3 IU/ml, IQR 0–1.25, Figure 1, Panel A). As the QFT-GIT assay incorporates TB7.7, which is Rv2654 of *Mycobacterium tuberculosis*, we included Rv2654c as a peptide pool in the initial 7-day diluted WBA. Using the 7 day diluted WBA with peptide pools as stimulants, 10/29 donors (34%) responded to at least one peptide pool. However, when recombinant proteins were used as stimulants, 27/29 (93%) donors responded to either recombinant ESAT-6 or CFP-10, and this difference was significant when compared to the peptides ($p<0.0001$ Wilcoxon matched pairs test, Figure 1, Panel B). The median IFN-gamma response to all ESAT-6, CFP-10 and Rv2654 peptide pools was 0 pg/ml, while the median response to the recombinant ESAT-6 and CFP-10 was 1878 (IQR 471–3863) and 305 pg/ml (IQR 118–1774) respectively ($p<0.0001$ for both comparisons). A significant correlation was observed between the QFT-GIT (IU/ml) and the 7 day WBA (pg/ml) response to the ESAT-6 peptide pool ($r=0.45$; $p=0.017$) and the recombinant CFP-10 ($r=0.59$; $p=0.001$) (Table 1).

We next compared the response between peptide pools and recombinant proteins in the overnight ELISpot assay using PBMC. As we had no recombinant Rv2654 available, these experiments were carried out using ESAT-6 and CFP-10 (peptides and recombinants) only. No difference in IFN-gamma induction was observed between peptide pools and recombinant proteins in the overnight ELISpot assay using PBMC from the same donors (Figure 2). Considering the low endotoxin content of these recombinant proteins, and the clear negative responses observed in some donors in the ELISpot assay, the possibility of contamination in the recombinant protein to induce higher IFN-gamma responses was ruled out. Additionally, experiments using PBMC from two donors stimulated with either peptide pools or the recombinant proteins for 1, 3, 5 or 7 days, showed similar concentrations of IFN-gamma secretion (Figure 3), indicating that peptides and recombinants have similar activity in prolonged PBMC based cultures.

SELDI-TOF MS Analysis of Whole Blood Assay Supernatants

Because peptides are degraded more easily by proteases present in whole blood than recombinant proteins [39,40], we hypothesized that insufficient peptide concentrations were present after 7

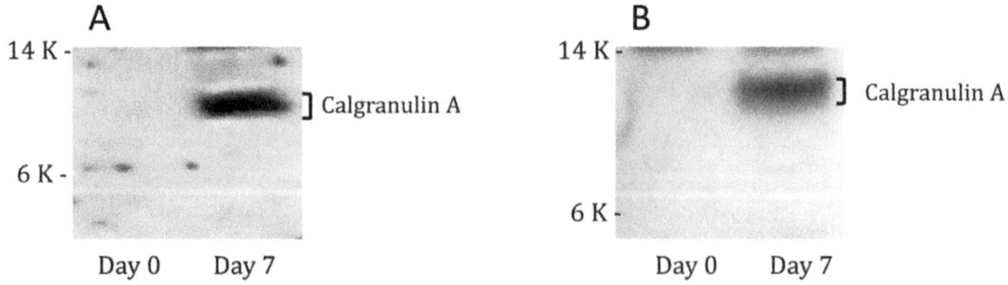

Figure 9. Detection of Calgranulin A (S100A8) by Western blot: Supernatant of whole blood stimulated with (A) rESAT-6 and (B) rCFP-10, was probed with Calgranulin A polycolonal antibody at Day 0 and after 7 days of incubation. The data indicates the presence of Calgranulin A after 7 days of culture, but not on Day 0.

days of culture to continue stimulating T cells. In order to detect possible peptide degradation, we compared the SELDI-TOF mass spectrometry profiles of ESAT-6 and CFP-10 peptide pools and recombinant proteins in the supernatants of whole blood assays from 2 donors collected at Day 0 and after 7 days of incubation. As shown in Figure 4, the ESAT-6 and CFP-10 peptide peaks were present at Day 0 but not detectable after 7 days of culture, suggesting that the peptide pools were degraded during the course of the experiment. Protein profiling of recombinant ESAT-6 and CFP-10 showed a clear peak for each antigen, corresponding to the predicted molecular weight at Day 0 that was still detectable at Day 7, although present in lower concentrations (Figures 5 and 6). Due to the low levels of rESAT-6 and rCFP-10 that were detectable by mass spectroscopy after 7 days of culture, we confirmed the presence of both recombinant proteins by western blot. The results shown in Figure 7 confirm, that both recombinant proteins are still present after 7 days of incubation.

We also observed new peaks of lower molecular weight in both the ESAT-6 and CFP-10 protein profiles, after 7 days of culture, in particular a highly expressed protein at 10.8 kDa. To determine whether protein cleavage of the 3′ C-terminal His-tag would account for the new peaks of lower molecular weight, Day 0 and Day 7 supernatants from recombinant ESAT-6 and CFP-10 stimulated whole blood were separated on a 1-D PAGE gel and probed with penta-anti-His antibody (Figure 8). These results showed that the His-tag was detectable at Day 0 and after 7 days of culture for both recombinant proteins, refuting that exopeptidase degradation of the C-terminus could account for the change in mass.

The peak observed at 10.8 kDa had a high intensity and appeared after stimulation with either rESAT-6 or rCFP-10 proteins but not with the peptide pools. A database search for proteins previously identified at an m/z of 10.8 kDa identified Calgranulin A (S100A8) as the potential protein. Calgranulin A is an S-100 calcium-binding protein that is expressed in a variety of cell types including macrophages and neutrophils during acute inflammation. In order to determine if there was an increase in Calgranulin A after long-term culture, rESAT-6 and rCFP-10 stimulated 7-day WBA supernatants were analysed by western blot. As shown in Figure 9, whole blood stimulated with either rESAT-6 or rCFP-10 contained Calgranulin A after 7 days of culture, but not at Day 0.

Discussion

In this study we investigated the secretion of IFN-γ in the commercially available QuantiFERON TB Gold in Tube (QFT-GIT) and long-term (7 day) whole blood assays, as well as the ELISpot assay in response to ESAT-6 and CFP-10. The performance of these two antigens was also compared in the form of synthetic overlapping peptide pools and recombinant proteins in 7 day WBA and PBMC based assays (both long term culture and ELISpot assay). Our data indicate a reduced response to peptide pools compared to recombinant proteins in the 7-day WBA, which has not been reported previously. The activity of peptide pools and recombinant proteins did not differ in PBMC cultures.

The IFN-gamma response in a 7-day culture results from proliferating cells. Therefore assays based on prolonged culture would depend on the presence of adequate amounts of antigenic stimuli (either peptide or recombinant protein) during the course of the culture period. Peptides have greater sensitivity than recombinant proteins to degradation by proteases present in cultured whole blood [39,40]. Our SELDI-TOF mass spectrometry data suggest that peptide degradation is most probably responsible for the reduced response observed, as the peptides are degraded by day 7 and not available to stimulate the proliferating T cells.

Our data also suggest that recombinant proteins are less abundant after 7 days of whole blood culture. However, we were clearly able to detect both recombinant proteins after 7 days of culture by Western blot. We did investigate the possibility that new peaks of lower molecular weight that have appeared after 7 days, were due to C-terminal degradation of the recombinant proteins. As we were able to detect that the His-tag was present in both Day 0 and Day 7 samples, we conclude that reduced abundance of the recombinant proteins is most likely due to uptake, processing and presentation by the antigen presenting cells during the culture period.

Interestingly, one of the lower molecular weight peaks that appeared after 7 days of incubation was observed at 10.8 kDa. A database search identified Calgranulin A (S100A8) as the potential candidate, and we confirmed that this molecule is indeed detectable after 7 days of stimulation with rESAT-6 and rCFP-10. Calgranulins are endogenously expressed at various levels in myeloid cells (neutrophils, macrophages, dendritic cells, monocytes), secreted upon cellular activation and involved in a number of cellular functions, including cell growth and differentiation [41]. The secreted form by neutrophils enhances CD11b expression on human monocytes, and is a potent chemotactic for leukocytes [42]. S100A8 detected by SELDI at day 7 could be the result of the maturation process that adherent whole blood derived monocytes go through during the incubation period, maturing into monocyte derived macrophages, and acting as antigen presenting cells. While this was not the focus of the present manuscript, the significance of calgranulin A needs to be explored in further studies.

Our results of recombinant proteins used in 7 day WBA as stimulants are in line with other published studies using the same assay. The majority of published studies use 7 day WBA with recombinant proteins and 24 hrs diluted/undiluted WBA with peptide pools. Two main recent antigen discovery papers from the Ottenhoff group use recombinant proteins [21,43]. A study looking at diagnosing leprosy with synthetic peptides [44] uses synthetic 20-mer peptides overlapping by 10 amino acids in PBMC cultured for 6 days, which is the assay system where we showed no difference between peptide pools and recombinant proteins. We are aware of one exception [45], which evaluates and describes novel MTB infection phase-dependent antigens in TB patients and household contacts. The authors included 8 peptide pools along with other recombinant proteins in a 7 day diluted whole blood assay similar to our assay. They describe the recognition of the 8 peptide pools as "poor" in both TB patients and household contacts.

While our data is preliminary with a number of limitations, and further confirmatory studies are warranted, we conclude that assessment of vaccine candidates in the form of peptides should be tested in short term assays, while long-term whole blood based assays should be based on recombinant protein formulations.

Author Contributions

Conceived and designed the experiments: KAW RJW PRL MSH HG. Performed the experiments: MSH HG KW RS KAW. Analyzed the data: KAW HG MSH RJW PRL. Contributed reagents/materials/analysis tools: RJW PRL CB. Wrote the paper: KAW MSH HG. Recruitment of volunteers: DP TO.

References

1. Hanekom WA, Dockrell HM, Ottenhoff TH, Doherty TM, Fletcher H, et al. (2008) Immunological Outcomes of New Tuberculosis Vaccine Trials: WHO Panel Recommendations. PLoS Med 5: e145.
2. Pai M, Zwerling A, Menzies D (2008) Systematic review: T-cell-based assays for the diagnosis of latent tuberculosis infection: an update. Ann Intern Med 149: 177–184.
3. Lalvani A, Pareek M (2009) A 100 year update on diagnosis of tuberculosis infection. Br Med Bull 93: 69–84.
4. Ravn P, Demissie A, Eguale T, Wondwosson H, Lein D, et al. (1999) Human T cell responses to the ESAT-6 antigen from *Mycobacterium tuberculosis*. J Infect Dis 179: 637–645.
5. Arend SM, Geluk A, van Meijgaarden KE, van Dissel JT, Theisen M, et al. (2000) Antigenic equivalence of human T-cell responses to *Mycobacterium tuberculosis*-specific RD1-encoded protein antigens ESAT-6 and culture filtrate protein 10 and to mixtures of synthetic peptides. Infect Immun 68: 3314–3321.
6. Tena-Coki NG, Scriba TJ, Peteni N, Eley B, Wilkinson RJ, et al. (2010) CD4 and CD8 T Cell Responses to Mycobacterial Antigens in African Children. Am J Respir Crit Care Med. 182: 120–129.
7. Soares AP, Scriba TJ, Joseph S, Harbacheuski R, Murray RA, et al. (2008) Bacillus Calmette-Guerin vaccination of human newborns induces T cells with complex cytokine and phenotypic profiles. J Immunol 180: 3569–3577.
8. Hanekom WA, Hughes J, Mavinkurve M, Mendillo M, Watkins M, et al. (2004) Novel application of a whole blood intracellular cytokine detection assay to quantitate specific T-cell frequency in field studies. J Immunol Methods 291: 185–195.
9. Scholvinck E, Wilkinson KA, Whelan AO, Martineau AR, Levin M, et al. (2004) Gamma interferon-based immunodiagnosis of tuberculosis: comparison between whole-blood and enzyme-linked immunospot methods. J Clin Microbiol 42: 829–831.
10. Bertholet S, Ireton GC, Kahn M, Guderian J, Mohamath R, et al. (2008) Identification of human T cell antigens for the development of vaccines against *Mycobacterium tuberculosis*. J Immunol 181: 7948–7957.
11. Wilkinson RJ, Wilkinson KA, De Smet KA, Haslov K, Pasvol G, et al. (1998) Human T- and B-cell reactivity to the 16 kDa alpha-crystallin protein of *Mycobacterium tuberculosis*. Scand J Immunol 48: 403–409.
12. Wilkinson RJ, Vordermeier HM, Wilkinson KA, Sjolund A, Moreno C, et al. (1998) Peptide-specific T cell response to *Mycobacterium tuberculosis*: clinical spectrum, compartmentalization, and effect of chemotherapy. J Infect Dis 178: 760–768.
13. Weir RE, Morgan AR, Britton WJ, Butlin CR, Dockrell HM (1994) Development of a whole blood assay to measure T cell responses to leprosy: a new tool for immuno-epidemiological field studies of leprosy immunity. J Immunol Methods 176: 93–101.
14. Todryk SM, Pathan AA, Keating S, Porter DW, Berthoud T, et al. (2009) The relationship between human effector and memory T cells measured by ex vivo and cultured ELISPOT following recent and distal priming. Immunology 128: 83–91.
15. Jurcevic S, Hills A, Pasvol G, Davidson RN, Ivanyi J, et al. (1996) T cell responses to a mixture of *Mycobacterium tuberculosis* peptides with complementary HLA-DR binding profiles. Clin Exp Immunol 105: 416–421.
16. Black GF, Weir RE, Floyd S, Bliss L, Warndorff DK, et al. (2002) BCG-induced increase in interferon-gamma response to mycobacterial antigens and efficacy of BCG vaccination in Malawi and the UK: two randomised controlled studies. Lancet 359: 1393–1401.
17. Beveridge NE, Price DA, Casazza JP, Pathan AA, Sander CR, et al. (2007) Immunisation with BCG and recombinant MVA85A induces long-lasting, polyfunctional *Mycobacterium tuberculosis*-specific CD4+ memory T lymphocyte populations. Eur J Immunol 37: 3089–3100.
18. Sidders B, Pirson C, Hogarth PJ, Hewinson RG, Stoker NG, et al. (2008) Screening of highly expressed mycobacterial genes identifies Rv3615c as a useful differential diagnostic antigen for the *Mycobacterium tuberculosis* complex. Infect Immun 76: 3932–3939.
19. Ewer K, Cockle P, Gordon S, Mansoor H, Govaerts M, et al. (2006) Antigen mining with iterative genome screens identifies novel diagnostics for the *Mycobacterium tuberculosis* complex. Clin Vaccine Immunol 13: 90–97.
20. Brock I, Weldingh K, Leyten EM, Arend SM, Ravn P, et al. (2004) Specific T-cell epitopes for immunoassay-based diagnosis of *Mycobacterium tuberculosis* infection. J Clin Microbiol 42: 2379–2387.
21. Black GF, Thiel BA, Ota MO, Parida SK, Adegbola R, et al. (2009) Immunogenicity of novel DosR regulon-encoded candidate antigens of *Mycobacterium tuberculosis* in three high-burden populations in Africa. Clin Vaccine Immunol 16: 1203–1212.
22. Agger EM, Brock I, Okkels LM, Arend SM, Aagaard CS, et al. (2003) Human T-cell responses to the RD1-encoded protein TB27.4 (Rv3878) from *Mycobacterium tuberculosis*. Immunology 110: 507–512.
23. Sutherland JS, de Jong BC, Jeffries DJ, Adetifa IM, Ota MO (2010) Production of TNF-alpha, IL-12(p40) and IL-17 can discriminate between active TB disease and latent infection in a West African cohort. PLoS One 5: e12365.
24. Leyten EM, Lin MY, Franken KL, Friggen AH, Prins C, et al. (2006) Human T-cell responses to 25 novel antigens encoded by genes of the dormancy regulon of *Mycobacterium tuberculosis*. Microbes Infect 8: 2052–2060.
25. Millington KA, Innes JA, Hackforth S, Hinks TS, Deeks JJ, et al. (2007) Dynamic relationship between IFN-gamma and IL-2 profile of *Mycobacterium tuberculosis*-specific T cells and antigen load. J Immunol 178: 5217–5226.
26. Aagaard C, Brock I, Olsen A, Ottenhoff TH, Weldingh K, et al. (2004) Mapping immune reactivity toward Rv2653 and Rv2654: two novel low-molecular-mass antigens found specifically in the *Mycobacterium tuberculosis* complex. J Infect Dis 189: 812–819.
27. Wilkinson RJ, Haslov K, Rappuoli R, Giovannoni F, Narayanan PR, et al. (1997) Evaluation of the recombinant 38-kilodalton antigen of *Mycobacterium tuberculosis* as a potential immunodiagnostic reagent. J Clin Microbiol 35: 553–557.
28. Schuck SD, Mueller H, Kunitz F, Neher A, Hoffmann H, et al. (2009) Identification of T-cell antigens specific for latent *Mycobacterium tuberculosis* infection. PLoS One 4: e5590.
29. Lin MY, Reddy TB, Arend SM, Friggen AH, Franken KL, et al. (2009) Cross-reactive immunity to *Mycobacterium tuberculosis* DosR regulon-encoded antigens in individuals infected with environmental, nontuberculous mycobacteria. Infect Immun 77: 5071–5079.
30. Chen J, Su X, Zhang Y, Wang S, Shao L, et al. (2009) Novel recombinant RD2- and RD11-encoded *Mycobacterium tuberculosis* antigens are potential candidates for diagnosis of tuberculosis infections in BCG-vaccinated individuals. Microbes Infect 11: 876–885.
31. Sallusto F, Lenig D, Forster R, Lipp M, Lanzavecchia A (1999) Two subsets of memory T lymphocytes with distinct homing potentials and effector functions. Nature 401: 708–712.
32. Godkin AJ, Thomas HC, Openshaw PJ (2002) Evolution of epitope-specific memory CD4(+) T cells after clearance of hepatitis C virus. J Immunol 169: 2210–2214.
33. Leyten EM, Arend SM, Prins C, Cobelens FG, Ottenhoff TH, et al. (2007) Discrepancy between *Mycobacterium tuberculosis*-specific gamma interferon release assays using short and prolonged in vitro incubation. Clin Vaccine Immunol 14: 880–885.
34. Beveridge NE, Fletcher HA, Hughes J, Pathan AA, Scriba TJ, et al. (2008) A comparison of IFNgamma detection methods used in tuberculosis vaccine trials. Tuberculosis (Edinb) 88: 631–640.
35. Rangaka MX, Diwakar L, Seldon R, van Cutsem G, Meintjes GA, et al. (2007) Clinical, immunological, and epidemiological importance of antituberculosis T cell responses in HIV-infected Africans. Clin Infect Dis 44: 1639–1646.
36. Connell TG, Shey MS, Seldon R, Rangaka MX, van Cutsem G, et al. (2007) Enhanced ex vivo stimulation of *Mycobacterium tuberculosis*-specific T cells in human immunodeficiency virus-infected persons via antigen delivery by the Bordetella pertussis adenylate cyclase vector. Clin Vaccine Immunol 14: 847–854.
37. Gideon HP, Wilkinson KA, Rustad TR, Oni T, Guio H, et al. (2010) Hypoxia induces an immunodominant target of tuberculosis specific T cells absent from common BCG vaccines. PLoS Pathog 6: e1001237.
38. Sinha S, Ambur OH, Langford PR, Tonjum T, Kroll JS (2008) Reduced DNA binding and uptake in the absence of DsbA1 and DsbA2 of Neisseria meningitidis due to inefficient folding of the outer-membrane secretin PilQ. Microbiology 154: 217–225.
39. Yeaman MR, Gank KD, Bayer AS, Brass EP (2002) Synthetic peptides that exert antimicrobial activities in whole blood and blood-derived matrices. Antimicrob Agents Chemother 46: 3883–3891.
40. Nusslein K, Arnt L, Rennie J, Owens C, Tew GN (2006) Broad-spectrum antibacterial activity by a novel abiogenic peptide mimic. Microbiology 152: 1913–1918.
41. Hofmann Bowman MA, Schmidt AM (2011) S100/calgranulins EN-RAGEing the blood vessels: implications for inflammatory responses and atherosclerosis. Am J Cardiovasc Dis 1: 92–100.
42. Donato R (2003) Intracellular and extracellular roles of S100 proteins. Microsc Res Tech 60: 540–551.
43. Commandeur S, van Meijgaarden KE, Prins C, Pichugin AV, Dijkman K, et al. (2013) An unbiased genome-wide *Mycobacterium tuberculosis* gene expression approach to discover antigens targeted by human T cells expressed during pulmonary infection. J Immunol 190: 1659–1671.
44. Geluk A, van der Ploeg J, Teles RO, Franken KL, Prins C, et al. (2008) Rational combination of peptides derived from different *Mycobacterium leprae* proteins improves sensitivity for immunodiagnosis of *M. leprae* infection. Clin Vaccine Immunol 15: 522–533.
45. Chegou NN, Black GF, Loxton AG, Stanley K, Essone PN, et al. (2012) Potential of novel *Mycobacterium tuberculosis* infection phase-dependent antigens in the diagnosis of TB disease in a high burden setting. BMC Infect Dis 12: 10.

17

Fusogenic-Oligoarginine Peptide-Mediated Delivery of siRNAs Targeting the CIP2A Oncogene into Oral Cancer Cells

Liliana Cantini, Christopher C. Attaway, Betsy Butler, Lourdes M. Andino, Melissa L. Sokolosky, Andrew Jakymiw*

Depatment of Craniofacial Biology and Center for Oral Health Research, Hollings Cancer Center, Medical University of South Carolina, Charleston, South Carolina, United States of America

Abstract

Despite a better understanding of the pathogenesis of oral cancer, its treatment outcome remains poor. Thus, there is a need for new therapeutic strategies to improve the prognosis of this disease. RNA interference (RNAi) appears to be a promising therapeutic tool for the treatment of many diseases, including oral cancer. However, an obstacle for RNAi-mediated therapies has been delivery, in particular, the retention of small interfering RNAs (siRNAs) in endosomes and their subsequent degradation in lysosomes, resulting in inefficient gene silencing. Thus, the current study examined the feasibility of designing and utilizing a peptide, termed 599, consisting of a synthetic influenza virus-derived endosome-disruptive fusogenic peptide sequence and a stretch of cationic cell-penetrating nona(D-arginine) residues, to deliver siRNAs into oral cancer cells and induce silencing of the therapeutic target, CIP2A, an oncoprotein overexpressed in various human malignancies including oral cancer. Increasing the 599 peptide-to-siRNA molar ratio demonstrated a higher binding capacity for siRNA molecules and enhanced siRNA delivery into the cytoplasm of oral cancer cells. In fact, quantitative measurements of siRNA delivery into cells demonstrated that a 50:1 peptide-to-siRNA molar ratio could deliver 18-fold higher amounts of siRNAs compared to cells treated with siRNA alone with no significant long-term cytotoxic effects. Most importantly, the 599 peptide-mediated siRNA delivery promoted significant CIP2A mRNA and protein silencing which resulted in decreased oral cancer cell invasiveness and anchorage-independent growth. Together, these data demonstrate that a chimeric peptide consisting of a fusogenic sequence, in combination with cell-penetrating residues, can be used to effectively deliver siRNAs into oral cancer cells and induce the silencing of its target gene, potentially offering a new therapeutic strategy in combating oral cancer.

Editor: Kin-Hang Kok, University of Hong Kong, Hong Kong

Funding: This work was supported by the NIDCR and NCRR grants R00DE018191 (AJ) and P20RR017696, respectively. This work was also funded by the Bankhead-Coley Florida Cancer Research Program grant 08BN-02 (AJ). Technical support provided by the Clemson Light Imaging Facility was made possible by the National Science Foundation Award #1126407 and a Clemson University Core Facility grant to Terri F. Bruce. The funders had no role in study design, data collection and analysis, decision to publish, or preparation of the manuscript.

Competing Interests: The authors have declared that no competing interests exist.

* E-mail: jakymiw@musc.edu

Introduction

It is estimated that about 40,000 new cases and approximately 8,000 deaths related to cancer of the oral cavity and pharynx will occur annually in the USA in 2012 [1]. Oral cavity cancer is currently ranked as the 6th most prevalent cancer globally, with squamous cell carcinomas of the oral mucosa being the most common type (~90%) [2,3]. Despite vast amounts of research and advances in the fields of oncology and surgery, the 5-year survival rate for oral cancer has only modestly improved in the last 30 years and its prognosis remains poorer compared to breast, colon, or prostate cancer [1]. Therefore, new therapeutic strategies are needed to improve the outcome of this disease.

RNA interference (RNAi) is a highly conserved post-transcriptional gene regulatory mechanism triggered by small, non-coding double-stranded RNA molecules that can specifically silence gene expression by either repressing translation and/or inducing mRNA degradation [4,5]. Short double-stranded RNA molecules, known as small interfering RNA (siRNA) are functional molecules that in association with the RNA-induced silencing complex (RISC) mediate sequence-specific mRNA target selection and cleavage [6,7,8,9]. The discovery that the introduction of chemically synthesized siRNAs into mammalian cells could efficiently induce sequence-specific inhibition of gene expression [6], made evident the therapeutic potential of harnessing RNAi as a means to specifically target and silence disease-causing genes. Subsequent preclinical experiments in animals and more recent clinical trials have further validated siRNAs as potent inhibitors of an assortment of disease-causing genes and as a promising new class of therapeutics [8,10,11].

Although the design of therapeutic-grade siRNAs has improved [8,10], delivery still remains the single greatest obstacle towards the pervasive use of siRNAs for therapeutic applications [8]. Because therapeutic macromolecules are generally delivered through endocytosis [12], one of the major limiting steps for many delivery approaches, including siRNA delivery, is endosomal entrapment and subsequent degradation of the therapeutic cargo in lysosomes [11,12,13]. Thus, to enhance the intracellular bioavailability of siRNAs, effective strategies for endosomal escape are needed.

To transport the viral genome into the cytoplasm, animal viruses that are internalized via receptor-mediated endocytosis, utilize proteins with endosome-disruptive fusion peptide domain sequences to mediate destabilization of the host cell endosomal membrane [14]. The method by which these viral proteins destabilize endosomal membranes occurs in an acidification-dependent manner and has been mimicked by synthetic peptides, termed fusogenic peptides [15,16]. In particular, several synthetic fusogenic peptides based on the N-terminal fusion domain of the HA2 subunit of the influenza virus hemagglutinin protein have proven to be effective at influencing gene transfer [15]. Among these fusogenic peptides, both the INF-7 peptide and its dimeric form diINF-7, have demonstrated their endosome disruptive capability by improving non-viral gene transfer systems and enhancing both the cytosolic delivery of immunoliposome-entrapped macromolecules and lipofectamine-mediated siRNA-induced gene silencing [15,16,17]. Moreover, co-incubation of chimeric peptides consisting of the HA2 endosome-disruptive peptide fused to cell-penetrating peptides (CPPs; cationic peptide vectors that can rapidly induce their own cellular internalization through different forms of endocytosis [18]), have also been demonstrated to enhance the endosomal escape of CPP-complexed cargo, consequently promoting stronger biological effects [12,18,19]. In one particular study, co-incubation of HA2-linked CPP peptides with CPP-complexed siRNAs was found to moderately improve the silencing of the targeted reporter gene [20].

Although the above described strategies have demonstrated improvements in siRNA-mediated silencing effects, several problems still remain that could limit the efficacy of fusogenic peptide-mediated cytosolic delivery of siRNAs. First, because the fusogenic or HA2-chimeric peptides were not complexed to the siRNAs, this approach did not guarantee that the peptides would co-localize within the same endocytic vesicles as the siRNA cargo [12]. Co-localization is required in order to release siRNA molecules from the endocytic vesicle [12]. Second, due to this lack of interaction between the peptides and siRNAs, it is possible that the peptides could escape the endosomes without seriously disrupting the endosomal membranes, ultimately leaving the siRNA cargo entrapped within the vesicle [12]. Thus, to circumvent these problems, this study aimed to design a chimeric peptide consisting of a fusogenic sequence linked to a CPP that would enable stable binding with siRNAs via electrostatic interactions and promote their intracellular delivery and endosomal escape, in order to induce the therapeutic silencing of a targeted oncogene in oral cancer cells. Because the INF-7 peptide is a more potent membrane-destabilizing peptide compared to the parent HA2 peptide and cationic nona(D-arginine) peptides are highly efficient CPPs capable of enhanced cellular uptake [12,21], as well as delivery of siRNAs into tumors and brains of mice [22,23], we designed a chimeric peptide, termed 599, that would harness both these properties within a single molecule to directly complex to and enhance the intracellular bioavailability of siRNAs. Here, we demonstrate that the 599 peptide effectively delivers siRNAs designed to target CIP2A, an oncoprotein overexpressed in human head and neck squamous cell carcinomas (HNSCCs), including oral squamous cell carcinomas (OSCCs) [24,25,26], into oral cancer cells, mediates efficient CIP2A silencing, and consequently inhibits oral cancer cell invasiveness and anchorage-independent growth.

Results

The 599 Peptide Effectively Binds and Delivers siRNAs into Oral Cancer Cells

In order to test whether the 599 peptide via its cationic nona(D-arginine) residues could effectively bind negatively charged siRNAs based on electrostatic interactions, an agarose gel shift assay was performed (**Fig. 1**). Using various amounts of the 599 peptide, ranging from 1 to 50-fold molar excess of siRNAs, it was demonstrated that the 599 peptide could indeed bind siRNA molecules designed to target CIP2A (siCIP2A) by retarding the siCIP2A starting at a peptide-to-siRNA molar (P:N) ratio of 20:1, with no free siRNAs detectable in the gel at 50:1. At P:N ratios of 10:1, the peptide only had moderate effects on siRNA binding, and at 1:1 it was completely ineffective.

Upon demonstrating that the 599 peptide could bind to siRNAs at specific P:N ratios, we next examined the ability of the peptide to deliver fluorescently-labeled siRNAs into oral cancer cells by fluorescence microscopy analysis (**Fig. 2A**). Using a fluorescently-labeled siRNA, DY547-conjugated to siCIP2A (D-siCIP2A), in complex with increasing amounts of 599 peptide, ranging from 1 to 50-fold molar excess of siRNAs, it was observed that increasing the P:N ratio enhanced the delivery of siRNA molecules into CAL 27 oral cancer cells after two hours incubation. More specifically, the 50:1 P:N ratio was demonstrated to have an increased ability at inducing siRNA uptake into cells compared to the 20:1 P:N ratio which only had a moderate effect. Conversely, cells treated with D-siCIP2A alone or in complex with 599 peptide at a 1:1 P:N

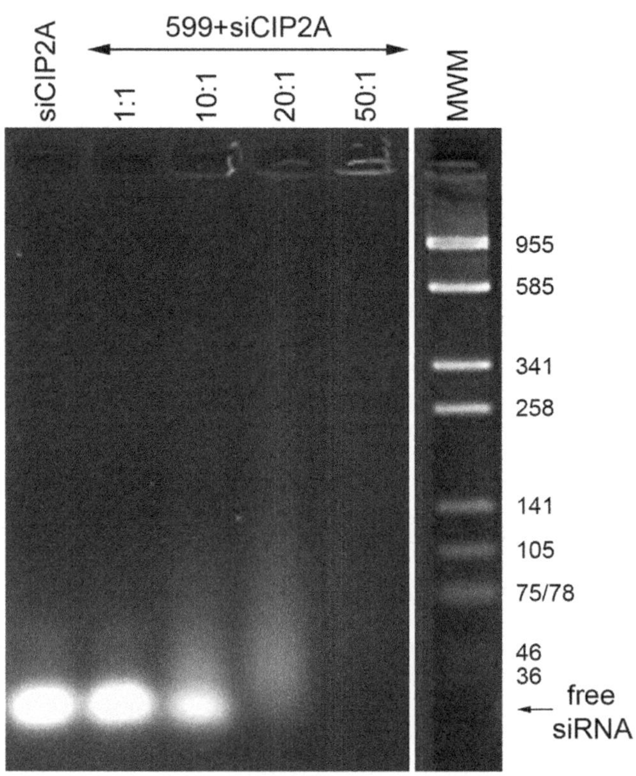

Figure 1. Optimization of 599 peptide binding to siRNAs. An ethidium bromide stained 4% agarose gel shift assay examining the ability of various amounts of the 599 peptide (ranging from 1 to 50-fold molar excess of siRNAs) to form complexes with siCIP2A. siCIP2A, siRNA targeting the CIP2A oncogene; MWM, molecular weight marker (the number of base pairs for each DNA fragment are shown).

ratio were incapable of effective siRNA uptake. Quantitative measurements of internalized D-siCIP2A into CAL 27 cells after 2.5 hours incubation also corroborated that increasing the 599 P:N ratios increased siRNA uptake into cells. In particular, the 50:1 and 100:1 P:N ratios delivered approximately 18 and 42-fold significantly higher amounts of D-siCIP2A, respectively, compared to cells treated with D-siCIP2A alone (**Fig. 2B**). As a comparison, the commercial transfection agent INTERFERin™ (IFN) only delivered approximately 2-fold higher amounts of D-siCIP2A, compared to cells treated with D-siCIP2A alone. Of note, although the 100:1 P:N ratio was capable of delivering the highest amount of siRNAs into cells, it induced significant long-term cytotoxic effects as measured by using a non-targeting siRNA (siNT) control (**Fig. 2C**). In particular, the 100:1 P:N ratio significantly reduced the proliferation of cells 48 hours post-treatment compared to untreated cells. Conversely, the 50:1 P:N ratio had no significant effects on long-term cytotoxicity. As a result, based on these findings the 50:1 P:N ratio was used for further characterization and experimentation.

Characterization of the 599 Peptide-siRNA Complex

To gain insight into the mechanism of cell entry of the 599/siRNA complex, the subcellular localization of the 599/D-siCIP2A complex was further examined in live cells by fluorescence microscopy coupled with phase contrast images (**Fig. 3A**). Upon closer inspection of the internalized D-siCIP2As, it appeared based on the extensive punctate uptake pattern of siRNAs that the 599/D-siCIP2A complex was being endocytosed. Co-staining of fixed cells with the early endosomal marker EEA1, confirmed the colocalization of D-siCIP2As with endosomal vesicles (**Fig. 3B**), thus indicating an endocytic internalization mode.

To assess the 599 peptide in terms of particle size and surface charge after complexation with siCIP2A both dynamic light scattering and zeta potential measurements were performed (**Fig. 4A**). Based on the measurements, 96% of the particles formed were centered at 74 nm and 4% at 220 nm by number-weighted size distribution, with the average particle size of the major population being 80±0.5 nm. The zeta potential of the 599/siCIP2A complex was determined to be positive with a value of 32.5±0.5 mV. The 599/siCIP2A complex was also visualized using darkfield-based optical illumination (**Fig. 4B**). Using this imaging technology the 599/siCIP2A complex was found to exhibit fairly uniform spherical structures.

To test the importance of the two core components (i.e. the endosome-disruptive and cationic cell-penetrating nona(D-arginine) residue portions) in conferring the ability of the 599 peptide to deliver siRNAs into cells, two additional peptides were synthesized consisting of either the endosome-disruptive sequence (616) or the cationic cell-penetrating nona(D-arginine) residues alone (9R). Upon treatment of CAL 27 cells with either the 599, 616, or 9R peptides in complex with D-siCIP2A at a 50:1 P:N ratio, fluorescence microscopy analyses revealed that only the intact 599 peptide with both the endosome-disruptive and cationic cell-penetrating nona(D-arginine) residue portions could mediate delivery of siRNAs into cells, whereas both the 616 and 9R peptides had no apparent effect on siRNA delivery 2 hours post-treatment (**Fig. 5**).

The failure of the 616 peptide to deliver siRNAs into cells was most likely due to its inability to bind siRNAs even at P:N ratios as high as 100:1 (data not shown). However, the result with the 9R peptide was somewhat surprising, since it was demonstrated to effectively bind siRNAs at a 50:1 P:N ratio based on an agarose gel shift assay (**Fig. S1A**) and could form particles with an average size of 58.2±0.2 nm and a zeta potential value of 25.8±0.5 mV (**Fig. S1B**). Because paraformaldehyde fixation of cells can result in the artifactual redistribution of $(Arg)_9$ peptides [27], it was possible that the cellular distribution of the 9R/D-siCIP2A complex may have been affected by this chemical fixative. Therefore, the uptake of D-siCIP2A was also assessed using the fixation-independent quantitative approach which similarly found that the 9R peptide, ranging from 0.1 to 100-fold molar excess of siRNAs, was incapable of mediating siRNA delivery into cells and was indistinguishable from cells treated with D-siCIP2A alone (**Fig. S1C**).

The 599 Peptide Mediates CIP2A Gene Silencing in Oral Cancer Cells

For the 599 peptide to be considered an effective delivery vehicle for siRNA-based therapeutics, it must be capable of delivering functional siRNAs into cells that can silence their intended gene target. To assess the functionality of the 599/siRNA complex, two oral cancer cell lines, CAL 27 and SCC-25, were treated with the 599 peptide complexed to either siCIP2A or a siNT control at a 50:1 P:N ratio, after which both the CIP2A mRNA and protein levels were assessed 48 hours post-treatment by real-time PCR and Western blot analyses, respectively. Quantitation by real-time PCR demonstrated a significant knockdown in CIP2A mRNA levels in both oral cancer cell lines treated with the 599/siCIP2A complex compared to control 599/siNT treated cells (**Fig. 6A**). More specifically, a ~60% and ~85% reduction in CIP2A mRNA levels was observed in CAL 27 and SCC-25 oral cancer cell lines, respectively. Furthermore, Western blot analyses confirmed the ability of the 599 peptide to mediate delivery of functional siRNAs in both cell lines by demonstrating the suppression of CIP2A protein levels 48 hours post-treatment with the 599/siCIP2A complex compared to control 599/siNT treated cells (**Fig. 6B**). Of note, the protein levels of the oncogenic transcription factor c-Myc were also assessed because CIP2A is known to regulate the stability of this protein [25]. Western blot analyses confirmed that silencing of CIP2A caused destabilization of c-Myc in both cell lines (**Fig. 6B**).

Characterization of the 599/siCIP2A Complex-Mediated RNAi Response

In an effort to further characterize 599 peptide-mediated silencing of CIP2A in oral cancer cells the duration of gene silencing, dose-dependent gene silencing, and activity of silencing in serum were also assessed. The successful implementation of RNAi as a form of therapy depends on these three key features. Thus, in an effort to determine the persistence of 599 peptide-mediated silencing of CIP2A in oral cancer cells over time, a Western blot analyses of CIP2A protein expression levels in CAL 27 and SCC-25 oral cancer cells 1, 3, 5, 7, and 9 days post-treatment with 599 peptide complexed to either siNT or siCIP2A at a 50:1 peptide-to-siRNA molar ratio were performed (**Fig. 7A**). After a single treatment with 599/siCIP2A, the silencing effect was observed to last for 5 days in CAL-27 cells and up to 9 days in SCC-25 cells with maximum silencing occurring for 3 and 7 days, respectively. In dose-response experiments in SCC-25 cells (**Fig. 7B**), the 599 peptide was observed to confer siCIP2A mediated silencing dose-dependently with siRNA concentrations ≥50 nM conferring ~100% CIP2A protein knockdown and siRNA concentrations as low as 20 nM still having measurable silencing effects. Similar results were observed in CAL 27 cells (data not shown), except that higher siRNA concentrations (≥80 nM) were needed to confer at best ~90% CIP2A protein

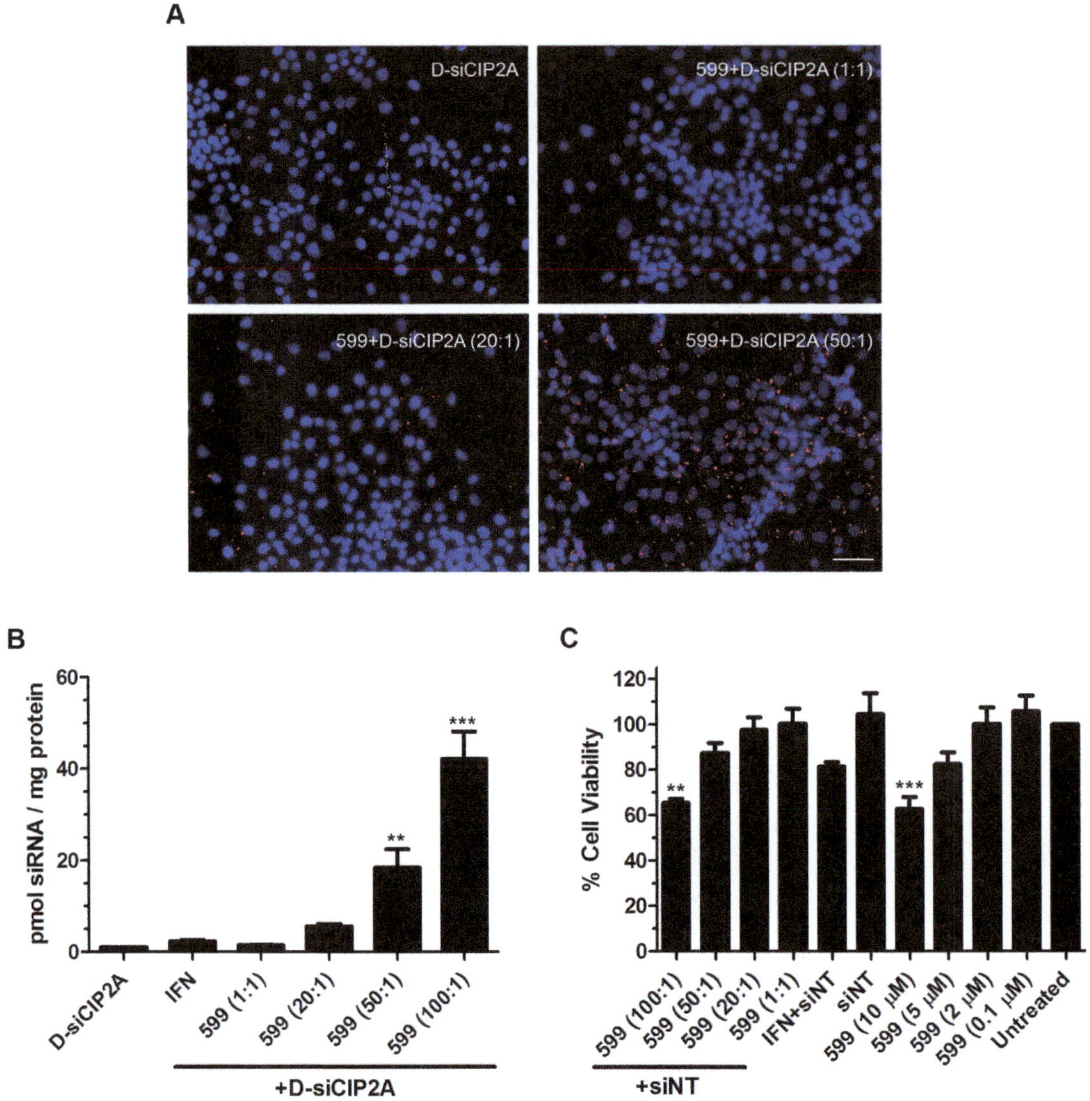

Figure 2. Optimization of 599 peptide-mediated delivery of siRNAs into oral cancer cells. (**A**) Fluorescence microscopy analysis of CAL 27 oral cancer cells incubated for 2 hours with DY547-conjugated siRNA targeting CIP2A (D-siCIP2A; red) alone or in complex with increasing amounts of 599 peptide (ranging from 1 to 50-fold molar excess of siRNAs). Nuclei (blue) were counterstained with DAPI. Scale bar: 50 μm. (**B**) CAL 27 cells incubated for 2.5 hours with D-siCIP2A alone or in complex with increasing amounts of 599 peptide (ranging from 1 to 100-fold molar excess of siRNAs). For comparison, the cells were also transfected using the commercial transfection reagent, INTERFERin^{TM} (IFN). The amount of siRNA delivered into cells in pmol per mg of protein is reported with each treatment normalized to D-siCIP2A alone. Data are mean ± SEM of four separate experiments, where ***$P<0.001$, **$P<0.01$ compared to D-siCIP2A alone treated cells (ANOVA, Dunnett's Multiple Comparison Test). (**C**) Assessment of long-term toxicity (as measured by a cell proliferation assay) of CAL 27 cells 48 hours post-treatment with either a non-targeting siRNA (siNT) alone, increasing concentrations of 599 peptide alone, or increasing amounts of 599 peptide (ranging from 1 to 100-fold molar excess of siRNAs) in complex with siNT. For comparison, the cells were also transfected with the commercial transfection reagent, IFN. Untreated cells were defined as 100% viable. Data are mean ± SEM performed in triplicate (n = 3), where ***$P<0.001$, **$P<0.01$ compared to untreated cells (ANOVA, Dunnett's Multiple Comparison Test).

silencing. Lastly, because the stability of siRNAs is an important concern in RNAi-therapy, 599 peptide-mediated CIP2A silencing was assessed in the absence and presence of serum and compared to several commercially available standard cationic lipid-based transfection reagents (**Fig. 7C**). Upon treating SCC-25 cells with 599/siCIP2A, ~95% CIP2A protein was knocked down in the absence of an initial input of serum and was comparable to CIP2A silencing observed for all cationic lipid-based transfection reagents tested (>99%) except HiPerFect® which exhibited only a ~30% reduction in protein levels. Of significance, however, is that even in the presence of serum, the 599/CIP2A complex could still produce a very effective RNAi response at ~70% CIP2A protein knockdown. The quantitation of these data were based on the densitometric analyses of three independent Western blot experiments using Image J software [28].

The 599 Peptide-Mediated Silencing of CIP2A Decreases Oral Cancer Cell Invasiveness and Anchorage-Independent Growth

Previously published literature has demonstrated that CIP2A silencing in cancer cells derived from different tissues, such as HNSCCs, renal cell carcinomas, and breast cancer can have

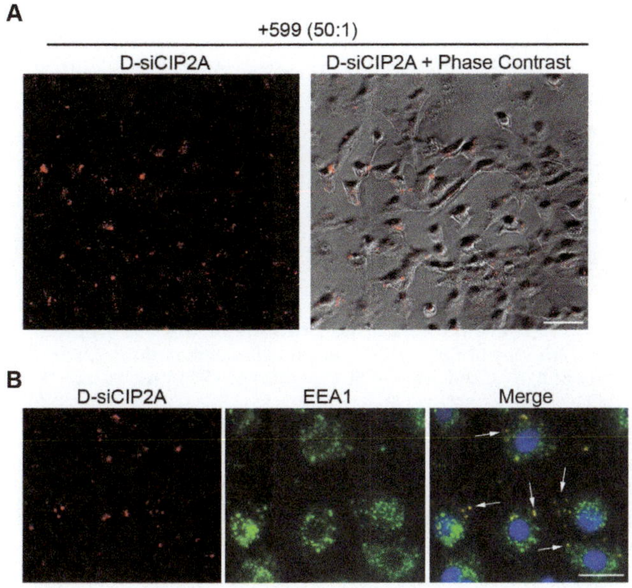

Figure 3. The 599 peptide mediates delivery of siRNAs into cells via endocytosis. (**A**) Fluorescence microscopy analysis and overlay of a phase contrast image of live CAL 27 oral cancer cells incubated for 2 hours with DY547-conjugated siRNA targeting CIP2A (D-siCIP2A; red) complexed with the 599 peptide at a 50:1 peptide-to-siRNA molar ratio. Scale bar: 50 μm. (**B**) Indirect immunofluorescence microscopy analysis of CAL 27 oral cancer cells incubated for 2 hours with D-siCIP2A (red) complexed with the 599 peptide at a 50:1 peptide-to-siRNA molar ratio. Endosomes (green) were stained using a rabbit monoclonal anti-EEA1 antibody. Nuclei (blue) were counterstained with DAPI. The co-localization of D-siCIP2A with endosomes (yellow) is observed in the merged panel and is emphasized by arrows. Scale bar: 25 μm.

effects on cancer cell invasion and anchorage-independent growth [25,29,30]. Thus, to further demonstrate the potential usefulness of the 599 peptide in siRNA-based therapeutics for oral cancer, we tested whether 599 peptide-mediated silencing of CIP2A could affect the invasiveness and anchorage-independent growth properties of oral cancer cells. Due to the fact that CAL 27 cells exhibited only a brief silencing effect and do not grow well in semi-solid medium [31], the 599 peptide-mediated CIP2A silencing effects on cell invasion and anchorage-independent growth were only tested in SCC-25 cells. Upon treatment of SCC-25 cells with the 599 peptide complexed to siCIP2A, we observed a significant ~33% inhibition in cell invasiveness, compared to control 599/siNT treated cells (**Fig. 8A**). Moreover, 599 peptide-mediated silencing of CIP2A in SCC-25 cells significantly reduced anchorage-independent growth by ~39% compared to control 599/siNT treated cells (**Fig. 8B**). Together, these results demonstrated the ability of the 599 peptide to function as an effective delivery vehicle for siRNA-based oral cancer therapeutics.

Discussion

For RNAi-based technologies to be effective therapies, they require the efficient intracellular delivery of therapeutic siRNAs into the cytoplasm of cells. However, because most delivery approaches traffic siRNA cargo into cells via endocytosis, the entrapment of these molecules within the endocytic vesicles has become a major limiting step [11,12] and thus a barrier towards the effective use of RNAi-based therapies. To circumvent this problem, we designed and tested a peptide, termed 599, that consisted in part the sequences of both a highly efficient CPP and the endosome-destabilizing INF-7 peptide that when combined would enable effective delivery of complexed siRNA cargo into cells. Results from our study demonstrated that the 599 peptide could increase the intracellular bioavailability of siRNAs in oral cancer cells and effectively mediate the silencing of the targeted CIP2A oncoprotein with consequent inhibition of oral cancer cell invasiveness and anchorage-independent growth.

Because covalent conjugation of siRNAs to CPPs can result in the inefficient delivery of functional siRNAs to cells [18,32,33], the 599 peptide was designed to include cationic nona(D-arginine) residues that would function not only as the CPP component, but also enable non-covalent binding of negatively charged siRNA molecules to the peptide via charged interactions. Agarose gel shift assays confirmed that the 599 peptide could complex to siRNA molecules and that the binding between the 599 peptide and siRNAs appeared to be dependent on the nona(D-arginine) residues. This was particularly evident based on the fact that the 616 peptide, which comprised only the highly anionic INF-7 peptide sequence, could not bind siRNAs.

Investigation into whether the 599 peptide-siRNA complex formation was sufficient to induce the internalization of the siRNAs into cells, revealed that increasing the P:N ratios would result in increased siRNA uptake into cells after two hours treatment compared to cells treated with INTERFERin™ or naked siRNA. These findings were consistent with several other studies that similarly found that increasing the concentrations of

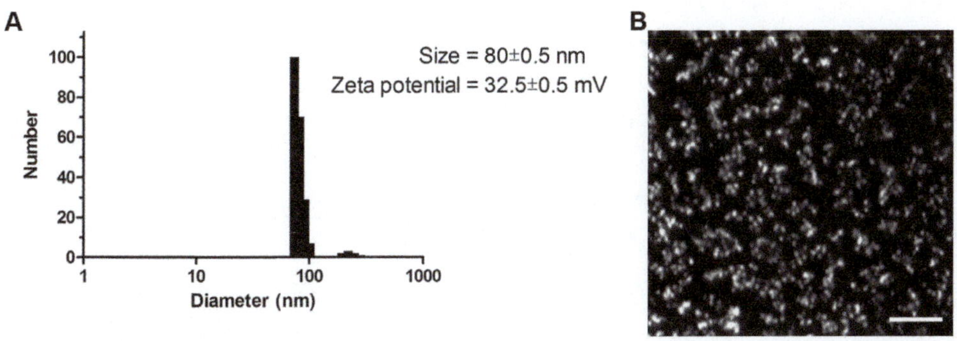

Figure 4. Particle characterization of the 599/siCIP2A complex. (**A**) Size distribution and zeta potential of the 599 peptide complexed with siCIP2A at a 50:1 peptide-to-siRNA molar ratio 20 minutes after formulation in water. (**B**) Darkfield-based optical microscopy image of the 599 peptide complexed with siCIP2A at a 50:1 peptide-to-siRNA molar ratio 20 minutes after formulation in water. Scale bar: 10,000 nm.

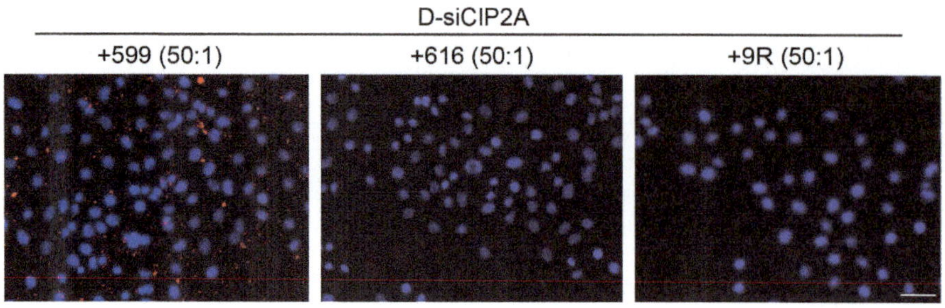

Figure 5. Both the endosome-disruptive and cationic cell-penetrating nona(D-arginine) residue portions of the 599 peptide are required for siRNA delivery into oral cancer cells. Fluorescence microscopy analyses of CAL 27 oral cancer cells incubated for 2 hours with DY547-conjugated siRNA targeting CIP2A (D-siCIP2A; red) complexed to peptides 599, 616, and 9R at a 50:1 peptide-to-siRNA molar ratio. Nuclei (blue) were counterstained with DAPI. Scale bar: 50 µm.

specific CPPs increased their siRNA uptake ability [20,23]. The increased uptake of siRNAs into cells most likely occurred at higher P:N ratios based on several factors. First, because the higher P:N ratios were more effective at binding free siRNAs in solution, as was evident through the agarose gel shift assays, this in-part would have contributed to larger quantities of siRNAs being internalized into cells. Second, because treatment of cells with high extracellular concentrations of CPPs has been reported to induce their internalization into cells via activation of endocytosis [34], the increased uptake of siRNAs into cells at higher P:N ratios may have also been attributed to the presence of higher amounts of the 599 peptide which would more effectively trigger endocytosis of the complex.

CPP-mediated internalization into cells is driven by cationic residues found within their sequence and this process has been reported to occur through various forms of endocytosis [18]. In our study, the 599 peptide was observed to mediate the internalization of siRNAs within the first two hours to endocytic vesicles and removal of the nona(D-arginines) from the 599 peptide, as represented by the 616 peptide, confirmed the requirement of cationic residues for the delivery of siRNA cargo. Surprisingly, the 9R peptide which consisted only of cationic nona(D-arginine) residues was similarly incapable of delivering siRNAs into cells, even though it could form complexes with siRNAs. One possible explanation for this contradictory effect could be based on a recent study that found that cells exhibit a preferential uptake of L- versus D-amino acid CPPs with chirality being an important determinant in the capacity of CPP endocytic uptake [35]. In fact, the study demonstrated that nona(D-arginine) peptides were less efficiently internalized into cells compared to their L-counterparts and appeared to accumulate more at the plasma membrane rather than intracellularly [35]. Thus, the 9R peptide in our study most likely could not deliver complexed siRNAs into cells because it was itself incapable of cellular uptake based on its D-amino acid stereochemistry and thus, could explain why in a study by Kim *et al.* [22] complexing siRNAs with nona(D-arginine) peptides alone could not induce silencing of their targeted gene. Interestingly, in this same study, Kim *et al.* also reported that attaching the hydrophobic lipid anchor cholesterol to nona(D-arginine) residues could both enhance the cellular uptake of plasmid DNA and induce silencing of their target gene [22]. In fact, several reports have theorized that combining hydrophobic moieties with cationic agents can produce synergistic effects that enhance the uptake of nucleic acid cargo by either increasing the adsorption of the nucleic acid/carrier complex to the cell surface thereby promoting a greater degree of internalization via endocytosis or causing the compaction of the nucleic acid/carrier complexes [22,36,37,38,39,40]. Although, the precise mechanism of how the 599 peptide mediates enhanced delivery of siRNAs still remains to be determined, it is interesting to note that

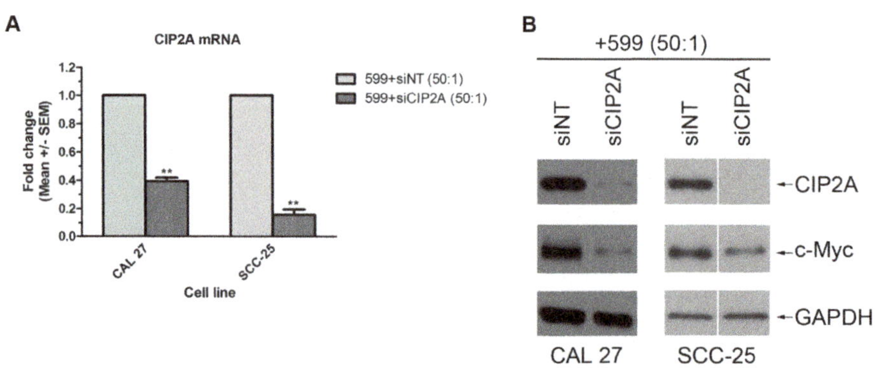

Figure 6. The 599 peptide mediates CIP2A gene silencing and subsequent destabilization of the c-Myc oncoprotein in oral cancer cells. (**A**) Real-time PCR analysis of CIP2A mRNA levels in CAL 27 and SCC-25 oral cancer cells 48 hours post-treatment with 599 peptide complexed at a 50:1 peptide-to-siRNA molar ratio to 100 nM of siRNA targeting CIP2A (siCIP2A) compared with control non-targeting siRNA (siNT). The CIP2A mRNA levels were normalized to 18S rRNA. Data are mean ± SEM of three separate experiments performed in triplicate, where **$P<0.01$ compared to siNT treated cells (Student's t test). (**B**) Western blot analyses of CIP2A and c-Myc protein expression levels in CAL 27 and SCC-25 oral cancer cells 48 hours post-treatment with 599 peptide complexed to either 100 nM of siNT or siCIP2A at a 50:1 peptide-to-siRNA molar ratio. GAPDH protein levels were monitored to ensure equal loading of samples.

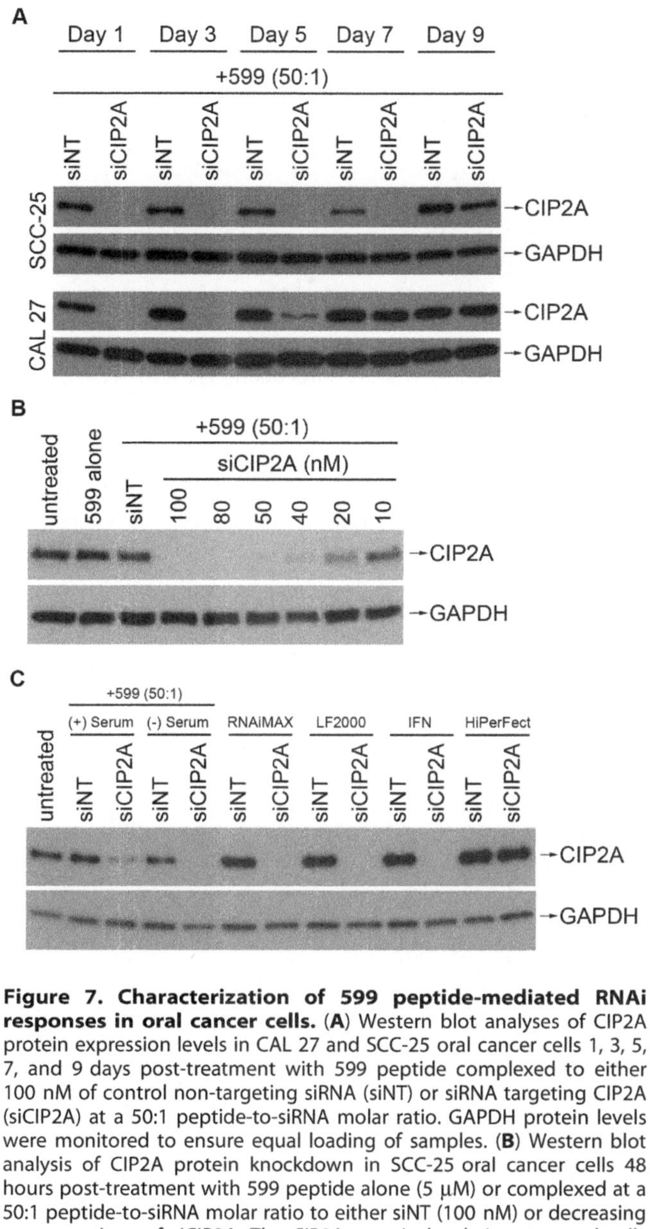

Figure 7. Characterization of 599 peptide-mediated RNAi responses in oral cancer cells. (**A**) Western blot analyses of CIP2A protein expression levels in CAL 27 and SCC-25 oral cancer cells 1, 3, 5, 7, and 9 days post-treatment with 599 peptide complexed to either 100 nM of control non-targeting siRNA (siNT) or siRNA targeting CIP2A (siCIP2A) at a 50:1 peptide-to-siRNA molar ratio. GAPDH protein levels were monitored to ensure equal loading of samples. (**B**) Western blot analysis of CIP2A protein knockdown in SCC-25 oral cancer cells 48 hours post-treatment with 599 peptide alone (5 μM) or complexed at a 50:1 peptide-to-siRNA molar ratio to either siNT (100 nM) or decreasing concentrations of siCIP2A. The CIP2A protein levels in untreated cells were also analyzed for comparison. GAPDH protein levels were monitored to ensure equal loading of samples. (**C**) Western blot analyses of CIP2A protein knockdown in SCC-25 oral cancer cells 48 hours post-treatment with 599 peptide complexed to either 100 nM of siNT or siCIP2A at a 50:1 peptide-to-siRNA molar ratio, in the presence (+) or absence (-) of serum, compared to untreated cells or cells treated with commercial transfection reagents Lipofectamine® RNAiMAX (RNAiMAX), Lipofectamine® 2000 (LF2000), INTERFERin™ (IFN), and HiPerFect®. GAPDH protein levels were monitored to ensure equal loading of samples.

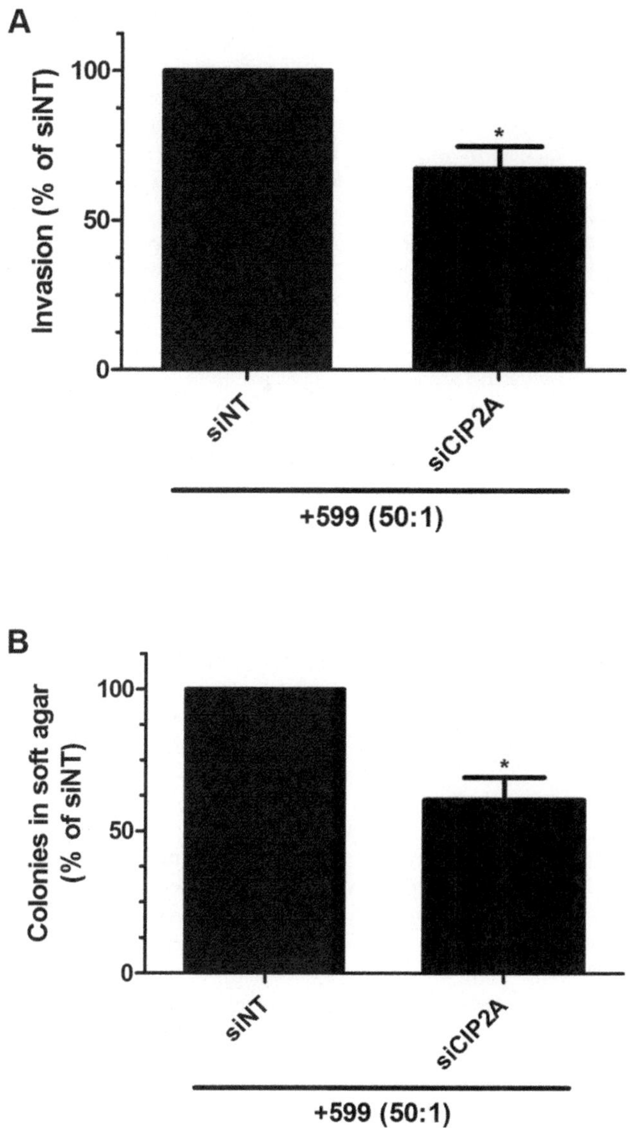

Figure 8. The 599 peptide-mediated silencing of CIP2A decreases oral cancer cell invasiveness and anchorage-independent growth. (**A**) Quantitation of the percentage of invading SCC-25 oral cancer cells after treatment with 599 peptide complexed at a 50:1 peptide-to-siRNA molar ratio to 100 nM of siRNA targeting CIP2A (siCIP2A) compared with control non-targeting siRNA (siNT). Data are mean ± SEM of four separate experiments, where $^{*}P<0.05$ compared to siNT treated cells (Student's t test). (**B**) Anchorage-independent growth of SCC-25 oral cancer cells after treatment with 599 peptide complexed at a 50:1 peptide-to-siRNA molar ratio to 100 nM of siCIP2A compared with control siNT. Data are mean ± SEM of four separate experiments, where $^{*}P<0.05$ compared to siNT treated cells (Student's t test).

the INF-7 peptide sequence built within the 599 peptide comprises 50% hydrophobic amino acid residues. Thus, one could speculate that the reason the 599 peptide has the ability to deliver siRNAs into cells is because the INF-7 peptide sequence not only has the ability to disrupt endosomal membranes [15], but it also comprises a hydrophobic component that when combined with cationic nona(D-arginines) could confer a synergistic effect that allows the 599 peptide the ability to more effectively deliver siRNA cargo into cells. This capacity to enhance the intracellular bioavailability of siRNAs by the 599 peptide was made evident by its ability to induce significant silencing of the targeted CIP2A oncogene with the consequent destabilization of the oncogenic transcription factor c-Myc 48 hours post-treatment.

The finding that the 599 peptide could mediate the delivery of functional siRNAs into cells was particularly significant because it revealed its potential applicability as a delivery vehicle for siRNA-based therapeutics. In fact, the measured significant inhibition in both cell invasiveness and anchorage-independent growth of SCC-25 cells upon treatment with the 599 peptide-siCIP2A complex compared to control treated cells demonstrated its usefulness for siRNA-based therapeutics, in particular for oral cancer. Although these studies were performed *in vitro*, the 599/siRNA complex does have the potential to also succeed in *in vivo* applications based on several factors. First, it is a simple formulation (a stable complex can be made within minutes in water by simply mixing the 599 peptide with siRNAs); therefore, it would not require time-consuming and expensive particle formulation procedures as observed in other effective *in vivo* siRNA delivery strategies that necessitate more complicated particle formulations [41]. Second, with the hydrodynamic mean diameter of the 599/siCIP2A complex being on average 80 nm, this size is well below particle size cutoffs required for efficient intracellular internalization via various endocytic routes [42], comparable to sizes of other efficient siRNA nanoparticle formulations (<200 nm) [43], and within the particle size range (20–200 nm) necessary for effective retention in tumor tissue [44]. Third, the 599/siCIP2A complex has no associated long-term cytotoxicities and its silencing activity is well preserved in serum. The fact that the 599 peptide could mediate a ~70% CIP2A knockdown even in the presence of serum with its simple peptide design, begs one to question what improvements could be made with only a few additional simple chemical modifications. For instance, it would be interesting to examine whether changing the chirality of the D-arginines within the 599 peptide to L-arginines would enhance the uptake of the 599/siRNA complex resulting in a stronger RNAi response in the presence of serum. Moreover, to further improve upon this design, one could also change the chirality of the INF-7 peptide sequence while retaining the nona(L-arginines) to make the peptide more protease resistant and thus more stable in the presence of serum. Additionally, because the INF-7 peptide is an amphipathic peptide [14], it would also be worthwhile to test whether adding a stearyl moiety to its N-terminus would improve its activity in serum. Recently, the addition of a stearic acid moiety to an amphipathic CPP was demonstrated to improve the activity of the CPP in serum and make more efficient the delivery of siRNAs in cell culture and systemically *in vivo* [43,45]. Of note, chemically modified siRNAs were not used in our studies; therefore, there is also the possibility that one could significantly improve the RNAi response even at lower siRNA doses by using optimized formulations of the 599 peptide complexed to chemically stabilized siRNAs. Lastly, because OSCCs are readily accessible for treatment within the oral cavity, one strategy to retain the very effective silencing observed for the 599/siCIP2A complex in the absence of serum that was in parity with the effects observed for well-established commercially available cationic lipid-based transfection reagents, would be to administer the complex via a direct intratumoral injection. The advantage of this form of administration is that this route of dosing does not require as much protection of siRNAs from nuclease degradation compared to systemic administrations [8], thus potentially limiting the loss of RNAi activity mediated by the 599/siCIP2A complex in an *in vivo* setting.

In conclusion, our data collectively demonstrate that the 599 peptide, designed to comprise both synthetic influenza virus-derived fusogenic peptide sequences and cationic CPP nona(D-arginine) residues, can be used to enhance the intracellular delivery and bioavailability of siRNAs via its direct complexation with these molecules. Moreover, the ability of the 599 peptide to mediate silencing of the CIP2A oncogene resulting in decreased oral cancer cell invasiveness and anchorage-independent growth demonstrate the therapeutic relevance of this delivery vehicle. However, future *in vivo* studies using mouse xenograft floor-of-mouth tumors will be needed to more accurately assess the therapeutic potential of the 599 peptide in mediating siRNA-based therapeutics for oral cancer and its prospective applicability in clinical settings.

Materials and Methods

Peptide Synthesis

Peptides 599 (GLFEAIEGFIENGWEGMIDG-WYGGGGRRRRRRRRRK), 616 (GLFEAIEGFIENGWEG-MIDGWYGK), and 9R (RRRRRRRRR) were synthesized and purified (>95% purity) by high-performance liquid chromatography at GenScript USA Inc. (Piscataway, NJ). The 599 and 616 peptides were also biotinylated on the lysine residue and 9R was biotinylated at the N-terminus. In peptides 599 and 9R, the nine arginine residues were D-arginine.

siRNAs

The siRNA targeting CIP2A (siCIP2A) was previously described (CIP2A.1 siRNA) [25] and synthesized by Thermo Fisher Scientific Dharmacon (Lafayette, CO). Additionally, a siCIP2A with a DY547 label at the 3′-end of the anti-sense strand (D-siCIP2A) was also synthesized and used for some experiments. The control siGENOME Non-Targeting siRNA #2 (siNT2) and #5 (siNT5) were purchased from Thermo Fisher Scientific Dharmacon.

Cell Culture

Human tongue squamous cell carcinoma (SCC) cell lines CAL 27 and SCC-25 were purchased from American Type Culture Collection (ATCC, Manassas, VA). Each of the cell lines was cultured in ATCC-specified complete growth media in a 37°C incubator with 5% CO_2.

Agarose Gel Shift Assay

30 pmol of siCIP2A was incubated with the 599 or 9R peptide at 50:1, 20:1, 10:1, and 1:1 molar ratios of peptide-to-siRNA at room temperature (RT) for 15 minutes. Afterwards, the samples were subjected to electrophoresis on a 4% agarose gel and stained with ethidium bromide. siCIP2A without added peptide served as a control. The Ambion® pUC19 DNA - Sau3A I digest DNA ladder was purchased from Life Technologies (Grand Island, NY).

Peptide-Mediated siRNA Delivery and CIP2A Silencing in Oral Cancer Cell Lines

100 nM of D-siCIP2A was incubated either alone or with the 599 peptide at 50:1, 20:1, and 1:1 molar ratios of peptide-to-siRNA for 15 minutes at RT in Opti-MEM® I Reduced Serum Media (Life Technologies). Concurrently, CAL 27 cells grown to a confluency of ~60% on Collagen Type I coated 8-chamber slides (BD Biosciences, San Jose, CA) were rinsed three times with Opti-MEM® media. Afterwards, the cells were incubated with either siRNA alone or peptide/siRNA solutions for 2 hours before being processed and examined by fluorescence microscopy. For experiments involving siRNA delivery comparisons between the 599, 616, and 9R peptides, 100 nM of D-siCIP2A was used at a 50:1 molar ratio of peptide-to-siRNA in Opti-MEM® media. Lastly, for CIP2A mRNA and protein silencing experiments, both 96 and 12-well plate formats were used, respectively. In these experiments, 125 nM of either siNT2 or siCIP2A were initially incubated with the 599 peptide at a 50:1 molar ratio of peptide-to-siRNA for 15 minutes at RT in Opti-MEM® media. Afterwards, the peptide/siRNA solutions were added to either CAL 27 or SCC-25 cells grown to a confluency of ~60% and incubated for 4 hours, prior

to adjusting the siRNA and fetal bovine serum (FBS) concentrations to 100 nM and 10%, respectively, by adding 50% FBS, Opti-MEM® media. For the dose-response experiments, peptide/siRNA solutions consisting of 12.5–125 nM of siCIP2A complexed at a 50:1 peptide-to-siRNA molar ratio in Opti-MEM® media were initially added to the cells and incubated for 4 hours, prior to adjusting the siRNA and FBS concentrations to 10–100 nM and 10%, respectively. For serum experiments, immediately after complex formation in serum-free media, 50% FBS, Opti-MEM® media was added to adjust the siRNA and FBS concentrations to 100 nM and 10%, respectively, after which the peptide/siRNA solutions were then added to cells. 48 hours post-treatment, the total RNA or protein was harvested for subsequent real-time PCR and Western blot analyses. Protein lysates were also collected 1, 3, 5, 7, and 9 days post-treatment to determine the duration of 599 peptide-mediated silencing by Western blot analysis. As controls for determining the efficacy of 599 peptide-mediated CIP2A silencing, cells were also transfected with 100 nM of siRNA using Lipofectamine® RNAiMAX (Life Technologies), Lipofectamine® 2000 (Life Technologies), HiPerFect® (Qiagen, Valencia, CA), and INTERFERin® (Polyplus-transfection, New York, NY), according to the manufacturer's instructions. Of note, siNT5 was used as the control non-targeting siRNA for the experiments involving serum stability and the commercially available transfection reagents described above.

Fluorescence Microscopy

Treated cells seeded on Collagen Type I coated 8-chamber slides (BD Biosciences, San Jose, CA) were rinsed with 1×phosphate-buffered saline (PBS), fixed in 3% paraformaldehyde at RT for 10 minutes and then either mounted immediately with a glass coverslip using VECTASHIELD Mounting Medium with 4′,6-diamidino-2-phenylindole (DAPI, VECTOR Laboratories, Burlingame, CA) or permeabilized with 0.5% Triton X-100 at RT for 5 minutes. For cells that were permeabilized, the slides were subsequently incubated with a rabbit monoclonal anti-EEA1 antibody (1:200, clone C45B10, Cell Signaling, Danvers, MA) at RT for 1 hour, washed with 1×PBS, and then incubated with the corresponding secondary antibody, Alexa Fluor® 488-conjugated goat anti-rabbit IgG (1:400, Life Technologies), at RT for 1 hour. Afterwards, coverslips were mounted onto the slides as described above. For live cell imaging, treated cells were washed three times with 1×PBS before being imaged in the presence of PBS. Fluorescence and phase contrast images were obtained using a Zeiss (Thornwood, NY) Axio Observer.D1 microscope. Images of fixed or live cells were taken using a LD A-Plan ×20/0.3 Ph1 objective.

Quantitative Uptake of Fluorescently Labeled siRNAs

This protocol was adapted from a previously described protocol [20]. Briefly, 60 pmol of D-siCIP2A was incubated either alone or with the 599 or 9R peptides at different peptide-to-siRNA molar ratios for 15 minutes in a total volume of 600 µl of Opti-MEM® media (Life Technologies). CAL 27 cells grown to a confluency of ~60% on a 24-well plate were then rinsed three times with Opti-MEM® media, after which the cells were incubated with either 500 µl of siRNA alone or peptide/siRNA solutions for 2.5 hours. As controls for the experiments involving the 599 peptide, the cells were also treated with peptide alone (the amount of peptide used was equivalent to that used at a 100:1 peptide-to-siRNA molar ratio) or transfected with 100 nM of D-siCIP2A using INTERFERin® (Polyplus-transfection), according to the manufacturer's instructions. After treatment, the cells were rinsed three times with 1×PBS, trypsinated for 10 minutes at 37°C, and then centrifuged at 4°C for 5 minutes at 1,000 g. After centrifugation, the cells were washed with ice cold 1×PBS, centrifuged again, and then lysed in 200 µl of ice cold 0.1 M NaOH. The cell lysates were centrifuged at 4°C for 5 minutes at 10,000 g to remove cell debris and then stored at −80°C. On the day of the fluorescence measurements, the cell lysates were thawed on ice and 100 µl per sample was transferred to a black 96-well plate to measure fluorescence at 530/590 nm using a BioTek (Winooski, VT) Synergy HT plate reader. Fluorescence measurements were converted to the amount of internalized siRNAs using a standard curve that was generated per experiment using known concentrations of D-siCIP2A ranging from 1.6–100 fmol/µl. Afterwards, the amount of internalized siRNA was normalized to the amount of protein which was quantitated using the Pierce BCA Protein Assay Kit (Thermo Fisher Scientific, Rockford, IL).

Particle Size and Zeta Potential Determination

The hydrodynamic diameter of the 599/siCIP2A and 9R/CIP2A particles formulated at a 50:1 peptide-to-siRNA molar ratio in water were measured at room temperature by dynamic light scattering using a 90Plus Nanoparticle Size Analyzer (Brookhaven Instruments, Holtsville, NY). Zeta potential was measured using the same instrument.

Darkfield-Based Optical Microscopy

After formulating the 599/siCIP2A particles at a 50:1 peptide-to-siRNA molar ratio in water, 2 µl was pipetted onto a glass slide, and the spot was allowed to air dry. Images of the particles were then collected using an Olympus (Center Valley, PA) upright BX51 microscope outfitted with a CytoViva (Auburn, AL) Hyperspectral Imaging System and oil immersion darkfield condenser. All images were collected using a 40×LUC PlanFL N air objective (NA 0.60).

Real-Time PCR

For CIP2A mRNA and 18S rRNA quantitation, total RNA was harvested and reverse-transcribed using the TaqMan® Fast Cells-to-CT™ kit (Life Technologies). Afterwards, quantitative real-time PCR was performed in an Applied Biosystems® StepOnePlus™ Real-Time PCR machine (Life Technologies) using predesigned TaqMan® Gene Expression Assays (Life Technologies) for CIP2A (Hs00405413_m1) and 18S (4310893E) combined with TaqMan® Fast Universal PCR Master Mix (Life Technologies), according to the manufacturer's instructions.

Western Blot Analysis

Treated cells were washed with 1×PBS and lysed using ice cold RIPA buffer (50 mM Tris-HCl, pH 7.4, 150 mM NaCl, 0.5% sodium deoxycholate, 0.1% SDS, and 1% NP-40) with protease inhibitor cocktail (Thermo Fisher Scientific). The protein lysates were then resolved by SDS-PAGE on a 10% gel and transferred to a PVDF membrane. The PVDF membrane was cut at the 50 and 75-kDa molecular weight protein markers into three pieces (for experiments not involving c-Myc analyses the membrane was only cut at the 75-kDa molecular weight protein marker) and blocked in 5% nonfat dried milk in Tris-HCl-buffered-saline (TBS)-0.1% Tween (pH 7.5) for 1 hour at RT. The top portion of the membrane containing the higher molecular weight proteins was incubated with mouse monoclonal anti-CIP2A antibody (1:500, clone 2G10-3B5, Santa Cruz Biotechnology, Santa Cruz, CA), the middle portion of the membrane containing proteins ranging from 50 to 75-kDa was incubated with rabbit monoclonal anti-c-Myc antibody (1:1,000, Cell Signaling), and the bottom portion of the membrane containing the lower molecular weight proteins was incubated with rabbit monoclonal anti-GAPDH antibody (1:10,000, Cell Signaling) for 1 hour at RT. The membranes were then washed four times with TBS-Tween and incubated with the corresponding secondary antibodies either horseradish peroxidase-conjugated goat anti-mouse (1:4,000) or anti-rabbit IgG (1:4,000 or 1:20,000, SouthernBiotech, Birmingham, AL) for 1 hour at RT. Immunoreactive bands were detected by the SuperSignal Chemiluminescent system (Thermo Fisher Scientific), according to the manufacturer's instructions.

Cell Viability Measurements

Long-term cell viability was assessed using the CellTiter 96® AQueous One Solution Cell Proliferation Assay (Promega, Madison, WI). Briefly, 15,000 cells were seeded on 96-well plates several days before treatment. Upon reaching a 50–60% cellular confluency, the cells were either treated in Opti-MEM® media with various concentrations of peptide alone, 125 nM siNT2 alone or in complex with 599 peptide (at different peptide-to-siRNA molar ratios), or transfected with 100 nM siNT2 using INTERFERin® (Polyplus-transfection). Of note, for cells treated with peptide alone, siRNA alone, or in complex with the 599 peptide, the cells were initially incubated for 4 hours, prior to adjusting the siRNA (if added) and FBS concentrations to 100 nM and 10%, respectively, by adding 50% FBS, Opti-MEM® media. After 48 hours treatment/transfection, the cell viability was assayed, according to the manufacturer's instructions. Absorbance at 490 nm was measured using a BioTek Synergy HT plate reader. Untreated cells were defined as 100% viable.

Cell Invasion Assay

Cell invasion properties were quantified using the CytoSelect™ 24-Well Cell Invasion Assay kit (Cell Biolabs, San Diego, CA). Briefly, 72 hours post-treatment, SCC-25 cells were trypsinized and serum starved in 0.1% (w/v) BSA, Opti-MEM® media for 24 hours prior to being resuspended in the same media at 2×10^5 cells/ml. Afterwards, 300 µl of cell suspension was added to the insert pre-coated with a basement membrane matrix layer and 500 µl of 10% FBS, Opti-MEM® media was added to the lower well of the invasion plate. After 48 hours, non-invasive cells were swabbed away from the interior of the inserts and invasive cells adhering to the underside were stained and quantified, according to the manufacturer's instructions. Absorbance at 560 nm was measured using a BioTek Synergy HT plate reader. Data were normalized to the amount of inputted cells grown on a separate plate for the same period of time.

Anchorage-Independent Growth Assay

Anchorage independent growth in semisolid media was assessed using the CytoSelect™ 96-Well Cell Transformation Assay (Soft Agar Colony Formation; Cell Biolabs). Briefly, 24 hours post-treatment, SCC-25 cells were resuspended using Versene (Life Technologies) at a concentration of 1.6×10^5 cells/ml. Following the manufacturer's protocol, the different semisolid layers were assembled. The top layer of growth media was replenished every 2 days. The assay was carried out for 9 days at which time the agar was solubilized, the cells were lysed, and growth was quantitated with a fluorescent dye. Fluorescence was measured at 485/528 nm using a BioTek Synergy HT plate reader.

Supporting Information

Figure S1 Characterization of 9R peptide binding and delivery of siRNAs. (**A**) An ethidium bromide stained 4% agarose gel shift assay examining the ability of various amounts of the 9R peptide (ranging from 1 to 50-fold molar excess of siRNAs) to form complexes with siCIP2A. siCIP2A, siRNA targeting the CIP2A oncogene; MWM, molecular weight marker (the number of base pairs for each DNA fragment are shown). (**B**) Size distribution and zeta potential of the 9R peptide complexed with siCIP2A at a 50:1 peptide-to-siRNA molar ratio 20 minutes after formulation in water. (**C**) CAL 27 cells incubated for 2.5 hours with DY547-conjugated siRNA targeting CIP2A (D-siCIP2A) alone or in complex with increasing amounts of 9R peptide (ranging from 0.1 to 100-fold molar excess of siRNAs). As a positive control, the cells were also treated with the 599 peptide complexed to D-siCIP2A at a 50:1 peptide-to-siRNA molar ratio. The amount of siRNA delivered into cells in pmol per mg of protein is reported with each treatment normalized to D-siCIP2A alone. Data are mean ± SEM of three separate experiments, where ***$P<0.001$ compared to D-siCIP2A alone treated cells (ANOVA, Dunnett's Multiple Comparison Test).

Acknowledgments

We thank Tilman Heise and Gunhild Sommer in the Department of Biochemistry and Molecular Biology at MUSC for the use of their Zeiss Axio Observer microscope. We also thank Frank Alexis and Thomas Moore in the Department of Bioengineering at Clemson University for the use of their Brookhaven 90Plus Nanoparticle Size Analyzer and for their technical assistance in particle size and zeta-potential analyses. Images from the Cytoviva microscope were collected by Rhonda R. Powell (Clemson Light Imaging Facility).

Author Contributions

Conceived and designed the experiments: LC CCA LMA AJ. Performed the experiments: LC CCA BB LMA MLS AJ. Analyzed the data: LC CCA LMA AJ. Contributed reagents/materials/analysis tools: AJ. Wrote the paper: LC AJ.

References

1. Siegel R, Naishadham D, Jemal A (2012) Cancer statistics, 2012. CA Cancer J Clin 62: 10–29.
2. Stewart SL, Cardinez CJ, Richardson LC, Norman L, Kaufmann R, et al. (2008) Surveillance for cancers associated with tobacco use–United States, 1999–2004. MMWR Surveill Summ 57: 1–33.
3. Campana JP, Meyers AD (2006) The surgical management of oral cancer. Otolaryngol Clin North Am 39: 331–348.
4. Rana TM (2007) Illuminating the silence: understanding the structure and function of small RNAs. Nat Rev Mol Cell Biol 8: 23–36.
5. Fire A, Xu S, Montgomery MK, Kostas SA, Driver SE, et al. (1998) Potent and specific genetic interference by double-stranded RNA in Caenorhabditis elegans. Nature 391: 806–811.
6. Elbashir SM, Harborth J, Lendeckel W, Yalcin A, Weber K, et al. (2001) Duplexes of 21-nucleotide RNAs mediate RNA interference in cultured mammalian cells. Nature 411: 494–498.
7. Hammond SM, Boettcher S, Caudy AA, Kobayashi R, Hannon GJ (2001) Argonaute2, a link between genetic and biochemical analyses of RNAi. Science 293: 1146–1150.
8. Rettig GR, Behlke MA (2012) Progress toward in vivo use of siRNAs-II. Mol Ther 20: 483–512.
9. Zamore PD, Tuschl T, Sharp PA, Bartel DP (2000) RNAi: double-stranded RNA directs the ATP-dependent cleavage of mRNA at 21 to 23 nucleotide intervals. Cell 101: 25–33.
10. Bramsen JB, Kjems J (2012) Development of Therapeutic-Grade Small Interfering RNAs by Chemical Engineering. Front Genet 3: 154.

11. Dominska M, Dykxhoorn DM (2010) Breaking down the barriers: siRNA delivery and endosome escape. J Cell Sci 123: 1183–1189.
12. El-Sayed A, Futaki S, Harashima H (2009) Delivery of macromolecules using arginine-rich cell-penetrating peptides: ways to overcome endosomal entrapment. Aaps J 11: 13–22.
13. Jarver P, Coursindel T, Andaloussi SEL, Godfrey C, Wood MJA, et al. (2012) Peptide-mediated Cell and *In Vivo* Delivery of Antisense Oligonucleotides and siRNA. Molecular Therapy-Nucelic Acids 1: e27.
14. Plank C, Zauner W, Wagner E (1998) Application of membrane-active peptides for drug and gene delivery across cellular membranes. Adv Drug Deliv Rev 34: 21–35.
15. Plank C, Oberhauser B, Mechtler K, Koch C, Wagner E (1994) The influence of endosome-disruptive peptides on gene transfer using synthetic virus-like gene transfer systems. J Biol Chem 269: 12918–12924.
16. Oliveira S, van Rooy I, Kranenburg O, Storm G, Schiffelers RM (2007) Fusogenic peptides enhance endosomal escape improving siRNA-induced silencing of oncogenes. Int J Pharm 331: 211–214.
17. Mastrobattista E, Koning GA, van Bloois L, Filipe AC, Jiskoot W, et al. (2002) Functional characterization of an endosome-disruptive peptide and its application in cytosolic delivery of immunoliposome-entrapped proteins. J Biol Chem 277: 27135–27143.
18. Meade BR, Dowdy SF (2008) Enhancing the cellular uptake of siRNA duplexes following noncovalent packaging with protein transduction domain peptides. Adv Drug Deliv Rev 60: 530–536.
19. Wadia JS, Stan RV, Dowdy SF (2004) Transducible TAT-HA fusogenic peptide enhances escape of TAT-fusion proteins after lipid raft macropinocytosis. Nat Med 10: 310–315.
20. Lundberg P, El-Andaloussi S, Sutlu T, Johansson H, Langel U (2007) Delivery of short interfering RNA using endosomolytic cell-penetrating peptides. Faseb J 21: 2664–2671.
21. Wender PA, Mitchell DJ, Pattabiraman K, Pelkey ET, Steinman L, et al. (2000) The design, synthesis, and evaluation of molecules that enable or enhance cellular uptake: peptoid molecular transporters. Proc Natl Acad Sci U S A 97: 13003–13008.
22. Kim WJ, Christensen LV, Jo S, Yockman JW, Jeong JH, et al. (2006) Cholesteryl oligoarginine delivering vascular endothelial growth factor siRNA effectively inhibits tumor growth in colon adenocarcinoma. Mol Ther 14: 343–350.
23. Kumar P, Wu H, McBride JL, Jung KE, Kim MH, et al. (2007) Transvascular delivery of small interfering RNA to the central nervous system. Nature 448: 39–43.
24. Bockelman C, Hagstrom J, Makinen LK, Keski-Santti H, Hayry V, et al. (2011) High CIP2A immunoreactivity is an independent prognostic indicator in early-stage tongue cancer. Br J Cancer 104: 1890–1895.
25. Junttila MR, Puustinen P, Niemela M, Ahola R, Arnold H, et al. (2007) CIP2A inhibits PP2A in human malignancies. Cell 130: 51–62.
26. Katz J, Jakymiw A, Ducksworth MK, Stewart CM, Bhattacharyya I, et al. (2010) CIP2A expression and localization in oral carcinoma and dysplasia. Cancer Biol Ther 10: 694–699.
27. Richard JP, Melikov K, Vives E, Ramos C, Verbeure B, et al. (2003) Cell-penetrating peptides. A reevaluation of the mechanism of cellular uptake. J Biol Chem 278: 585–590.
28. Abramoff MD, Magelhaes PJ, Ram SJ (2004) Image processing with ImageJ. Biophotonics Int 11: 36–42.
29. Come C, Laine A, Chanrion M, Edgren H, Mattila E, et al. (2009) CIP2A is associated with human breast cancer aggressivity. Clin Cancer Res 15: 5092–5100.
30. Ren J, Li W, Yan L, Jiao W, Tian S, et al. (2011) Expression of CIP2A in renal cell carcinomas correlates with tumour invasion, metastasis and patients' survival. Br J Cancer 105: 1905–1911.
31. Gioanni J, Fischel JL, Lambert JC, Demard F, Mazeau C, et al. (1988) Two new human tumor cell lines derived from squamous cell carcinomas of the tongue: establishment, characterization and response to cytotoxic treatment. Eur J Cancer Clin Oncol 24: 1445–1455.
32. Moschos SA, Jones SW, Perry MM, Williams AE, Erjefalt JS, et al. (2007) Lung delivery studies using siRNA conjugated to TAT(48–60) and penetratin reveal peptide induced reduction in gene expression and induction of innate immunity. Bioconjug Chem 18: 1450–1459.
33. Turner JJ, Jones S, Fabani MM, Ivanova G, Arzumanov AA, et al. (2007) RNA targeting with peptide conjugates of oligonucleotides, siRNA and PNA. Blood Cells Mol Dis 38: 1–7.
34. Jiao CY, Delaroche D, Burlina F, Alves ID, Chassaing G, et al. (2009) Translocation and endocytosis for cell-penetrating peptide internalization. J Biol Chem 284: 33957–33965.
35. Verdurmen WP, Bovee-Geurts PH, Wadhwani P, Ulrich AS, Hallbrink M, et al. (2011) Preferential uptake of L- versus D-amino acid cell-penetrating peptides in a cell type-dependent manner. Chem Biol 18: 1000–1010.
36. Khalil IA, Futaki S, Niwa M, Baba Y, Kaji N, et al. (2004) Mechanism of improved gene transfer by the N-terminal stearylation of octaarginine: enhanced cellular association by hydrophobic core formation. Gene Ther 11: 636–644.
37. Claffey DJ, Meyer JD, Beauvais R, Brandt T, Shefter E, et al. (2000) Long chain arginine esters: a new class of cationic detergents for preparation of hydrophobic ion-paired complexes. Biochem Cell Biol 78: 59–65.
38. Furgeson DY, Chan WS, Yockman JW, Kim SW (2003) Modified linear polyethylenimine-cholesterol conjugates for DNA complexation. Bioconjug Chem 14: 840–847.
39. Yockman JW, Maheshwari A, Han SO, Kim SW (2003) Tumor regression by repeated intratumoral delivery of water soluble lipopolymers/p2CMVmIL-12 complexes. J Control Release 87: 177–186.
40. Blessing T, Remy JS, Behr JP (1998) Monomolecular collapse of plasmid DNA into stable virus-like particles. Proc Natl Acad Sci U S A 95: 1427–1431.
41. Hatakeyama H, Ito E, Akita H, Oishi M, Nagasaki Y, et al. (2009) A pH-sensitive fusogenic peptide facilitates endosomal escape and greatly enhances the gene silencing of siRNA-containing nanoparticles in vitro and in vivo. J Control Release 139: 127–132.
42. Rejman J, Oberle V, Zuhorn IS, Hoekstra D (2004) Size-dependent internalization of particles via the pathways of clathrin- and caveolae-mediated endocytosis. Biochem J 377: 159–169.
43. Andaloussi SE, Lehto T, Mager I, Rosenthal-Aizman K, Oprea, II, etal. (2011) Design of a peptide-based vector, PepFect6, for efficient delivery of siRNA in cell culture and systemically in vivo. Nucleic Acids Res 39: 3972–3987.
44. Danhier F, Feron O, Preat V (2010) To exploit the tumor microenvironment: Passive and active tumor targeting of nanocarriers for anti-cancer drug delivery. J Control Release 148: 135–146.
45. Mae M, El Andaloussi S, Lundin P, Oskolkov N, Johansson HJ, et al. (2009) A stearylated CPP for delivery of splice correcting oligonucleotides using a non-covalent co-incubation strategy. J Control Release 134: 221–227.

18

Panning of a Phage Display Library against a Synthetic Capsule for Peptide Ligands That Bind to the Native Capsule of *Bacillus anthracis*

Michael Beer, Chun-Qiang Liu*

Human Protection and Performance Division, Defence Science and Technology Organisation, Fishermans Bend, Victoria, Australia

Abstract

Bacillus anthracis is the causative agent of anthrax with the ability to not only produce a tripartite toxin, but also an enveloping capsule comprised primarily of γ-D-glutamic acid residues. The purpose of this study was to isolate peptide ligands capable of binding to the native capsule of *B. anthracis* from a commercial phage display peptide library using a synthetic form of the capsule consisting of 12 γ-D-glutamic acid residues. Following four rounds of selection, 80 clones were selected randomly and analysed by DNA sequencing. Four clones, each containing a unique consensus sequence, were identified by sequence alignment analysis. Phage particles were prepared and their derived 12-mer peptides were also chemically synthesized and conjugated to BSA. Both the phage particles and free peptide-BSA conjugates were evaluated by ELISA for binding to encapsulated cells of *B. anthracis* as well as a *B. anthracis* capsule extract. All the phage particles tested except one were able to bind to both the encapsulated cells and the capsule extract. However, the peptide-BSA conjugates could only bind to the encapsulated cells. One of the peptide-BSA conjugates, with the sequence DSSRIPMQWHPQ (termed G1), was fluorescently labelled and its binding to the encapsulated cells was further confirmed by confocal microscopy. The results demonstrated that the synthetic capsule was effective in isolating phage-displayed peptides with binding affinity for the native capsule of *B. anthracis*.

Editor: Patrick M. Schlievert, University of Iowa Carver College of Medicine, United States of America

Funding: All funding for this research was provided by the Australian Government Department of Defence. The funders had no role in study design, data collection and analysis, decision to publish, or preparation of the manuscript.

Competing Interests: The authors have declared that no competing interests exist.

* E-mail: chun.liu@dsto.defence.gov.au

Introduction

Bacillus anthracis is capable of infecting many mammal species (including humans), and the acute form of infection, commonly referred to anthrax, is often fatal [1]. Infections arise from the highly durable and dormant form of *B. anthracis* known as endospores, which are formed from the metabolically active vegetative form in response to nutrient deficiency and/or environmental stress. Entry of endospores into an animal host can occur through cutaneous contact, ingestion or inhalation; the latter two routes of infection having the greatest chance of developing systemic infection and therefore highest mortality. Endospores germinate into metabolically active vegetative bacilli after being enveloped by macrophages and transported to lymph nodes, where they rapidly multiply. Vegetative bacilli are then released into the bloodstream, where they produce anthrax toxin, which is comprised of three proteins known as edema factor (EF), lethal factor (LF), and protective antigen (PA), encoded by the pXO1 plasmid. After binding to cell surface receptors, PA undergoes protease cleavage which then enables it to form a heptamer complex. This complex, known as a prepore, can bind both EF and LF. The tripartite toxin enters the host cell *via* endocytosis, resulting in edema (EF) and kinase inactivation (LF) [1].

The ability of *B. anthracis* to generate a capsule is also an essential component of its capacity to induce anthrax [2,3]. The capsule of *B. anthracis* is comprised almost exclusively of a polymer of γ-D-glutamic acid residues (γPGA), which is resistant to protease activity [4]. The number of glutamic acid residues ranges from 1000 to 2500 [4], and the genes for capsule synthesis located on the pXO2 plasmid have also been well characterised [5]. γPGA is negatively charged, which enables the vegetative cells to evade phagosytosis by macrophages [6]. Given its crucial role in virulence, γPGA has been targeted for both active and passive immunization against anthrax infection [7,8,9]. Further, γPGA may also serve as a biomarker for early diagnosis of anthrax infection, because measurable levels of γPGA were produced as early as 24 hours after infection in a mouse model [7].

Several monoclonal antibodies have been produced for use in passive immunotherapy against anthrax [7,8,9] and immunoassays for γPGA in early diagnosis of anthrax [7,10]. However, due to the poor immunogenicity, the capsule had to be pre-conjugated to either an immunogenic protein carrier [9,11,12,13] or CD40 mAb [7,8]. Most of the antibodies showed varying levels of protection when tested in a murine model of pulmonary anthrax [8]. Clearly, there is a need for alternative types of binding molecules that target the capsule of *B. anthracis*.

Phage display technology has proven highly effective in discovering peptides (including antibody fragments) with affinities to virtually any target. As drug candidates, peptides are often preferred to antibodies because of their lower manufacturing costs, greater stability and better organ or tumour penetration [14]. To date, several oligopeptides [15] and human antibody-like fragments have been isolated from phage display libraries and shown to have high affinity for *B. anthracis* endospores [16,17], LF [18], and PA [19], or multiple/associated toxin components [20,21,22].

Here, we report for the first time the isolation and initial characterisation of phage-displayed peptides that bind to the capsule of vegetative cells of *B. anthracis*. This work represents a first step towards the development of novel therapeutic agents that allow selective killing of *B. anthracis* cells.

Methods

Phage Display Library and Panning Procedure

The Ph.D.-12^{TM} phage display peptide library was purchased from New England Biolabs (Ipswich, MA, USA). The library displays random 12-mer peptides fused to the minor coat protein (pIII) of M13 phage. The following oligo-peptides were synthesized by AnaSpec (CA, USA): CL-1 (Biotin-LC-(γ-D-Glu)$_{12}$-NH_2), CL-2 (Biotin-LC-Gly-Gly-Pro-His-Ser-Gly- NH_2), CL-3 (BSA-Cys-(γ-D-Glu)$_{12}$-NH_2), CL-4 (Acetyl-(γ-D-Glu)$_{12}$- NH_2) and CL-5 (Acetyl-Leu-Gly-Thr-Pro-His-Ser-Gly-Thr-Arg-Leu-Ser-Glu-NH_2). LC is a 6-aminohexanoic acid spacer molecule used to minimize potential steric hindrance effects on the peptide. CL-1 and CL-3 are synthetic capsules, each consisting of 12 D-glutamic acid residues linked to biotin and BSA, respectively. CL-4 is a non-biotinylated version of CL-1. CL-2 and CL-5 are non-capsule control peptides.

Panning of the phage display library was carried out as per manufacturer's instructions with modifications. Briefly, the synthetic capsule peptide CL-1 was captured via strepavidin conjugated magnetic beads and the non-biotinylated control peptide CL-5 was added in excess at the beginning of the each cycle to help eliminate non-specific binders. The screening stringency was increased by decreasing the concentration of the coated target peptide CL-1 and increasing the detergent percentage in the washing buffer. From the 1st to the 3rd round of selection, acidic elution was carried out to collect more binders. However in the 4th round, competitive elution with a high concentration of CL-4 (the non-biotinylated target peptide) was performed to enrich specific binders. A 10-fold serial dilution of the phage particles from the 4th round elute was prepared in Luria Broth, and added to a freshly prepared culture of *E. coli* ER2738 (New England Biolabs). The culture was incubated at room temperature for 10 minutes before plating out. Random single plaques were inoculated into *E. coli* ER2738 and incubated at 37°C for 5 hours.

PCR and Nucleotide Sequencing

The DNA sequence of the 12-mer peptide in each phage clone was amplified by PCR using primers M13KE-fl (5′-GTTGTTGTCATTGTCGGCGCA-3′) and M13KE-r2 (5′-GCCCTCATAGTTAGCGTAACG-3′). The temperature cycling consisted of an initial step of 5 minutes at 94°C, followed by 30 cycles of 96°C for 10 seconds, 50°C for 5 seconds, and 60°C for 4 minutes, and then held at 4°C. The resulting PCR product was purified with QIAquick PCR Purification Kit (Qiagen GmbH, Hilden Germany) and quantified with a NanoDrop spectrophotometer (Thermo Scientific, Wilmington, DE, USA). The DNA sequencing reactions were prepared using BigDye Terminator Ready Mix v3.1 (Roche), and sequencing reactions were analysed on an ABI PRISM 3100 Genetic Analyzer (Applied Biosystems, Mulgrave VIC, AUSTRALIA).

Extraction of Capsular Material

The capsular material was extracted from ΔAmes-1, a non-toxigenic, encapsulated strain of *B. anthracis* which harbours the pXO2 plasmid responsible for capsule synthesis, but lacks the pOX1 plasmid encoding the three toxin components (Welkos, Little et al. 2001). The ΔAmes-1 strain was kindly provided by Dr J. Bates (Queensland Health Forensic and Scientific Services, Australia). A nutrient agar plate (3 g/L beef extract, 5 g/L peptone and 15 g/L agar) containing 0.8% $NaHCO_3$ (added as an 8% (w/v) solution to molten agar at 50°C) was inoculated with *B. anthracis* ΔAmes-1, and incubated overnight at 37°C. A single colony from the nutrient agar plate was used to inoculate 4 mL of nutrient broth containing 0.8% (w/v) $NaHCO_3$ in sterile 50 mL tube. The tube was shaken at 175 rpm for 24 hours at 37°C. The culture was spread onto 10 nutrient agar plates (50 µL each), and the plates incubated in an anaerobic jar containing about 15% CO_2 that was generated by a Gas Generating Pouch (Oxoid, Melbourne, Australia). Plates were incubated for 48 hours at 37°C, after which growth was harvested with a cell scraper (Sigma, Melbourne, Australia) and the bacteria resuspended in 25 mL 1× PBS (8 g/L NaCl, 0.2 g/L KCl, 1.44 g/L Na a_2HPO_4, 0.24 g KH_2PO_4, pH 7.4) containing 4% (v/v) formaldehyde (Sigma), and incubated for 48 hours at 37°C. Two and a half grams of solid sodium acetate (Sigma) was added to the remaining suspension and agitated until dissolved. Glacial acetic acid (250 µL) (Sigma) was then added and the suspension mixed thoroughly. Cells were pelleted by centrifugation at 2 500× g for 5 minutes and the supernatant filtered through a 0.2 µm filter (MilliPore, North Ryde NSW, Australia). Two volumes of absolute ethanol were added to the filtrate, and mixed thoroughly. Capsular material was left to precipitate for 30 minutes at room temperature, before being collected by centrifugation at 5 500× g for 15 minutes at room temperature. The supernatant was discarded and the precipitate resuspended in 1.0 mL of sterile MilliQ water. The amount of capsular material was quantified with a NanoDrop Spectrophotometer (ThermoFisher Scientific).

Enzyme-linked Immunosorbent Assay (ELISA)

Formaldehyde treated *B. anthracis* ΔAmes-1 cells were diluted to an OD_{600} of 0.5 in coating buffer (0.05 M $NaHCO_3$ pH 9.6), and extracted capsular material diluted to 1 µg/mL in coating buffer. They were then added separately to wells (100 µL each) of a Nunc MaxiSorp F96 microplate (ThermoFisher Scientific), the plate then incubated overnight at 4°C. Wells were washed 3 times with

Table 1. Amino acid sequences of the 12-mer peptides carried by selected phage clones.

Phage clone[a]	Translated amino acid sequence[b]	Peptide name[c]
P01(26/72)	DSS*R***IPM**Q**W***H***P**Q	G1
P04(15/72)	**Y**TMQ**YSLI**N*H***P**Q	G2
P26(3/72)	S*H*S**AMY**E**Y**D**A**ST	G3
P50(3/72)	*HSH***P**S*K***FPALAY**	G4

[a]The designation of each phage clone is given along with their representation within the sequenced clones shown in parentheses.
[b]The translated amino acid sequence, determined from the Ph.D-12 insert library sequence, is also given (underlined - acidic residues, *italics* - basic residues, **bold** - neutral hydrophobic residues, normal – neutral hydrophilic residues).
[c]Peptide name designates the representative synthesised peptide.

PBST, and blocked for 1 hour at room temperature with PBST (PBS buffer with 0.2% v/v Tween20) containing 1.5%w/v BSA. Phage preparation (1.5×10^{12} pfu/mL) was diluted 1:10 in PBST, added to designated wells (100 µL per well), incubated for 1 hour at room temperature, and washed three times in PBST. Mouse anti-M13 bacteriophage antibody (Sigma) was diluted 1:2000 in PBST and added at 100 µL to each well. The plate was incubated for 1 hour at room temperature, and then washed three times in PBST. To each well, 100 µL of a 1:2000 dilution of Anti-mouse IgG conjugated to Alkaline Phosphatase (Sigma) in PBST were added, incubated for 1 hour at room temperature, and washed three times in PBST. To each well, 100 µL of 1 mg/mL of 4-nitrophenyl phosphate disodium salt hexahydrate in 0.2 M Tris buffer (Sigma) was added. The plate was protected from light and shaken at room temperature for 1 hour. Absorbance was read at a wavelength of 405 nm with a Synergy HT Multi-Mode Microplate Reader (Bio-Tek, Winooski, VT, United States).

The synthetic peptides were manufactured and conjugated to BSA by ANASpec. The conjugated peptides were each diluted to 1.0 µg/mL in coating buffer (0.05 M $NaHCO_3$ pH 9.6), and 100 µL added to a Nunc MaxiSorp F96 microplate (Thermo-Fisher Scientific) in triplicate for each target, and incubated at 4°C overnight. Plates were washed three times in PBST (PBS buffer with 0.2% v/v Tween20), after which 350 µL of BSA-PBST (0.25%w/v BSA in PBST) was added to each well and incubated for 1 hour at room temperature with gentle shaking (about 50 rpm). Each well was washed three times with 350 µL of PBST. Washed formaldehyde fixed *B. anthracis* ΔAmes-1 cells were diluted to an OD_{600} of 0.5 in BSA-PBST, and extracted capsular material diluted to 1.0 µg/mL in BSA-PBST. 100 µL of these preparations were added separately to triplicate wells coated with each peptide. The plate was incubated at room temperature for 1 hour with gentle shaking and washed three times in PBST. Monoclonal mouse anti-spore antigen (Hytest, Turku Finland) was added to the wells to which encapsulated cells were bound, and 100 µL of a single domain fragment expression culture lysate (diluted 1:10 in PBST) known to bind to capsule extract (unpublished) was added to the wells to which extract was bound. The plate was incubated at room temperature for 1 hour with gentle shaking and washed three times in PBST. 100 µL of a 1:2000 dilution of goat anti-mouse antibody conjugated to alkaline phosphatase (Sigma) in PBST were added to each well to which BA cells were previously bound, and 100 µL of a 1:2000 dilution of Rabbit Anti-E tag Polyclonal Antibody (Thermo) in PBST were added to each well to which capsule extract was previously bound. The plate was incubated for 1 hour at room temperature, and then washed three times in PBST. 100 µL of a 1:2000 dilution of anti-Rabbit IgG conjugated to Alkaline Phosphatase (Sigma) in PBST were added to each well to which capsule extract was previously bound and incubated for 1 hour at room temperature, and washed three times in PBST. After each well was washed once with 100 µL of distilled water, 100 µL of 1 mg/mL of 4-nitrophenyl phosphate disodium salt hexahydrate in 0.2 M Tris buffer (Sigma) was added to each well, and the plate was shaken at room temperature for 1 hour. Absorbance was read at 405 nm in a Synergy HT Multi-Mode microplate reader (Bio-Tek). All appropriate controls were carried out during the ELISA experiment, including a background control from which the primary binding (encapsulated cells/capsule extract) was omitted.

Confocal Microscopy

Peptide-BSA conjugates were labelled with AlexaFluor488 using the AlexaFluor488 Microscale Labelling Kit (Molecular Probes), as per the manufacturers instruction. BSA conjugated forms of the peptides G1 and CL3 (AnaSpecUSA) were diluted to 1.0 mg/mL in sterile 1×PBS, and the dye added at a ratio of 1:30 (protein: AlexaFluor288). Absorbances at 280 nm and 494 nm were determined using the NanoDrop Spectrophotometer (Thermo), and the number of dye molecules per protein molecule was approximately 2.2 and 1.8 for G1-BSA and CL3-BSA respectively. Two *B. anthracis* strains were used: ΔAmes-1 and Sterne 34F2. The latter is a vaccine strain that is devoid of the capsule plasmid pXO2 (Green, Battisti et al. 1985). The two strains were cultured in nutrient media for 24 h at 37°C, and treated with 4% formaldehyde as detailed previously. Cells for each culture were collected by centrifugation (3500× g for 5 mins), and washed three times in 1× PBS. Cell suspensions were adjusted to an OD_{600} of 0.40, and diluted 1:10 with 1× PBS. The suspended cells (20 µL) were spotted onto a glass cover slip, and allowed to dry at 20°C for 20 minutes. Slides were blocked by immersion into PBTS solution containing 0.25% (w/v) BSA for 30 minutes at room temperature. Cover slips were washed with 1× PBS, and stained with 50 µL of a 0.8 mg/mL solution containing one of the Alexifluor488 labelled peptides (G1-BSA conjugate or CL3-BSA conjugate). After 5 minutes the cover slips were washed twice in 1×PBS. Fluorescence confocal microscopy was performed using a Leica TCS SP5

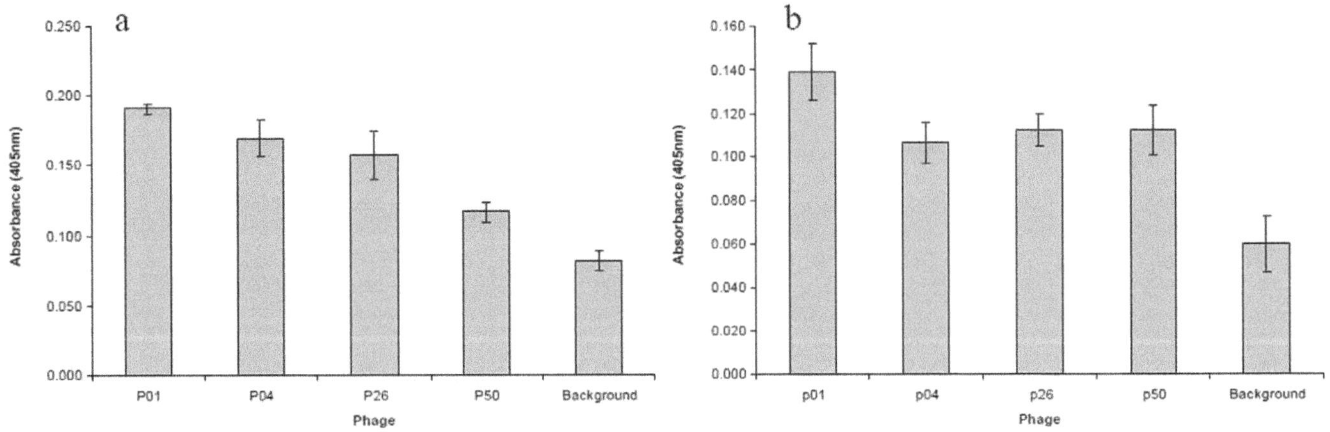

Figure 1. Binding assays for phage particles. Affinity between phage particles with encapsulated *B. anthracis* cells(a) and extracted capsule (b) was assessed by ELISA. The error bars indicate a confidence interval of 95%.

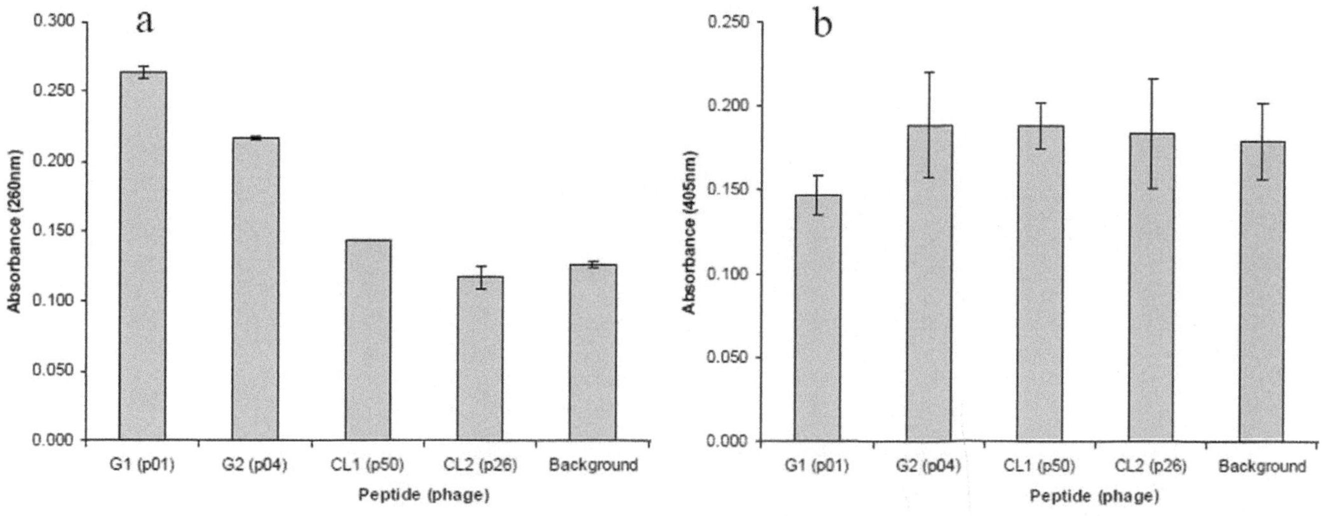

Figure 2. Binding assays for peptides detached from phage particles. Affinity was assayed by ELISA for the synthetic peptides derived from the variable region of phage clones with encapsulated *B. anthracis* cells (a) and extracted capsule (b). The error bars indicate a confidence interval of 95%.

spectral confocal microscope based on a DMI600 inverted microscope. A 63× oil immersion lens with a numerical aperture of 1.4 was used for all images. A 488 nm line from an Ar laser was used for excitation of AlexaFluor488, and emission detected with a standard fluorescein filter (522±35 nm). Image manipulations of confocal microscope data were performed using Leica Applications Suite, and were limited to brightness, contrast, intensity and tone curve adjustments.

Results

Panning of Phage Display Peptide Library Ph.D.-12

Following four rounds of panning, 80 phage clones were randomly selected and the region flanking the DNA insert in the gene III of each clone was amplified using primers M13KE-f1 and M13KE-r2. No PCR product was obtained on eight clones. Of the remaining 72 clones sequenced, clone p01 was found to be the most represented, comprising 36% of all the sequenced clones. Another clone which had a similar motif to p01 (three identical terminal residues) and representation in both screenings (about 21%) was p04. Phage clones p26 and p50 were represented three times and also warranted further investigation. The remaining clones had variable sequences which were only represented once, and were thus considered to have a lesser γPGA binding potential. The amino acid sequence of the peptides p01, p04, p26 and p50 and their representation among the sequenced clones, are given in Table 1.

Binding of Phage Particles to *B. anthracis* Capsule Extract and Encapsulated Cells

The capsule was extracted from the vegetative cells of *B. anthracis* ΔAmes-1 as described in the Materials and Methods. A sample of the capsule extract was analysed on an SDS-PAGE gel. A high molecular weight band (greater than 250 kDa) stained positive with methylene blue, but negative for coomassie blue indicating the presence of γPGA (data not shown). Phage particles were prepared from the clones listed in Table 1 and assayed by ELISA for binding to the capsule and the vegetative cells of *B. anthracis*, respectively. Phage particles were first adsorbed to immobilised encapsulated cells or capsule extract, and then detected with an antibody to M13 bacteriophage. All the phage particles except p50, produced comparable signals against encapsulated cells (Fig. 1a). Clone p01 resulted in the greatest signal against the capsule extract, while the other clones gave a lower but comparable result to one another (Fig. 1b).

Binding of Phage Particle-free Peptides to *B. anthracis* Capsule Extract and Encapsulated Cells

Five peptides were synthesised, four of which were based on the sequence information in Table 1, and the fifth, named G5 (SDYPWKNYHEAQ), was from the insert of a random phage clone that did not bind to the capsule of *B. anthracis*. A short spacer (GGGC) was also added at the 3′ end of each peptide. The five synthetic peptides were each conjugated *via* the cysteine residue to BSA. The binding of BSA-peptide conjugate to synthetic PGA was confirmed by ELISA, in which biotinylated-PGA was adsorbed to plate bound peptides and the former then detected with streptavidin-AP conjugate (data not shown). Encapsulated cells or capsule extract was adsorbed to plate bound peptides, the former then detected with an antibody generated to target capsule (unpublished). The ELISA results for binding to encapsulated cells are shown in Fig. 2a. G1, the peptide derived from phage clone p01, had the greatest ELISA signal, with reduced signals obtained for peptides G2 (p04) and G4 (p50). The random peptide G5 did not provide a signal significantly greater then the background control. However, no binding was detected between the capsule extract and synthetic peptides (Fig, 2b), as all peptide signals were indistinguishable from the background control. It is worth noting that these ELISA data (Fig. 1 and 2, a and b) are not quantitatively comparable to one another as no other no characterised antibody which targets capsule was readily available to us.

Surface Plasmon Resonance (SPR) was employed to examine the binding between the G1 peptide and extracted capsule, but experiments were not successful (data not shown). As a consequence, confocal microscopy was carried out to visualise of the binding to encapsulated cells, as detailed below.

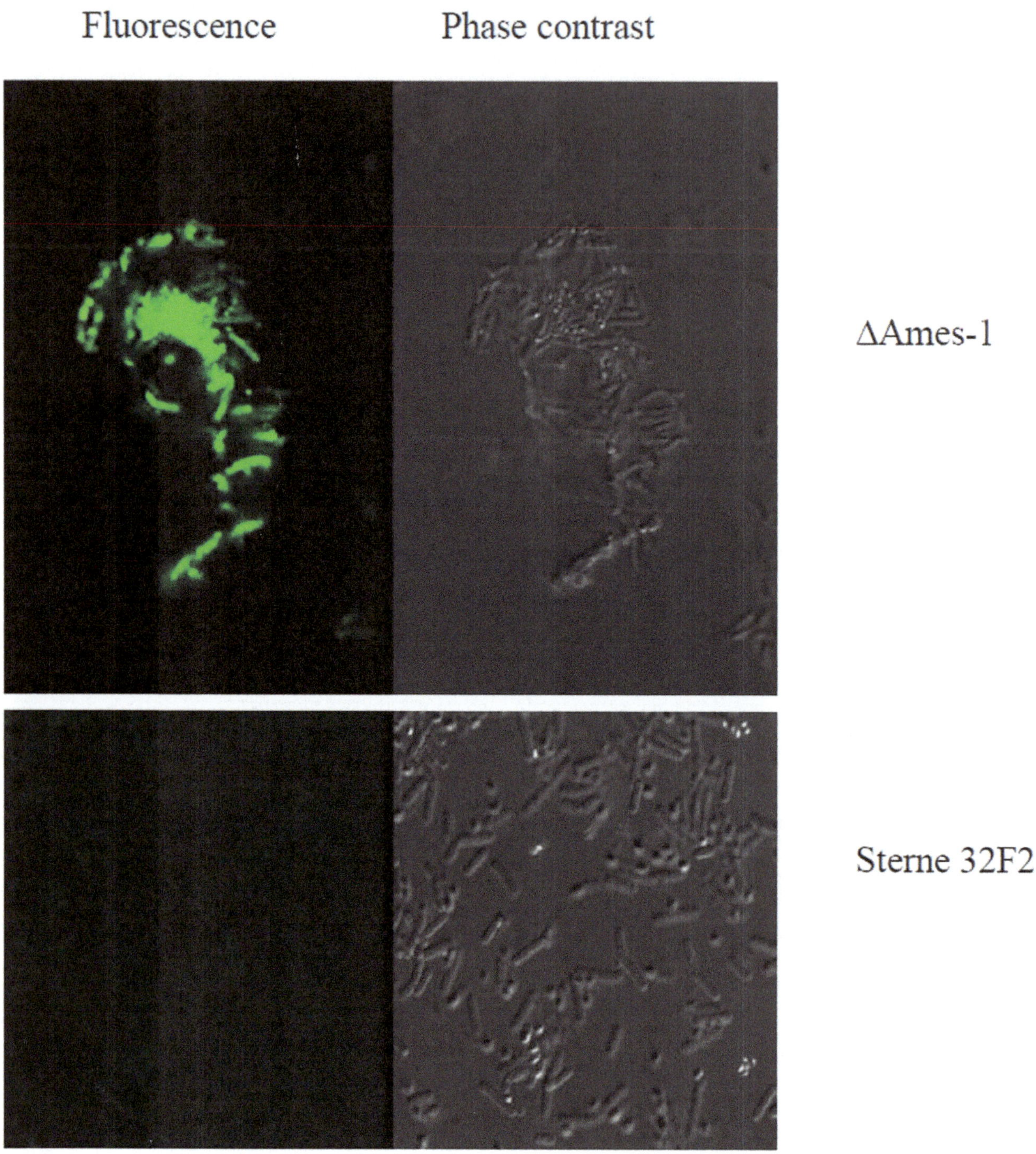

Figure 3. Confocal microscopy and image analysis. Binding of peptide G1 to *B. anthracis* encapsulated cells was determined by confocal laser scanning microscopy and digital image analysis showing selective binding of AlexaFluo488 labelled peptide G1 to encapsulated cells of *B. anthracis* (ΔAmes-1) over unencapsulated cells of *B. anthracis* (Sterne 32F2). Scale bar = 5 μm.

Confirmation of Peptide Binding to *B. anthracis* Encapsulated Cells by Confocal Microscopy

Peptide G1 was fluorescently labelled with AlexaFluo488 and then mixed with encapsulated *B. anthracis* (ΔAmes-1) cells and unencapsulated *B. anthracis* (Sterne 32F2) cells. Although cells of both strains exhibited auto-fluorescence, the signal obtained with the encapsulated strain was readily discernible from that of the unencapsulated strain (Fig. 3). In another experiment, the two strains were mixed, separately, with the control peptide G5 labelled with AlexaFluo488, however no signal was detectable by confocal microscopy (data not shown). The results clearly demonstrated that the G1 peptide was able to bind to the native capsule of vegetative cells of *B. anthracis*.

Discussion

In this work, we used a synthetic capsule peptide of 12 γ-D-glutamic acid residues to pan a phage display library for peptide ligands that bound directly to the native capsule of *B. anthracis*. Even though the phage display library could have been panned against vegetative cells of *B. anthracis* or its capsule extract, the synthetic peptide was chosen for the following reasons. First, neither encapsulated cells nor its capsule extract was free from non-γPGA antigens, which could result in the selection of phage particles with an affinity for the non-γPGA antigens. In contrast, the synthetic peptide could be readily synthesized and purified. Second, the preparation of encapsulated cells or its capsule extract involves handling of pathogenic *B. anthracis*, with stringent requirements for biosafety and physical containment. Finally and most importantly, similar oligo-γPGA-peptides (5, 10, 15 and 20 monomers) had been used successfully to raise antibodies with affinity for the native capsule in animals including guinea pigs, rabbits and mice [9,12].

A peptide with high affinity to the capsule of *B. anthracis* may be used potentially to deliver a therapeutic agent to the surface of the vegetative cells of *B. anthracis*. Examples of such agents include those that degrade capsular material, such as capsule depolymerase [3,23], immunostimulatory components, and bioreactive peptides that reduce microbial function, such as antimicrobial peptides (AMPs) [24,25]. Over the last few years, we and others have identified a number of AMPs that are potent against *B. anthracis* [26]. Targeting of such AMPs, through conjugation with a capsule-binding peptide, to the vegetative cells of *B. anthracis* would be highly beneficial, as most AMPs discovered to date have a broad range of activity killing both pathogenic and beneficial bacteria. As demonstrated for *Pseudomonas mendocina* [27], the presence of such a target-binding peptide would allow the accumulation of a specific AMP on the surface of the target cell resulting in enhanced antimicrobial activity. Of the four peptides identified in this study, G1 was the strongest binder to the encapsulated cells of *B. anthracis*, and hence it represents the most promising peptide for use as a delivery agent that targets *B. anthracis* cells. The affinity of the G1 peptide for the vegetative cell could be increased through application of *in vitro* affinity maturation (e.g. error-prone PCR), potentially increasing its effectiveness in therapeutic applications.

The reason that the G1 peptide bound only to encapsulated cells but not the capsule extract is not clear. Our confocal microscopy result clearly demonstrated that the G1 peptide targeted specifically the encapsulated ΔAmes-1 but not the un-encapsulated Sterne 32F2. Therefore, it is unlikely that the failure of the G1 to bind the capsule extract was caused by the loss of a non-capsule compound to which the G1 might bind during extraction. As a potential delivery agent for clinical applications, the ability of a peptide to bind the encapsulated cells (rather than the capsule extract) is essential to achieve targeted killing of vegetative cells of *B. anthracis*.

Acknowledgments

The authors thank Ms Trinh Vo and Dr David Thirkettle-Watts for technical assistance and Dr Peter Gray for critical reading of the manuscript.

Author Contributions

Conceived and designed the experiments: CQL MB. Performed the experiments: MB CQL. Analyzed the data: MB CQL. Contributed reagents/materials/analysis tools: CQL MB. Wrote the paper: MB CQL.

References

1. Brossier F, Mock M (2001) Toxins of *Bacillus anthracis*. Toxicon 39: 1747–1755.
2. Drysdale M, Heninger S, Hutt J, Chen Y, Lyons CR, et al. (2005) Capsule synthesis by *Bacillus anthracis* is required for dissemination in murine inhalation anthrax. EMBO J 24: 221–227.
3. Scorpio A, Chabot DJ, Day WA, Hoover TA, Friedlander AM (2010) Capsule depolymerase overexpression reduces *Bacillus anthracis* virulence. Microbiology 156: 1459–1467.
4. Candela T, Fouet A (2006) Poly-gamma-glutamate in bacteria. Mol Microbiol 60: 1091–1098.
5. Candela T, Mock M, Fouet A (2005) CapE, a 47-amino-acid peptide, is necessary for *Bacillus anthracis* polyglutamate capsule synthesis. J Bacteriol 187: 7765–7772.
6. Ezzell JW, Welkos SL (1999) The capsule of *Bacillus anthracis*, a review. J Appl Microbiol 87: 250.
7. Kozel TR, Murphy WJ, Brandt S, Blazar BR, Lovchik JA, et al. (2004) mAbs to *Bacillus anthracis* capsular antigen for immunoprotection in anthrax and detection of antigenemia. Proc Natl Acad Sci U S A 101: 5042–5047.
8. Kozel TR, Thorkildson P, Brandt S, Welch WH, Lovchik JA, et al. (2007) Protective and immunochemical activities of monoclonal antibodies reactive with the *Bacillus anthracis* polypeptide capsule. Infect Immun 75: 152–163.
9. Lee DY, Chun JH, Ha HJ, Park J, Kim BS, et al. (2009) Poly-gamma-d-glutamic acid and protective antigen conjugate vaccines induce functional antibodies against the protective antigen and capsule of *Bacillus anthracis* in guinea-pigs and rabbits. FEMS Immunol Med Microbiol 57: 165–172.
10. AuCoin DP, Sutherland MD, Percival AL, Lyons CR, Lovchik JA, et al. (2009) Rapid detection of the poly-gamma-D-glutamic acid capsular antigen of *Bacillus anthracis* by latex agglutination. Diagn Microbiol Infect Dis 64: 229–232.
11. Joyce J, Cook J, Chabot D, Hepler R, Shoop W, et al. (2006) Immunogenicity and protective efficacy of *Bacillus anthracis* poly-gamma-D-glutamic acid capsule covalently coupled to a protein carrier using a novel triazine-based conjugation strategy. J Biol Chem 281: 4831–4843.
12. Schneerson R, Kubler-Kielb J, Liu TY, Dai ZD, Leppla SH, et al. (2003) Poly(gamma-D-glutamic acid) protein conjugates induce IgG antibodies in mice to the capsule of *Bacillus anthracis*: a potential addition to the anthrax vaccine. Proc Natl Acad Sci U S A 100: 8945–8950.
13. Wang TT, Lucas AH (2004) The capsule of *Bacillus anthracis* behaves as a thymus-independent type 2 antigen. Infect Immun 72: 5460–5463.
14. Ladner RC, Sato AK, Gorzelany J, de Souza M (2004) Phage display-derived peptides as therapeutic alternatives to antibodies. Drug Discov Today 9: 525–529.
15. Brigati J, Williams DD, Sorokulova IB, Nanduri V, Chen IH, et al. (2004) Diagnostic probes for *Bacillus anthracis* spores selected from a landscape phage library. Clin Chem 50: 1899–1906.
16. Turnbough CL Jr (2003) Discovery of phage display peptide ligands for species-specific detection of *Bacillus* spores. J Microbiol Methods 53: 263–271.
17. Zhou B, Wirsching P, Janda KD (2002) Human antibodies against spores of the genus *Bacillus*: a model study for detection of and protection against anthrax and the bioterrorist threat. Proc Natl Acad Sci U S A 99: 5241–5246.
18. Pelat T, Hust M, Laffly E, Condemine F, Bottex C, et al. (2007) High-affinity, human antibody-like antibody fragment (single-chain variable fragment) neutralizing the lethal factor (LF) of *Bacillus anthracis* by inhibiting protective antigen-LF complex formation. Antimicrob Agents Chemother 51: 2758–2764.
19. Laffly E, Danjou L, Condemine F, Vidal D, Drouet E, et al. (2005) Selection of a macaque Fab with framework regions like those in humans, high affinity, and ability to neutralize the protective antigen (PA) of *Bacillus anthracis* by binding to the segment of PA between residues 686 and 694. Antimicrob Agents Chemother 49: 3414–3420.
20. Basha S, Rai P, Poon V, Saraph A, Gujraty K, et al. (2006) Polyvalent inhibitors of anthrax toxin that target host receptors. Proc Natl Acad Sci U S A 103: 13509–13513.
21. Gujraty K, Sadacharan S, Frost M, Poon V, Kane RS, et al. (2005) Functional characterization of peptide-based anthrax toxin inhibitors. Mol Pharm 2: 367–372.
22. Mourez M, Kane RS, Mogridge J, Metallo S, Deschatelets P, et al. (2001) Designing a polyvalent inhibitor of anthrax toxin. Nat Biotechnol 19: 958–961.
23. Scorpio A, Chabot DJ, Day WA, O'Brien DK, Vietri NJ, et al. (2007) Poly-gamma-glutamate capsule-degrading enzyme treatment enhances phagocytosis and killing of encapsulated *Bacillus anthracis*. Antimicrob Agents Chemother 51: 215–222.
24. Dawson RM, Liu CQ (2008) Properties and applications of antimicrobial peptides in biodefense against biological warfare threat agents. Crit Rev Microbiol 34: 89–107.
25. Gordon YJ, Romanowski EG, McDermott AM (2005) A review of antimicrobial peptides and their therapeutic potential as anti-infective drugs. Curr Eye Res 30: 505–515.

26. McGillivray SM, Ebrahimi CM, Fisher N, Sabet M, Zhang DX, et al. (2009) ClpX contributes to innate defense peptide resistance and virulence phenotypes of *Bacillus anthracis*. J Innate Immun 1: 494–506.

27. Eckert R, Brady KM, Greenberg EP, Qi F, Yarbrough DK, et al. (2006) Enhancement of antimicrobial activity against *Pseudomonas* aeruginosa by coadministration of G10KHc and tobramycin. Antimicrob Agents Chemother 50: 3833–3838.

In Vitro and *In Vivo* Studies for Assessing the Immune Response and Protection-Inducing Ability Conferred by *Fasciola hepatica*-Derived Synthetic Peptides Containing B- and T-Cell Epitopes

Jose Rojas-Caraballo[1,2], Julio López-Abán[1], Luis Pérez del Villar[1], Carolina Vizcaíno[2], Belén Vicente[1], Pedro Fernández-Soto[1], Esther del Olmo[3], Manuel Alfonso Patarroyo[2,4], Antonio Muro[1]*

1 Parasite and Molecular Immunology Laboratory, Tropical Disease Research Centre, Universidad de Salamanca (IBSAL-CIETUS), Salamanca, Spain, **2** Molecular Biology and Immunology Department, Fundación Instituto de Inmunología de Colombia (FIDIC), Bogotá, Colombia, **3** Pharmaceutical Chemistry Department, Tropical Disease Research Centre, Universidad de Salamanca (IBSAL-CIETUS), Salamanca, Spain, **4** Basic Sciences Department, School of Medicine and Health Sciences, Universidad del Rosario, Bogotá, Colombia

Abstract

Fasciolosis is considered the most widespread trematode disease affecting grazing animals around the world; it is currently recognised by the World Health Organisation as an emergent human pathogen. Triclabendazole is still the most effective drug against this disease; however, resistant strains have appeared and developing an effective vaccine against this disease has increasingly become a priority. Several bioinformatics tools were here used for predicting B- and T-cell epitopes according to the available data for *Fasciola hepatica* protein amino acid sequences. BALB/c mice were immunised with the synthetic peptides by using the ADAD vaccination system and several immune response parameters were measured (antibody titres, cytokine levels, T-cell populations) to evaluate their ability to elicit an immune response. Based on the immunogenicity results so obtained, seven peptides were selected to assess their protection-inducing ability against experimental infection with *F. hepatica* metacercariae. Twenty-four B- or T-epitope-containing peptides were predicted and chemically synthesised. Immunisation of mice with peptides so-called B1, B2, B5, B6, T14, T15 and T16 induced high levels of total IgG, IgG1 and IgG2a ($p<0.05$) and a mixed Th1/Th2/Th17/Treg immune response, according to IFN-γ, IL-4, IL-17 and IL-10 levels, accompanied by increased CD62L^{+} T-cell populations. A high level of protection was obtained in mice vaccinated with peptides B2, B5, B6 and T15 formulated in the ADAD vaccination system with the AA0029 immunomodulator. The bioinformatics approach used in the present study led to the identification of seven peptides as vaccine candidates against the infection caused by *Fasciola hepatica* (a liver-fluke trematode). However, vaccine efficacy must be evaluated in other host species, including those having veterinary importance.

Editor: Patricia Talamas-Rohana, Centro de Investigacion y de Estudios Avanzados del Instituto Politecnico Nacional, Mexico

Funding: This study was supported by Fundación Ramón Areces (Madrid, Spain). Reference: XV-2010-2013 (www.fundacionareces.es). The funders had no role in study design, data collection and analysis, decision to publish, or preparation of the manuscript.

Competing Interests: The authors have declared that no competing interests exist.

* Email: ama@usal.es

Introduction

Fasciolosis is one of the most important helminthiasis worldwide affecting grazing livestock due its widespread geographical distribution and resulting economic loss; it is caused by the common liver fluke *Fasciola hepatica,* along with the related species *Fasciola gigantica* [1]. Besides being a well-known veterinary problem, fasciolosis has also recently become considered as an emerging parasitic human disease, having a significant impact on public health, causing millions of people to be at risk of infection. Reports have indicated its increase in many Latin-American, African, European and Asian countries [2,3]. Taking its impact on human health and wide emergence into account, human fasciolosis has been recently included in the World Health Organization's (WHO) list of priorities related to Neglected Tropical Diseases [4].

It is well-known that methodological and technical difficulties related to diagnosis have limited progress in combating human fasciolosis globally, including drawbacks in diagnosing infection and assessing drug efficacy and resistance, mainly concerning triclabendazole which is still the most effective drug for combating the disease. Indeed, no commercial vaccine is currently available and developing vaccines for controlling animal and human fasciolosis thus represents a tremendous research opportunity. Many candidate proteins have been tested for a long time now as target antigens in vaccination assays against fluke, including fatty acid-binding proteins, glutathione S-transferases, cathepsin proteases, leucine aminopeptidase, fluke haemoglobin and thioredoxin peroxidase. However, no consensus regarding the factors required for immunological protection has yet emerged and there has been no report to date of a successful field trial concerning a liver fluke vaccine [5,6]. Many factors may be responsible for the failure of these vaccines when tested; vaccine formulation, choice of adjuvant and delivery route and dosage will affect the way in which different animals and breeds will respond to different vaccines and, possibly, the choice of target antigen. The aforementioned research challenge must thus involve identifying new target antigens for obtaining an effective vaccine against *F. hepatica* [7].

Public access to an increasing number of pathogen genomes which have been totally or partially sequenced, along with the use of powerful *in silico* analysis, currently relies on rapidly screening a large number of expressed pathogen proteins for their ability to induce a protective immune response; vaccine candidates based on genome information has thus become possible [8]. Synthetic peptide-based vaccines, in which small peptides derived from known target epitopes are used to induce an immune reaction, have thus attracted interest as a promising approach to treating several infectious diseases and tumours, since they have several advantages over other forms of vaccine, particularly regarding safety, ease of production, reproducibility, low cost and ensuring a more effective antigen-specific immune response to a particular cell type [9].

As epitope-based vaccines only contain small sequences derived from an entire protein known to bind to various major histocompatibility complex (MHC) molecules, predicting peptide-MHC binding and mapping epitopes are crucial in their design [10,11]. This approach has led to identifying specific binding motifs for effectively predicting both B- and T-cell epitopes. There are several online-based tools for predicting the MHC-peptide interaction available for researchers, although B-cell epitope mapping algorithms have lagged behind T-cell ones and only a few B-cell epitope mapping algorithms are in current use [11]; this is because there are still several obstacles to developing B-cell epitope prediction for peptide-based vaccine design [12].

Synthetic peptides have been examined as potential prophylactic vaccines against viral, bacterial and parasitic diseases for many years now [13,14] and as therapeutic vaccines for chronic infections and non-infectious diseases, as well as cancer [15]. Despite such a large number of potential synthetic peptides having been identified, none are currently being marketed for human use [16] and few studies reported to date have used synthetic peptides as anti-helminth vaccines, including *Echinococcus granulosus* [17], *Trichinella spiralis* [18], *Brugia malayi* [19], *Taenia solium* [20], *Schistosoma mansoni* [21] and *F. gigantica* [22]. Regarding *F. hepatica*, synthetic peptides have been used in diagnosing human infections. In the search for selecting optimal vaccine candidate proteins expressed by *F. hepatica* and trigger an immune response induced by previously reported candidate proteins, our group has focused on the rational identification of B- and T-cell epitopes by *in silico* mapping.

Several peptides have thus been chemically-synthesised and then assessed using *in vitro* and *in vivo* assays to evaluate the induced immune response and their inducing-protection ability. Our trials have involved using a murine model prepared with an adjuvant/adaptation (ADAD) vaccination system [23] and then immunised with a chosen peptide antigen, a natural immunomodulator extracted from the rhizome of the fern *Phlebodium pseudoaureum* (PAL) or a chemically-synthesised aliphatic diamine immunomodulator AA0029 [24] and a non-haemolytic adjuvant containing *Quillaja saponaria* (QS) saponins to form an emulsion with a non-mineral oil in a 70/30 oil/water ratio. This study was aimed at selecting peptides containing B- and T-cell epitopes, assessing their immunogenicity, and testing the protection-inducing ability against experimental infection with *F. hepatica* metacercariae of the highly immunogenic ones.

Materials and Methods

Ethics statement and experimental animals

The animal procedures in this study complied with Spanish (Real Decreto RD53/2013) and European Union (European Directive 2010/63/EU) guidelines regarding animal experimentation for the protection and humane use of laboratory animals, and were conducted at the University of Salamanca's accredited Animal Experimentation Facilities (Register number: PAE/SA/001). University of Salamanca's Ethics Committee approved procedures used in the present study (protocol approval number 48531). Seven-week-old female BALB/c and CD1 mice (Charles River Laboratories, Barcelona, Spain) weighing 20 to 22 g were used for the experiments. Animals were kept in plastic boxes with food and water *ad libitum* at the University of Salamanca's Animal Experimentation Facilities. Animal care involved regular 12 h light–dark periods at 20°C. All efforts to minimise suffering were made.

Fasciola hepatica sequence selection

Due to a lack of data regarding the *F. hepatica* complete genome sequence, only partial information concerning several genes and protein sequences was available in pertinent databases at the beginning of the present study. Each sequence obtained was individually predicted for signal peptide (SP) and trans-membrane (TM) domains to select secreted proteins. SP was predicted with the SignalP 3.0 server [25] available at (http://www.cbs.dtu.dk/services/SignalP/) and the TM domain was predicted using the TMHMM v.2.0 server (http://www.cbs.dtu.dk/services/TMHMM/). Only proteins showing a SP and no TM domains were finally selected and grouped into families. All selected sequences were subjected to multiple sequence alignment using ClustalW (http://npsa-pbil.ibcp.fr) and conserved or semi-conserved fragments were chosen for B- and T-cell epitope prediction.

B- and T-cell epitope prediction

The BepiPred method was used for predicting linear B-cell epitopes [26]; the server (BepiPred 1.0) and training datasets are publicly available at http://www.cbs.dtu.dk/services/BepiPred. This method involves each amino acid receiving a prediction score based on Hidden Markov Model (HMM) profiles for known antigens and incorporates propensity scale methods based on hydrophilicity and secondary structure prediction. Predicted linear B-cell epitopes were then compared to the values given for each amino acid using ANTHEPROT 3D software (http://antheprot-pbil.ibcp.fr) regarding several physical-chemical profiles, such us antigenicity, hydrophobicity, flexibility and solvent accessibility [27]. The values obtained for each profile were averaged in groups of 20 amino acids. Only regions showing the best probability for each protein (score ≥0.5, according to the HMM result) were selected as promising linear B-cell epitopes.

The SYFPEITHI database [28], freely accessible at http://www.syfpeithi.de/was used as the source of MHC class II binding peptides. SYFPEITHI allows predicting peptide binding to a defined MHC type and predictions were made for H2-Ad murine MHC class-II ligands. The analysis was performed choosing 15-mer (15 amino acids) for MHC type II as prediction parameter. The resulting peptides showing the highest scores in the dataset capable of binding to the murine class-II molecules described above were selected as candidate epitopes.

Table 1. Selected *F. hepatica* proteins containing signal peptide (SP) but no transmembrane domain (TM).

Epitope	Synthesised peptide sequence	Position	GenBank	Description	Protein length (aa)
B1	KGAGSSQDACIKFIQYEVDG	63–82	AAB02579.1	Amoebapore homologue	102
B2	KGAGSSQDATIKFIQYEVDG	63–82	AAB02579.1	Amoebapore homologue	102
B3	FASFDVPSKQPTIDIDLCDI	14–33	AAF88069.1	Amoebapore-like protein	101
B4	FASFDVPSKQPTIDIDLTDI	14–33	AAF88069.1	Amoebapore-like protein	101
B5	ISEIRDQSSTSSTWAVSSAS	102–121	ABU62951.1	Cathepsin B	337
B6	GVENGVKYWLIANSWNEGWG	293–312	ABU62951.1	Cathepsin B	337
B7	QTCSPLRVNHAVLAVGYGTQ	260–279	AAB41670.2	Secreted cathepsin L1	326
		260–279	AAA29136.1	Cathepsin	326
		264–283	AAP49831.1	Cathepsin L	326
		260–279	Q24940.1	Cathepsin L-like proteinase	326
B8	QTTSPLRVNHAVLAVGYGTQ	260–279	AAB41670.2	Secreted cathepsin L1	326
		260–279	AAA29136.1	Cathepsin	326
		264–283	AAP49831.1	Cathepsin L	326
		260–279	Q24940.1	Cathepsin L-like proteinase	326
B9	YTEPRSVTPEERSVFQPMIL	27–46	AAV68752.1	Cystatin	116
B10	FVPLYSSKSATSVGTPTRVS	95–114	AAV68752.1	Cystatin	116
B11	VTTNGPPNGKHNDKHTYVEC	350–369	CAC85636	Legumain-like	419
B12	VTTNGPPNGKHNDKHTYVET	350–369	CAC85636	Legumain-like	419
T13	TVNLVKRLLQNSVVE	37–51	AAB02579.1	Amoebapore homologue	102
T14	DYIIDHVDQHNATEI	80–94	AAF88069.1	Amoebapore-like protein	101
T15	DRNTQRQTVRYSVSE	69–83	ABU62925.1	Cathepsin B	337
T16	FYMFEDFLVYKSGIY	260–274	ABU62925.1	Cathepsin B	337
T17	KYLTEMSRASDILSH	83–97	AAB41670.2	Secreted cathepsin L1	326
		83–97	ABQ95351.1	Secreted cathepsin L2	326
		83–97	AAA29136.1	Cathepsin	326
		83–97	AAP49831.1	Cathepsin L	326
		83–97	AAR99518.1	Cathepsin L protein	326
		83–97	Q24940.1	Cathepsin L-like proteinase	326
T18	ISFSEQQLVDTSGPW	153–167	AAB41670.2	Secreted cathepsin L1	326
		153–167	AAA29136.1	Cathepsin	326
		153–167	AAP49831.1	Cathepsin L	326
		153–167	BAA23743.1	Cathepsin L	325
		153–167	AAR99518.1	Cathepsin L protein	326
		153–167	AAT76664.1	Cathepsin L1 proteinase	326
		153–167	Q24940.1	Cathepsin L-like proteinase	326
T19	ENAYEYLKHNGLETE	178–192	AAC47721.1	Secreted cathepsin L2	326
		178–192	ABN50361.2	Cathepsin L	326
		178–192	CAA80446.1	Cathepsin L-like protease	326
		53–67	CAA80445.1	Cathepsin L-like protease	166
T20	LDPYFNLVSPEVYNY	29–43	BAE44988.1	Cytochrome oxidase subunit I	145
		29–43	BAE45005.1	Cytochrome oxidase subunit I	145
T21	DLNLPRLNALSAWLL	76–90	BAE44988.1	Cytochrome oxidase subunit I	145
		76–90	BAE45005.1	Cytochrome oxidase subunit I	145
T22	FAGHGKAYLHGSFDK	56–70	AAA29144.1	Vitelline protein B2	272
T23	YEKYEDDYARETPYD	254–268	AAA29144.1	Vitelline protein B2	272
		254–268	AAA29143.1	Vitelline protein B1	272
T24		101–115	AAA31753.2	NADH dehydrogenase subunit 3	118
		101–115	Q34522.1	NADH dehydrogenase subunit 3	118
		101–115	AAG13152.2	NADH dehydrogenase subunit 3	118

GenBank accession number and amino acid sequences from each B- and T-cell chemically synthesised epitope are also indicated.

Peptide synthesis

All derived peptides selected on the basis of B- or T-cell epitope predictions for each protein were chemically synthesised (Fundación Instituto de Inmunología, FIDIC, Colombia) by the solid-phase peptide synthesis according to the methodology first described by Merrifield [29] and subsequently modified by Houghten [30] using the t-Boc strategy and a BHA (benzyhydrylamine) resin (0.7 meq/mg). One cysteine and one glycine residue were added at both amino and carboxyl-terminal extremes to allow their polymerisation via oxidisation. Peptides were purified by reverse phase high performance liquid chromatography (to at least >90% purity), characterised by MALDI-TOF mass spectrometry, lyophilised and quantified. Freeze-dried synthesised peptides were used in the ensuing experiments.

Cytotoxicity evaluation

The J774.2 mouse peritoneal macrophage cell line was used in this study; it was grown in RPMI-1640 culture medium supplemented with 10% FBS, 2 mM L-glutamine, 100 U/mL penicillin, and 100 μg/mL streptomycin, at 37°C in humidified 95% air and 5% CO_2. J774.2 peritoneal macrophage cells were plated in complete RPMI-1640 culture medium at 1×10^6 cells/well concentration in 12-well culture plates (Costar, Cambridge, MA), and left to adhere for 2 h at 37°C in 5% CO_2. Non-adhering cells were removed by gentle washing with complete RPMI-1640 culture medium. Adherent J774.2 cells were incubated with each synthetic peptide at different concentrations (1–100 μg/mL). After 48 h incubation at 37°C in 5% CO_2, supernatants were removed and cell viability was measured on adhered cells by MTT (3-(4,5-dimethylthiazole-2-yl)-2,5-diphenyltetrazolium bromide) assay, measuring the absorbance at 540 nm. Controls for checking solvent cytotoxicity were also included.

Vaccine formulation and immunisation trial

One hundred and sixty-eight female BALB/c mice (56 groups of 3 mice per group) were used in this study. The immune response in mice was studied in two separate experiments (A and B). Experiment A: group 1 (untreated control group; n = 6), group 2 (ADAD and natural immunomodulator PAL [ADADn]; n = 6), groups 3–14 (ADADn together with B1–B12 B-epitope-containing peptides; n = 36) and groups 15–26 (ADADn together with T13–T24 T-epitope-containing peptides; n = 36). Experiment B: group 1 (untreated control group; n = 6), group 2 (ADAD and synthetic immunomodulator AA0029 [ADADs]; n = 6), groups 3–14 (ADADs together with B1–B12 B-epitope-containing peptides; n = 36) and groups 15–26 (ADADs together with T13–T24 T-epitope-containing peptides; n = 36). The mice were subcutane-

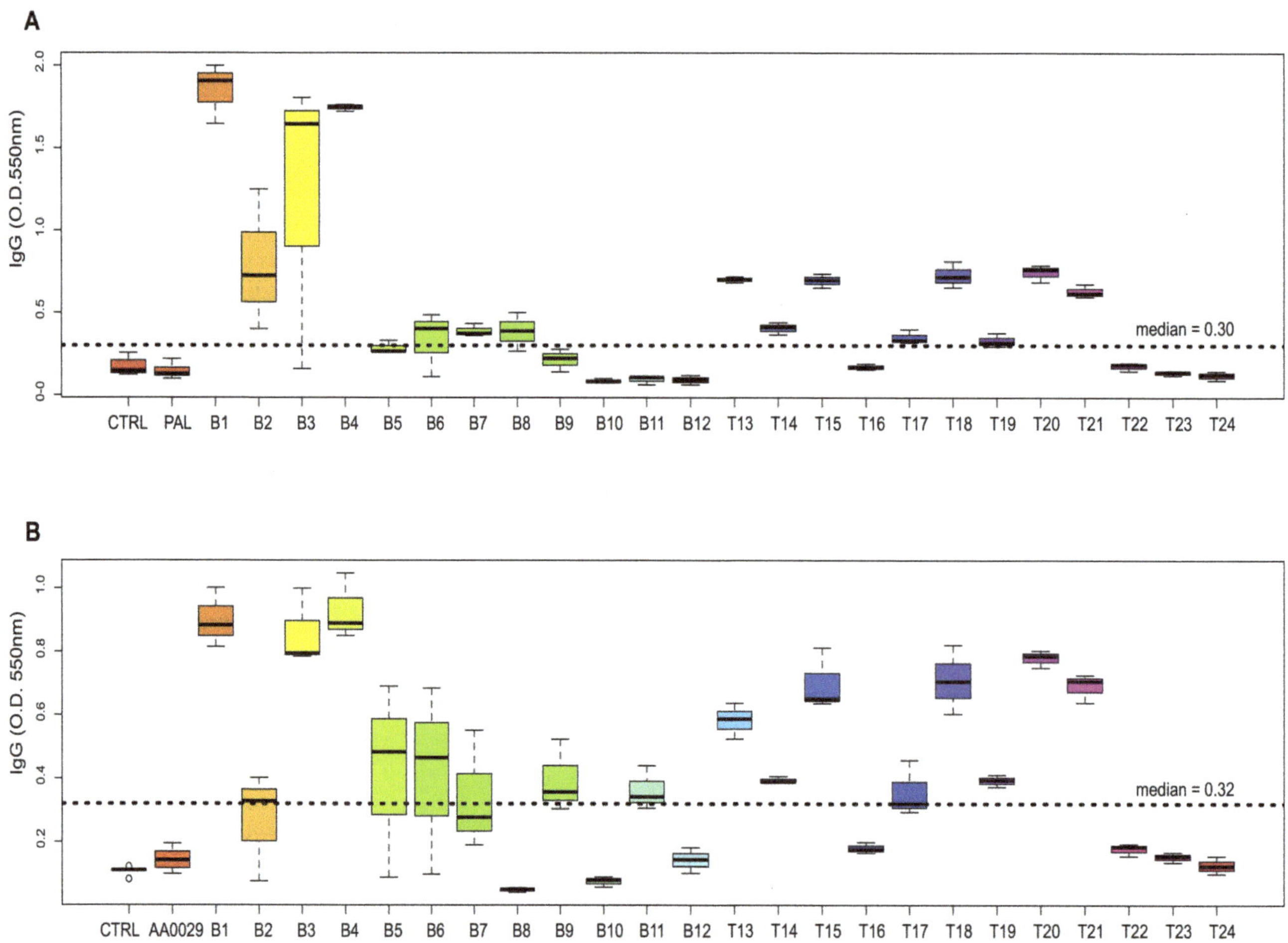

Figure 1. Boxplot showing IgG anti-peptide antibody levels in mice immunised with the synthetic peptides formulated in the ADAD vaccination system. The bottom and the top of the box indicate the 25th and 75th percentiles, respectively. A). Mice immunised using PAL. B). Mice immunised using AA0029.

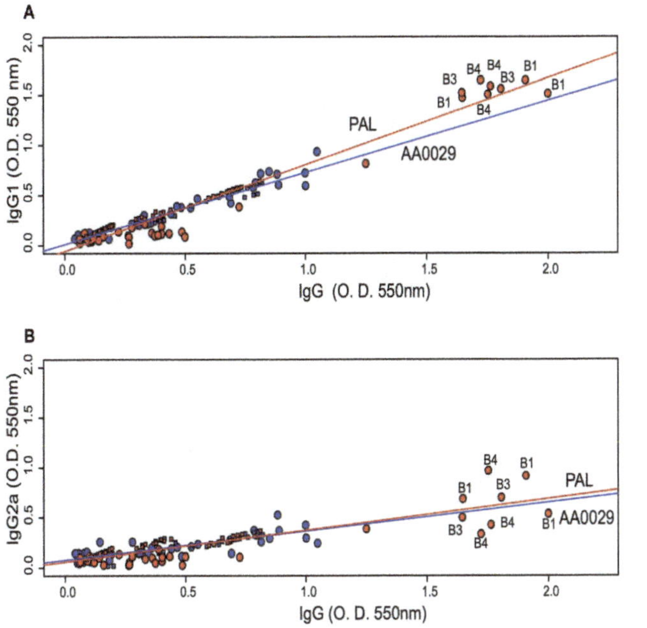

Figure 2. Linear regression comparing the effect of PAL and AA0029 immunomodulators on IgG subtype levels. A). IgG1 related to IgG. B). IgG2a related to IgG. Red indicates the PAL immunomodulator and blue AA0029. Circles and squares represent B- and T-peptides, respectively.

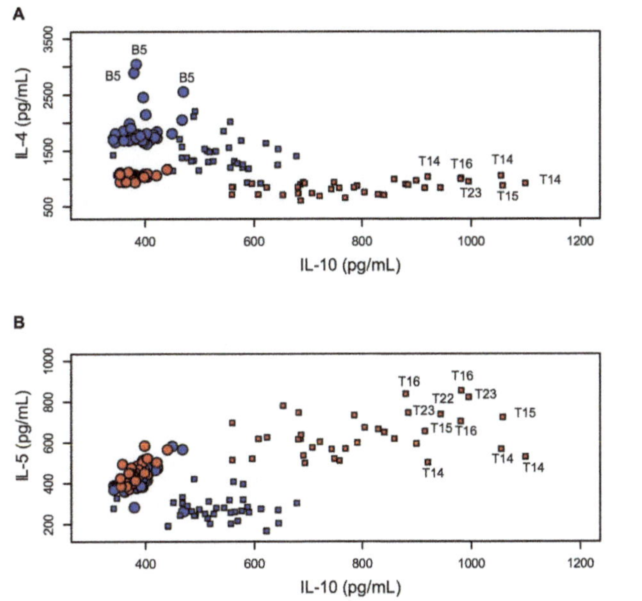

Figure 3. Scatterplot comparing the effect of PAL and AA0029 immunomodulators on cytokine levels. A). IL-4 related to IL-10. B). IL-5 related to IL-10. Red indicates PAL and blue AA0029. Circles and squares represent B- and T-peptides, respectively.

ously immunised using an adjuvant adaptation (ADAD) system [23].

Briefly, the ADAD vaccination system included the vaccine antigen, an immunomodulator (natural or chemically-synthesised), together with non-haemolytic adjuvant *Quillaja saponaria* (QS) saponins to form an emulsion with non-mineral oil in a 70/30 oil/water ratio. Vaccination with this system included a set of 2 subcutaneous injections. The first, also called adaptation, contained QS and the immunomodulator emulsified in non-mineral oil, but without the vaccine antigen; the second injection was administered 5 days after adaptation and contained the vaccine antigen with QS and the immunodulator in the emulsion oil. The individual doses injected during mice immunisation were formulated as follows: 600 μg natural immunodulator PAL (ASAC Pharmaceutical International, Alicante, Spain) or 100 μg chemically synthesised aliphatic diamine immunomodulator AA0029 [24] together with 20 μg *Q. saponaria* (QS) and, when evaluated, 10 μg of each peptide. A final 100 μL/injection volume was emulsified with non-mineral oil (Montanide ISA763A, SEPPIC, Paris, France) in a 70/30 oil/water ratio. Mice were immunised on day 0 and two booster doses with 100 μL of the preparations mentioned above were administered on days 14 and 28.

Sample collection

Mice were humanely euthanised 2 weeks after third immunisation. The spleen was aseptically removed during necropsy to obtain splenocytes for *in vitro* assays. Spleen cell suspensions were collected by spleen perfusion by passing sterile phosphate buffered solution (PBS) through individual spleens according to the methodology described elsewhere by [31]. Blood samples for each mouse were obtained before each immunisation and also during necropsy for serological studies.

Measuring antibody responses

Sera from mice immunised with the aforementioned formulations were analysed by ELISA for measuring total IgG, IgE and IgM levels as well as IgG1 and IgG2a antibody isotype levels. Briefly, 96-well polystyrene microplates (Costar, Corning Costar Corp, Cambridge, Mass) were coated with 1 μg of each peptide in carbonate buffer pH 9.6 (100 μL per well) and incubated overnight at 4°C. The plates were then washed thrice for 5 min with PBS containing 0.05% Tween 20 (PBST). The plates were blocked with 5% skimmed milk (SM) in PBST (200 μL per well) for 1 h at 37°C and then washed again thrice, as described above. Sera samples were appropriately diluted at 1:100 in dilution buffer (5% SM and PBST) and added to the wells (100 μL per well) in duplicate. After 1 h incubation at 37°C the plates were washed as described above and, according to each assay, goat peroxidase-conjugated anti-mouse IgG, IgE, IgM, IgG1 and IgG2a (1:1,000 in dilution buffer, 100 μL per well; Sigma) were incubated for 1 h at 37°C. After washing as above, the bound antibodies were detected using H_2O_2 (0.012%) and ortho-phenylenediamine (0.04%) in 0.1 M citrate/phosphate buffer (100 μL per well). The enzyme reaction was stopped after 15–20 min by adding 3N H_2SO_4 (100 μL per well) and optical density was measured at 550 nm (OD_{550}) on an Ear400FT ELISA reader (STL Lab Instruments, Groding, Austria). The mean absorbance values for each mouse serum from each group were determined and included in each data point.

Cytokine determination

The frequencies of antigen specific IFN-γ, IL-1α, IL-2, IL-4, IL-5, IL-6, IL-10, IL-17, TNF-α producing T-cells in the spleens were quantified by using a flow cytometry-based methodology. Individual mouse splenocytes were cultured in 6-well plates at 1×10^6 cells per well concentration in complete medium (RPMI 1640 medium containing 10% heat-inactivated foetal bovine serum and antibiotics, 100 U/mL penicillin and 100 μg/mL streptomycin) and stimulated with each synthetic peptide at final

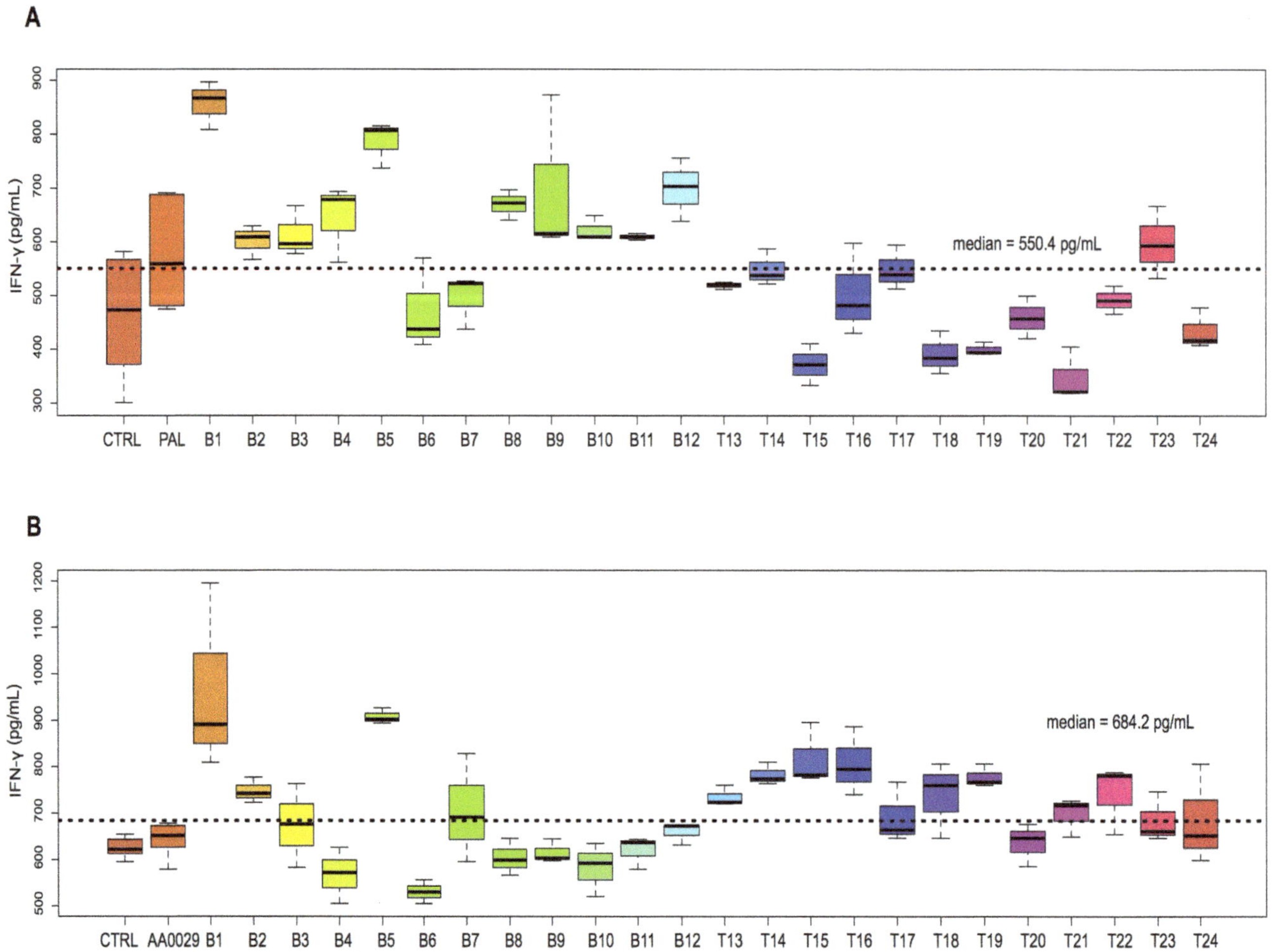

Figure 4. Boxplot showing IFN-γ cytokine levels in splenocyte mouse cell culture immunised with synthetic peptides. A). Mice immunised with PAL. B). Mice immunised with AA0029.

10 μg/mL concentration for 72 h at 37°C in a humidified atmosphere with 5% CO_2. Control wells were prepared containing untreated mouse splenocytes. After the incubation period, splenocyte culture supernatants were recovered for cytokine determination. A FlowCytomix Mouse Th1/Th2 10plex kit (Bender MedSystems GmbH, Vienna, Austria) was used, according to the manufacturer's instructions. Briefly, different sized fluorescent beads coated with capture antibodies specific for the cytokines mentioned above were incubated with splenocyte supernatant cell culture samples to form sandwich complexes with phycoerythrin (PE)-conjugated secondary antibodies. Flow cytometry data were collected using a FACSCalibur flow cytometer (BD Biosciences) at the University of Salamanca's Flow Cytometry Central Service. A total of 8,000 events were collected gated by forward and side scatter and data were analysed using FlowCytomix Pro 3.0 software (Bender MedSystems, Vienna, Austria). Each cytokine concentration was determined from standard curves using known concentrations of mouse recombinant cytokines.

Experimental identification of linear B- and T-cell epitopes

An ELISA assay was used for evaluating the presence of each B-cell epitope using sera from infected mice. Briefly, plates were coated with 1 μg of either each synthetic peptide or *F. hepatica* excretory/secretory antigen and incubated overnight at 4°C. After blocking and washing steps, as previously described, sera samples from mice infected with *F. hepatica* metacercariae or immunised with each synthetic peptide were appropriately diluted at 1:100 dilution and added to the wells in duplicate. Serial dilutions of sera from infected mice were made to calculate the specific titre. The plates were washed after 1 h incubation at 37°C and goat peroxidase-conjugate anti-mouse IgG added to the plates (1:1,000 in dilution buffer, 100 μL per well; Sigma) which were then incubated for 1 h at 37°C. After washing as above, bound antibodies were detected using H_2O_2 (0.012%) and ortho-phenylenediamine (0.04%) in 0.1 M citrate/phosphate buffer (100 μL per well).

A flow cytometry analysis was performed on stained and fixed splenocytes to investigate T-cell populations responding to the immunisation of mice with T-cell epitope-containing peptides. Regarding immunofluorescence staining, 5×10^5 cells were incubated with fluorescein isothyosanate (FITC)-conjugated mouse monoclonal antibodies (mAb) against CD4, or with phycoerythrin (PE)-conjugated mouse mAb against CD8, or with allophycocyanin (APC)-conjugated mouse mAb against CD62L. All samples were incubated with anti CD16/CD32 blocking monoclonal antibody for 5 min at room temperature. Each specific antibody

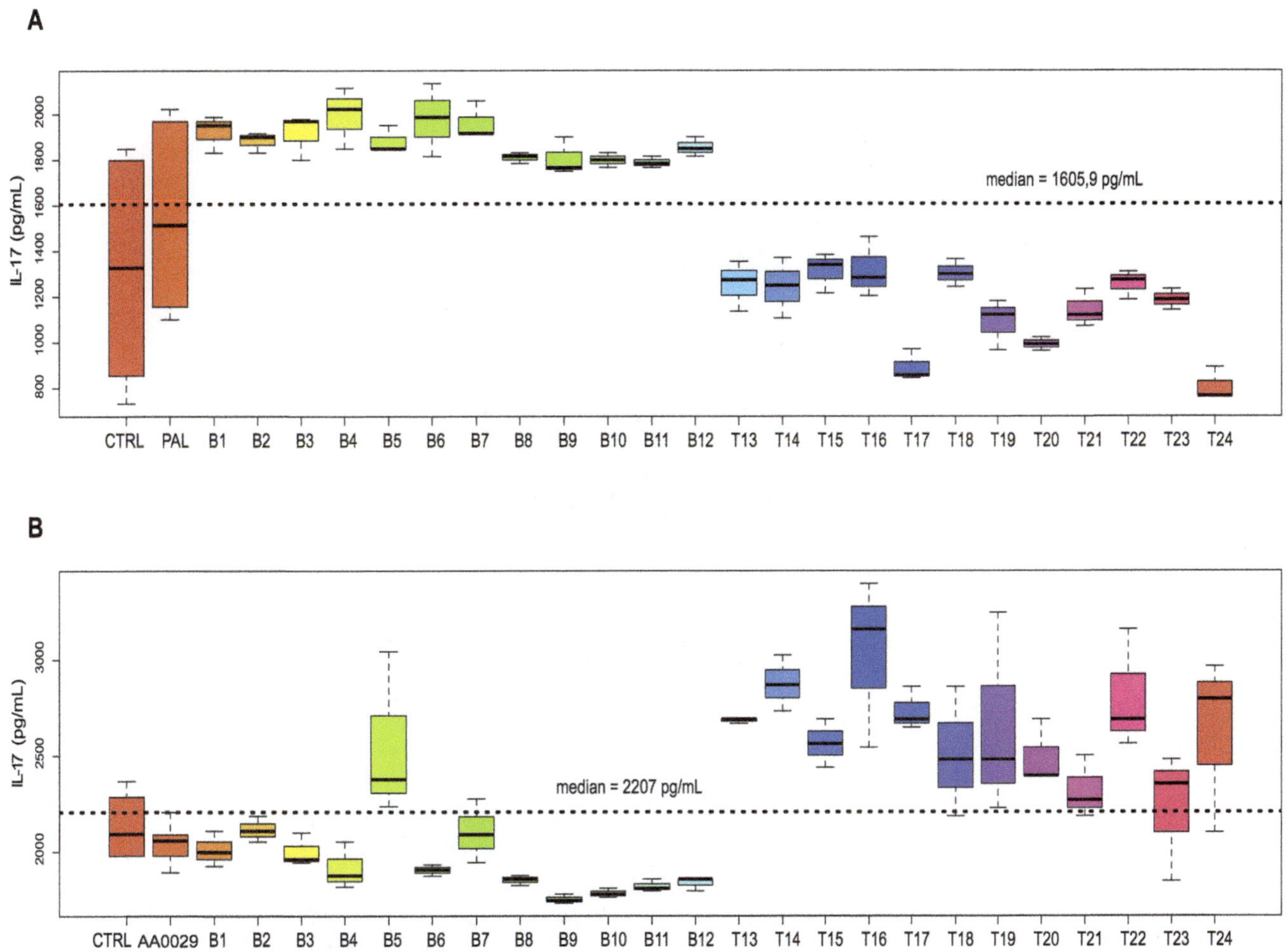

Figure 5. Boxplot showing IL-17 cytokine levels in splenocyte mouse cell culture immunised with synthetic peptides. A). Mice immunised with PAL. B). Mice immunised with AA0029.

(BD Biosystems) was incubated in 1/50 dilution factor in PBS plus 2% foetal calf serum (PBS-FCS) for 30 min at 4°C. The cells were washed with PBS-FCS after the incubation period, spun at 1,200 rpm for 5 min and the supernatant discarded. Splenocytes were fixed with 100 µL of a solution containing 2% p-formaldehyde in PBS for no longer than 12 h at 4°C before data acquisition. Data was collected using a FACSCalibur flow cytometer (BD Biosciences) at the University of Salamanca's Flow Cytometry Central Service. A total of 30,000 events were collected (gated by forward and side scatter) and data were analysed using Gatelogic Flow Cytometry Analysis Software (Inivai Technologies Pty Ltd).

In vivo protection studies: antigens, vaccine formulation and immunisation trials

Based on the immune response induced by each of the peptides assayed, those inducing the following immunological patterns were selected for the *in vivo* protection studies: peptides inducing a high -but not unique- Th1 response; peptides inducing a strong Th2 response; peptides inducing a Treg response, peptides inducing a mixed Th1/Th2 response and peptides inducing a combination of Th1/Th2/Treg/Th17 response. The aforementioned peptides were selected as vaccine candidates due to their high immunogenicity, but taking also into account that immune mechanisms associated with protection against *F. hepatica* are not well understood. Concerning the immunomodulator, the one inducing the highest overall immune response was selected. Seventy mice were divided into 10 groups as follows: group 1 consisted of untreated and uninfected controls, group 2 untreated and infected controls, group 3 adjuvant-administered and infected controls, group 4 those immunised with peptide B1, group 5 mice immunised with peptide B2, group 6 immunised with peptide B5, group 7 immunised with peptide B6, group 8 immunised with peptide T14, group 9 immunised with peptide T15 and group 10 immunised with peptide T16. The mice were subcutaneously immunised using the adjuvant adaptation (ADAD) system as previously described.

Experimental infection and protection assessment

All the animals included in this study (except untreated controls) were orally infected with 7 *F. hepatica* metacercariae two weeks after the last immunisation. *F. hepatica* metacercariae were provided by Ridgeway Research Ltd (Gloucestershire, U. K.) and were stored at 4°C on 0.4% carboxymethylcellulose until use. Metacercariae viability was confirmed by microscope observation before infection. Human endpoints were used when an evidence of severe pain, excessive distress, suffering or an impending death was observable in any of the animals and then euthanised with an intraperitoneal injection of pentobarbital at 60 mg/kg using 30 g

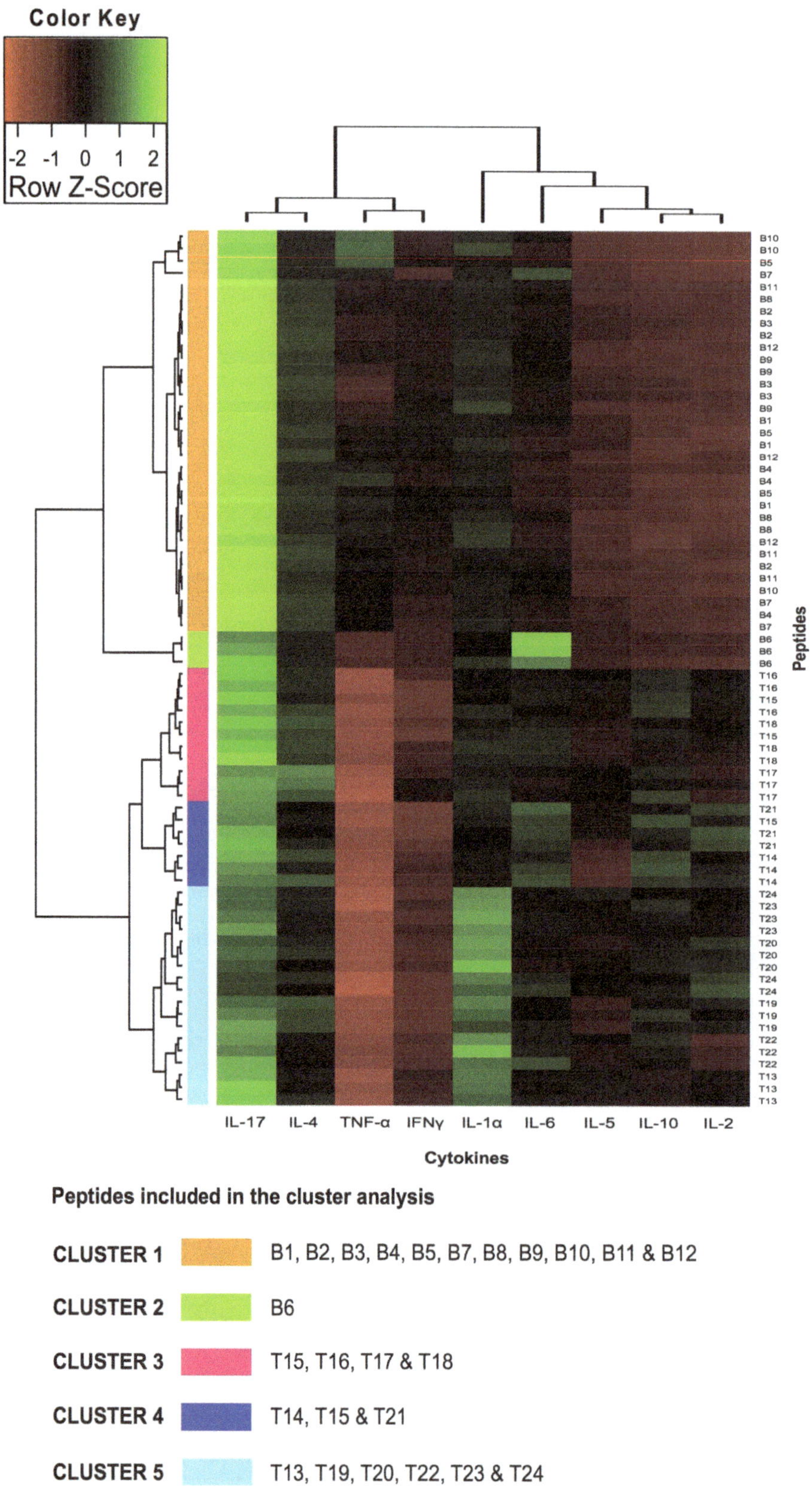

Figure 6. Individual cytokine expression levels are represented by shades of green to red in the central heatmap (highest values shown in green and lowest in dark red). The right-hand margin provides the names of peptide sets. Rows and columns represent clusters of interleukins and peptides having a similar immunological response. A list of cytokines grouped within each cluster is also provided. Groupings having shorter distances (as indicated by the distance to k-means nearest group) had greater similarity. Bicluster analysis for B- and T-cell peptides formulated with the PAL immunomodulator. Five major clusters can be discerned (1, 2, 3, 4 and 5), encompassing peptides having similar cytokine levels.

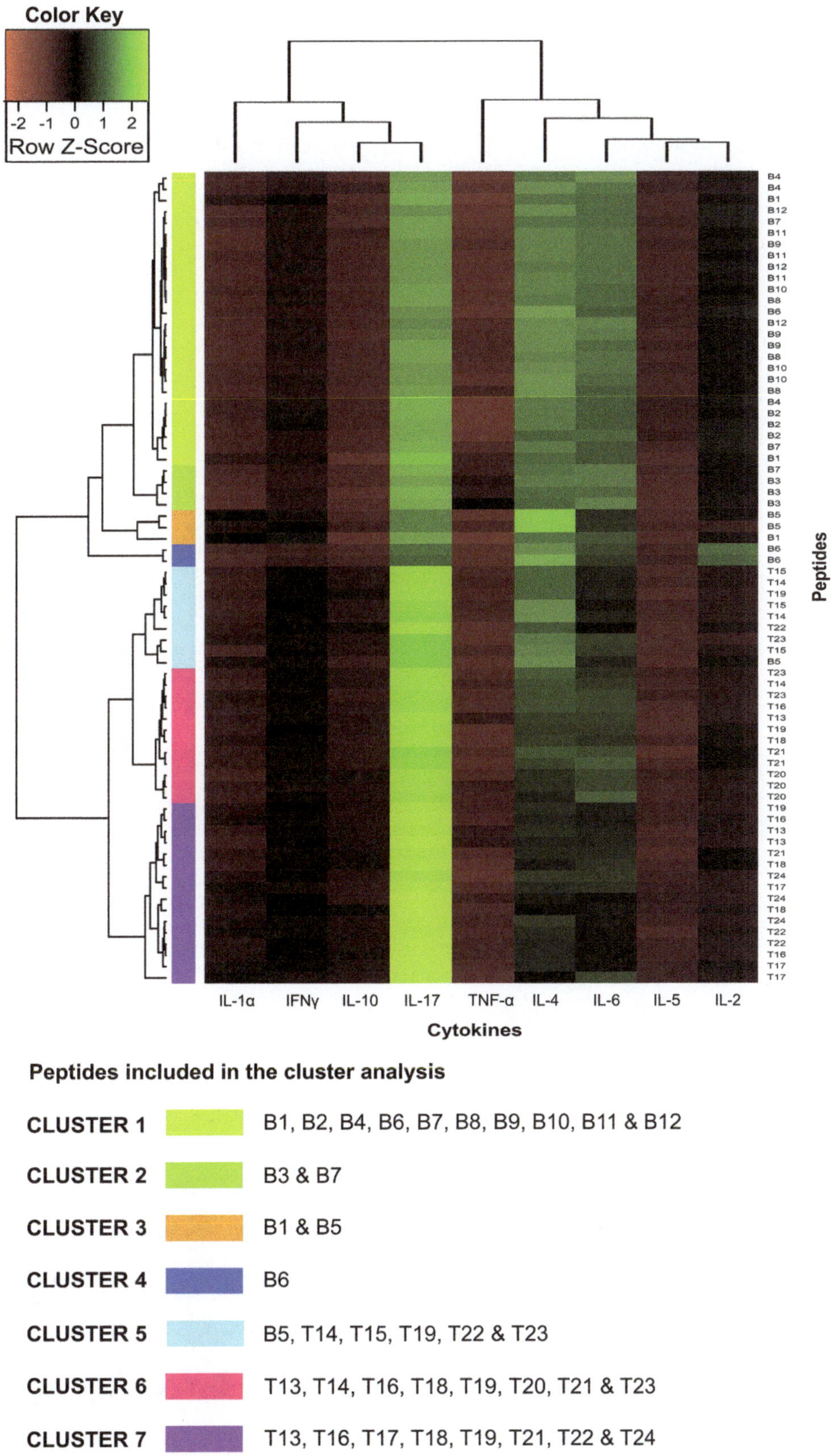

Figure 7. Individual cytokine expression levels are represented by shades of green to red on the central heatmap (highest values shown in green and lowest in dark red). The right-hand margin provides the name of peptide sets. Rows and columns represent clusters of interleukins and peptides having a similar immunological response. A list of cytokines grouped within each cluster is also provided. Groupings having shorter distances (as indicated by the distance to k-means nearest group) had greater similarity. Bicluster analysis for B- and T-cell peptides formulated with the AA0029 immunomodulator. Seven major clusters can be discerned (1, 2, 3, 4, 5, 6 and 7), encompassing peptides having similar cytokine levels.

needles. On day 42 post infection mice remaining alive were then humanely euthanised and necropsied to recover flukes from liver, score hepatic damage and study the survival rates. Mice welfare was evaluated daily according to behaviour and appearance, physiological indicators and other general clinical signs. Two experienced pathologists independently evaluated liver lesions without knowing which group the livers belonged to. Changes regarding size, colour, consistency concerning blood vessels, bile ducts and surface wounds were evaluated; a score was assigned to each feature: 0 points if no lesions was observed, 1 point if a liver lobe was affected, 2 if an entire lobe was affected and 3 if more than 1 lobe was affected. No lesions was assigned when the sum was 0 points, mild (+) for 1–5 points, moderate (++) for 6–10 points and severe (+++) for 11–14 points. Survival rate percentage in mice infected was then calculated as the ratio of the number of surviving experimental mice on day 42 and the total of experimental mice in each group.

Statistical analysis

An initial descriptive analysis was made of each group of mice, exploring cytokine and immunoglobulin patterns and cell expression. A linear model with interaction was used to evaluate cytokine and cell expression, including two factors: epitope type (B or T) and the immunomodulator used (AA0029 or PAL). Interactions were plotted for representing this linear model using Plotrix package in R. Furthermore, groups of peptides were compared using a Kruskall Wallis test followed by a Dunn multiple comparisons test. Differences having $p<0.05$ were considered statistically significant. The results were reported as the mean of each group and the standard deviation (SD). SPSS 20.0 software (SPSS Inc., USA) was used for data analysis. Hierarchical clustering was used to identify sets of cytokines whose expression levels correlated among peptides with B- and T-cell epitopes. Centring and scaling are the previous data transformation steps. The complete linkage clustering method was used, based on a similarity matrix derived from Pearson (rows) and Spearman (columns) moment correlations. The heatmap was visualised using the heatmap.2 function of gplots package in R. Kaplan-Meier survival curves were used for evaluating survival rates.

Results

Selecting candidate proteins, epitope prediction and peptide synthesis

A total of 269 reported *F. hepatica* protein sequences were accessible at and downloaded from NCBI; these proteins were individually predicted for both SP and TM domains for selecting those likely to be secreted. According to such criteria, 21 proteins were finally chosen, aligned and grouped into six different families. Most were found to be cathepsins and cathepsin-like proteases. Once selected, the proteins were then analysed for determining B- and T-cell epitopes. Twenty-four linear peptides were chemically

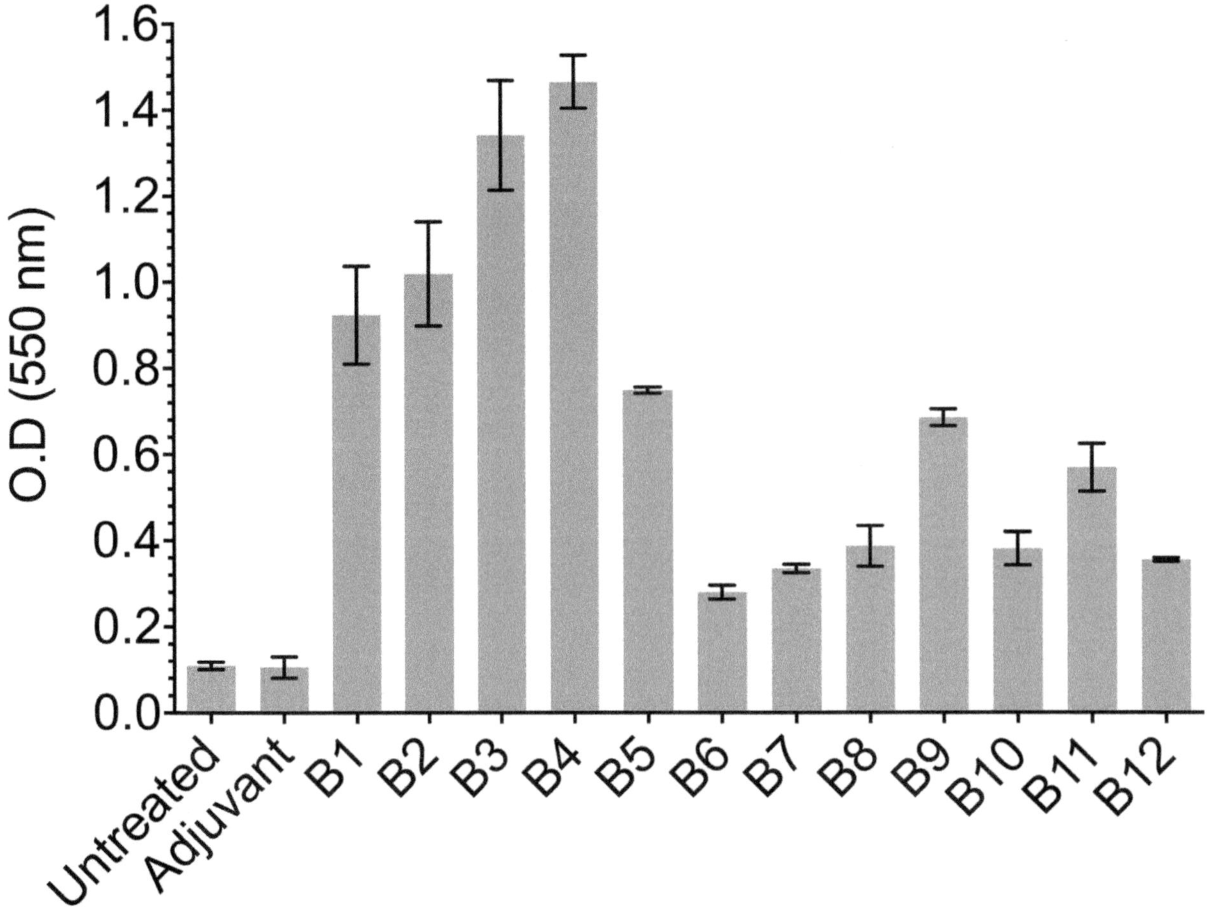

Figure 8. ELISA showing B-cell epitope-containing synthetic peptide reactivity against a pool of sera from mice infected with *F. hepatica* metacercariae.

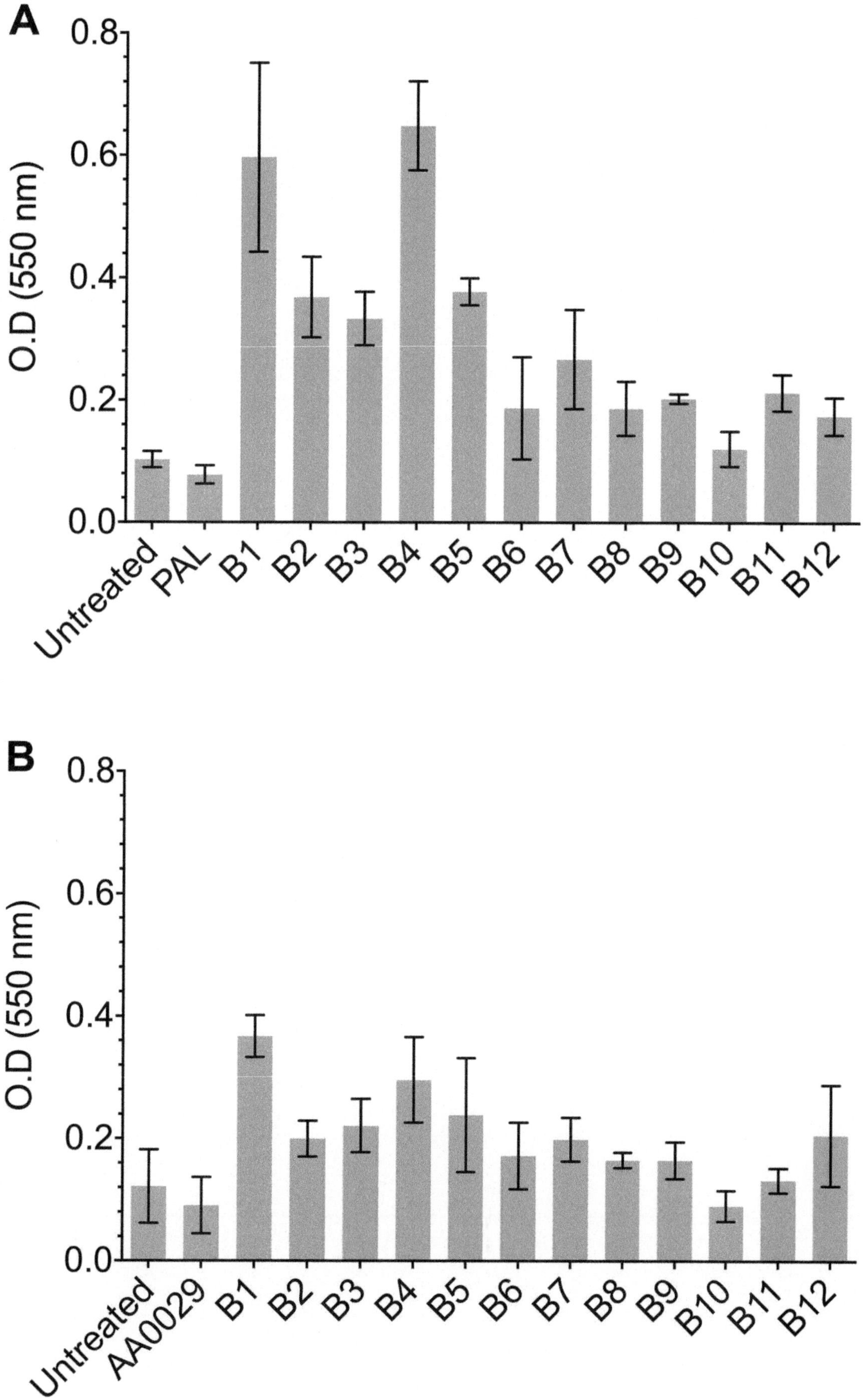
A
0.8
0.6
0.4
0.2
0.0
O.D (550 nm)
Untreated
PAL
B1
B2
B3
B4
B5
B6
B7
B8
B9
B10
B11
B12
B
0.8
0.6
0.4
0.2
0.0
O.D (550 nm)
Untreated
AA0029
B1
B2
B3
B4
B5
B6
B7
B8
B9
B10
B11
B12

Figure 9. ***F. hepatica*** **excretory/secretory antigen reactivity against hyper-immune sera from mice immunised with synthetic peptides containing B-cell epitopes as assessed by ELISA.** A). B-cell peptides formulated with PAL. B). B-cell peptides formulated with AA0029.

synthesised according to *in silico* epitope prediction: twelve 20-amino-acid-long peptides for B-cell epitopes (peptides B1 to B12) and twelve 15-amino-acid-long peptides for T-cell epitopes (peptides T13 to T24). Four peptide pairs (B1–B2, B3–B4, B7–B8 and B11–B12) were synthesised in which a cysteine residue was replaced by a threonine in one peptide of the pair. Taking into account that peptides used for immunisation were synthesised adding a glycine and a cysteine at both termini and then polymerised inducing the formation of disulphide bridges, the presence of any additional cysteine residue in a monomer-synthesised peptide could lead to several polymerised forms being obtained from the peptide. To avoid this issue, and to obtain a well-defined polymerised form, we replaced all cysteine residues in the peptide sequence by threonine ones (a residue sharing similar biochemical properties). More than 5 mg of each purified peptide was obtained and its purity was found to be higher than 90%. Table 1 shows the amino acid sequence of each B- or T-cell epitope, indicating which proteins belonged to each peptide.

In vitro peptides cytotoxicity evaluation

Each peptide was assayed in four different concentrations ranging from 1 to 100 µg/mL for *in vitro* cytotoxicity evaluation. Some peptides' hydrophobic nature resulted in them having low solubility in PBS buffer. Such peptides needed DMSO to be added at a concentration no higher than 2%. Controls for checking solvent cytotoxicity were added to the study (data not shown). Results showed that none of the peptides were toxic for macrophages below 50 µg/mL and more than 90% were still viable after treatment (*Fig. S1*). Some T-peptides showed decreased cell viability at the highest concentration (100 µg/mL)

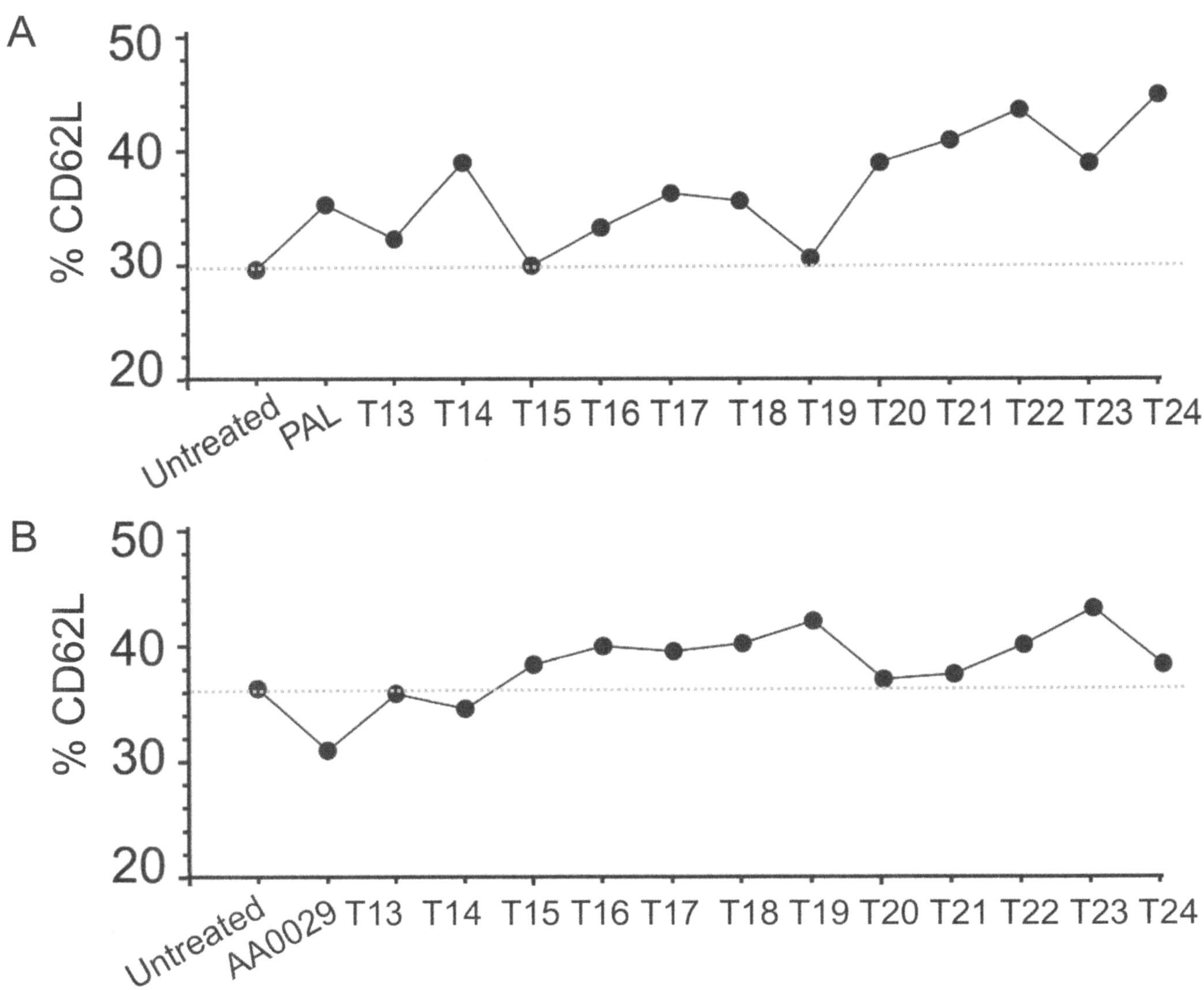

Figure 10. Flow cytometry analysis for quantitation of T-lymphocyte population through CD62L$^+$ T-cell memory marker in mice immunised with T-cell epitope-containing synthetic peptides. A). T-cell peptides formulated with PAL. B). T-cell peptides formulated with AA0029.

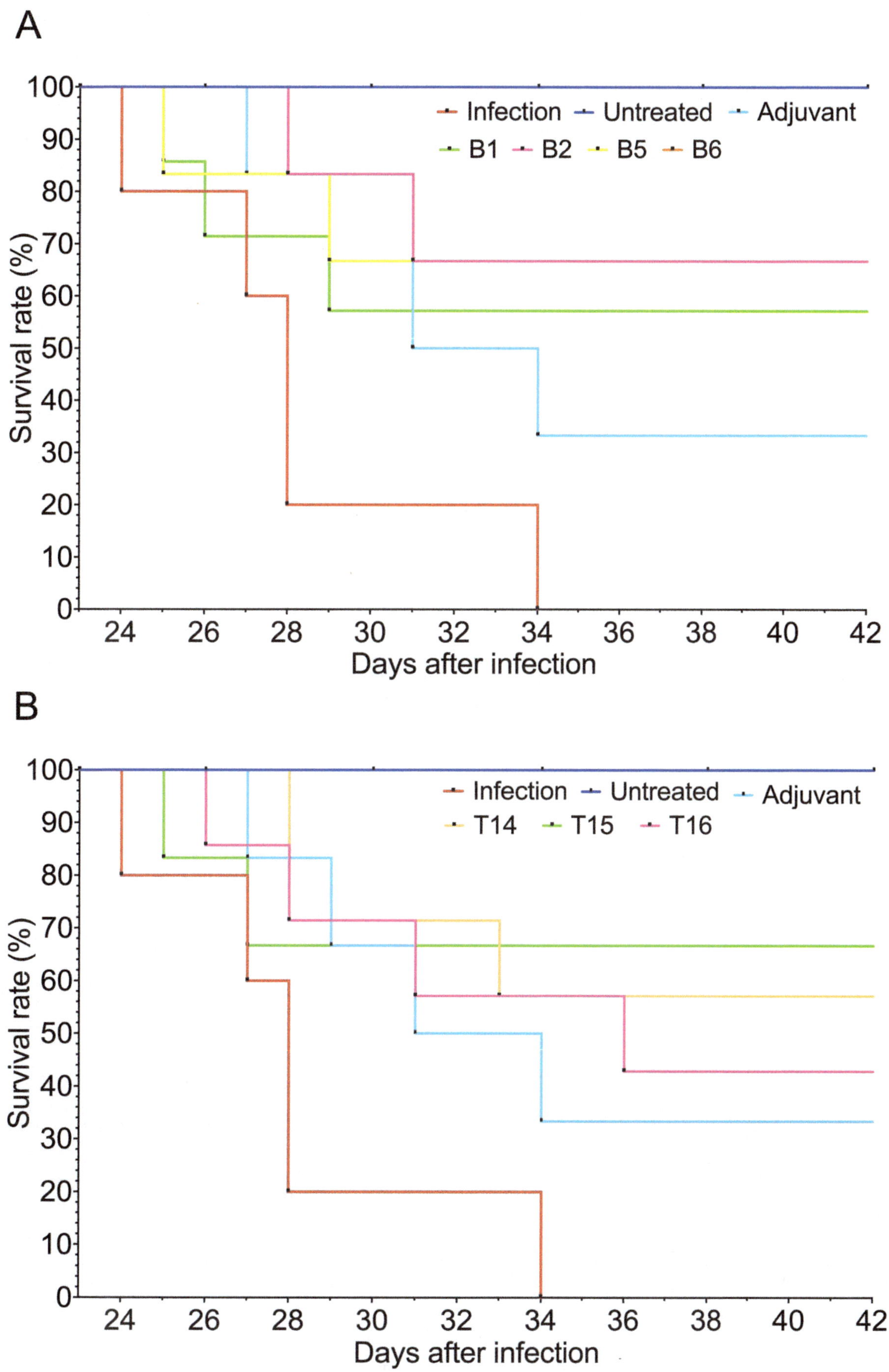
A
Survival rate (%)
Days after infection
Infection
Untreated
Adjuvant
B1
B2
B5
B6
B
Survival rate (%)
Days after infection
Infection
Untreated
Adjuvant
T14
T15
T16

Figure 11. Kaplan-Meier curves depicting survival rates in mice immunised with single peptides and infected with *F. hepatica* metacercariae. A). Mice immunised with single peptides containing B-epitopes. B). Mice immunised with single peptides containing T-epitopes. Survival rates of mice belonging to both untreated and infected controls are also represented. Human endpoint was established when an indicator of severe pain, excessive distress, suffering or an impending death was observed in any of the animals and then euthanised with an intraperitoneal injection of pentobarbital at 60 mg/kg using 30 g needles.

and only 80% of the macrophage population was viable. However, the maximum peptide concentration reached in *in vivo* immunisation assays was no higher than 10 μg/mL.

The effect on immune response induced by immunising mice with synthetic peptides: antibody response and cytokine levels

Immunomodulator (PAL and AA0029), epitope (B or T) and peptide effect on antibody levels were studied. Regarding the immunomodulator effect, it was observed that using PAL induced higher IgG levels than AA0029 ([0.5267] *cf* [0.4332] $p = 0.003$). Significantly higher total IgG levels were obtained when B-epitope-containing peptides were used compared to T-epitope-containing peptides, regardless of adjuvant effect ([0.5258] *cf* [0.4332] $p = 0.05$). The boxplot in *Fig. 1A*, showed that peptides B1, B2, B3, B4, T18 and T20 induced the highest levels of IgG after being administered in mice using the PAL compared to the control group ($p<0.05$), whilst peptides B1, B3, B4, B5, B6, T13, T15, T18, T20 and T21 induced the highest levels after being injected using the AA0029 compared to the control group ($p< 0.05$) (*Fig. 1B*). Concerning those peptides with similar amino acid sequences (different in one amino acid only, where a cysteine was replaced by a threonine), significant differences in IgG levels were only observed in the B1–B2 peptide pair, being the peptide B1 more immunogenic than peptide B2 when using any of the immunomodulators. Regarding IgG subclasses, the following relationships were described through linear regression adjustment: IgG1 with total IgG (*Fig. 2A*) and IgG2a with total IgG (*Fig. 2B*). It was observed that PAL induced significantly higher IgG1 levels than AA0029 ($p<0.05$) in immunised mice (*Fig. 2A*). IgG2a levels were not influenced by the immunomodulator used, since there were no statistical differences between the slopes (*Fig. 2B*). Peptides B1, B3 and B4 induced the highest IgG1 and IgG2a levels with the PAL immunomodulator.

There were no statistical differences regarding IgE levels when either of the immunomodulators was used. Furthermore, B-epitope-containing peptides showed a stronger IgE response than T-epitope-containing peptides ([0.1588] *cf* [0.0952] $p<0.01$). The same pattern was observed in anti-peptide IgM antibodies, as there were no statistical differences when comparing immunomodulator effect; however, B-epitope-containing peptides had a stronger response than T-epitope-containing ones ([0.1161] *cf* [0.1014] $p<0.05$). The boxplot showing the IgE and IgM anti-peptide for each peptide has been included in *Figs. S2 & S3*.

The type of immunomodulator and epitope used in the immunisation trial influenced cytokine levels, as can be seen in the three-dimensional scatterplots in *Figs S4 & S5*. Specifically, using AA0029 with B-epitope-containing peptides led to obtaining high levels of IFN-γ, IL-4, IL-10 and IL-17, whilst using peptides containing T-epitopes and the AA0029 immunomodulator produced higher IFN-γ, IL-4 and IL-17 levels. Changes in cytokine level may thus have resulted from the interaction between the immunomodulator (PAL or AA0029) and the epitope (B or T) being used (*Fig. S6*).

Each peptide was also individually analysed for cytokine quantitation after being injected into the mice using the ADAD vaccination system. The following relationships were described: IL-4 and IL-10 levels (*Fig. 3A*) and IL-5 and IL-10 levels (*Fig. 3B*) in all the mice used in this study. Figure 3A and *Fig. 3B* show that most T-epitope-containing peptides induced high IL-10 and IL-5 levels, but lower IL-4 levels. Increased IL-4 levels were observed in mice immunised with peptides B5, B6, T14 and T15 compared to the AA0029 adjuvant-vaccinated group and also peptides B6 and B12 when the PAL immunomodulator was used. Peptides T16, T22 and T14 induced changes in IL-10 levels when these peptides were formulated with AA0029, whilst T15, T16, T22 and T23 did so with PAL. Concerning IL-5 levels, peptides B1, B2, B3, B4, B6, B7 and T17 produced the highest IL-5 levels with AA0029, whereas peptides T13, T15, T16, T17 and T20 did so with PAL. Regarding Th1 cytokine profile, the boxplot in *Fig. 4* shows IFN-γ levels; these became increased for peptides B1, B5, T15 and T16 compared to the AA0029 adjuvant-vaccinated group, whilst peptides B1, B5, B8, B9 and B12 produced changes in IFN-γ when PAL was used. Concerning peptides with similar amino acid sequence (Cys by Thr substitutions), higher IFN-γ levels were obtained with peptide B1 (compared to peptide B2), using either immunomodulator. Peptide B8 also induced higher levels of IFN-γ than peptide B7 but only when using the PAL immunomodulator. Changes in IL-17 levels were produced by peptides T14, T16 and T22 when AA0029 was used and by peptides B1, B3, B4, B6 and B7 when using PAL (*Fig. 5*). These results clearly showed that some peptides stimulated the production of cytokines associated with different T-helper cell profiles. It has been observed that the AA0029 immunomodulator has induced a more powerful, but not unique, Th2-like immune response. B-peptides only had a Th1-type immune response for peptide B1 whilst T-peptides elicited a mixed Th1/Th2 immune response when the natural PAL immunomodulator was used.

Although analysing individual cytokines identified peptides associated with different cytokine level patterns, such associations did not represent the relationship between the peptides or cytokines included in this study. Two-dimensional cluster analysis was therefore performed to identify sets of cytokines that might have been coordinately expressed induced by immunisation with peptides having a low immune response (compared to a high one) and more strongly correlate them with an effective immune response. Bicluster analysis led to a comprehensive representation of splenocyte state throughout their response to peptides used in this study. The expressed cytokines' functional concordance gave biological significance to the broad patterns seen in images like the biclusters in *Figs. 6 & 7* for PAL and AA0029, respectively. Peptides containing B- and T-cell epitopes clustered separately: cluster 1–2 and cluster 3–5 for the PAL adjuvant (*Fig. 6*) and cluster 1–4 and cluster 5–7 for adjuvant AA0029 (*Fig. 7*). It was also observed that cytokine correlation depended on the adjuvant used in vaccine formulation. PAL-adjuvant formulation clustered IL-17, IL-4, IFN-γ and TNF-α separately from IL-1α, IL-6, IL-5, IL-10 and IL-2 cytokines. AA0029-adjuvant formulations induced different cytokine co-expression pattern; IL-1α, IFN-γ, IL-10 and IL-17 clustered separately from TNF-α, IL-4, IL-6, IL-5 and IL-2. This result confirmed the strong immunomodulator effect conferred by the adjuvant being used.

Analysis of B-epitopes clustered with the PAL adjuvant (*Fig. 6*) revealed that cluster 1 was able to cluster each B-epitope-

Table 2. Recovered flukes and assessment of macroscopic hepatic lesions in CD1 mice immunised with synthetic peptides containing B- and T-cell epitopes using the ADAD vaccination system and challenged with 7 *F. hepatica* metacercariae.

Group	Treatment	Number of flukes in individual mice	Worm recovery (Mean ± SEM)	Reduction (%)	Hepatic lesion in individual mice	Lesion score (Mean ± SEM)	Reduction (%)
1	Untreated uninfected						
2	Infected	3, 1, 2, 2, 2	2.0±0.3		14, 14, 12, 10, 10	12.0±1.5	
3	Adjuvant treated	0, 2, 0, 2, 2, 0	1.0±0.4	50	10, 14, 5, 11, 14, 2	9.3±1.8	22
4	B1	1, 0, 1, 2, 1, 1, 1	1.0±0.2	50	10, 12, 6, 11, 6, 9, 11	9.3±1.5	23
5	B2	0, 1, 1, 1, 1, 1	0.8±0.2	58	6, 14, 5, 9, 9, 5	8.0±1.3	33
6	B5	1, 0, 1, 0, 2, 0	0.7±0.3	67*	4, 1, 12, 12, 7, 8	7.3±2.1	39*
7	B6	2, 0, 2, 1, 0, 1	1.0±0.3	50	13, 2, 8, 10, 14, 13	10.0±1.7	17
8	T14	0, 1, 2, 1, 1, 0, 1	0.9±0.3	57	4, 7, 10, 13, 14, 9, 10	9.6±1.8	20
9	T15	0, 2, 0, 1, 2, 0	0.8±0.4	58	6, 14, 5, 4, 14, 10	8.8±1.7	26
10	T16	1, 0, 0, 1, 2, 1, 1	0.9±0.3	57	10, 10, 11, 5, 13, 12 14	10.7±1.1	11

*$p<0.05$ compared to infected controls.

containing peptide, except peptide B6, while cluster 2 represented the most interesting cluster in *Fig. 6* because it clustered the three B6 peptide replicates which induced high IL-4, IL-6 and IL-1α levels. Regarding B-peptides clustered with AA0029 (*Fig. 7*), a single cluster (cluster 4) also represented two of the three B6 peptide replicates, which also induced high IL-4 and IL-5 levels. Cluster 3 in *Fig. 7* included peptide B1 that induced high IL-5, IFN-γ and TNF-α levels and peptide B5 that also induced high IL-4 and IFN-γ levels, suggesting different immunological effects induced by such B-peptides regarding the groups that clustered T-epitopes using the PAL adjuvant (*Fig. 7*). Peptides T14, T15 and T16 were included in clusters 3 and 4; they were able to induce high levels of regulatory cytokine IL-10, IL-5 and proinflammatory cytokine IL-2. Regarding T-epitope clustered groups using AA0029 (*Fig. 7*), the peptides were extremely disseminated throughout the different clusters, although T14 and T15 both seemed to group in cluster 5, inducing high IL-4 and IFN-γ levels.

Experimental identification of linear B- and T-cell epitopes

An ELISA assay was used for evaluating the recognition of peptides containing predicted B-cell epitopes using sera from mice infected with *F. hepatica*. Figure 8 shows that peptides containing predicted B-cell epitopes had some reactivity against sera from infected mice, thus indicating the presence of linear epitopes within the synthetic peptides. *F. hepatica* amoebapore and cathepsin B protein-derived B1, B2, B3, B4 and B5 peptides had the highest reactivity, reaching 1:3,200 specific titre values (data not shown). The ELISA depicted in *Fig. 9* shows the reactivity of sera raised against predicted B-cell epitopes formulated in either PAL or AA0029 immunomodulators against *F. hepatica* excretory/secretory antigens. Higher reactivity was observed when using hyper-immune sera from mice immunised with the B-cell epitopes formulated with the PAL immunomodulator (*Fig. 9A*). The highest reactivity was shown by B1, B2, B3, B4 and B5 peptides in both cases, thereby confirming the presence of immunogenic regions within peptides containing predicted B-cell epitopes (*Figs. 9A & 9B*).

Three lymphocyte subsets (LT $CD4^+$, LT $CD8^+$ and LT $CD62L^+$) were analysed to better evaluate the cellular immune response of T-cell epitope-containing synthetic peptides formulated with both the PAL and/or AA0029 immunomodulators. No statistically significant differences were detected in either LT $CD4^+$ or LT $CD8^+$ immunophenotypes in mice immunised with the T-cell epitope-containing peptides. However, differences in the $CD62L^+$ immunophenotype were detected in mice immunised with peptides containing T-cell epitopes formulated in either of the immunomodulators (PAL or AA0029) when compared to the non-immunised control group. Different lymphocyte expression patterns were observed when the peptides were individually analysed. Figure 10 shows the LT $CD62L^+$ levels reached by each group of mice immunised with T-cell epitope-containing peptides formulated with either the PAL (*Fig. 10A*) or AA0029 immunomodulator (*Fig. 10B*). Using the PAL immunomodulator led to higher LT $CD62L^+$ levels being obtained in mice immunised with T-cell epitopes, except those immunised with peptides T13, T15 and T19 (*Fig. 10A*). Concerning the AA0029 immunomodulator, its use led to an increase in the LT $CD62L^+$ population in mice immunised with the T-cell epitopes, except those immunised with peptides T13, T14, T20 and T21 (*Fig. 10B*). The results obtained here confirmed the presence of immunodominant T-cell epitopes in peptides selected using a bioinformatics approach.

In vivo protection assessment

Immunising mice with the synthetic peptides enhanced their survival rate, compared to the unimmunised and infected control group. All mice in the unimmunised and infected control group died between 24 and 34 days pi; mice which died before day 42 pi were considered unprotected in our study. Figure 11 shows that all the groups of mice immunised with the synthetic peptides had some degree of protection against experimental infection with *F. hepatica* metacercariae. The highest survival rate (66.7% in each group) was obtained with the peptides B2, B5, B6 and T15 (groups 5, 6, 7 and 9, respectively), according to Kaplan-Meier estimates (*Figs. 11A & 11B*). Mice immunised with single peptides B1 and T14 (groups 4 and 8, respectively) had a 57.1% survival rate (*Figs. 11A & 11B*) and mice immunised with the single peptide T16 (group 10) had a survival rate of 42.9% (*Fig. 11B)*.

Immunising mice with the peptide B5 (group 6) produced the lowest fluke burden compared to the unimmunised and infected control group ([0.7±0.3] *cf* [2.0±0.3]) ($p<0.05$). Mice immunised with the single peptides B2 and T15 decreased worm burden by 58% compared to the unimmunised and infected control group. Concerning liver damage score, mice immunised with the single peptide B5 showed the lowest hepatic damage when compared to the infected control group ([7.3±2.1] *cf* [12.0±01.5]) ($p<0.05$). Mice immunised with the peptide B2 also reduced the liver damage ([8.0±1.3] *cf* [2.0±0.3]) ($p = 0.05$). These results suggest a direct correlation between the survival percentage with lower liver fluke burdens and liver damage scores (Table 2).

Discussion

The present study was designed to bioinformatically predict and experimentally assess both the immune response and protection-inducing ability of *F. hepatica* protein-derived peptides containing B- and T-cell epitopes. The newly-emergent discipline of bioinformatics, immunoinformatics and rational vaccine design were used here, involving a reverse vaccinology strategy which has been successfully used in malaria and some helminthic infections [11,14,17–21]. According to reports in the pertinent literature, cysteine proteases released by *F. hepatica* play a key role in parasite feeding, migration through host tissues and immune evasion and they are considered good vaccine candidates [32]. Cysteine proteases have been used as vaccine candidates in immunisation trials involving experimental models such as cattle and sheep and using different adjuvants, resulting in up to 79% protection [33–38]. The rationale behind selecting proteins that could be secreted by *F. hepatica* has been based on the fact that these proteins are exposed to a host's immune system, making them easily recognisable. Most epitopes included in this study belonged to either the cathepsin L or cathepsin B family, eliciting high activation of the immune response. Cathepsin B is predominantly released during the juvenile life-cycle stage whilst cathepsin L is released throughout the whole cycle, thereby representing an important issue when designing a multi-stage, multi-epitope, subunit-based vaccine against *F. hepatica* [32]. Other studies have also shown the importance of two *F. hepatica* cathepsins (FheCB and FheCL), expressed in the infective stage of newly excysted juveniles (NEJs), in acquiring infection. Silencing the expression of both these proteins by using the interference RNA (iRNA) methodology induced less movement and penetration of NEJs to the gut [39]. The present study has demonstrated that immunisation with peptides containing B- and T-cell epitopes formulated with the AA0029 immunomodulator induced a strong immune response -based on high IL-4 and IFN-γ cytokine levels-, these being important biomarkers in Th2 and Th1 differentiation [40]. This pattern could be essential for using these immunomodulator in vaccination trials where a more powerful Th1 or Th2 immune response would be required for protection against disease. Furthermore, AA0029 was able to stimulate some pro-inflammatory and Th17 cytokines, such as IL-6 and IL-17, while PAL previously tested in other immunisation experiments was associated with a down-regulated Th2 immune response [31]. PAL, together with T-epitopes, was able to induce high IL-10 and IL-5 cytokine levels and PAL plus B-epitopes seemed to better stimulate IgG1, the antibody subtype classically belonging to the Th2 immune response profile. However, other components in the ADAD vaccination system used here could modify or highly modulate the immune response, such as the adjuvant Qs21 and the non-mineral oil Montanide. These components could increase antigen immunogenicity, could be used to enhance immune response speed and duration and could stimulate cell-mediated immunity [41]. According to reports in the pertinent literature, Qs21 acts as an immunostimulatory adjuvant involved in Th1 cytokine (i.e. IL-2 and IFN-γ) and IgG2a isotype antibody induction [42]. It is well-known that cattle are highly susceptible to primary infection but are resistant to reinfection, thus generating chronically and silent *F. hepatica* infection [43]. The mechanisms associated with such tolerance are mediated by host's humoral and cellular Th2 immune responses.

A prolonged Th2 immune response to *F. hepatica* infection is thus not protective and can lead to the parasite residing within the liver or passageways conducting bile. According to the literature, the mechanism behind chronic *F. hepatica* infection development involves IgG1 antibody generation and little or no IgG2a [44,45]. However, an antibody response against *F. hepatica* cathepsin L as vaccine candidate involves generating both IgG1 and IgG2a antibodies, suggesting a mixed Th1/Th2 protective immune response [34,44]. The results obtained here showed that most peptides produced both IgG1 and IgG2a antibodies (having higher IgG1 titres in all cases), showing a mixed Th1/Th2 immune response, whilst induced IgE and IgM immunoglobulin levels resulted in very low levels for all the synthetic peptides used in this study. Helminth infection has been associated with IgE response; however, its role in protective immunity is not well understood [46]. There is some evidence for a positive correlation between IgE level and protection in human populations infected with *Schistosoma mansoni* and *Schistosoma haematobium* [47]. Despite the low IgE levels obtained by immunisation with synthetic peptides in the present study, some peptides were able to induce higher levels compared to the adjuvant-vaccinated group.

Cytokine quantitation in stimulated-splenocyte supernatant cell cultures has indicated that B- or T-cell epitope-containing peptides have been able to induce high levels of Th1 and Th2-associated immune response. Peptides inducing high IL-17 levels were also found in the present study. These results were consistent with those from other immunisation trials, resulting in the generation of cytokines involved in a mixed Th1/Th2 immune response [44,48,49]. IL-4 producing T-cell differentiation is an important step in developing an effective protective immune response. IL-4 can directly mediate worm expulsion mechanisms and is required for Th2 cell amplification and IL-5 also plays an import role in mediating host protection against helminth parasites [50]. The present results showed that some peptides such as B5, B6 and T14–17 plus AA0029, induced high IL-4 and IL-5 levels. It is worth noting that the protection induced by many vaccines against helminths, particularly *S. mansoni* experimental infection, has been associated with high IFN-γ and TNF-α production [51,52]. TNF-α might play a role in accelerated worm expulsion through Th2 immune response enhancement [53] and IFN-γ production suppression may mediate parasite survival in fasciolosis [54]. The present study also found that the B1 peptide stimulated both TNF-α and IFN-γ, whilst peptides B5, T15 and T16 induced high IFN-γ levels, classically corresponding to a Th1-like immune response. This study strongly suggested that Th1 cytokines play a central role as biomarkers and should be used for measuring vaccination effectiveness. Some peptides evidently stimulated both IFN-γ and IL-4 cytokines, such as B5 and T15, suggesting their potential as vaccine candidates. Peptides B5 and T16 were the only ones having high IL-17 levels; however, the role of IL-17 has not yet been completely understood in *F. hepatica* infection. IL-17 has been associated with a severe form of the disease in experimental schistosomiasis models [55]. According to the data obtained in the present study, it can be hypothesised that immunological mechanisms participating in conferring protection against infection caused by *F. hepatica* involve Th1, Th2 and Th17 immune responses; the most protection-inducing peptides -according to survival rates, fluke burden and hepatic damage-, stimulated the production of high IFN-γ, IL-4 and IL-17 levels, respectively, after their administration with the AA0029 immunomodulator.

The specific role of $CD4^+$ and $CD8^+$ T-cells in protection against *F. hepatica* infection remains to be clearly elucidated. However, immunosuppression is one of the main mechanisms leading to liver fluke survival in a chronically-infected host, inducing a significant decrease in peripheral blood $CD4^+$ and $CD8^+$ T cells. Previous studies have demonstrated that *F. hepatica* cathepsin L has down-regulated $CD4^+$ from the surface of human T-cells [56]. The more susceptible sheep host has shown reduced T-lymphocyte proliferation during *F. hepatica* infection [57]. Other authors have suggested lymphocyte migration, including $CD4^+$ and $CD8^+$ lymphocytes, from mesenteric lymph nodes to the antigen exposure site. $CD8^+$ T-cells have been particularly evident in fibrotic areas of the liver during chronic infection [58]. It is well-known that helminth parasites are able to live for long periods in a host, immunoregulation being a key factor in this [59]. Some memory T-cells, including central memory T-cells and effector memory T-cells expressing CD197 (CCR7) and $CD62L^+$, respectively, are involved in such immunoregulation [60]. However, it is not yet well understood how immunoregulation works in *F. hepatica*. Nevertheless, other studies have suggested the importance of central memory T-cells in establishing long-term immunity; central memory T-cells are maintained when the parasite is eliminated, but effector memory T-cells are not [61]. Vaccine effectiveness combines the action of both humoral and cellular immune responses; this is why central memory T-cells were also considered in the present study as a key target for peptide screening in the search for vaccine candidates. High $CD62L^+$ expression levels have been shown to be protective in various infectious diseases [61,62]. Importantly, higher $CD62L^+$ percentages were obtained with T-epitopes in this study using either of the immunomodulators.

Most recent work aimed at developing an effective vaccine against *F. hepatica* has been based on recombinant antigens, parasite purified proteins, excretory/secretory released proteins and DNA-based vaccines. However, little is known concerning the development of subunit-based, chemically-synthesised anti-*F. hepatica* vaccines [63,64]. This article has demonstrated that immunising CD1 mice with subunit-based and chemically-synthesised peptides containing B- or T-epitopes formulated in the ADAD vaccination system using the AA0029 immunomodulator led to obtaining protection against infection caused by *F. hepatica*, based on survival rates, reduced fluke burden and hepatic lesion score. Despite the use of a murine model for evaluating vaccine candidates against fasciolosis is still controversial, as mice are highly susceptible to the infection and animals could die during experimentation, the survival rate was here considered as a clear protection-inducing indicator, since all mice belonging to the unimmunised and infected control group, died between days 24 and 34 post-infection.

The present study's findings have highlighted the immunoprophylactic potential of B- and T-cell epitope-containing synthetic peptides (B2, B5, B6 and T15) formulated in the ADAD vaccination system in a murine model. However, it should be noted that a deeper knowledge of the host-parasite interactions, and understanding the molecular and immunological mechanisms involved in inducing a protective response, might be crucial for better selecting the most appropriate vaccine formulation. Further studies aimed at studying the protection-inducing ability of the aforementioned peptides when formulated in combination, and tested in natural fasciolosis models, are in need.

Supporting Information

Figure S1 *In vitro* cell viability using MTT assay after 48 h culture of mouse peritoneal macrophages cell line (J774.2) with B- and T-cell epitope-containing synthetic peptides. Synthetic peptides were assayed in a range from 1 to 100 µg/mL.

Figure S2 IgE antibody level detection in mice immunised with the synthetic peptides throughout the immunisation schedule. Data presented as box plots with the bottom and the top of the box indicating the 25th and 75th percentiles, respectively. A). Peptides formulated with the PAL immunomodulator. B). Peptides formulated with AA0029.

Figure S3 IgM antibody level detection in mice immunised with synthetic peptides throughout the immunisation schedule. Data presented as box plots with the bottom and the top of the box indicating the 25th and 75th percentiles, respectively. A). Peptides formulated with the PAL immunomodulator. B). Peptides formulated with AA0029.

Figure S4 Three-dimensional scatterplots represents cytokine levels induced by immunisation of mice with peptides containing B-cell epitopes. The Z-axis represents IFN-γ, IL-4, IL-10 and IL-17 levels for Figure A, B, C and D, respectively. The x and y axis represent CD197 and CD27 memory T-lymphocytes for each Figure. Blue indicates the use of AA0029 and green indicates PAL.

Figure S5 Three-dimensional scatterplots represents cytokine levels induced by immunisation of mice with peptides containing T-cell epitopes. The Z-axis represents IFN-γ, IL-4, IL-10 and IL-17 levels for Figure A, B, C and D, respectively. The x and y axis represent CD197 and CD27 memory T-lymphocytes for each Figure. Blue indicates the use of AA0029 and green indicates PAL.

Figure S6 Interaction plot for regulatory (A. IL-10), Th2 (B. IL-5, C. IL-4), Th17 (D. IL-17), Th1 (E. IL-2, F. IFN-γ) and innate inflammatory cytokine levels (G. IL-6, H. TNFα, I. IL-1α) elicited by epitope effect (B & T) and immunomodulator effect (AA0029 & PAL).

Acknowledgments

We would like to thank Magnolia Vanegas (Fundación Instituto de Inmunología de Colombia - FIDIC) for peptide synthesis, Dr Arturo San

Feliciano (Universidad de Salamanca) for kindly providing us with AA0029, Dr Antonio Martínez-Fernández (Universidad Complutense de Madrid) for supplying PAL and Montanide, and Jason Garry for correcting the manuscript.

Author Contributions

Conceived and designed the experiments: JRC JLA LPV CV BV PFS MAP AM. Performed the experiments: JRC JLA BV. Analyzed the data: JRC JLA LPV MAP AM EO. Contributed reagents/materials/analysis tools: JRC JLA BV EO. Wrote the paper: JRC JLA PFS MAP AM.

References

1. Kaplan RM (2001) Fasciola hepatica: a review of the economic impact in cattle and considerations for control. Vet Ther 2: 40–50.
2. Mas-Coma S (2005) Epidemiology of fascioliasis in human endemic areas. J Helminthol 79: 207–216.
3. Mas-Coma S, Valero MA, Bargues MD (2009) Climate change effects on trematodiases, with emphasis on zoonotic fascioliasis and schistosomiasis. Vet Parasitol 163: 264–280.
4. WHO (2011) Initiative to estimate the Global burden of Foodborne Diseases.
5. McManus DP, Dalton JP (2006) Vaccines against the zoonotic trematodes Schistosoma japonicum, Fasciola hepatica and Fasciola gigantica. Parasitology 133 Suppl: S43–61.
6. Hillyer GV (2005) Fasciola antigens as vaccines against fascioliasis and schistosomiasis. J Helminthol 79: 241–247.
7. Fairweather I (2011) Reducing the future threat from (liver) fluke: realistic prospect or quixotic fantasy? Vet Parasitol 180: 133–143.
8. Sette A, Rappuoli R (2010) Reverse vaccinology: developing vaccines in the era of genomics. Immunity 33: 530–541.
9. Purcell AW, McCluskey J, Rossjohn J (2007) More than one reason to rethink the use of peptides in vaccine design. Nat Rev Drug Discov 6: 404–414.
10. De Groot AS, Berzofsky JA (2004) From genome to vaccine–new immunoinformatics tools for vaccine design. Methods 34: 425–428.
11. De Groot AS (2006) Immunomics: discovering new targets for vaccines and therapeutics. Drug Discov Today 11: 203–209.
12. Caoili SE (2010) Benchmarking B-cell epitope prediction for the design of peptide-based vaccines: problems and prospects. J Biomed Biotechnol 2010: 910524.
13. Arnon R, Ben-Yedidia T (2003) Old and new vaccine approaches. Int Immunopharmacol 3: 1195–1204.
14. Patarroyo ME, Cifuentes G, Bermudez A, Patarroyo MA (2008) Strategies for developing multi-epitope, subunit-based, chemically synthesized anti-malarial vaccines. J Cell Mol Med 12: 1915–1935.
15. Vlieghe P, Lisowski V, Martinez J, Khrestchatisky M (2010) Synthetic therapeutic peptides: science and market. Drug Discov Today 15: 40–56.
16. Hans D, Young PR, Fairlie DP (2006) Current status of short synthetic peptides as vaccines. Med Chem 2: 627–646.
17. Read AJ, Casey JL, Coley AM, Foley M, Gauci CG, et al. (2009) Isolation of antibodies specific to a single conformation-dependant antigenic determinant on the EG95 hydatid vaccine. Vaccine 27: 1024–1031.
18. Robinson K, Bellaby T, Chan WC, Wakelin D (1995) High levels of protection induced by a 40-mer synthetic peptide vaccine against the intestinal nematode parasite Trichinella spiralis. Immunology 86: 495–498.
19. Madhumathi J, Prince PR, Anugraha G, Kiran P, Rao DN, et al. (2010) Identification and characterization of nematode specific protective epitopes of Brugia malayi TRX towards development of synthetic vaccine construct for lymphatic filariasis. Vaccine 28: 5038–5048.
20. de Aluja AS, Villalobos NM, Nava G, Toledo A, Martinez JJ, et al. (2005) Therapeutic capacity of the synthetic peptide-based vaccine against Taenia solium cysticercosis in pigs. Vaccine 23: 4062–4069.
21. Arnon R, Tarrab-Hazdai R, Steward M (2000) A mimotope peptide-based vaccine against Schistosoma mansoni: synthesis and characterization. Immunology 101: 555–562.
22. Jezek J, El Ridi R, Salah M, Wagih A, Aziz HW, et al. (2008) Fasciola gigantica cathepsin L proteinase-based synthetic peptide for immunodiagnosis and prevention of sheep fasciolosis. Biopolymers 90: 349–357.
23. Martinez-Fernandez AR, Nogal-Ruiz JJ, Lopez-Aban J, Ramajo V, Oleaga A, et al. (2004) Vaccination of mice and sheep with Fh12 FABP from Fasciola hepatica using the new adjuvant/immunomodulator system ADAD. Vet Parasitol 126: 287–298.
24. del Olmo E, Plaza A, Muro A, Martinez-Fernandez AR, Nogal-Ruiz JJ, et al. (2006) Synthesis and evaluation of some lipidic aminoalcohols and diamines as immunomodulators. Bioorg Med Chem Lett 16: 6091–6095.
25. Bendtsen JD, Nielsen H, von Heijne G, Brunak S (2004) Improved prediction of signal peptides: SignalP 3.0. J Mol Biol 340: 783–795.
26. Larsen JE, Lund O, Nielsen M (2006) Improved method for predicting linear B-cell epitopes. Immunome Res 2: 2.
27. Deleage G, Combet C, Blanchet C, Geourjon C (2001) ANTHEPROT: an integrated protein sequence analysis software with client/server capabilities. Comput Biol Med 31: 259–267.
28. Rammensee H, Bachmann J, Emmerich NP, Bachor OA, Stevanovic S (1999) SYFPEITHI: database for MHC ligands and peptide motifs. Immunogenetics 50: 213–219.
29. Merrifield RB (1969) Solid-phase peptide synthesis. Adv Enzymol Relat Areas Mol Biol 32: 221–296.
30. Houghten RA (1985) General method for the rapid solid-phase synthesis of large numbers of peptides: specificity of antigen-antibody interaction at the level of individual amino acids. Proc Natl Acad Sci U S A 82: 5131–5135.
31. Lopez-Aban J, Andrade Mdo A, Nogal-Ruiz JJ, Martinez-Fernandez AR, Muro A (2007) Immunomodulation of the response to excretory/secretory antigens of Fasciola hepatica by Anapsos in Balb/C mice and rat alveolar macrophages. J Parasitol 93: 428–432.
32. Dalton JP, Neill SO, Stack C, Collins P, Walshe A, et al. (2003) Fasciola hepatica cathepsin L-like proteases: biology, function, and potential in the development of first generation liver fluke vaccines. Int J Parasitol 33: 1173–1181.
33. Dalton JP, McGonigle S, Rolph TP, Andrews SJ (1996) Induction of protective immunity in cattle against infection with Fasciola hepatica by vaccination with cathepsin L proteinases and with hemoglobin. Infect Immun 64: 5066–5074.
34. Mulcahy G, O'Connor F, McGonigle S, Dowd A, Clery DG, et al. (1998) Correlation of specific antibody titre and avidity with protection in cattle immunized against Fasciola hepatica. Vaccine 16: 932–939.
35. Piacenza L, Acosta D, Basmadjian I, Dalton JP, Carmona C (1999) Vaccination with cathepsin L proteinases and with leucine aminopeptidase induces high levels of protection against fascioliasis in sheep. Infect Immun 67: 1954–1961.
36. Wedrychowicz H, Kesik M, Kaliniak M, Kozak-Cieszczyk M, Jedlina-Panasiuk L, et al. (2007) Vaccine potential of inclusion bodies containing cysteine proteinase of Fasciola hepatica in calves and lambs experimentally challenged with metacercariae of the fluke. Vet Parasitol 147: 77–88.
37. Jayaraj R, Piedrafita D, Dynon K, Grams R, Spithill TW, et al. (2009) Vaccination against fasciolosis by a multivalent vaccine of stage-specific antigens. Vet Parasitol 160: 230–236.
38. Villa-Mancera A, Mendez-Mendoza M (2012) Protection and antibody isotype responses against Fasciola hepatica with specific antibody to pIII-displayed peptide mimotopes of cathepsin L1 in sheep. Vet J 194: 108–112.
39. McGonigle L, Mousley A, Marks NJ, Brennan GP, Dalton JP, et al. (2008) The silencing of cysteine proteases in Fasciola hepatica newly excysted juveniles using RNA interference reduces gut penetration. Int J Parasitol 38: 149–155.
40. Agnello D, Lankford CS, Bream J, Morinobu A, Gadina M, et al. (2003) Cytokines and transcription factors that regulate T helper cell differentiation: new players and new insights. J Clin Immunol 23: 147–161.
41. Singh M, O'Hagan D (1999) Advances in vaccine adjuvants. Nat Biotechnol 17: 1075–1081.
42. Kensil CR (1996) Saponins as vaccine adjuvants. Crit Rev Ther Drug Carrier Syst 13: 1–55.
43. Haroun ET, Hillyer GV (1986) Resistance to fascioliasis–a review. Vet Parasitol 20: 63–93.
44. Mulcahy G, O'Connor F, Clery D, Hogan SF, Dowd AJ, et al. (1999) Immune responses of cattle to experimental anti-Fasciola hepatica vaccines. Res Vet Sci 67: 27–33.
45. Hoyle DV, Taylor DW (2003) The immune response of regional lymph nodes during the early stages of Fasciola hepatica infection in cattle. Parasite Immunol 25: 221–229.
46. Negrao-Correa D (2001) Importance of immunoglobulin E (IgE) in the protective mechanism against gastrointestinal nematode infection: looking at the intestinal mucosae. Rev Inst Med Trop Sao Paulo 43: 291–299.
47. Hagan P (1993) IgE and protective immunity to helminth infections. Parasite Immunol 15: 1–4.
48. Espino AM, Osuna A, Gil R, Hillyer GV (2005) Fasciola hepatica: humoral and cytokine responses to a member of the saposin-like protein family following delivery as a DNA vaccine in mice. Exp Parasitol 110: 374–383.
49. Espino AM, Rivera F (2010) Quantitation of cytokine mRNA by real-time RT-PCR during a vaccination trial in a rabbit model of fascioliasis. Vet Parasitol 169: 82–92.
50. Gause WC, Urban JF Jr, Stadecker MJ (2003) The immune response to parasitic helminths: insights from murine models. Trends Immunol 24: 269–277.
51. Cardoso FC, Macedo GC, Gava E, Kitten GT, Mati VL, et al. (2008) Schistosoma mansoni tegument protein Sm29 is able to induce a Th1-type of immune response and protection against parasite infection. PLoS Negl Trop Dis 2: e308.
52. Farias LP, Cardoso FC, Miyasato PA, Montoya BO, Tararam CA, et al. (2010) Schistosoma mansoni Stomatin like protein-2 is located in the tegument and induces partial protection against challenge infection. PLoS Negl Trop Dis 4: e597.
53. Artis D, Humphreys NE, Bancroft AJ, Rothwell NJ, Potten CS, et al. (1999) Tumor necrosis factor alpha is a critical component of interleukin 13-mediated protective T helper cell type 2 responses during helminth infection. J Exp Med 190: 953–962.

54. Flynn RJ, Mulcahy G (2008) The roles of IL-10 and TGF-beta in controlling IL-4 and IFN-gamma production during experimental Fasciola hepatica infection. Int J Parasitol 38: 1673–1680.
55. Rutitzky LI, Smith PM, Stadecker MJ (2009) T-bet protects against exacerbation of schistosome egg-induced immunopathology by regulating Th17-mediated inflammation. Eur J Immunol 39: 2470–2481.
56. Prowse RK, Chaplin P, Robinson HC, Spithill TW (2002) Fasciola hepatica cathepsin L suppresses sheep lymphocyte proliferation in vitro and modulates surface CD4 expression on human and ovine T cells. Parasite Immunol 24: 57–66.
57. Zimmerman GL, Kerkvliet NI, Brauner JA, Cerro JE (1983) Modulation of host immune responses by Fasciola hepatica: responses by peripheral lymphocytes to mitogens during liver fluke infections of sheep. J Parasitol 69: 473–477.
58. McCole DF, Doherty ML, Baird AW, Davies WC, McGill K, et al. (1999) T cell subset involvement in immune responses to Fasciola hepatica infection in cattle. Parasite Immunol 21: 1–8.
59. Taylor MD, van der Werf N, Maizels RM (2012) T cells in helminth infection: the regulators and the regulated. Trends Immunol 33: 181–189.
60. Totte P, Duperray C, Dedieu L (2010) CD62L defines a subset of pathogen-specific bovine CD4 with central memory cell characteristics. Dev Comp Immunol 34: 177–182.
61. Zaph C, Uzonna J, Beverley SM, Scott P (2004) Central memory T cells mediate long-term immunity to Leishmania major in the absence of persistent parasites. Nat Med 10: 1104–1110.
62. Bustamante JM, Bixby LM, Tarleton RL (2008) Drug-induced cure drives conversion to a stable and protective CD8+ T central memory response in chronic Chagas disease. Nat Med 14: 542–550.
63. Harmsen MM, Cornelissen JB, Buijs HE, Boersma WJ, Jeurissen SH, et al. (2004) Identification of a novel Fasciola hepatica cathepsin L protease containing protective epitopes within the propeptide. Int J Parasitol 34: 675–682.
64. Vilar MM, Barrientos F, Almeida M, Thaumaturgo N, Simpson A, et al. (2003) An experimental bivalent peptide vaccine against schistosomiasis and fascioliasis. Vaccine 22: 137–144.

MHC Class I-Presented T Cell Epitopes Identified by Immunoproteomics Analysis Are Targets for a Cross Reactive Influenza-Specific T Cell Response

James S. Testa, Vivekananda Shetty, Julie Hafner, Zacharie Nickens, Shivali Kamal, Gomathinayagam Sinnathamby, Ramila Philip*

Immunotope, Inc., Doylestown, Pennsylvania, United States of America

Abstract

Influenza virus infection and the resulting complications are a significant global public health problem. Improving humoral immunity to influenza is the target of current conventional influenza vaccines, however, these are generally not cross-protective. On the contrary, cell-mediated immunity generated by primary influenza infection provides substantial protection against serologically distinct viruses due to recognition of cross-reactive T cell epitopes, often from internal viral proteins conserved between viral subtypes. Efforts are underway to develop a universal flu vaccine that would stimulate both the humoral and cellular immune responses leading to long-lived memory. Such a universal vaccine should target conserved influenza virus antibody and T cell epitopes that do not vary from strain to strain. In the last decade, immunoproteomics, or the direct identification of HLA class I presented epitopes, has emerged as an alternative to the motif prediction method for the identification of T cell epitopes. In this study, we used this method to uncover several cross-specific MHC class I specific T cell epitopes naturally presented by influenza A-infected cells. These conserved T cell epitopes, when combined with a cross-reactive antibody epitope from the ectodomain of influenza M2, generate cross-strain specific cell mediated and humoral immunity. Overall, we have demonstrated that conserved epitope-specific CTLs could recognize multiple influenza strain infected target cells and, when combined with a universal antibody epitope, could generate virus specific humoral and T cell responses, a step toward a universal vaccine concept. These epitopes also have potential as new tools to characterize T cell immunity in influenza infection, and may serve as part of a universal vaccine candidate complementary to current vaccines.

Editor: Geetha P. Bansal, Tulane University, United States of America

Funding: This work is supported by National Institutes of Health SBIR grant 1 R43 AI091232-01. The funders had no role in study design, data collection and analysis, decision to publish, or preparation of the manuscript.

Competing Interests: Provisional United States patent application: #US 61/548,985 Title: MHC class I associated peptides for prevention and treatment of Influenza virus infection. All authors are employed by Immunotope, Inc. There are no further patents, products in development or marketed products to declare.

* E-mail: rphilip@immunotope.com

Introduction

Influenza virus is a significant public health problem internationally, causing three to five million cases of severe illness, and an estimated 250,000 to 500,000 deaths annually [1]. Influenza virus is a member of orthomixovirdae and its genome is comprised of eight segments of negative single stranded RNA [2]. Viral strains are divided into A, B, and C viruses and differ serologically only between the HA and NA proteins [3]. Influenza constantly modifies these glycoproteins by implementing antigenic drift and shift [4], which is the main reason for influenza pandemics and the requirement for seasonal vaccines. The immune response to influenza is governed by both innate and adaptive immunity and has been well-studied. The humoral arm of the adaptive immune response utilizes secretory IgA and IgM to provide protection against the establishment of initial infection, while IgG acts to neutralize newly replicating virus [5,6]. Improving humoral immunity to influenza is the target of current conventional influenza vaccines, however, they are generally not cross-protective [7]. Cell-mediated immunity, on the other hand, as elicited by major histocompatibility complex (MHC) class I-restricted CD8+ cytotoxic T lymphocytes (CTLs), plays a central role in controlling influenza virus infection [8–11] [12]. Indeed, cell-mediated immunity generated by primary influenza infection provides substantial protection against serologically distinct viruses due to the recognition of cross-reactive epitopes, often from internal viral proteins conserved between viral subtypes [13–15].

Tremendous efforts are underway to develop a universal flu vaccine that would work against all types of influenza. Such a universal vaccine should target conserved influenza virus antibody and T cell epitopes that do not vary from strain to strain [16]. Unfortunately, most conserved viral proteins lie within the virus, out of reach of antibodies. With a focus on antibody-mediated protection, attempts are being made to use a part of the external M2 protein (M2e) in addition to the HA stalk region, both of which are highly conserved among human influenza type A viruses [17–20]. Primarily, motif prediction methodology is used to identify shared T cell epitopes [21], however, there is evidence in the literature that a high number of predicted epitopes are not processed and presented by infected cells [22]. In the last decade, immunoproteomics, or the direct identification of HLA class I presented epitopes from infected cells, has emerged as an alternative to the motif prediction method [23–26]. These analyses are generally based on the isolation of the HLA-peptide complexes, elution of bound peptides from HLA molecules, and examination using mass spectrometry [25,27–31].

In this study, we have identified T cell epitopes naturally presented on influenza A-infected cells using immunoproteomics. By surveying the MHC/peptide complexes present on the surface of influenza A-infected cells, we have identified novel conserved epitopes from various surface and internal influenza proteins. These shared T cell epitopes, when combined with a cross-reactive antibody epitope, such as the M2e peptide, generate cross-strain specific cell mediated and humoral immunity, which would be a positive step towards a potential universal vaccine for influenza infection.

Methods

Ethics Statement

All of the animal experiments were conducted according to the recommendations in the Guide for the Care and Use of Laboratory Animals of the National Institutes of Health. The protocol was approved by the Lampire Biologicals Institutional Animal Care and Use Committee (IACUC Permit - Core Rodent Immunotope 2012 Approved 5-31-2012).

Mice

Six to eight-week old female HLA-A2 transgenic mice were purchased from Taconic (Strain HLA-A2.1, CB6F1-Tg(HLA-A*0201/H2-Kb)A*0201).

Virus

Non-irradiated, sucrose-purified influenza virus strains A/PR/8/34(H1N1) (PR8), and A X:31, A Aichi/68(H3N2) (X31) were purchased from Charles River laboratories. Influenza strain A2/Japan/305/57(H2N2) (JAP) was a generous gift from Laurence Eisenlohr and used as unpurified allantoic fluid.

Cells

HepG2, hepatoma cells, JY, EBV transformed lymphoblastoid B cells, and T2, lymphoblasts, were obtained from ATCC. HepG2 were maintained in DMEM:F12 medium while JY, and T2 were maintained in RPMI 1640 (Mediatech). All culture medium was supplemented with 10% fetal bovine serum, L-glutamine (300 mg/mL), non-essential amino acids (1× concentration), 0.5 mM sodium pyruvate, penicillin and streptomycin (1× concentration, supplements were purchased from Mediatech) [complete medium]. All cell lines were maintained at 37°C in a humidified incubator with 5% CO2.

Dendritic cells (DC) were generated from leukophereses obtained from HLA-A2+ healthy donors (Research Blood Components, LLC) as previously described [32]. Briefly, peripheral blood mononuclear cells (PBMC) were purified according to standard methods and, after overnight culture, were treated with 100 ng/mL IL-4 and 25 ng/mL Granulocyte Monocyte Colony Stimulating Factor (GM-CSF) (both cytokines from Peprotech or eBiosciences) for six days prior to their infection with influenza virus.

Analysis of Infection

To verify infection, HepG2, JY, and monocyte-derived human DC were pulsed with 1000HAU/1×10^6 cells of purified virus (PR8, X31) or 200HAU/1×10^6 cells of unpurified virus (JAP) in PBS+.1% BSA for approximately 1 hr at 37°C. 16 hrs post-infection, cells were stained for intracellular nucleoprotein (NP) and were analyzed using a Guava flow cytometer (Millipore). Cells were stained with the primary antibody, HB65 (a generous gift from Dr. Laurence Eisenlohr), followed by FITC labeled anti-mouse secondary antibody (Invitrogen). The stained cells were analyzed using the Guava flow cytometer accompanying software.

Isolation, Purification and Fractionation of MHC Class I Bound Peptides

JY cells were grown to about 1×10^9 cells. Monocyte-derived DCs were generated over a period of six days to about 5×10^8 cells. These cells were then infected with PR8 at 1000HAU/1×10^6 cells. Influenza-infected cells were then processed as previously described [33]. Briefly, cell lysates were prepared in buffer containing 1% NP40 (IBI-Scientific), and MHC/peptide complexes were isolated by immunoaffinity chromatography using W6/32 antibody- (monoclonal antibody recognizing pan MHC class I molecule) coated protein A/G beads (UltraLink Immobilized Protein A/G, Pierce). The bound MHC/peptide complexes were eluted from the beads by the addition of 10% acetic acid. The eluate was heated at 85°C for 15 min to dissociate the bound peptides from the MHC molecules. Peptides were separated from the antibody by centrifugation using Amicon Ultra-3 kDa molecular mass cutoff membrane filters (Millipore). The purified peptide mixture was fractionated using a C-18 reversed phase (RP) column on an offline ultimate 3000 HPLC (Dionex).

Mass Spectrometry Analysis

Mass spectrometry experiments were carried out using an Orbitrap (Thermo) instrument interfaced with nano ultimate HPLC (Dionex). RP-HPLC purified peptide fractions were injected individually into the LC–MS/MS system to identify the sequences of the peptides. The peptides were analyzed using a Data-Dependent method and the acquired spectra data were searched against the influenza protein database using Proteome Discoverer (Thermo) to interpret data and derive peptide sequences. For validation, synthetic peptides were obtained from Genscript Corporation and subjected to LC-MS/MS analysis under identical experimental conditions. Their sequences were confirmed based on their MS/MS data with that of their synthetic analogs.

Generation of CTL

Peptide-specific T cells were generated as previously described [32]. Briefly, Peripheral blood mononuclear cells (PBMC) were purified from heparinized blood obtained from healthy HLA-A2+ donors (Research Blood Components, LLC.) using lymphocyte separation medium (Mediatech). PBMC were pulsed with 50 µg/mL synthetic peptide and 1.5 µg/mL human β2-microglobulin (EMD Biosciences) overnight followed by stimulation with IL-7 at 5 ng/mL, Keyhole Limpet Hemocyanin (KLH, Sigma-Aldrich) at 5 µg/mL, GM-CSF at 25 ng/mL and IL-4 at 50 ng/mL (all cytokines and growth factors were purchased from Peprotech or eBiosciences). Two weeks after initial stimulation, T cells were restimulated with autologous CD4+/CD8+ T cell-depleted PBMCs pulsed with synthetic peptide at 10 µg/mL and 1.5 µg/mL human β2-microglobulin in complete medium containing 5 ng/mL IL-15, 12.5 ng/mL GM-CSF, and 10 U/mL IL-2 for 7 days. Restimulation was repeated three times at the indicated time intervals prior to use in ELISpot assays.

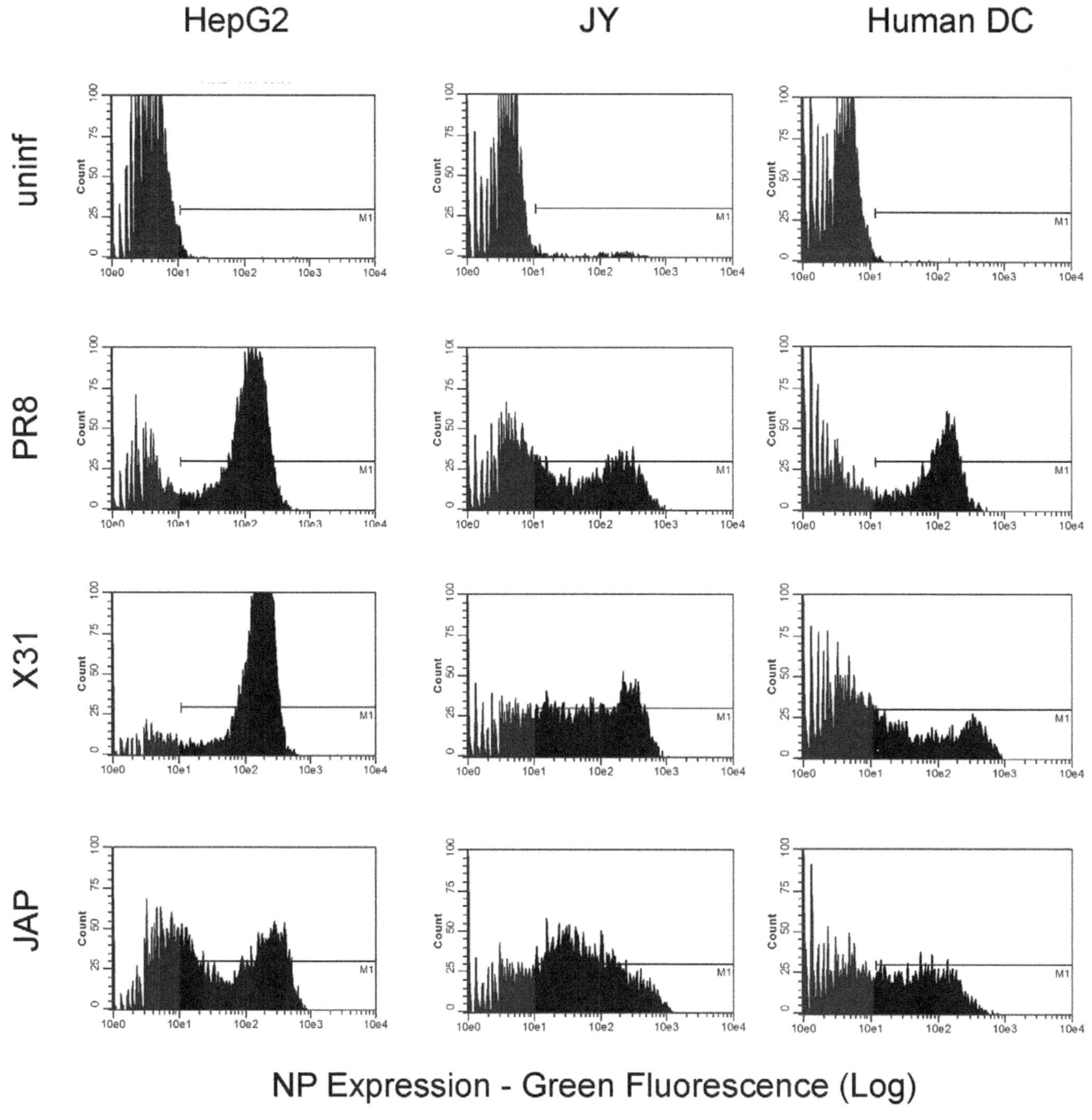

Figure 1. Influenza strain infection analysis. HepG2, JY, and monocyte-derived human DC were pulsed with 1000HAU/1 $\times 10^6$ cells of purified virus (PR8, X31) or 200HAU/1 $\times 10^6$ cells of unpurified virus (JAP) in PBS+.1% BSA for approximately 1 hr at 37°C. Cells were fixed and intracellularly stained for influenza NP after overnight incubation.

ELISpot Assay

Antigen stimulated interferon-γ (IFN-γ) release as a measure of CTL activation was assayed using an ELISpot assay kit (BD-Pharmingen) according to the manufacturer's instructions. Typically, a fixed number of various target cells and effector cells at an effector to target ratio of 100:1 for PBMC or 10:1 for murine splenocytes were cultured in replicate wells overnight. Spots were quantitated using an ELISpot Reader System (AID). Results are presented as the number of interferon-γ producing cells per 1×10^6 PBMCs/splenocytes. Each assay was performed with PBMCs from at least three different healthy HLA-A2+ donors. Error bars represent SEM of experimental replicates.

In vivo Analysis

HLA-A2 transgenic mouse experiments were conducted at Lampire Biologicals. Their initial inoculation consisted of a mixture of pooled free peptide in PBS plus Montanide ISA 51 (Seppic) (50:50 emulsion). In addition to the T cell epitopes, an antibody epitope derived from the ectodomain of influenza Matrix 2 protein (pM2e, MSLLTEVETPTRNEWESRSSDSSD) [17,18]

Table 1. List of MHC class I presented influezna virus specific peptides identified by immunoproteomics analysis.

Peptide Number	Peptide	Protein	Influenza strain	Accession ID	Motif	m/z	Charge	Xcorr	Score	
									Bimas[†]	SYFPEITHI*
P1	YINTALLNA	polymerase PA	H5N1	gi168805480	A2	496.8	2	1.8	7.2	20
		polymerase PA	H1N1	A3DRP7						
		polymerase PA	H3N2	O91742						
		polymerase PA	H3N8	P13169						
		polymerase PA	H4N6	P13172						
P2	TVIKTNMI	polymerase PB1	H5N1	gi172053048	A2	460.3	2	2.1	-	-
		PB1 polymerase subunit	H9N2	Q9Q0S3						
		RNA-directed RNA polymerase catalytic subunit	H7N3	E9P6L9						
		RNA-directed RNA polymerase catalytic subunit	H1N1	A6M777						
P3	PVAGGTSSIYI	polymerase PB2	H3N2	gi215480628	A2	603.3	2	2.4	1.4	19
		polymerase PB2	H1N2	Q75TA1						
P4	MTIIFLILM	hemagglutinin	H2N3	gi257123295	A2	1094.6	1	1.7	1.0	15
P5	AIMDKNIIL	nonstructural protein 1	H1N1	gi138898	A2	515.8	2	2.1	18.3	23
		nonstructural protein 1	H3N8	Q77ZM3						
		nonstructural protein 1	H2N8	O57276						
		nonstructural protein 1	H5N9	O41649						
		nonstructural protein 1	H4N6	Q6LDH2						

[†]Calcluated score based on halftime for dissociation of the peptide from HLA molecule [51].
*Binding affinity scores calculated using SYFPEITHI database [52].

at 10 μg each was selectively injected. Mice were injected at two sites; i.d. near the base of the tail and s.c. on the flank. Injections were repeated two more times at 10-day intervals. One week after the third inoculation, mice were bled, sacrificed, and spleens were harvested for analysis.

Spleens were homogenized and RBCs lysed. Next, cells were restimulated *in vitro* with peptides as previously described [34] and used in an ELISpot assay at an E:T ratio of 10:1 with uninfected or infected HepG2 or JY cells and peptide-loaded T2 cells as targets.

Flow cytometry was also carried out on *in vitro* stimulated splenocytes as previously described [35,36]. After overnight incubation, cells were stained with directly conjugated antibodies against murine CD8 and CD107a (eBiosciences). Samples were analyzed using a Guava flow cytometer and accompanying software by gating CD8+ cells and graphing the mean fluorescence of CD107a for each group.

Anti-M2e Analyses

Serum samples from terminal bleeds were analyzed for IgG specific for M2e peptide using standard ELISA techniques. Additionally, serum samples were used to measure the presence of the M2e antibody epitope on the surface of infected cells. HepG2 cells were infected with PR8, X31, and JAP viruses as described above. After overnight incubation, cells were stained with serum samples at a 1:50 dilution followed by FITC-labeled anti-mouse IgG (Invitrogen) secondary antibody. Samples were analyzed using a Guava flow cytometer and accompanying software.

The neutralization function of anti-M2e antibodies in the serum of peptide-immunized mice was measured by adding serum to influenza infected HepG2 cells. HepG2 cells were infected with a suboptimal dose of PR8 (30HAU/1×10^6 cells) in the presence of 1:50 serum. After the initial pulse, serum was added into the culture at a 1:50 dilution for an overnight incubation. The next day, samples were intracellularly stained for NP as described above.

Results

Productive Infection with Influenza Virus for Epitope Discovery

To uncover influenza specific epitopes by immunoproteomics analysis, we chose JY cells and monocyte-derived human DCs as professional APCs, and nonprofessional HepG2 cells, which express the most globally prevalent MHCI molecule, HLA-A*0201 [37] (www.allelefrequency.net). [33]. We first established the infectivity of these cell types *in vitro* by infecting the cells with infectious viral strains A/PR/8/34(H1N1) (PR8), A X-31, Aichi/68(H3N2) (X31), or A2/Japan/305/57(H2N2) (JAP) and assessed nucleoprotein expression [14] (Figure 1). As demonstrated, protein levels were very high, and would prove ideal for immunoproteomic probing and discovery of naturally processed and presented influenza T cell epitopes derived from viral proteins.

Identification of Influenza-specific MHC Class I Peptides by Nano-LC/MS/MS Analysis

Employing an immunoproteomics strategy, we identified 6 A2, 10 B44, and 6 B7 motif containing peptides (Table S1). The data including their sequences, virus strains and proteins, HLA-A2, HLA-B44, and HLA-B7 motif anchor residues, and the binding affinity/half time dissociation scores are given in Table S1. Prior

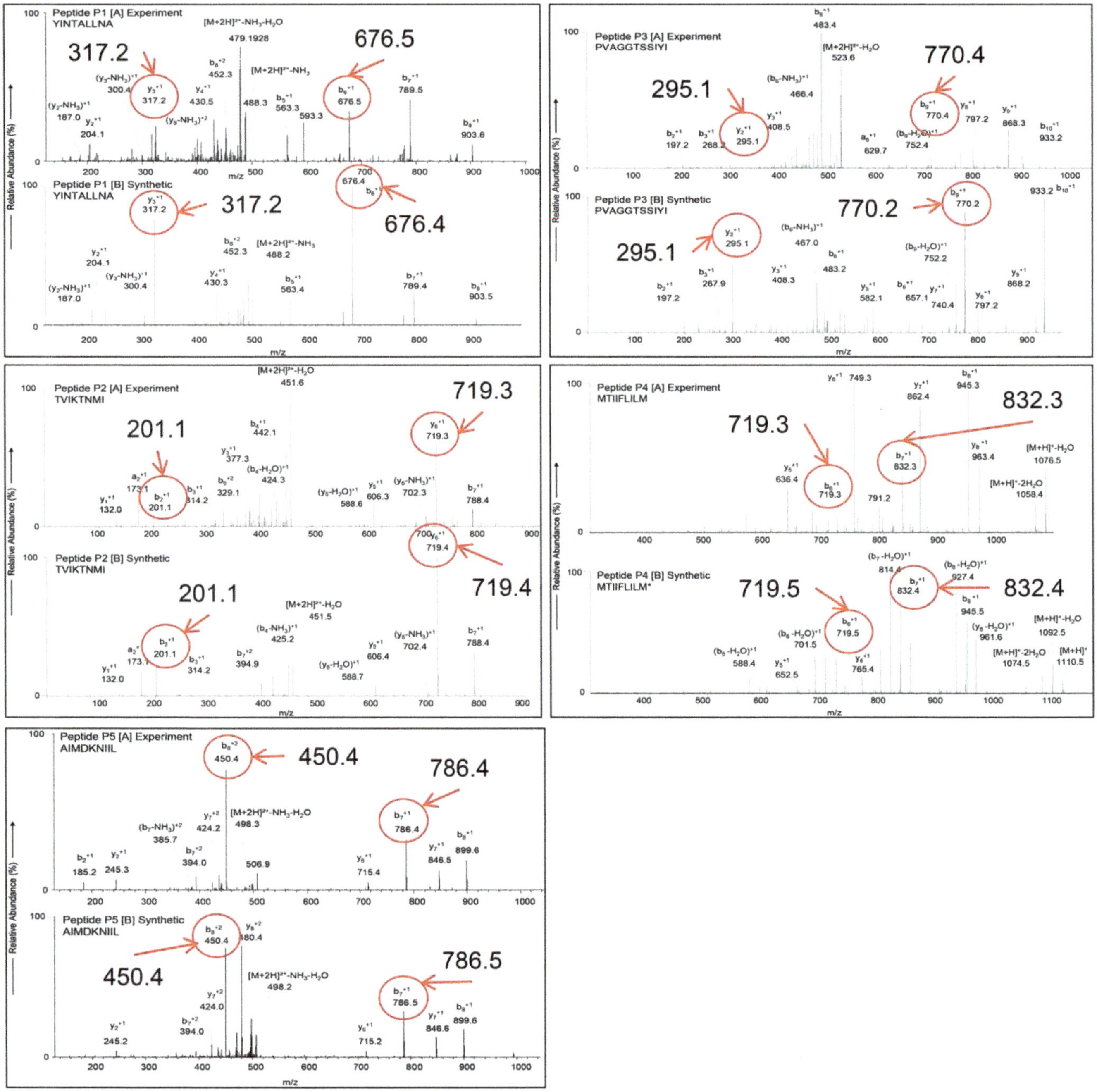

Figure 2. Synthetic Validation of Peptides. MS/MS spectra of experimentally identified MHC peptides [P1A–P5A] and their synthetic analogs [P1B–P5B]. Example fragment masses that match are denoted (circled).

to CTL characterization experiments, we confirmed the authenticity of 5 HLA-A2 specific peptides, P1–P5 (Table 1) using their synthetic peptide analogs. As illustrated in Figure 2, most of the fragment ions in the ms/ms spectra of experimentally identified peptides (Figure 2: P1-A–P5-A) matched with spectra of their corresponding synthetic peptides as indicated by the denoted masses (Figure 2: P1-B–P5-B). In the case of P4 peptide, because of the methionine oxidative modification in the synthetic peptide, the corresponding ms/ms spectrum show a mass shift of 16 Da in all the 'y' fragment ions (y_5 and y_6), precursor ion $[(M+H)^+]$, and the ions $[(M+H)^+\text{-}H_2O]$ and $[(M+H)^+\text{-}2H_2O]$, due to water losses from the precursor ion (Table 1 P4 [B]). In addition, we further verified the ms/ms spectra manually to confirm the identity of all the experimentally observed peptides.

CTL Characterization of Influenza Peptides

To verify the presentation of these epitopes by infected cells, CTLs specific for each of the 5 peptides were generated using PBMCs from healthy HLA-A2+ donors and synthetic peptides corresponding to the identified epitopes. In ELISpot assays, CTL functionality was measured by detection of antigen specific IFNγ secretion. As illustrated in Figure 3*A*, PR8-infected JY and HepG2 cells stimulated all five of the influenza epitope-specific T cells.

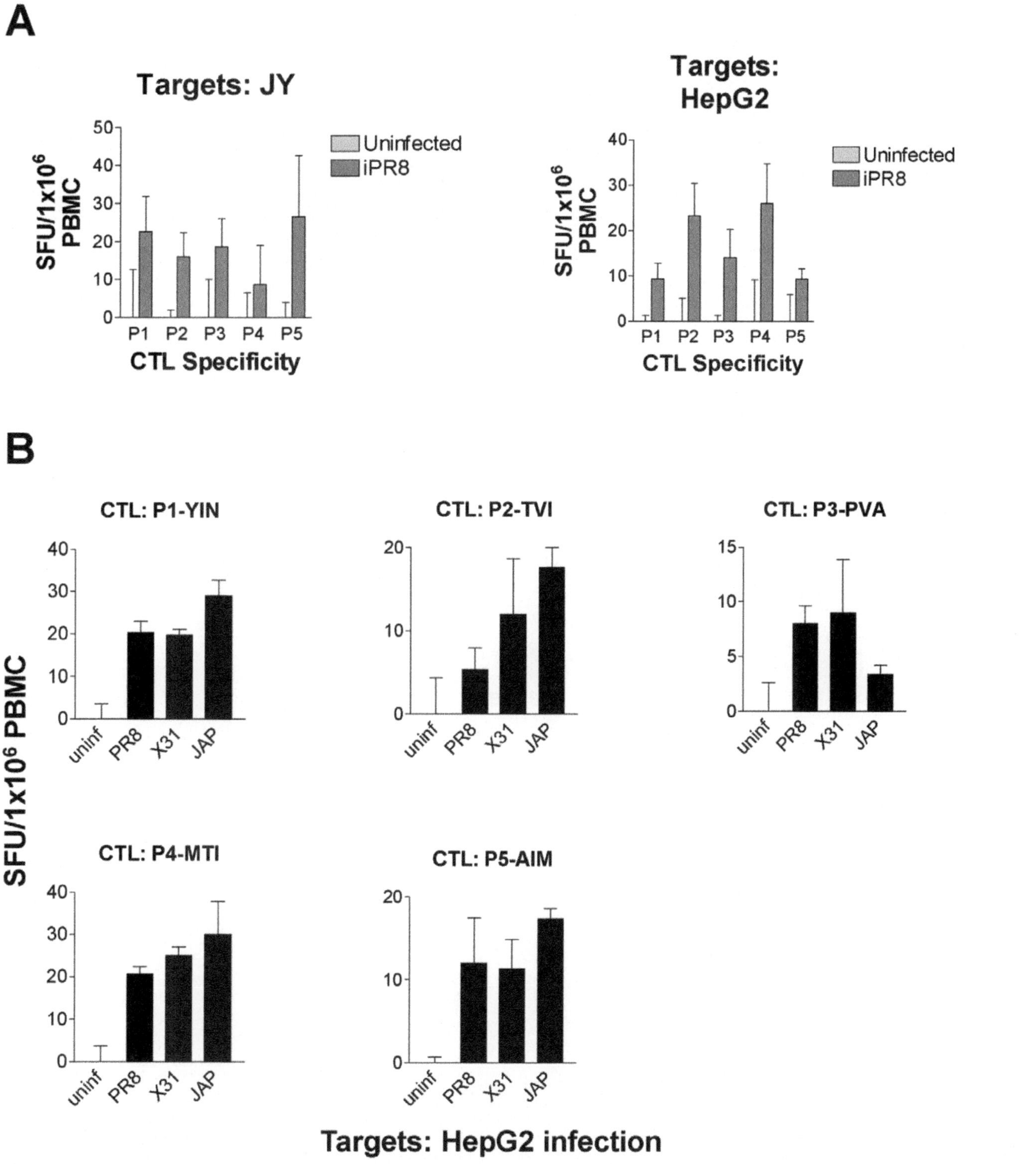

Figure 3. CTLs generated *in vitro* with influenza epitopes are specific and cross-reactive. (A) HepG2 and JY cells were left untreated or infected with PR8 and used as targets in an ELISpot assay with CTLs that were generated from HLA-A2+ PBMCs against specific peptides. (B) HepG2 cells were infected with PR8, X-31, or JAP and used as targets in an ELISpot assay. Results were normalized against uninfected controls.

Additionally, cross-reactivity to other strains was demonstrated using HepG2 target cells infected with various influenza A strains (X31, H3N2 and JAP, H2N2), indicating the presentation of these epitopes in various influenza strain-infected cells (Figure 3*B*).

In vivo Characterization of Influenza Peptides

To further characterize the immune response generated by these epitopes *in vivo*, we immunized HLA-A2+ transgenic mice with a mixture of the aforementioned five epitopes. Immunizations were carried out using these peptides in the presence of Montanide

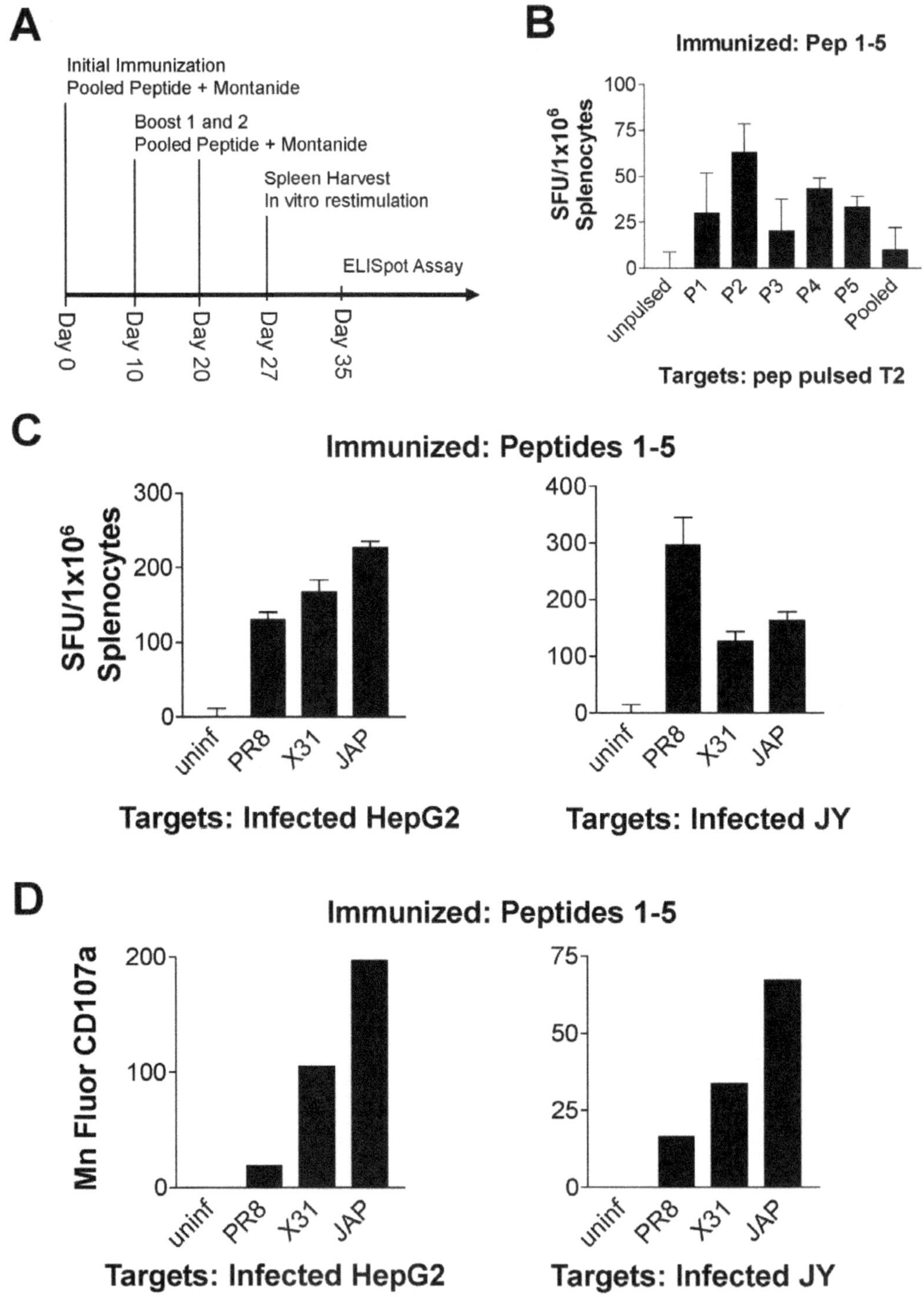

Figure 4. CTLs generated with influenza epitopes *in vivo* using humanized mice are specific and cross-reactive. (A) Immunization scheme for peptide injections. (B) T2 cells were pulsed with peptide and used as targets in an ELISpot assay with CTLs that were generated from humanized mice immunized with influenza-specific (P1-5) peptides. (C) HepG2 and JY cells were infected with PR8, X-31, or JAP and used as targets in an ELISpot assay. (D) Additionally, cells were stained for CD8 and CD107a after overnight co-culture and results are given as the mean fluorescence intensity of CD107a gated on CD8+ cells. Results were normalized against unpulsed or uninfected controls.

ISA 51 as an adjuvant (Figure 4*A*). We determined the influenza-specific T cell response by measuring murine IFNγ secretion in an ELISpot assay. Using T2 pulsed with individual peptides 1–5, we observed a response to all 5 peptides when mice were immunized with the mixture (Figure 4*B*). In conjunction with above *in vitro* results, *in vivo*-generated CTLs specific for these peptides were stimulated equally well when HepG2 and JY cells infected with various strains of influenza were used as targets (Figure 4*C*) indicating that these epitopes are presented by various influenza strain infected cells.

In addition to IFNγ release, we also measured the phenotypic changes of CD8+ T cells from splenocytes with regards to

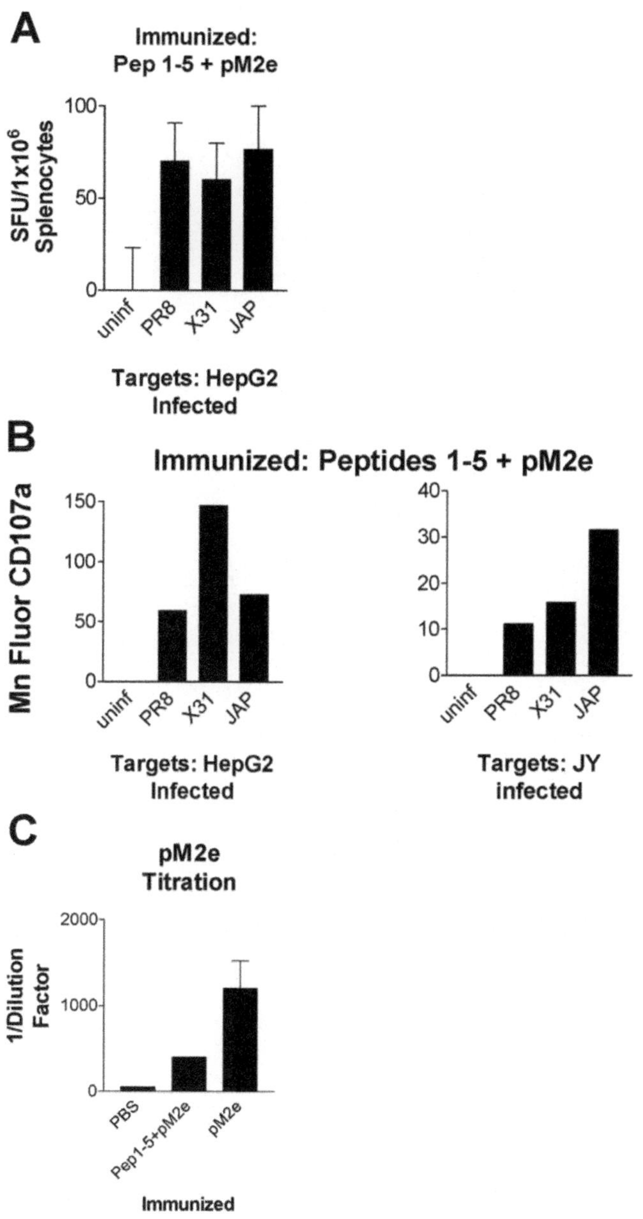

Figure 5. CTLs generated *in vivo* with influenza epitopes in addition to pM2e are specific and cross-reactive. HepG2 and JY cells were infected with PR8, X-31, or JAP and used as targets in an (A) ELISpot assay or (B) FACS analysis using CTLs generated from humanized mice immunized with influenza-specific (P1-5) peptides plus pM2e. (C) Serum titers of mice immunized with PBS, P1-5+pM2e, or pM2e alone were analyzed using ELISA. Plates were coated with pM2e and serum was added and serially diluted. Serum IgG was then detected using anti-mouse IgG secondary reagents. The dilution where the signal was reduced to background was measured and graphed using the reciprocal of the dilution factor (1/DF).

CD107a, an activation marker present on granulating effector CTLs [35,36]. As illustrated in Figure 4*D*, splenocytes incubated with infected target cells displayed a higher staining density for CD107a when gated on CD8+ T cells.

Functional Antibodies to Influenza M2e Peptide

Evaluation of both humoral and T cell immunity simultaneously was accomplished by the injection of multiple T cell epitopes, which drive a strong cellular response, combined with a shared antibody epitope from influenza matrix 2 protein (M2). To this end, we immunized a group of mice with MHCI peptides 1–5 in addition to a peptide from the ectodomain of M2 (pM2e) [17,18]. To ensure that the T cell response was at least the same as the mice immunized with only P1-5, we repeated our IFNγ ELISpot assay (Figure 5*A*) and CD107a (Figure 5*B*) flow cytometric analysis with the splenocytes from mice immunized with MHCI peptides 1–5+ M2e peptide and observed a comparable T cell response. The concentration of circulating M2e-specific antibody was then measured by a standard ELISA using serum collected from the terminal bleeds of immunized mice. As illustrated in Figure 5*C*, mice immunized with the M2e peptide generated a robust and antigen specific IgG response.

Functionality of the M2e antisera for cross reactivity was then characterized by infecting HepG2 cells with all three strains of influenza virus that we have thus far tested. Using flow cytometry, we labeled the cells with the M2e-specific antisera and demonstrated the binding of the antibody specifically on infected cells indicating the presence of this epitope in various strains of influenza infected cells (Figure 6*A*). Lastly, the neutralizing ability of these antibodies was determined by adding antisera to HepG2 cells infected with PR8 virus. As illustrated in Figure 6*B*, infection levels were lower when the serum containing anti-M2e antibody was used indicating the functional ability of the M2e specific antibodies.

Discussion

In this study, we attempted to evaluate the immune responses to cross reactive T-cell and antibody epitopes, which together, can work towards the development of a potential universal vaccine. Using an immunoproteomics method, we identified and characterized five naturally presented epitopes that are conserved amongst various strains of influenza virus. Furthermore, we examined the efficacy of these epitopes *in vivo* in combination with a known universal antibody epitope, M2e. We demonstrate a strong T cell response, in addition to neutralization, across several strains. Together, these epitopes could potentially form the backbone of a universal influenza vaccine.

Although current influenza vaccines target humoral immunity, there is evidence that T-cell responses are extremely important for protection against influenza. Importantly, in addition to the role of CTLs in mediating viral clearance [38,39], CD8+ T cells in humans were shown to have cross-reactive acute [21,40,41] and memory responses [42] to different subtypes of influenza A virus. Although our work primarily focused on HLA-A2 supertype-specific T cell epitopes, we have also indicated that other major HLA supertype specific peptides can be identified using the same immunoproteomics approach, which could lead to the generation of a multi-epitope universal vaccine. Goodman *et al* recently demonstrated a multi-epitope DNA prime/pox virus boost recombinant vaccine protected against influenza virus infection in a mouse influenza model [43] indicating this is a valid and effective approach for the generation of a cross-reactive vaccine.

Although a vast number of predicted and immunologically characterized epitopes are reported in the literature (IEDB-AR, http://tools.immuneepitope.org), successful vaccines based on these predicted epitopes have not yet been developed. Generally, motif-based predicted epitopes may activate CTL in the context of *in vitro* assays, however, not all of the activated CTLs will recognize naturally processed antigenic peptides on infected cells due to differences between motif-predicted epitopes and those that are endogenously presented [44]. Furthermore, comparison of the

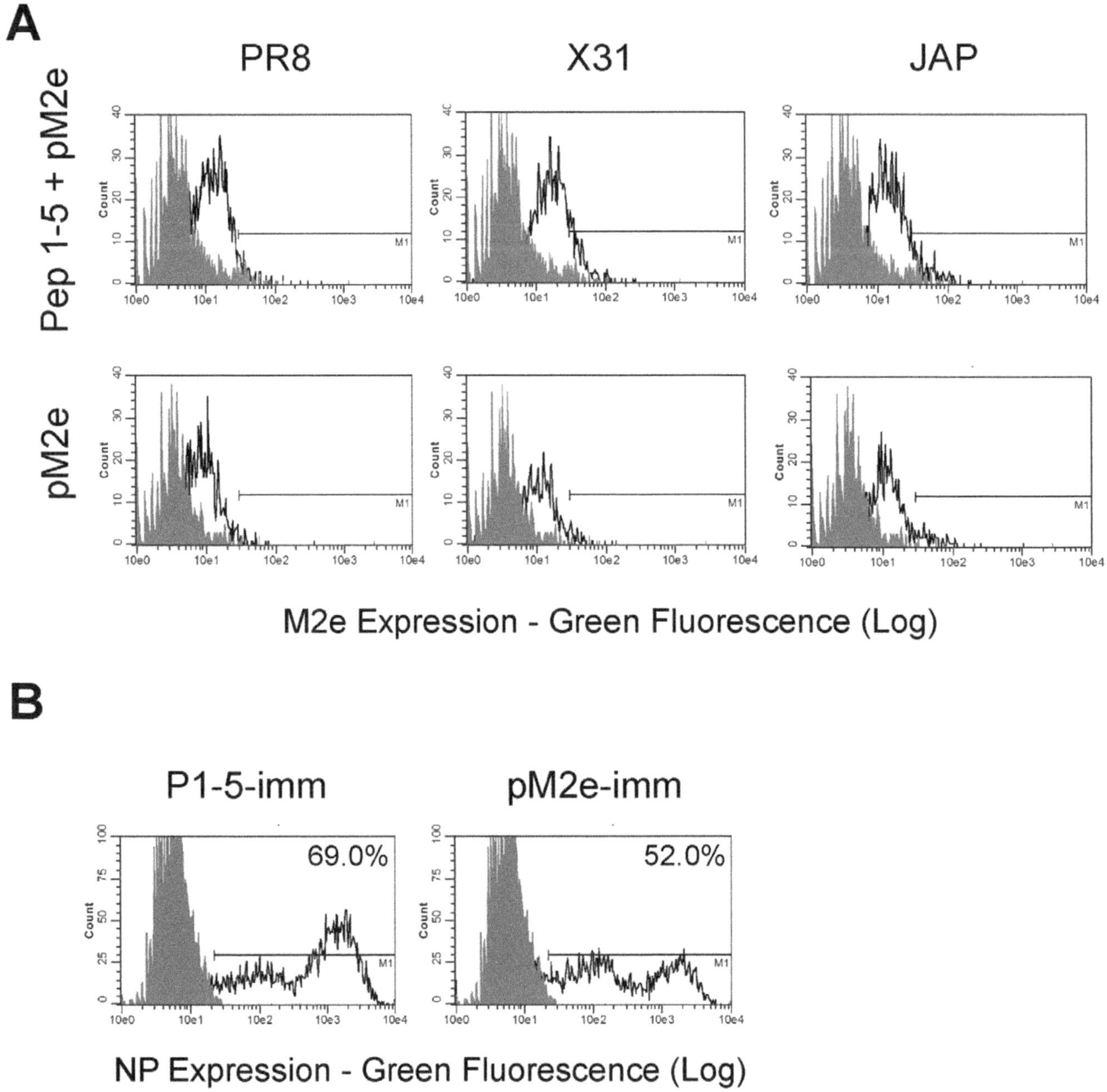

Figure 6. Antibodies against M2e are cross-reactive and neutralizing. (A) HepG2 cells were infected with PR8, X-31, or JAP. After overnight incubation, cells were stained using serum from P1-5+pM2e or pM2e immunized mice at a 1:50 dilution followed by anti-mouse IgG secondary reagents. (B) HepG2 cells were pulsed with a low dose of PR8 in the presence of serum from P1-5, P1-5+pM2e or pM2e immunized mice at a 1:50 dilution for 1 hr. Following overnight incubation with the same levels of serum, cells were fixed and intracellularly stained for influenza NP. Percent positive cells are depicted. Gray-filled histograms are uninfected controls.

motif prediction method with direct mass spectrometry analysis of endogenously presented epitopes from virus infected cells revealed a high number of predicted epitopes were not processed and presented [22]. Interestingly, we identified a previously reported epitope derived from NS1 protein (P5) [45], which was identified using motif-prediction and *ex vivo* CTL analysis. While this study demonstrates that obtaining an epitope with this approach is possible, it is unlikely to work in a high throughput setting due to the numerous predicted peptide sequences, which vary in their binding affinities for the MHC molecule. Natural presentation of these peptides by infected cells would be difficult to examine without actually analyzing the MHC repertoire.

It is interesting to note that our peptide analysis of influenza infected cells did not yield a well studied motif predicted epitope derived from M1 protein (GILGFVFTL, $M1_{58-66}$) [46]. This is concerning, although informative, and is most likely be due to the fact that this epitope is not well-presented on infected cells. Since epitopes presented by infected cells are critical for generating a clinically relevant protective T cell response, it is likely that this is the reason cellular immunity to seasonal influenza is limited and essentially non-protective [47]. In addition, our epitope specific

CTL analysis using PBMCs from healthy donors did not show an enhanced secondary response as one would expect from an influenza-exposed and vaccinated population. Specifically, if memory in the immunized population existed to the peptide epitopes we uncovered, a strong CTL response would have been illustrated. However, the response was moderate at best suggesting no preexisting immunity, i.e. CTL memory, specific for the identified epitopes was present in this population. Our results may also provide reason to believe that epitopes presented by infected cells may be subdominant epitopes [48,49] and do not activate a strong CTL response as opposed to the dominant epitopes that generate a memory CTL response during infection or vaccination.

The data generated by in silico approaches may overestimate the number of conserved epitopes and may not necessarily identify all immunoreactive naturally presented epitopes. In addition, the motif prediction method may be limited in identifying subdominant epitopes, which are reported to activate T cells in secondary influenza virus specific responses [44]. Furthermore, considering motif predicted epitopes are validated by screening circulating CTLs from virus infected individuals, it is important to note a study by Thomas et al [44] where they demonstrated a CD8+ T cell immunodominance hierarchy that was suggested to be dependent on the concentration of the presented epitope, size of the available CD8+ T cell repertoire, activation after initial priming, and competition and cooperation between different epitope-specific responses. Additionally, Zhong et al. [22] reported a large number of naturally processed and in vivo presented viral CTL epitopes in a mouse model system. In fact, comparison of the motif prediction method with direct mass spectrometry analysis of endogenously presented epitopes isolated from virus infected cells revealed a high number of predicted epitopes were not processed and presented by infected cells [22]. These findings illustrate the complexity of CD8+ T cell based screening of functional epitopes and how this approach may miss hidden subdominant epitopes. Our results, in accordance with these previous studies, further emphasize the need to identify epitopes presented by infected cells and the potential benefit of immunizing individuals with those peptides in a vaccine formulation to gain clinically relevant broad protection.

M2e-specific antibodies have the potential to provide broad protective immunity across influenza A strains. Mice immunized intranasally with combined M2e and T cell determinants exhibited significant production of M2e-specific antibodies as well as decreased morbidity after influenza virus infection [50]. In our study, we have demonstrated strong humoral and cell mediated immunity against influenza virus by immunization with both M2e and five conserved T cell epitopes. We did, however, observe a slight decrease in antibody response when the M2e peptide is immunized in combination with the T cell epitopes. This may be due to the competition of peptide uptake by the APCs. Ultimately, the most promising universal influenza vaccine candidate may come from such a combination of antigens.

In the viral pathogen field, due to the lack of knowledge of vital T cell epitopes, T cell responses to viral infection including influenza have not been routinely assessed in infected patients or in vaccinated individuals. Although several influenza specific T cell epitopes identified by motif prediction and overlapping peptide library screening methodologies exist, a comprehensive analysis of naturally presented T cell epitopes from infected cells, such as our study, has never been undertaken. We believe that these epitopes may serve as part of a universal vaccine candidate complementary to current vaccines and, if successful, could lead to a paradigm shift in how prophylactic and therapeutic antiviral vaccines are formulated.

This report is the first ever on naturally presented conserved influenza specific T cell epitopes identified by direct analysis of infected cells. Additionally, we have demonstrated that conserved epitope-specific CTLs could recognize multiple strain infected target cells and, when combined with a universal antibody epitope, could be useful in a vaccine formulation. Any universal vaccine that produces broad immunity against influenza would be a significant improvement over the current strain-specific vaccines that must be produced anew each year. Moreover, an epitope based vaccine would be produced synthetically, a major improvement over the current method of growing influenza vaccine in fertilized eggs. Therefore, it not only would generate a more effective immune response, but would also prove more economically sensible. The primary objective of this study is an attempt to create such a vaccine product that will complement the universal antibody based vaccines currently in development to combat influenza infection and pandemic.

Acknowledgments

We would like to thank Dr. Laurence Eisenlohr for the generous gifts of the A2/Japan/305/57(H2N2) (JAP) influenza strain as well as the HB65 anti-nucleoprotein antibody. Additionally, we would like to acknowledge the excellent work done at Lampire Biologicals as it related to our mouse studies.

Author Contributions

Conceived and designed the experiments: RP JT VS. Performed the experiments: JT JH ZN GS SK. Analyzed the data: JT VS RP. Wrote the paper: JT VS RP.

References

1. (2012) WHO Influenza (Seasonal). World Health Organization.
2. Lamb RA (1983) The influenza virus RNA segments and their encoded proteins. In: Palese P, Kingsbury DW, editors. Genetics of Influenza Viruses. New York: Springer-Verlag. 21–69.
3. Schild GC, Dowdle WR (1975) Influenza virus characterization and diagnostic serology. In: Kilbourne ED, editor. The Influenza Viruses and Influenza. New York: Academic Press. 315–372.
4. Chen J, Deng YM (2009) Influenza virus antigenic variation, host antibody production and new approach to control epidemics. Virol J 6: 30.
5. Murphy BR, Nelson DL, Wright PF, Tierney EL, Phelan MA, et al. (1982) Secretory and systemic immunological response in children infected with live attenuated influenza A virus vaccines. Infect Immun 36: 1102–1108.
6. Burlington DB, Clements ML, Meiklejohn G, Phelan M, Murphy BR (1983) Hemagglutinin-specific antibody responses in immunoglobulin G, A, and M isotypes as measured by enzyme-linked immunosorbent assay after primary or secondary infection of humans with influenza A virus. Infect Immun 41: 540–545.
7. Johansson BE, Bucher DJ, Kilbourne ED (1989) Purified influenza virus hemagglutinin and neuraminidase are equivalent in stimulation of antibody response but induce contrasting types of immunity to infection. J Virol 63: 1239–1246.
8. Doherty PC, Allan W, Eichelberger M, Carding SR (1992) Roles of alpha beta and gamma delta T cell subsets in viral immunity. Annu Rev Immunol 10: 123–151.
9. Eichelberger M, Allan W, Zijlstra M, Jaenisch R, Doherty PC (1991) Clearance of influenza virus respiratory infection in mice lacking class I major histocompatibility complex-restricted CD8+ T cells. J Exp Med 174: 875–880.
10. Epstein SL, Lo CY, Misplon JA, Bennink JR (1998) Mechanism of protective immunity against influenza virus infection in mice without antibodies. J Immunol 160: 322–327.

11. Graham MB, Braciale TJ (1997) Resistance to and recovery from lethal influenza virus infection in B lymphocyte-deficient mice. J Exp Med 186: 2063–2068.
12. Subbarao K, Murphy BR, Fauci AS (2006) Development of effective vaccines against pandemic influenza. Immunity 24: 5–9.
13. Rimmelzwaan GF, Osterhaus AD (1995) Cytotoxic T lymphocyte memory: role in cross-protective immunity against influenza? Vaccine 13: 703–705.
14. Yewdell JW, Bennink JR, Smith GL, Moss B (1985) Influenza A virus nucleoprotein is a major target antigen for cross-reactive anti-influenza A virus cytotoxic T lymphocytes. Proc Natl Acad Sci U S A 82: 1785–1789.
15. Tan PT, Khan AM, August JT (2011) Highly conserved influenza A sequences as T cell epitopes-based vaccine targets to address the viral variability. Hum Vaccin 7: 402–409.
16. Doherty PC, Kelso A (2008) Toward a broadly protective influenza vaccine. J Clin Invest 118: 3273–3275.
17. Fiers W, De Filette M, Birkett A, Neirynck S, Min Jou W (2004) A "universal" human influenza A vaccine. Virus Res 103: 173–176.
18. Grandea AG, 3rd, Olsen OA, Cox TC, Renshaw M, Hammond PW, et al. (2010) Human antibodies reveal a protective epitope that is highly conserved among human and nonhuman influenza A viruses. Proc Natl Acad Sci U S A 107: 12658–12663.
19. Kang SM, Song JM, Compans RW (2011) Novel vaccines against influenza viruses. Virus Res 162: 31–38.
20. Li C, Jaentschke B, Song Y, Wang J, Cyr TD, et al. (2010) A simple slot blot for the detection of virtually all subtypes of the influenza A viral hemagglutinins using universal antibodies targeting the fusion peptide. Nat Protoc 5: 14–19.
21. Greenbaum J (2009) http://iedb.zendesk.com/forums/45499/entries/35037 Knowledgebase and Forums/Epitope analysis in emerging H1N1 swine flu viruses/Analysis version 1.0). Accessed 2012 Jul.
22. Zhong W, Reche PA, Lai CC, Reinhold B, Reinherz EL (2003) Genome-wide characterization of a viral cytotoxic T lymphocyte epitope repertoire. J Biol Chem 278: 45135–45144.
23. Pleschka S, Jaskunas R, Engelhardt OG, Zurcher T, Palese P, et al. (1996) A plasmid-based reverse genetics system for influenza A virus. J Virol 70: 4188–4192.
24. Wiesmuller KH, Fleckenstein B, Jung G (2001) Peptide vaccines and peptide libraries. Biol Chem 382: 571–579.
25. Berzofsky JA, Ahlers JD, Belyakov IM (2001) Strategies for designing and optimizing new generation vaccines. Nat Rev Immunol 1: 209–219.
26. Purcell AW, Gorman JJ (2004) Immunoproteomics: Mass Spectrometry-based Methods to Study the Targets of the Immune Response. Mol Cell Proteomics 3: 193–208.
27. Hunt DF, Henderson RA, Shabanowitz J, Sakaguchi K, Michel H, et al. (1992) Characterization of peptides bound to the class I MHC molecule HLA-A2.1 by mass spectrometry. Science 255: 1261–1263.
28. di Marzo Veronese F, Arnott D, Barnaba V, Loftus DJ, Sakaguchi K, et al. (1996) Autoreactive cytotoxic T lymphocytes in human immunodeficiency virus type 1-infected subjects. J Exp Med 183: 2509–2516.
29. van Els CA, Herberts CA, van der Heeft E, Poelen MC, van Gaans-van den Brink JA, et al. (2000) A single naturally processed measles virus peptide fully dominates the HLA-A*0201-associated peptide display and is mutated at its anchor position in persistent viral strains. Eur J Immunol 30: 1172–1181.
30. Hickman HD, Luis AD, Bardet W, Buchli R, Battson CL, et al. (2003) Cutting edge: class I presentation of host peptides following HIV infection. J Immunol 171: 22–26.
31. Lemmel C, Weik S, Eberle U, Dengjel J, Kratt T, et al. (2004) Differential quantitative analysis of MHC ligands by mass spectrometry using stable isotope labeling. Nat Biotechnol 22: 450–454.
32. Shetty V, Sinnathamby G, Nickens Z, Shah P, Hafner J, et al. (2011) MHC class I-presented lung cancer-associated tumor antigens identified by immunoproteomics analysis are targets for cancer-specific T cell response. J Proteomics 74: 728–743.
33. Testa JS, Shetty V, Sinnathamby G, Nickens Z, Hafner J, et al. (2012) Conserved MHC class I-presented dengue virus epitopes identified by immunoproteomics analysis are targets for cross-serotype reactive T-cell response. J Infect Dis 205: 647–655.
34. Motomura Y, Ikuta Y, Kuronuma T, Komori H, Ito M, et al. (2008) HLA-A2 and -A24-restricted glypican-3-derived peptide vaccine induces specific CTLs: preclinical study using mice. Int J Oncol 32: 985–990.
35. Betts MR, Brenchley JM, Price DA, De Rosa SC, Douek DC, et al. (2003) Sensitive and viable identification of antigen-specific CD8+ T cells by a flow cytometric assay for degranulation. J Immunol Methods 281: 65–78.
36. Mittendorf EA, Storrer CE, Shriver CD, Ponniah S, Peoples GE (2005) Evaluation of the CD107 cytotoxicity assay for the detection of cytolytic CD8+ cells recognizing HER2/neu vaccine peptides. Breast Cancer Res Treat 92: 85–93.
37. Middleton D, Menchaca L, Rood H, Komerofsky R (2003) New allele frequency database: http://www.allelefrequencies.net.Tissue Antigens 61: 403–407. Accessed 2012 July.
38. Yap KL, Braciale TJ, Ada GL (1979) Role of T-cell function in recovery from murine influenza infection. Cell Immunol 43: 341–351.
39. Yap KL, Ada GL, McKenzie IF (1978) Transfer of specific cytotoxic T lymphocytes protects mice inoculated with influenza virus. Nature 273: 238–239.
40. McMichael AJ, Gotch FM, Noble GR, Beare PA (1983) Cytotoxic T-cell immunity to influenza. N Engl J Med 309: 13–17.
41. De Groot AS, Ardito M, McClaine EM, Moise L, Martin WD (2009) Immunoinformatic comparison of T-cell epitopes contained in novel swine-origin influenza A (H1N1) virus with epitopes in 2008–2009 conventional influenza vaccine. Vaccine 27: 5740–5747.
42. Jameson J, Cruz J, Ennis FA (1998) Human cytotoxic T-lymphocyte repertoire to influenza A viruses. J Virol 72: 8682–8689.
43. Goodman AG, Heinen PP, Guerra S, Vijayan A, Sorzano CO, et al. (2011) A human multi-epitope recombinant vaccinia virus as a universal T cell vaccine candidate against influenza virus. PLoS One 6: e25938.
44. Thomas PG, Brown SA, Keating R, Yue W, Morris MY, et al. (2007) Hidden epitopes emerge in secondary influenza virus-specific CD8+ T cell responses. J Immunol 178: 3091–3098.
45. Man S, Newberg MH, Crotzer VL, Luckey CJ, Williams NS, et al. (1995) Definition of a human T cell epitope from influenza A non-structural protein 1 using HLA-A2.1 transgenic mice. Int Immunol 7: 597–605.
46. Man S, Ridge JP, Engelhard VH (1994) Diversity and dominance among TCR recognizing HLA-A2.1+ influenza matrix peptide in human MHC class I transgenic mice. J Immunol 153: 4458–4467.
47. Webby RJ, Andreansky S, Stambas J, Rehg JE, Webster RG, et al. (2003) Protection and compensation in the influenza virus-specific CD8+ T cell response. Proc Natl Acad Sci U S A 100: 7235–7240.
48. Crowe SR, Miller SC, Woodland DL (2006) Identification of protective and non-protective T cell epitopes in influenza. Vaccine 24: 452–456.
49. Dominguez MR, Silveira EL, de Vasconcelos JR, de Alencar BC, Machado AV, et al. (2011) Subdominant/cryptic CD8 T cell epitopes contribute to resistance against experimental infection with a human protozoan parasite. PLoS One 6: e22011.
50. Mozdzanowska K, Feng J, Eid M, Kragol G, Cudic M, et al. (2003) Induction of influenza type A virus-specific resistance by immunization of mice with a synthetic multiple antigenic peptide vaccine that contains ectodomains of matrix protein 2. Vaccine 21: 2616–2626.
51. Parker KC, Bednarek MA, Coligan JE (1994) Scheme for ranking potential HLA-A2 binding peptides based on independent binding of individual peptide side-chains. J Immunol 152: 163–175.
52. Rammensee H, Bachmann J, Emmerich NP, Bachor OA, Stevanovic S (1999) SYFPEITHI: database for MHC ligands and peptide motifs. Immunogenetics 50: 213–219.

Cleavage Specificity Analysis of Six Type II Transmembrane Serine Proteases (TTSPs) Using PICS with Proteome-Derived Peptide Libraries

Olivier Barré[1], Antoine Dufour[1], Ulrich Eckhard[1], Reinhild Kappelhoff[1], François Béliveau[2], Richard Leduc[2], Christopher M. Overall[1,3]*

1 Centre for Blood Research, Department of Oral Biological & Medical Sciences, University of British Columbia, Vancouver, BC, Canada, 2 Department of Pharmacology, Faculty of Medicine and Health Sciences, Université de Sherbrooke, Sherbrooke, QC, Canada, 3 Department of Biochemistry and Molecular Biology, University of British Columbia, Vancouver, BC, Canada

Abstract

Background: Type II transmembrane serine proteases (TTSPs) are a family of cell membrane tethered serine proteases with unclear roles as their cleavage site specificities and substrate degradomes have not been fully elucidated. Indeed just 52 cleavage sites are annotated in MEROPS, the database of proteases, their substrates and inhibitors.

Methodology/Principal Finding: To profile the active site specificities of the TTSPs, we applied Proteomic Identification of protease Cleavage Sites (PICS). Human proteome-derived database searchable peptide libraries were assayed with six human TTSPs (matriptase, matriptase-2, matriptase-3, HAT, DESC and hepsin) to simultaneously determine sequence preferences on the N-terminal non-prime (P) and C-terminal prime (P′) sides of the scissile bond. Prime-side cleavage products were isolated following biotinylation and identified by tandem mass spectrometry. The corresponding non-prime side sequences were derived from human proteome databases using bioinformatics. Sequencing of 2,405 individual cleaved peptides allowed for the development of the family consensus protease cleavage site specificity revealing a strong specificity for arginine in the P1 position and surprisingly a lysine in P1′ position. TTSP cleavage between R↓K was confirmed using synthetic peptides. By parsing through known substrates and known structures of TTSP catalytic domains, and by modeling the remainder, structural explanations for this strong specificity were derived.

Conclusions: Degradomics analysis of 2,405 cleavage sites revealed a similar and characteristic TTSP family specificity at the P1 and P1′ positions for arginine and lysine in unfolded peptides. The prime side is important for cleavage specificity, thus making these proteases unusual within the tryptic-enzyme class that generally has overriding non-prime side specificity.

Editor: Alexander Wlodawer, NCI-Frederick, United States of America

Funding: This study was funded by Proteolytic Signatures in Breast Cancer C.I.H.R. Grant M.O.P. 111058. O.B. was supported by the Swiss National Science Foundation (SNF) and a post-doctoral fellowship by the Canadian Institutes of Health Research (CIHR). A.D. and U.E. were supported by post-doctoral fellowships from CIHR and the Michael Smith Foundation for Health Research (MSFHR), respectively. A Canada Research Chair in Metalloproteinase Proteomics and Systems Biology supports C.M.O. This work was supported by grants to C.M.O. and R.L. from the CIHR and the Canadian Cancer Society. The University of British Columbia, Centre for Blood Research Mass Spectrometry Suite is supported by the Canada Foundation for Innovation (CFI) and MSFHR. The funders had no role in study design, data collection and analysis, decision to publish, or preparation of the manuscript.

Competing Interests: The authors have declared that no competing interests exist.

* Email: chris.overall@ubc.ca

These authors contributed equally to this work.

Introduction

Pericellular proteolysis is involved in many important cellular processes such as the transduction of signals across the cell membrane, the release of bioactive growth factors, cytokines and peptide hormones, as well as the interactions with other cells, basement membrane and extracellular matrix proteins [1–3]. The increasing relevance of these processes at the cell surface has focused attention on membrane-associated proteolytic systems, including the family of type II transmembrane serine proteases (TTSPs) [4,5]. The TTSP family is composed of more than 20 members that share a number of structural features including an N-terminal cytoplasmic domain, a transmembrane domain, a central region containing various domains potentially involved in protein-protein interaction, and a C-terminal extracellular serine protease domain (Figure 1A). TTSPs are synthesized as inactive single chain proenzyme zymogens and enzyme activation proceeds by cleavage after a basic amino acid residue in a conserved activation motif N-terminal the catalytic domain. However, the catalytic domain remains membrane-associated after activation because of a disulfide bond that links the prodomain and catalytic domain [6]. Although a few of the TTSPs are expressed across several tissue and cell types, in general these enzymes demonstrate relatively restricted expression patterns, indicating tissue specific regulation or functions [1,4,5,7–11].

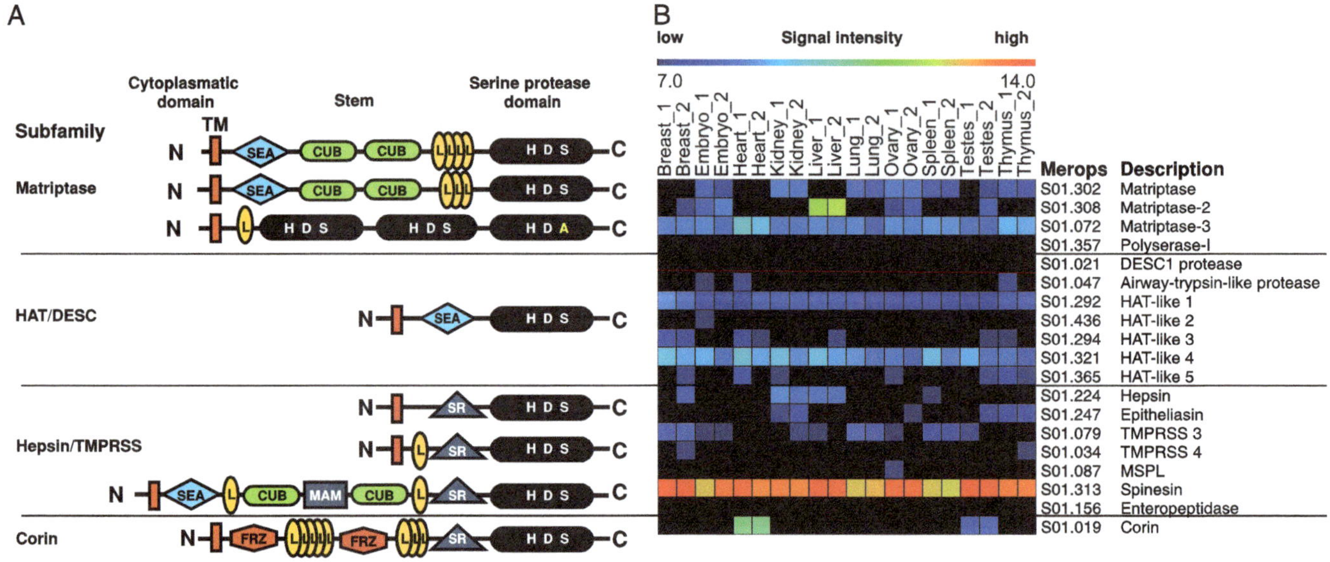

Figure 1. A. Schematic representation of the protein structures and arrangements of the four TTSP subfamilies: HAT/DESC, hepsin/TMPRSS, matriptase and corin. All TTSPs contain a N-terminal transmembrane signal anchor domain (**TM**) and a C-terminal serine protease domain (**H D S**). In case of polyserase-1 one of the 3 catalytic domains is inactive (**H D A**). The stem region of the TTSPs contains 1–6 different domains: the sea urchin sperm protein/enteropeptidase/agrin domain (**SEA**), group A scavenger receptor domain (**SR**), low-density lipoprotein receptor class A domain (**L**), Cls/CLr, urchin embryonic growth factor, bone morphogenetic protein-1 domain (**CUB**) and Corin contains two frizzled (FRIZ) domain. **B.** Murine CLIP-CHIP RNA expression profile and its distribution of 19 members of the TTSP family in 10 murine tissues in duplicates according to their average signal intensity.

Based on phylogenetic analysis of the serine protease domains and the domain structure of the extracellular stem region, the TTSPs have been divided into four subfamilies (Figure 1A). The largest is the HAT/DESC subfamily, with currently 7 proteases: human/murine airway trypsin-like (HAT/MAT), differentially expressed in squamous cells carcinoma (DESC)1, and HAT-like 1–5 protease. HAT-like 2 and 3 are only expressed in rodents [12]. HAT is predominantly expressed in the trachea [13,14] whereas DESC1 is restricted to the epithelia of the skin and oral cavity [15,16]. TTSPs of the hepsin/TMPRSS/enteropeptidase subfamily include hepsin, mosaic serine protease large form (MSPL), type II transmembrane serine protease (TMPRSS) 2, 3, 4, 5 and enteropeptidase. These TTSPs are expressed predominantly in fetal liver and kidney [1], prostate [17] and on the brush-border of the duodenum, respectively [18]. The matriptase subfamily contains three highly homologous proteases; matriptase, matriptase-2, and matriptase-3, as well as a protein with an atypical mosaic structure, polyserase-1. Matriptase was originally identified from a human breast cancer cell line and shows the most ubiquitous pattern of expression of the TTSP family [9–11]. Matriptase-2 plays an important role in iron homeostasis [19–21] and is expressed primarily in the liver, where it suppresses hepcidin expression [22]. Matriptase-3 is expressed in brain, ovary, testis and salivary gland [23], whereas polyserase-1 is mainly restricted to skeletal muscle, liver, placenta and heart [24]. The fourth subfamily consists of a single TTSP, corin, which is mainly expressed in cardiomyocytes [25] and is involved in the activation of proatrial natriuretic peptide, a cardiac hormone that regulates blood pressure and cardiac function by promoting natriuresis, dieresis, and vasodilatation [26].

All TTSPs contain a highly conserved catalytic serine protease domain (Figure 1A) of the trypsin (S1) fold subfamily. The large variability of the central modular region together with the diverse expression patterns of TTSP family members suggest that these enzymes may play diverse roles, but few of these functions have been identified. Therefore, substrate discovery is essential to establish *in vivo* functions [27]. As a first step, quenched fluorescence peptide assays combined with positional scanning synthetic combinatorial libraries and phage display approaches have identified a limited number of candidate peptide substrates with their cleavage sites [15,23,28–31]. However, the protease database MEROPS (http://merops.sanger.ac.uk/) shows that the number of known physiological cleavage sites for the TTSPs is only 52 (+92 synthetic peptide cleavages), and the peptide substrates used only extend from P4 to P4′ [32]. Further, most of these studies do not provide any information of prime side specificity corresponding to the sequence of the carboxyl cleavage product.

By knowing the cleavage site and preferred amino acids in recognition sequences, more specific and selective inhibitors can be developed. To overcome these challenges, we used a proteomics approach to determine the cleavage specificity of six different TTSPs belonging to three subfamilies. PICS (Proteome-wide Identification of Cleavage site Specificity) is a high throughput mass spectrometric approach that allows for the simultaneous identification of the amino acid preference at the cleavage site on both the prime and non-prime sides [33,34]. This technique using proteome-wide peptide libraries has the possibility to precisely identify several hundred cleaved peptides by tandem mass spectrometry and searchable databases. Since 2008, thousands of new cleavage sites have been both discovered or confirmed using PICS [35–39]. In the present study, the amino acid preferences of six human TTSPs, from P6 to P6′, were established from 2,405 cleavage sites and a rational interpretation of family-wide peptide substrate specificity was obtained by linking their specificities to 3D models of their catalytic domains.

Materials and Methods

Murine CLIP-CHIP microarray

The expression profile of a panel of 10 commercial murine RNAs (Life Technologies) was analyzed with the murine CLIP-CHIP microarray. mRNA sample preparation for the hybridization to the microarray chip and its analysis were conducted as previously described [40,41].

Proteases expression and purification

pRSET5a containing the cDNA of human matripase (UniProt ID Q9Y5Y6) was a generous gift of Dr. U. Bodendorf, Novartis, Switzerland. Human matriptase-2 (Q8IU80) cDNA was a generous gift from Dr. C. Lopez-Otin (Universidad de Oviedo, Oviedo, Spain). Human hepsin (P05981) cDNA was cloned from a human liver cDNA library from Life Technologies. Human DESC1 (Q9UL52) cDNA was a generous gift from Dr. D. E. Schuller (Ohio State University, OH, USA). Human HAT (O60235) cDNA was purchased from GeneCopoeia (Rockville, MD, USA).

Proteases were cloned, expressed and purified as described [29]. In brief, cDNAs corresponding to amino acids 78–811 of matriptase-2, 45–417 of hepsin, 44–423 of DESC1 and 45-418 of HAT were amplified by PCR and ligated into the pMT/BiP/V5-His vector (Life Technologies, Burlington, Canada). These constructs each contained a C-terminal V5-His tag for affinity purification using immobilized metal–chelate affinity chromatography. Drosophila Schneider 2 (S2) cells (Life Technologies) were grown in six-well plates to a density of $2–4\times10^6$ cells/mL and transfected with 19 μg cDNAs in pMT/BiP/V5-His and 1 μg pCoBlast selection vector (Life Technologies) using calcium phosphate transfection kits (Life Technologies). The calcium phosphate solution was removed 16 h post-transfection and fresh medium was added. Cells were grown for an additional 2 days. Blasticidin (20 μg/mL) was added to the medium, and the cells were incubated for a 14-day selection period. Cells were then inoculated into 1 L of selection medium and grown to a density of 3×10^6 cells/mL. Recombinant TTSP expression was induced by adding 500 μM CuSO4. Cells were removed by centrifugation at 6,000×g. Supernatants were loaded on HisTrap FF columns (GE, Mississauga, Canada) and bound TTSPs were eluted using 500 mM imidazole. Fractions were analyzed by immunoblotting and fractions containing TTSPs were pooled and for imidazole removal dialyzed for 16 h at 4 °C against 50 mM Tris-HCl (pH 8.5), 10% glycerol and 250 mM NaCl. Dialyzed pooled fractions were then loaded on HiTrap Benzamidine FF affinity columns (GE, Mississauga, Canada) and eluted with 10 mM HCl (pH 2.0), 0.5 M NaCl and then buffered to 50 mM HEPES (pH 9.0) with 20% glycerol. The four purified TTSPs were active-site titrated with the burst titrant 4-methylumbelliferyl-*p*-guanidino benzoate.

Spodoptera frugiperda Sf9 cells expressing mouse matriptase-3 (Q7RTY8) cDNA encoding the serine protease domain (amino acids 581–829) were a kind gift from Dr. T. Bugge (NIH, Bethesda, MD, USA). Matriptase-3 protease was purified as previously described [1,4,23] with slight modifications. Conditioned medium was collected from Blasticidin-resistant Sf9 cells, and sodium phosphate buffer (pH 7.5) with NaCl were added to a final concentration of 20 mM and 500 mM respectively. The solution was filtered through a 0.22 μm filter and applied to a HisTrap column using an AKTA explorer system. Chelated proteins were eluted with a linear gradient of 20–400 mM imidazole, and TTSP-positive fractions were identified by immunoblotting using the anti-V5 antibody (Life Technologies). The positive fractions were dialysed overnight at 4 °C against 200 volumes of 50 mM Tris-HCl (pH 8.0), 1 mM $CaCl_2$ and 0.1% Tween 20 using dialysis membranes of 10 kDa MWCO.

PICS analysis

To identify TTSP cleavage sites, we prepared two overlapping proteome-derived peptide libraries. Endoproteinases GluC or chymotrypsin were used to digest HEK293 cell (Life Technologies) proteomes. Peptides were reduced, cysteines alkylated and primary amines protected by dimethylation [33]. The two peptide libraries were then separately incubated with the individual TTSPs after which the N-termini of prime-side cleavage products were tagged with biotin, affinity enriched and identified by liquid chromatography tandem mass spectrometry (LC-MS/MS), performed on a Packings capillary LC system (Dionex) coupled to a quadrupole time-of-flight mass spectrometer (QSTAR XL; Applied Biosystems, Foster City, CA) at the UBC Centre for Blood Research Mass Spectrometry Suite. In brief, samples were resuspended in 5% (v/v) acetonitrile, 3% formic acid and loaded onto a column packed with Magic C18 resin (Michrom Bioresources, Auburn, CA). Peptides were eluted using a 7–40% gradient of organic phase over 95 min. Buffer A was 2% acetonitrile with 0.1% formic acid and buffer B was 98% acetonitrile, 0.1% formic acid. Mass spectrometry data were acquired automatically using Analyst QS software (Applied Biosystems) for data dependent acquisition based on a 1 s mass spectrometry survey scan followed by up to two (QSTAR XL) MS/MS scans of 3 s each. MS and MS/MS data were acquired in centroid mode. QSTAR data were converted to the mzXML format with mzwiff (version 4.3.1). These sequences represent the prime side products of TTSP cleavage. The corresponding non-prime side sequences were derived from human proteome databases using bioinformatics searches. To increase the confidence of our mass spectrometry queries both search engines Mascot (version 2.2.04) [42] and X! Tandem (version 2007.07.01.2) [43] were used and the results were combined. Peptides were analyzed with PeptideProphet [44] and iProphet [45], both implemented within the Trans Proteomic Pipeline (Institute for Systems Biology, Seattle, WA, USA) [46]. Peptides were analyzed at a global false discovery rate of 5% at peptide level by using a decoy search strategy. This allowed for the selection of peptides having a minimum threshold probability corresponding to this false discovery rate percentage [46]. Search parameters included a mass tolerance of 0.4 Da for parental ions and 0.2 Da for fragment ions, allowing up to three missed cleavages, and for the following fixed peptide modifications: carbamidomethylation of cysteine residues (+57.02 Da), dimethylation of lysine amines (+28.03 Da), and variable methionine oxidation (+15.99 Da), variable thioacylation (+88.00 Da) and dimethylation (+28.03 Da) of peptide N-termini. N-terminally thioacylated peptides identified by MS/MS represent prime-side cleavage products of the assayed proteases. The complete cleavage sites were reconstructed by bioinformatic determination of the non-prime side sequences from the human database IPI v3.42 (http://ebi.edu.au/ftp/databases/IPI/last_release/msipi/old/HUMAN/) using WebPICS [39], available at http://clipserve.clip.ubc.ca/pics/. Subsite positions with ambiguous information coming from different protein isoforms were omitted and replaced by X, the one-letter notation for an unknown amino acid. Reconstructed cleavage sites were aligned and summarized as heat maps and iceLogos (University Ghent, Belgium) [7].

Peptide cleavage site confirmation

In order to confirm the PICS results for the cleavage specificity, mass spectrometry peptide cleavage assays were performed.

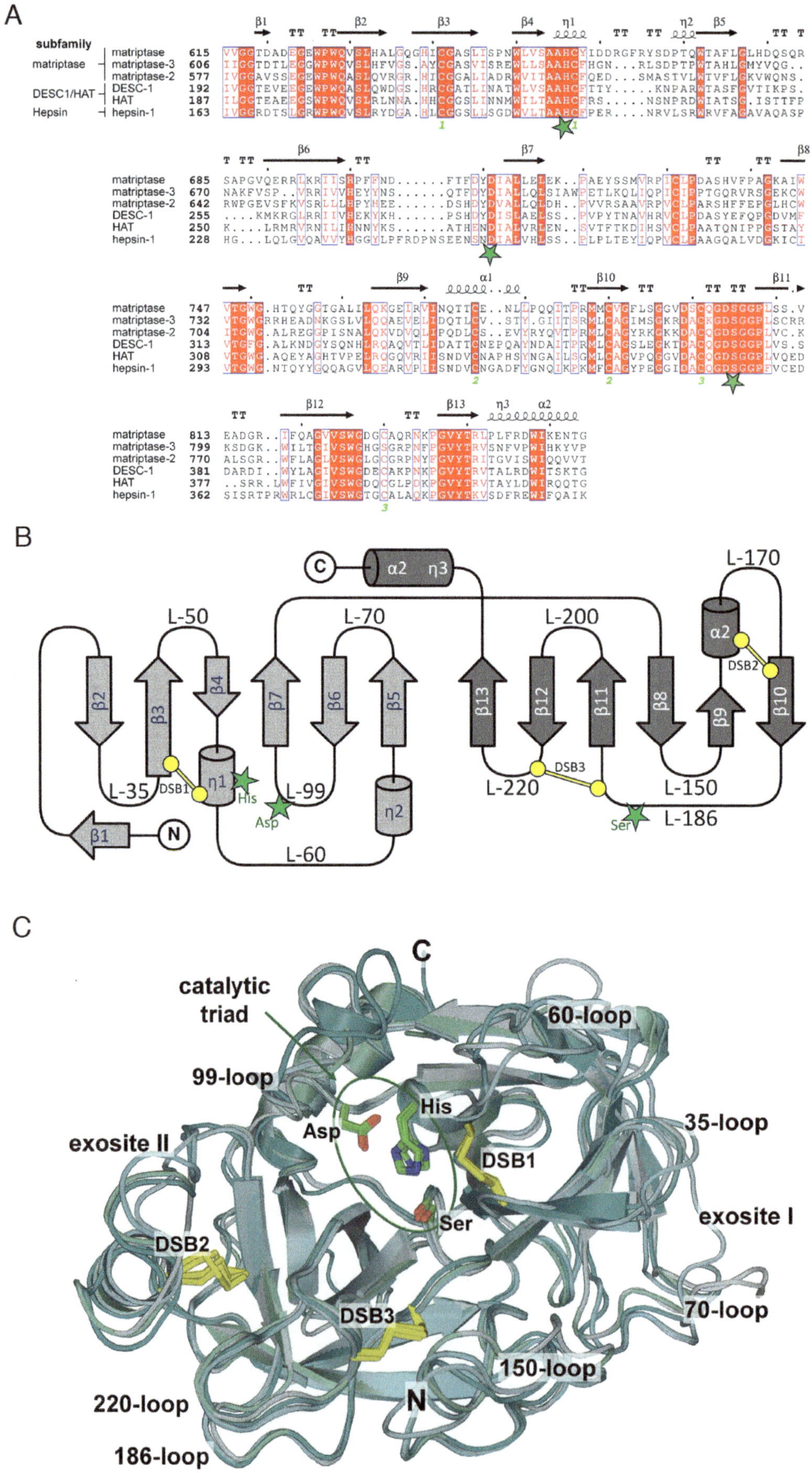
A
subfamily
matriptase
DESC1/HAT
Hepsin
matriptase
matriptase-3
matriptase-2
DESC-1
HAT
hepsin-1
B
C
L-50
L-70
L-200
L-170
L-35
L-99
L-60
L-220
L-150
L-186
DSB1
DSB2
DSB3
His
Asp
Ser
α2 η3
catalytic triad
60-loop
99-loop
35-loop
exosite II
exosite I
DSB1
DSB2
DSB3
70-loop
150-loop
220-loop
186-loop
N

Figure 2. A. Multiple sequence alignment of the catalytic domains of six TTSPs was performed with Clustal Omega [59] **and displayed using ESPript 2.2** (http://esprit.ibcp.fr/ESPript/cgi-bin/ESPript.cgi [63]**.** Green stars indicate the residues of the catalytic triad; conserved residues are shaded in red. Blue boxes indicate amino acid with similar physico-chemical properties. Secondary structure elements are represented with flat arrows for β sheets and helices for α- and 3_{10}- (η) helices. Three conserved disulfide bridges are indicated in orange numerals below the alignment. Sequence numbering is according to the respective UniProt entries. **B.** Topology diagram of a TTSP catalytic domain. Residues of the catalytic triad are indicated as green stars, disulfide bonds (DSB) are shown in yellow. Loops (L-) are labeled corresponding to the thrombin nomenclature suggested by Bode et al. [64]. **C.** Superposition of the catalytic domains of matriptase, matriptase-2, matriptase-3, hepsin, DESC1 and HAT, including both known and modeled structures, in traditional serine protease standard orientation [65]. Catalytic triad and conserved disulfide bridges are shown in ball and stick representation. Important loops and exosites are indicated and labeled.

Scanning through the peptides identified, we selected one peptide sequence that was found for each of the TTSPs analyzed, namely AVIGRKFGDP. The peptide (theoretical mass of 1059.22 Da) was synthesized (GenScript) (>90% purity). For the dimethylated peptide, 30 mM formaldehyde, 30 mM of cyanoborohydride (ALD coupling solution, Sterogene, Carlsbad, CA) and 10 mM HEPES pH 7.4 were added to the peptide and incubated overnight (18 h) at room temperature to generate (dm)A-VIGR(dm)KFGDP. To assay for cleavage, 200 µg of AVIGRKFGDP or (dm)AVIGR(dm)KFGDP were added to 2 µg of matriptase, matriptase-2, HAT and hepsin in 100 mM HEPES, pH 7.4, 150 mM NaCl and incubated at 37°C for 18 h. 20 µL of the reaction mixtures were used and buffer exchanged using C18 OMIX 10 µL tips (Agilent, Mississauga, Canada) using the manufacturer's instructions. Samples (0.8 µL) were mixed with 0.5 µL of 10 mg/mL of CHCA matrix solution (Sigma, Oakville, Canada) and spotted on a stainless steel MALDI plate. Spectra were acquired in reflectron positive mode and 2,175 random shots were performed and analyzed on a 4700 MALDI-TOF/TOF mass spectrometer (Applied Biosystems) with a mass range of 300-1,200 Da. Data analysis was conducted using protein data software provided by Applied Biosystems.

Structural analysis

For matriptase, hepsin, and DESC1, the crystal structures 3p8g [47], 1z8g [48] and 2oq5 [49] were used, respectively. For matriptase-2, -3 and HAT, structural models were generated using the interactive mode of the modeling server SWISS-MODEL [50,51], and crystal structures of closely related enzymes identified by the FFAS03 server [52,53]. We used DESC1 (PDB code 2oq5) [49] for matriptase-2 and HAT, and matriptase (PDB code 3p8g) [47] for matriptase-3. Sequence numbering was corrected and secondary structures assigned to all structures and models using the UCSF Chimera package [44]. The topology representation was created using TopDraw [54]. Peptide docking with the structural model of matriptase-2 was performed using the Rosetta FlexPepDock web server [55,56], allowing full flexibility to the peptide (AEAALR(P1).(P1′)KLLEVA), and full side-chain flexibility to the protease. Initial models were based on available complex structures of matriptase with aprotinin (PDB code 1eaw) [47] and trypsin inhibitor 1 (PDB entry 3p8f) [57], and our PICS P6 to P6′ sequence results. Molecular graphic figures were made using the molecular visualization system PyMOL (The PyMOL Molecular Graphics System, Version 1.5.03, Schrödinger, LLC) [58].

Results

Tissue distribution of TTSPs

The tissue distribution of TTSP family members was determined by transcript analysis of ten murine tissues using the dedicated murine CLIP-CHIP [1–3], with probes for all murine proteases, inactive homologues, and inhibitors (Figure 1B). The expression profile reveals that most of the TTSPs are very low or non expressed (average signal intensity below 7.5) in the analyzed tissues with the exception of the high expression of matriptase-2 in liver and corin in heart, both of which have been previously described as predominantly expressed in these tissues [25,28]. Hepsin shows significant expression in liver and kidney, TMPRSS3 in breast and lung, matriptase in kidney and lung. Matriptase-3, Hat-like 1, Hat-like 4 and spinesin show expression in all 10 tissues. However, polyserase-1, DESC1 and enteropeptidase were not detected in these ten tested tissues. This suggests potential tissue specificity for some of the TTSPs whereas other TTSPs might be involved in more general processes due to their ubiquitous expression.

Sequence alignment and structural analysis of the catalytic domain of TTSPs

To better understand the cleavage specificity of the TTSP family, we first compared the sequences of their catalytic domains using Clustal Omega [59] and ESPript 3.0 [57] (Figure 2A). The catalytic domain of matriptase showed the highest sequence identity with matriptase-2 (45% identity) and matriptase-3 (31% identity). HAT showed the highest sequence identity with DESC1 (50% identity). From the three subfamilies, the conserved regions within the catalytic domain are clearly visible (indicated in red). The catalytic triad, indicated as green stars, consists of histidine 656, aspartate 711 and serine 805, following the numbering of matriptase (top row). Interestingly, strictly conserved regions are often observed in stretches that adopt a β-sheet structure (Figure 2A). A topology diagram of a typical TTSP catalytic domain is shown in Figure 2B. Conservative substitutions, for example such as leucine by valine and serine by threonine were identified in the β4-strand close to the active site. This suggests that a specific chemical environment is necessary for the catalytic domain to perform catalysis.

Such conservation of key regions and three disulfide bridges led us to hypothesize that the catalytic domain of all the TTSPs had structural similarities. Crystal structures of matriptase [47], DESC1 [49] and hepsin [48,60] have been solved previously. Therefore, matriptase-2, matriptase-3 and HAT were modeled and analyzed by superimposition of all the catalytic domains (Figure 2C), and typical structural elements of serine proteases of the trypsin fold are depicted. The first striking aspect observed was that the structures of all the TTSP catalytic domains are similar and highly superimposable with few major differences. One of these differences occurs in matriptase-3 where an extended loop in its catalytic domain following the β7 strand that is diverging from the other family members. This loop may interact with substrates or effectors that could induce a conformational change in the matriptase-3 catalytic domain and modulate substrate binding and presentation to the active site for catalysis. Another example is the extended 60-loop between the two 3_{10}-helices (η1 and η2) of matriptase, which potentially provides additional substrate specificity and possibly flip over the active site upon substrate binding, blocking further access until cleavage occurs. However, these differences in structure do not affect the active site geometry of the

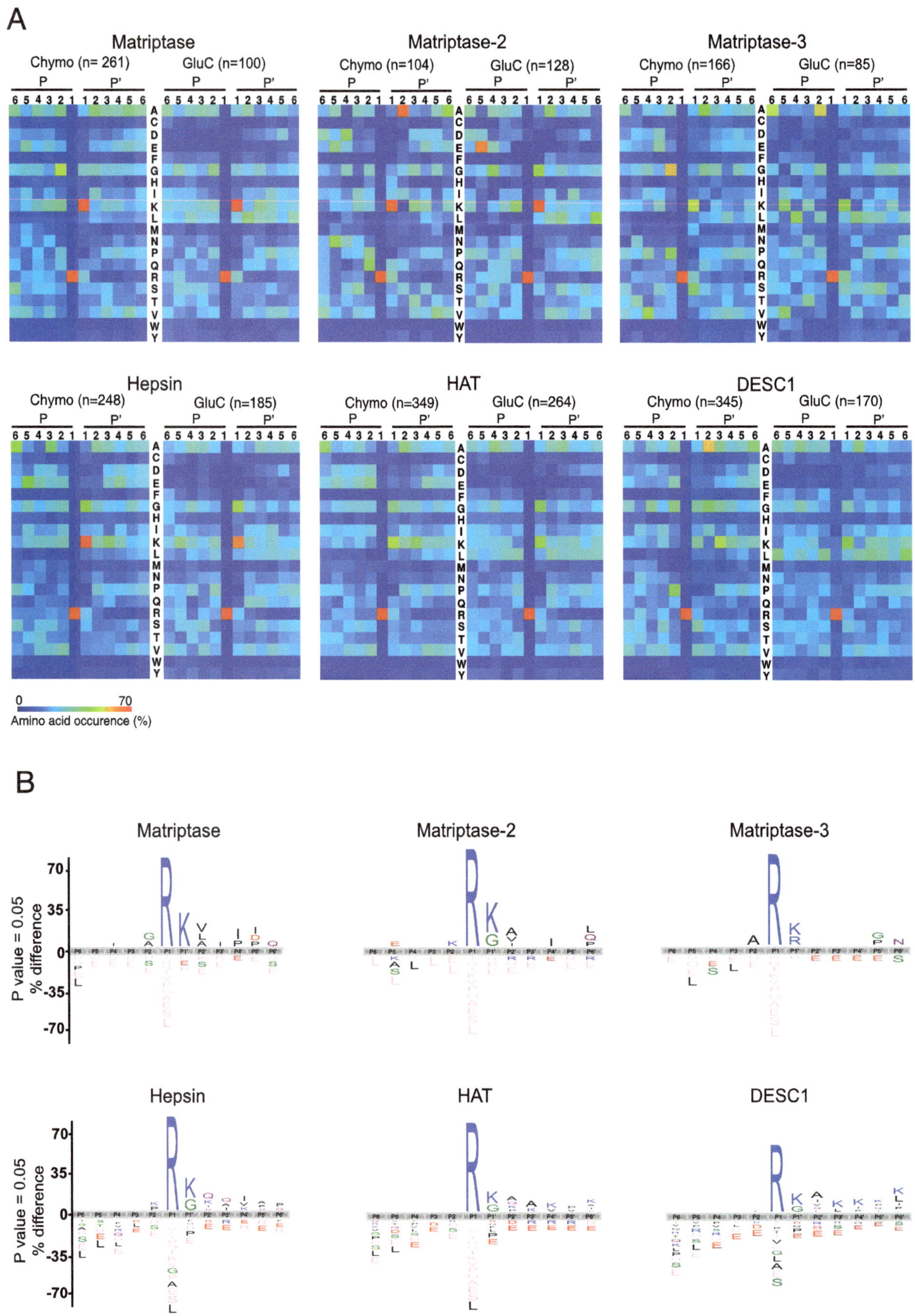
A
Matriptase
Chymo (n= 261)
GluC (n=100)
Matriptase-2
Chymo (n=104)
GluC (n=128)
Matriptase-3
Chymo (n=166)
GluC (n=85)
Hepsin
Chymo (n=248)
GluC (n=185)
HAT
Chymo (n=349)
GluC (n=264)
DESC1
Chymo (n=345)
GluC (n=170)
P
P'
6 5 4 3 2 1
1 2 3 4 5 6
A C D E F G H I K L M N P Q R S T V W Y
0
70
Amino acid occurence (%)
B
Matriptase
Matriptase-2
Matriptase-3
Hepsin
HAT
DESC1
P value = 0.05
% difference
70
35
0
-35
-70
P6 P5 P4 P3 P2 P1 P1' P2' P3' P4' P5' P6'

Figure 3. A. Matriptase, matriptase-2, matriptase-3, hepsin, DESC1, an HAT cleavage site specificities obtained by PICS using two different peptide libraries: GluC and chymotryptic. For each class of peptide library, the amino acid occurrences in P6–P6′ are shown as heat maps. Above each heat map, the number of identified cleaved peptides is specified. **B.** Taking into account the relative natural abundances of amino acids, iceLogo representations of the different cleavage site specificities obtained from the GluC library are shown. Peptide sequences were aligned on the scissile bond between P1 and P1′. Statistically significant amino acid residue occurrences present (P<0.01) were plotted and amino acid heights are indicative for the degree of conservation at the indicated position. Residues that were completely absent in the identified terminal peptides are shown below in pink.

catalytic triad in these TTSPs, and their specificity for peptidic substrates, but may explain differences in native substrate cleavage or in comparing peptide cleavage specificities with native protein cleavage specificities.

PICS analysis represented by heat maps and iceLogo

Using PICS, 2,405 cleavage sites were identified using both chymotryptic and GluC-generated proteome-derived peptide libraries, greatly expanding upon the known 144 cleavage sites annotated in MEROPS (Tables S1–S12 in File S1). These peptide

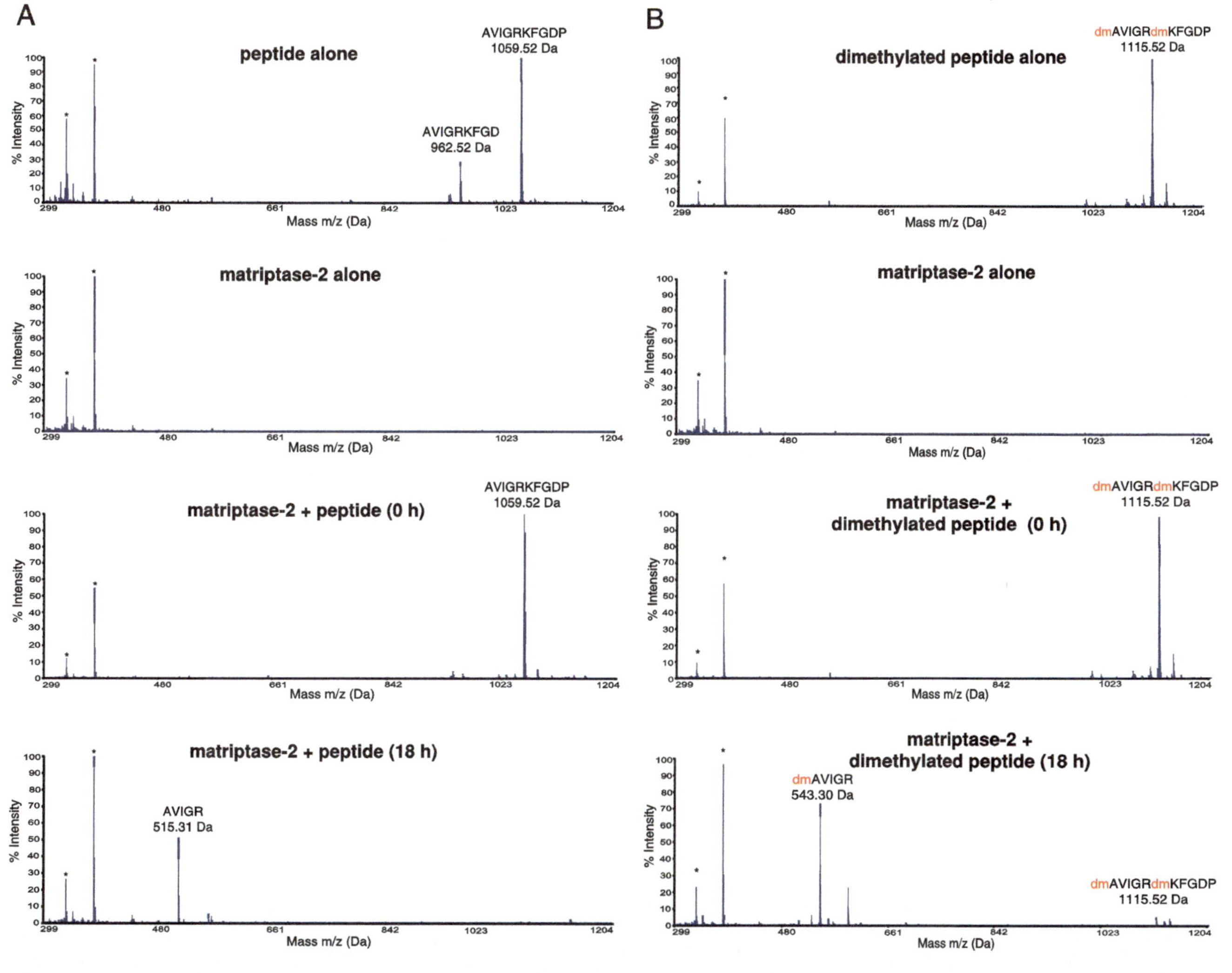

Figure 4. A. ***Upper panel.*** MALDI-TOF spectrum of the synthetic peptide AVIGRKFGDP. The sample contained a minor synthesis contaminant of AVIGRKFGD. The experimental determined $[M+H]^+$ is 1059.52 Da (predicted m/z is 1059.22 Da). The sequence is presented by the spectra. The two asterix peaks represent the MALDI matrix peaks which are found in all the spectra in this m/z range. ***Second panel.*** MALDI-TOF spectrum of matriptase-2. ***Third panel.*** MALDI-TOF spectrum of the synthetic peptide added to matriptase-2 at 0 h. ***Lower panel.*** MALDI-TOF spectrum of the assay reaction products generated after incubation of the synthetic peptide with matriptase-2 for 18 h. The spectral peak at 1059.52 m/z disappeared and a new peak at 515.31 Da appeared corresponding to the cleavage product $[AVIGR+H]^+$ (predicted m/z is 514.96 Da). **B.** ***Upper panel.*** MALDI-TOF spectrum of the dimethylated (dm) synthetic peptide (dm)AVIGR(dm)KFGDP. The experimental determined $[M+H]^+$ is 1115.52 Da. The sequence is presented on the spectral peak. The two asterix peaks represent the MALDI matrix peaks and they can be found in all the spectra. ***Second panel.*** MALDI-TOF spectrum of matriptase-2. ***Third panel.*** MALDI-TOF spectrum of the reaction products after incubation of the dimethylated synthetic peptide with matriptase-2 added at 0 h. ***Lower panel.*** MALDI-TOF spectrum of the reaction products after incubation of the dimethylated synthetic peptide with matriptase-2 for 18 h. The peak at 1115.52 Da disappeared and a new peak at 543.30 Da appeared, corresponding to $[dAVIGR+H]^+$. Red (dm) is dimethylation.

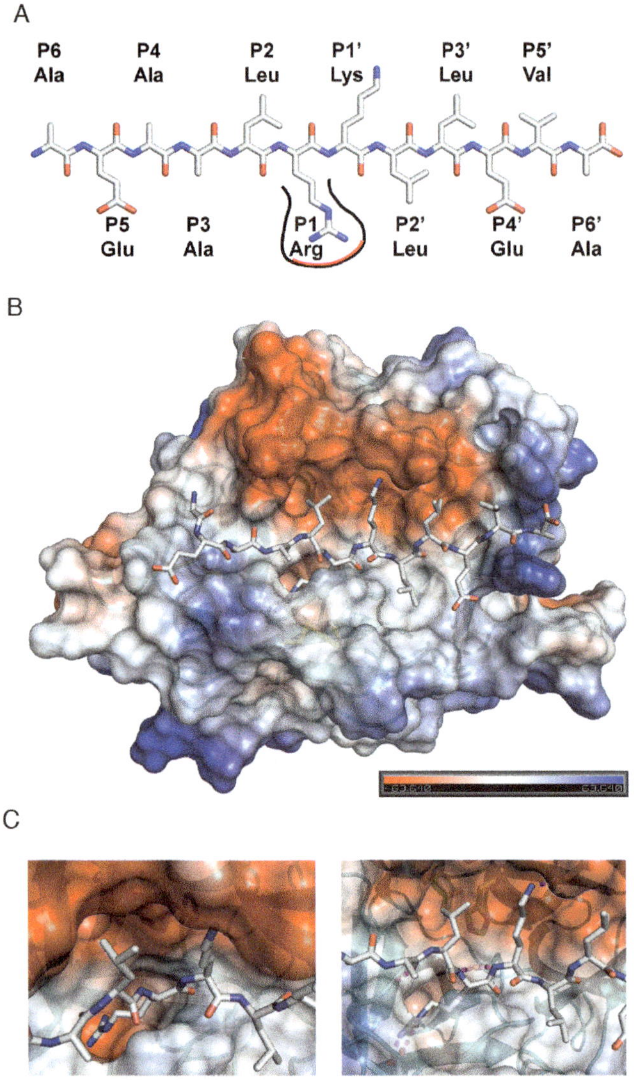

Figure 5. A Schematic representation of the dodecapeptide (AEAALR↓KLLEVA) used for active site docking. The deep TTSP S1 subsite, accommodating the P1-Arg, is shown. **B.** Structural model of matriptase-2 with the peptide substrate (AEAALRKLLEVA) identified by PICS docked into the active site. Matriptase-2 was colored according to its surface charge distribution, and is shown in standard orientation with the modeled peptide occupying subsites S6 to S6′ (from left to right). **C.** Close-up views of the active site illustrating the deep and negatively charged S1 pocket (left) and key stabilizing interaction between the protease and the modeled peptide (right).

libraries were selected because the TTSP family belongs to the trypsin-like family and so have a lysine or arginine specificity in P1. Thus, peptide libraries generated by trypsin for use with the PICS technique will contain peptides displaying an arginine or lysine at the C-terminus and hence will not contain internal basic residues. In such cases alternate peptide digestive proteases are easily used to generate appropriate peptide libraries. Here we used chymotryptic and GluC peptide libraries in which the C-terminal residues were aromatic/hydrophobic or negatively charged, respectively. These two libraries are overlapping and both libraries display internal lysine and arginine residues and complement each other by displaying acidic and hydrophobic residues within the peptide sequences. Using two or more libraries also allows for experimental validation of results obtained from one library with the other.

From the heat maps generated for the six human TTSPs, a common feature was apparent—arginine is preferred in the P1 position, ~17-fold more than the natural occurrence of arginine in proteins, and lysine is preferred in P1′ (Figure 3A). Note in the PICS workflow lysine is dimethylated. For several TTSPs, the P1′ position has glycine as an alternative amino acid preference, which is strongly favored in the case of hepsin. Similar specificities also occur within subfamilies. For example, for matriptase and matriptase-3, glycine or alanine is important at P2, whereas matriptase-2 prefers glutamate in P5 and alanine in P2′. Another example is across the HAT/DESC1 subfamily, where alanine and lysine are preferred in the P2′ and P3′ positions, respectively, for DESC1, but these are less preferred for HAT. HAT requires an aspartate in P2, a feature that was shown by Wysocka et al [31] using positional scanning libraries.

The iceLogo representation (Figure 3B) also revealed the least preferred amino acid preferences at each position. For example, since arginine is the only preferred amino acid at P1, this means that all other residues are thereby under represented. Similar negative preferences result from the lysine or glycine in P1′. Least preferred residues also reinforces the differences among members of the same subfamily. For example, matriptase and matriptase-3 prefer a glycine in P2, but matriptase-2 does not. Conversely, matriptase-2 favors a glutamate in P5, but this is not the case for matriptase and matriptase-3. This also points out the utility of PICS as it is easy to profile 6 amino acid residues in both the non-prime and prime sides, which other techniques generally do not do cover. Another interesting finding was the preference for alanine in the P2′ position for most of the TTSPs, especially important for matriptase-2 and DESC1, suggesting a specific feature for these two subfamilies.

To confirm the identified cleavage specificities of the six studied TTSPs we used synthetic peptides. We first examined the PICS mass spectrometry results in more detail and found that the AVIGRKFGDP peptide ($[M+H]^+$ = 1059.52) was cleaved by all TTSPs. This peptide was synthesized and incubated with matriptase-2 (Figure 4), matriptase, HAT and hepsin (Figures S1–S3), and the cleavage site was identified by MALDI-TOF MS. The N-terminal nonprime cleavage product AVIGR (515.30 Da) was identified after cleavage between Arg and Lys confirming the unusual P1′ specificity.

In PICS, all primary amines within the peptide library are dimethylated resulting in a dimethylated lysine side chain with altered charge, raising the question about the involvement of lysine in P1′. We found that the dimethylated (dm) version of the peptide (dm)AVIGR(dm)KFGDP was also cleaved by matriptase-2 (Figure 4B) and all other tested TTSPs with similar kinetics, validating the identified PICS cleavage sites for these TTSPs. Overall, our data reveal that the prime side is also important for the cleavage specificity of the TTSPs where assayed using unstructured peptide libraries. This makes the TTSPs unique within the tryptic-enzyme class that are dominated by P1 interactions.

Interpreting cleavage specificity by structural modeling of matriptase-2

Based on the known crystal structures, a matriptase-2 3D homology model was generated (Figure 2C and Figure 5), and a PICS determined model peptide, AEAALR↓KLLEVA (Figure 5A) was incorporated in the active site by peptide docking. An electrostatic surface charge representation of matriptase-2 including the docked peptide is shown in Figure 5B, illustrating

Table 1. Known protein substrates obtained from MEROPS [32] and UniProt [66] showing the P4-P4′ cleavage specificity in natural substrates of matriptase, matriptase-2, matriptase-3, HAT, hepsin and corin where cleaved as native proteins *versus* denatured peptides in PICS.

	Cleavage site								
Matriptase substrates	**P4**	**P3**	**P2**	**P1**	**P1′**	**P2′**	**P3′**	**P4′**	**References**
CDCP1	Lys	Gln	Ser	Arg	Lys	Phe	Val	Pro	[67]
fibronectin	Val	Thr	Gly	Arg	Gly	Asp	Ser	Pro	[68]
filaggrin	Arg	Lys	Arg	Arg	Gly	Ser	Arg	Val	[11]
filaggrin	Thr	Gly	His	Gly	Gln	Ala	Ser	Ser	[69]
hepatocyte growth factor	Lys	Gln	Leu	Arg	Val	Val	Asn	Gly	[11]
insulin-like growth factor-binding protein 1	Lys	Ala	Leu	His	Val	Thr	Asn	Ile	[11]
insulin-like growth factor-binding protein 7	Arg	Lys	Gly	Lys	Ala	Gly	Ala	Ala	[70]
laminin, alpha 1-like	Arg	Val	Val	Arg	Lys	Gly	Val	Ser	[68]
matriptase	Arg	Gln	Ala	Arg	Val	Val	Gly	Gly	[11]
prostasin	Ile	Gln	Pro	Arg	Ile	Thr	Gly	Gly	[71]
proteinase-activated receptor 2	Ser	Lys	Gly	Arg	Ser	Leu	Ile	Gly	[11]
stromelysin-1	Arg	Lys	Pro	Arg	Cys	Gly	Val	Pro	[72]
urokinase-type plasminogen activator	Pro	Arg	Phe	Lys	Ile	Ile	Gly	Gly	[11]
Matriptase-2 substrates									
matriptase-2 precursor	Pro	Ser	Ser	Arg	Ile	Val	Gly	Gly	[73]
matriptase-2	Pro	Gly	Val	Arg	Val	His	Tyr	Gly	[73]
hemojuvelin	Ile	Ile	Ile	Arg	Gln	Thr	Ala	Gly	[74]
Matriptase-3 substrates									
alpha-2-antiplasmin	Ala	Met	Ser	Arg	Met	Ser	Leu	Ser	[23]
antithrombin-III	Ile	Ala	Gly	Arg	Ser	Leu	Asn	Pro	[23]
heparin cofactor 2	Phe	Met	Pro	Leu	Ser	Thr	Gln	Val	[23]
plasma serine protease inhibitor	Phe	Thr	Phe	Arg	Ser	Ala	Arg	Leu	[23]
plasminogen activator inhibitor-1	Val	Ser	Ala	Arg	Met	Ala	Pro	Glu	[23]
Hepsin substrates									
coagulation factor IX	Asp	Phe	Thr	Arg	Val	Val	Gly	Gly	[60]
coagulation factor VII	Pro	Gln	Gly	Arg	Ile	Val	Gly	Gly	[75]
coagulation factor XII	Ser	Met	Thr	Arg	Val	Val	Gly	Gly	[60]
hepatocyte growth factor	Lys	Gln	Leu	Arg	Val	Val	Asn	Gly	[60]
hepsin precursor	Pro	Val	Asp	Arg	Ile	Val	Gly	Gly	[60]
laminin subunit beta-3	Ser	Gln	Leu	Arg	Leu	Gln	Gly	Ser	[61]
urokinase-type plasminogen activator	Pro	Arg	Phe	Lys	Ile	Ile	Gly	Gly	[76]
Corin substrates									

Table 1. Cont.

				Cleavage site					
Matriptase substrates	**P4**	**P3**	**P2**	**P1**	**P1′**	**P2′**	**P3′**	**P4′**	**References**
atrial natriuretic peptide-converting enzyme	Gly	Asp	Gln	Arg	Cys	Leu	Tyr	Asn	[77]
pro-atrial natriuretic peptide	Thr	Ala	Pro	Arg	Ser	Leu	Arg	Arg	[18]
HAT substrates									
urokinase plasminogen activator surface receptor	Asn	Ser	Gly	Arg	Ala	Val	Thr	Tyr	[78]
urokinase plasminogen activator surface receptor	Thr	Tyr	Ser	Arg	Ser	Arg	Tyr	Leu	[78]

the negative charge distribution of the upper half of the active site including subsite S2 and S1′, and the slightly positive subsites S5 and S4′, explaining their opposed preferences for charged residues. The deep S1 pocket with Asp756 forming the negative base and main interactor of the P1-Arg is depicted in Figure 5C. Both the groove and the electrostatic environment suggest that a positively charged amino acid composed of a long side chain can be readily accommodated and stabilized in S1. Indeed, this would explain why an arginine is highly favored at the P1 position. Noteworthy, both backbone nitrogens of the P1 arginine and the P1′ lysine are hold in position by the active site Ser762. Specificity for lysine in P1′ is mainly provided by the negatively charged side-chain of Glu621 at the top of the S1 subsite (Figure 5C). We suggest that these features are the key players for the preferred cleavage between arginine in P1 and lysine in P1′.

Glycine 784 of the protease backbone connects to the P2 position of the peptide through hydrogen bonding and stabilizes this atom at the non-prime site of the peptide. The slight preference in this position for glycine might be guided by the high flexibility of its α carbon, which may compensate for less-preferred amino acids in other subsites. Subsite P3′, which favors small hydrophobic residues for most of the investigated TTSPs, is stabilized by anti-parallel backbone-backbone interaction of the peptide to the protease.

Discussion

By utilizing diverse natural peptides of variable length and amino acid composition, PICS identifies the cleavage specificity of a protease and the exact sequence on both sides of the scissile bond by analysis of hundreds of cleaved peptides in each experiment [33,34]. The mass spectrometric determination of the active site specificity of six important members of the TTSP subfamilies using PICS revealed several important features. First, it is an unusual finding for tryptic proteases for their specificity to be determined by both the P1 and P1′ positions. Second it is also unusual for tryptic proteases to have specificity exerted through extended features on the prime side. The specificity for arginine in P1 is in agreement with other recent reports using other techniques such as positional scanning libraries and/or fluorescent substrates [15,23,29,31,60,61]. Furthermore, almost all of the known protein substrates of the assayed TTSPs contain an arginine in P1 position (Table 1). However, we cannot rule out that native substrates with a lysine in P1 might be cleavable since all lysines in PICS are dimethylated to protect against NHS-biotin reactivity, which is used to selectively tag the neo-N terminal α-amine for enrichment after cleavage. Third, the specificity for lysine in P1′ is shown for the first time using a peptide library technique in our study and is consistent with the natural substrate specificity of matriptase, which cleaves for example CDCP1 and laminin, alpha 1-like between arginine and lysine (Table 1). Consistent with our findings, Beliveau *et al* [29] demonstrated that negatively charged amino acids are the least preferred at P1′ position as shown by having a very low k_{cat}/K_M.

PICS showed that if the nature of the peptide substrate was negatively charged, non-polar, or hydrophobic, then these characteristics affected the amino acid preferences accommodated in the TTSP active sites depending on which library was used. For example, using a chymotryptic library, a preference for aspartate and glutamate was identified at the P3, P4 and P5 positions for hepsin, matriptase, matriptase-3 and DESC1, whereas peptides generated by GluC had a leucine and/or lysine preferred at P3 and P4. Since aspartate and glutamate are displayed internally in the chymotryptic peptides these libraries better reflect the true specificities for acidic residues than can be derived from GluC libraries, which are internally depleted for negatively charged residues. This reaffirms the need for using at least two libraries in PICS to accurately interpret cleavage specificity preferences for proteases [33,34].

It was notable that the prime side of a peptide is important for TTSP cleavage specificity. Thus alanine or other small hydrophobic residues are required in the P2′ position, as well as nonpolar amino acids such as leucine and valine at the other prime positions. This feature can also be observed in natural substrates of matriptase (hepatocyte growth factor), matriptase-3 (plasminogen activator-inhibitor-1) and hepsin (coagulation factor IX and VII) (Table 1).

The application of PICS was shown to be highly successful by the identification of 2,405 cleavage sites for the six TTSP family members studied. These data greatly expand the 144 known cleavages sites for these proteases. The specific amino acid residue features found in our study of three subfamilies of the TTSPs provide new insight that might be used to design specific inhibitors and peptide substrates, such as AEAALR ↓ KLLEVA that we identified here and validated. It is important to keep in mind that the overall similarities in specificity for the TTSPs will likely render small molecule drug development challenging. The presence of several domains outside the catalytic domain suggests that these exosites could be critical for substrate binding and recognition [62] and undoubtedly influence native substrate binding and cleavage of intact proteins. Indeed many native protein substrates do not have Lys in P1′. Native protein structures often mask even highly preferred cleavage sites rendering these unavailable for protease access and cleavage. However, this can be levered to advantage in designing small molecule inhibitors. Thus future studies of exosite interactions, and possibly designing exosite inhibitors, will be insightful in revealing TTSP specificity for natural substrates *in vivo* and hence *in vivo* roles.

Acknowledgments

O.B. was supported by the Swiss National Science Foundation (SNF) and a post-doctoral fellowship by the Canadian Institutes of Health Research (CIHR). A.D. and U.E. were supported by post-doctoral fellowships from CIHR and the Michael Smith Foundation for Health Research (MSFHR), respectively. A Canada Research Chair in Metalloproteinase Proteomics and Systems Biology supports C.M.O. This work was supported by grants to C.M.O. and R.L. from the CIHR and the Canadian Cancer Society. We thank Wei Chen, Suzanne Perry and Jason Rogalski for LC-MS/MS measurements. The University of British Columbia, Centre for Blood Research Mass Spectrometry Suite is supported by the Canada Foundation for Innovation (CFI) and MSFHR. We also thank Ursula Bodendorf, Thomas Bugge and Roman Szabo for their generous donation of plasmids and the cell line expressing Matriptase-3.

Author Contributions

Conceived and designed the experiments: OB FB RL CMO. Performed the experiments: OB AD UE RK. Analyzed the data: OB AD UE RK CMO. Contributed reagents/materials/analysis tools: OB AD UE RK FB RL CMO. Wrote the paper: OB AD RK UE CMO. Prepared all the figures: OB AD RK UE. Produced and purified the TTSPs: FB.

References

1. Leytus SP, Loeb KR, Hagen FS, Kurachi K, Davie EW (1988) A novel trypsin-like serine protease (hepsin) with a putative transmembrane domain expressed by human liver and hepatoma cells. Biochemistry 27: 1067–1074.
2. Rodríguez D, Morrison CJ, Overall CM (2010) Matrix metalloproteinases: What do they not do? New substrates and biological roles identified by murine models and proteomics. Biochimica et Biophysica Acta (BBA) - Molecular Cell Research 1803: 39–54. doi:10.1016/j.bbamcr.2009.09.015.
3. Overall CM, Blobel CP (2007) In search of partners: linking extracellular proteases to substrates. Nat Rev Mol Cell Biol 8: 245–257. doi:10.1038/nrm2120.
4. Netzel-Arnett S, Hooper JD, Szabo R, Madison EL, Quigley JP, et al. (2003) Membrane anchored serine proteases: a rapidly expanding group of cell surface proteolytic enzymes with potential roles in cancer. Cancer Metastasis Rev 22: 237–258.
5. Szabo R, Bugge TH (2008) Type II transmembrane serine proteases in development and disease. Int J Biochem Cell Biol 40: 1297–1316. doi:10.1016/j.biocel.2007.11.013.
6. Hooper JD, Clements JA, Quigley JP, Antalis TM (2001) Type II transmembrane serine proteases. Insights into an emerging class of cell surface proteolytic enzymes. J Biol Chem 276: 857–860. doi:10.1074/jbc.R000020200.
7. Colaert N, Helsens K, Martens L, Vandekerckhove J, Gevaert K (2009) Improved visualization of protein consensus sequences by iceLogo. Nat Methods 6: 786–787. doi:10.1038/nmeth1109-786.
8. Wu F, Yan W, Pan J, Morser J, Wu Q (2002) Processing of pro-atrial natriuretic peptide by corin in cardiac myocytes. J Biol Chem 277: 16900–16905. doi:10.1074/jbc.M201503200.
9. Bugge TH, List K, Szabo R (2007) Matriptase-dependent cell surface proteolysis in epithelial development and pathogenesis. Front Biosci 12: 5060–5070.
10. List K, Haudenschild CC, Szabo R, Chen W, Wahl SM, et al. (2002) Matriptase/MT-SP1 is required for postnatal survival, epidermal barrier function, hair follicle development, and thymic homeostasis. Oncogene 21: 3765–3779. doi:10.1038/sj.onc.1205502.
11. Uhland K (2006) Matriptase and its putative role in cancer. Cell Mol Life Sci 63: 2968–2978. doi:10.1007/s00018-006-6298-x.
12. Sales KU, Hobson JP, Wagenaar-Miller R, Szabo R, Rasmussen AL, et al. (2011) Expression and Genetic Loss of Function Analysis of the HAT/DESC Cluster Proteases TMPRSS11A and HAT. PLoS ONE 6: e23261. doi:10.1371/journal.pone.0023261.
13. Yamaoka K, Masuda K, Ogawa H, Takagi K, Umemoto N, et al. (1998) Cloning and characterization of the cDNA for human airway trypsin-like protease. J Biol Chem 273: 11895–11901.
14. Yasuoka S, Ohnishi T, Kawano S, Tsuchihashi S, Ogawara M, et al. (1997) Purification, characterization, and localization of a novel trypsin-like protease found in the human airway. Am J Respir Cell Mol Biol 16: 300–308. doi:10.1165/ajrcmb.16.3.9070615.
15. Hobson JP, Netzel-Arnett S, Szabo R, Réhault SM, Church FC, et al. (2004) Mouse DESC1 is located within a cluster of seven DESC1-like genes and encodes a type II transmembrane serine protease that forms serpin inhibitory complexes. J Biol Chem 279: 46981–46994. doi:10.1074/jbc.M403299200.
16. Lang JC, Schuller DE (2001) Differential expression of a novel serine protease homologue in squamous cell carcinoma of the head and neck. Br J Cancer 84: 237–243. doi:10.1054/bjoc.2000.1586.
17. Dhanasekaran SM, Barrette TR, Ghosh D, Shah R, Varambally S, et al. (2001) Delineation of prognostic biomarkers in prostate cancer. Nature 412: 822–826. doi:10.1038/35090585.
18. Kitamoto Y, Yuan X, Wu Q, McCourt DW, Sadler JE (1994) Enterokinase, the initiator of intestinal digestion, is a mosaic protease composed of a distinctive assortment of domains. Proc Natl Acad Sci USA 91: 7588–7592.
19. Folgueras AR, de Lara FM, Pendás AM, Garabaya C, Rodríguez F, et al. (2008) Membrane-bound serine protease matriptase-2 (Tmprss6) is an essential regulator of iron homeostasis. Blood 112: 2539–2545. doi:10.1182/blood-2008-04-149773.
20. Silvestri L, Pagani A, Nai A, De Domenico I, Kaplan J, et al. (2008) The serine protease matriptase-2 (TMPRSS6) inhibits hepcidin activation by cleaving membrane hemojuvelin. Cell Metab 8: 502–511. doi:10.1016/j.cmet.2008.09.012.
21. Du X, She E, Gelbart T, Truksa J, Lee P, et al. (2008) The serine protease TMPRSS6 is required to sense iron deficiency. Science 320: 1088–1092. doi:10.1126/science.1157121.
22. Ramsay AJ, Hooper JD, Folgueras AR, Velasco G, López-Otín C (2009) Matriptase-2 (TMPRSS6): a proteolytic regulator of iron homeostasis. Haematologica 94: 840–849. doi:10.3324/haematol.2008.001867.

23. Szabo R, Netzel-Arnett S, Hobson JP, Antalis TM, Bugge TH (2005) Matriptase-3 is a novel phylogenetically preserved membrane-anchored serine protease with broad serpin reactivity. Biochem J 390: 231–242. doi:10.1042/BJ20050299.
24. Cal S, Quesada V, Garabaya C, López-Otín C (2003) Polyserase-I, a human polyprotease with the ability to generate independent serine protease domains from a single translation product. Proc Natl Acad Sci USA 100: 9185–9190. doi:10.1073/pnas.1633392100.
25. Hooper JD, Scarman AL, Clarke BE, Normyle JF, Antalis TM (2000) Localization of the mosaic transmembrane serine protease corin to heart myocytes. Eur J Biochem 267: 6931–6937.
26. Chan JCY, Knudson O, Wu F, Morser J, Dole WP, et al. (2005) Hypertension in mice lacking the proatrial natriuretic peptide convertase corin. Proc Natl Acad Sci USA 102: 785–790. doi:10.1073/pnas.0407234102.
27. López-Otín C, Overall CM (2002) Protease degradomics: a new challenge for proteomics. Nat Rev Mol Cell Biol 3: 509–519. doi:10.1038/nrm858.
28. Lee SL, Dickson RB, Lin CY (2000) Activation of hepatocyte growth factor and urokinase/plasminogen activator by matriptase, an epithelial membrane serine protease. J Biol Chem 275: 36720–36725. doi:10.1074/jbc.M007802200.
29. Béliveau F, Désilets A, Leduc R (2009) Probing the substrate specificities of matriptase, matriptase-2, hepsin and DESC1 with internally quenched fluorescent peptides. FEBS Journal 276: 2213–2226. doi:10.1111/j.1742-4658.2009.06950.x.
30. Velasco G, Cal S, Quesada V, Sánchez LM, López-Otín C (2002) Matriptase-2, a membrane-bound mosaic serine proteinase predominantly expressed in human liver and showing degrading activity against extracellular matrix proteins. J Biol Chem 277: 37637–37646. doi:10.1074/jbc.M203007200.
31. Wysocka M, Spichalska B, Lesner A, Jaros M, Brzozowski K, et al. (2010) Substrate specificity and inhibitory study of human airway trypsin-like protease. Bioorg Med Chem 18: 5504–5509. doi:10.1016/j.bmc.2010.06.059.
32. Rawlings ND, Waller M, Barrett AJ, Bateman A (2014) MEROPS: the database of proteolytic enzymes, their substrates and inhibitors. Nucleic Acids Res 42: D503–D509. doi:10.1093/nar/gkt953.
33. Schilling O, Overall CM (2008) Proteome-derived, database-searchable peptide libraries for identifying protease cleavage sites. Nat Biotechnol 26: 685–694. doi:10.1038/nbt1408.
34. Schilling O, Huesgen PF, Barré O, auf dem Keller U, Overall CM (2011) Characterization of the prime and non-prime active site specificities of proteases by proteome-derived peptide libraries and tandem mass spectrometry. Nat Protoc 6: 111–120. doi:10.1038/nprot.2010.178.
35. Starr AE, Bellac CL, Dufour A, Goebeler V, Overall CM (2012) Biochemical characterization and N-terminomics analysis of leukolysin, the membrane-type 6 matrix metalloprotease (MMP25): chemokine and vimentin cleavages enhance cell migration and macrophage phagocytic activities. J Biol Chem 287: 13382–13395. doi:10.1074/jbc.M111.314179.
36. Eckhard Ulrich, Huesgen PF, Brandstetter H, Overall CM (2014) Proteomic protease specificity profiling of clostridial collagenases reveals their intrinsic nature as dedicated degraders of collagen. J Proteomics 100: 102–114. doi:10.1016/j.jprot.2013.10.004.
37. Marino G, Huesgen PF, Eckhard Ulrich, Overall CM, Schröder WP, et al. (2014) Family-wide characterization of matrix metalloproteinases from Arabidopsis thaliana reveals their distinct proteolytic activity and cleavage site specificity. Biochem J 457: 335–346. doi:10.1042/BJ20130196.
38. Becker-Pauly C, Barré O, Schilling O, auf dem Keller U, Ohler A, et al. (2011) Proteomic analyses reveal an acidic prime side specificity for the astacin metalloprotease family reflected by physiological substrates. Mol Cell Proteomics 10: M111.009233. doi:10.1074/mcp.M111.009233.
39. Schilling O, auf dem Keller U, Overall CM (2011) Factor Xa subsite mapping by proteome-derived peptide libraries improved using WebPICS, a resource for proteomic identification of cleavage sites. Biol Chem 392: 1031–1037. doi:10.1515/BC.2011.158.
40. Kappelhoff R, Overall C (2001) The CLIP-CHIP Oligonucleotide Microarray: Dedicated Array for Analysis of All Protease, Nonproteolytic Homolog, and Inhibitor Gene Transcripts in Human and Mouse. Hoboken, NJ, USA: John Wiley & Sons, Inc. doi:10.1002/0471140864.ps2119s56.
41. Kappelhoff R, auf dem Keller U, Overall CM (2010) Analysis of the degradome with the CLIP-CHIP microarray. Methods Mol Biol 622: 175–193. doi:10.1007/978-1-60327-299-5_10.
42. Perkins DN, Pappin DJ, Creasy DM, Cottrell JS (1999) Probability-based protein identification by searching sequence databases using mass spectrometry data. Electrophoresis 20: 3551–3567. doi:10.1002/(SICI)1522-2683(19991201)20:18<3551:AID-ELPS3551>3.0.CO;2-2.
43. Craig R, Beavis RC (2004) TANDEM: matching proteins with tandem mass spectra. Bioinformatics 20: 1466–1467. doi:10.1093/bioinformatics/bth092.
44. Keller A, Nesvizhskii AI, Kolker E, Aebersold R (2002) Empirical statistical model to estimate the accuracy of peptide identifications made by MS/MS and database search. Anal Chem 74: 5383–5392.
45. Shteynberg D, Deutsch EW, Lam H, Eng JK, Sun Z, et al. (2011) iProphet: multi-level integrative analysis of shotgun proteomic data improves peptide and protein identification rates and error estimates. Mol Cell Proteomics 10: M111.007690. doi:10.1074/mcp.M111.007690.
46. Deutsch EW, Mendoza L, Shteynberg D, Farrah T, Lam H, et al. (2010) A guided tour of the Trans-Proteomic Pipeline. Proteomics 10: 1150–1159. doi:10.1002/pmic.200900375.
47. Friedrich R, Fuentes-Prior P, Ong E, Coombs G, Hunter M, et al. (2002) Catalytic domain structures of MT-SP1/matriptase, a matrix-degrading transmembrane serine proteinase. J Biol Chem 277: 2160–2168. doi:10.1074/jbc.M109830200.
48. Katz BA, Luong C, Ho JD, Somoza JR, Gjerstad E, et al. (2004) Dissecting and designing inhibitor selectivity determinants at the S1 site using an artificial Ala190 protease (Ala190 uPA). J Mol Biol 344: 527–547. doi:10.1016/j.jmb.2004.09.032.
49. Kyrieleis OJP, Huber R, Ong E, Oehler R, Hunter M, et al. (2007) Crystal structure of the catalytic domain of DESC1, a new member of the type II transmembrane serine proteinase family. FEBS J 274: 2148–2160. doi:10.1111/j.1742-4658.2007.05756.x.
50. Bordoli L, Kiefer F, Arnold K, Benkert P, Battey J, et al. (2009) Protein structure homology modeling using SWISS-MODEL workspace. Nat Protoc 4: 1–13. doi:10.1038/nprot.2008.197.
51. Arnold K, Bordoli L, Kopp J, Schwede T (2006) The SWISS-MODEL workspace: a web-based environment for protein structure homology modelling. Bioinformatics 22: 195–201. doi:10.1093/bioinformatics/bti770.
52. Xu D, Jaroszewski L, Li Z, Godzik A (2014) FFAS-3D: improving fold recognition by including optimized structural features and template re-ranking. Bioinformatics 30: 660–667. doi:10.1093/bioinformatics/btt578.
53. Jaroszewski L, Li Z, Cai X-H, Weber C, Godzik A (2011) FFAS server: novel features and applications. Nucleic Acids Res 39: W38–W44. doi:10.1093/nar/gkr441.
54. Bond CS (2003) TopDraw: a sketchpad for protein structure topology cartoons. Bioinformatics 19: 311–312.
55. London N, Raveh B, Cohen E, Fathi G, Schueler-Furman O (2011) Rosetta FlexPepDock web server—high resolution modeling of peptide-protein interactions. Nucleic Acids Res 39: W249–W253. doi:10.1093/nar/gkr431.
56. Raveh B, London N, Schueler-Furman O (2010) Sub-angstrom modeling of complexes between flexible peptides and globular proteins. Proteins 78: 2029–2040. doi:10.1002/prot.22716.
57. Yuan C, Chen L, Meehan EJ, Daly N, Craik DJ, et al. (2011) Structure of catalytic domain of Matriptase in complex with Sunflower trypsin inhibitor-1. BMC Struct Biol 11: 30. doi:10.1186/1472-6807-11-30.
58. DeLano WL (2005) The case for open-source software in drug discovery. Drug Discov Today 10: 213–217. doi:10.1016/S1359-6446(04)03363-X.
59. Sievers F, Wilm A, Dineen D, Gibson TJ, Karplus K, et al. (2011) Fast, scalable generation of high-quality protein multiple sequence alignments using Clustal Omega. Mol Syst Biol 7: 539. doi:10.1038/msb.2011.75.
60. Herter S, Piper DE, Aaron W, Gabriele T, Cutler G, et al. (2005) Hepatocyte growth factor is a preferred in vitro substrate for human hepsin, a membrane-anchored serine protease implicated in prostate and ovarian cancers. Biochem J 390: 125–136. doi:10.1042/BJ20041955.
61. Tripathi M, Nandana S, Yamashita H, Ganesan R, Kirchhofer D, et al. (2008) Laminin-332 is a substrate for hepsin, a protease associated with prostate cancer progression. J Biol Chem 283: 30576–30584. doi:10.1074/jbc.M802312200.
62. Overall CM (2002) Molecular determinants of metalloproteinase substrate specificity: matrix metalloproteinase substrate binding domains, modules, and exosites. Mol Biotechnol 22: 51–86. doi:10.1385/MB:22:1:051.
63. Gouet P, Robert X, Courcelle E (2003) ESPript/ENDscript: Extracting and rendering sequence and 3D information from atomic structures of proteins. Nucleic Acids Res 31: 3320–3323.
64. Bode W, Mayr I, Baumann U, Huber R, Stone SR, et al. (1989) The refined 1.9 A crystal structure of human alpha-thrombin: interaction with D-Phe-Pro-Arg chloromethylketone and significance of the Tyr-Pro-Pro-Trp insertion segment. EMBO J 8: 3467–3475.
65. Huber R, Kukla D, Bode W, Schwager P, Bartels K, et al. (1974) Structure of the complex formed by bovine trypsin and bovine pancreatic trypsin inhibitor. II. Crystallographic refinement at 1.9 A resolution. J Mol Biol 89: 73–101.
66. Magrane M, Consortium U (2011) UniProt Knowledgebase: a hub of integrated protein data. Database (Oxford) 2011: bar009. doi:10.1093/database/bar009.
67. Bhatt AS, Erdjument-Bromage H, Tempst P, Craik CS, Moasser MM (2005) Adhesion signaling by a novel mitotic substrate of src kinases. Oncogene 24: 5333–5343. doi:10.1038/sj.onc.1208582.
68. Satomi S, Yamasaki Y, Tsuzuki S, Hitomi Y, Iwanaga T, et al. (2001) A Role for Membrane-Type Serine Protease (MT-SP1) in Intestinal Epithelial Turnover. Biochemical and Biophysical Research Communications 287: 995–1002. doi:10.1006/bbrc.2001.5686.
69. Darragh MR, Bhatt AS, Craik CS (2008) MT-SP1 proteolysis and regulation of cell-microenvironment interactions. Front Biosci 13: 528–539.
70. Ahmed S, Jin X, Yagi M, Yasuda C, Sato Y, et al. (2006) Identification of membrane-bound serine proteinase matriptase as processing enzyme of insulin-like growth factor binding protein-related protein-1 (IGFBP-rP1/angiomodulin/mac25). FEBS Journal 273: 615–627. doi:10.1111/j.1742-4658.2005.05094.x.
71. Netzel-Arnett S, Currie BM, Szabo R, Lin CY, Chen LM, et al. (2006) Evidence for a Matriptase-Prostasin Proteolytic Cascade Regulating Terminal Epidermal Differentiation. Journal of Biological Chemistry 281: 32941–32945. doi:10.1074/jbc.C600208200.
72. Jin X, Yagi M, Akiyama N, Hirosaki T, Higashi S, et al. (2006) Matriptase activates stromelysin (MMP-3) and promotes tumor growth and angiogenesis. Cancer Science 97: 1327–1334. doi:10.1111/j.1349-7006.2006.00328.x.
73. Stirnberg M, Maurer E, Horstmeyer A, Kolp S, Frank S, et al. (2010) Proteolytic processing of the serine protease matriptase-2: identification of the cleavage sites

required for its autocatalytic release from the cell surface. Biochem J 430: 87–95. doi:10.1042/BJ20091565.

74. Maxson JE, Chen J, Enns CA, Zhang AS (2010) Matriptase-2- and Proprotein Convertase-cleaved Forms of Hemojuvelin Have Different Roles in the Down-regulation of Hepcidin Expression. Journal of Biological Chemistry 285: 39021–39028. doi:10.1074/jbc.M110.183160.
75. Kazama Y, Hamamoto T, Foster DC, Kisiel W (1995) Hepsin, a putative membrane-associated serine protease, activates human factor VII and initiates a pathway of blood coagulation on the cell surface leading to thrombin formation. J Biol Chem 270: 66–72.
76. Moran P, Li W, Fan B, Vij R, Eigenbrot C, et al. (2006) Pro-urokinase-type plasminogen activator is a substrate for hepsin. J Biol Chem 281: 30439–30446. doi:10.1074/jbc.M605440200.
77. Jiang J, Wu S, Wang W, Chen S, Peng J, et al. (2011) Ectodomain shedding and autocleavage of the cardiac membrane protease corin. Journal of Biological Chemistry 286: 10066–10072. doi:10.1074/jbc.M110.185082.
78. Beaufort N, Leduc D, Eguchi H, Mengele K, Hellmann D, et al. (2007) The human airway trypsin-like protease modulates the urokinase receptor (uPAR, CD87) structure and functions. AJP: Lung Cellular and Molecular Physiology 292: L1263–L1272. doi:10.1152/ajplung.00191.2006.

22

Reproducibility of an HPLC-ESI-MS/MS Method for the Measurement of Stable-Isotope Enrichment of *in Vivo*-Labeled Muscle ATP Synthase Beta Subunit

Sarah Everman[1,2], Zhengping Yi[1,2,3], Paul Langlais[1,2], Lawrence J. Mandarino[1,2], Moulun Luo[1,2], Christine Roberts[1,2], Christos S. Katsanos[1,2]*

1 Center for Metabolic and Vascular Biology, School of Life Sciences, Arizona State University, Tempe, Arizona, United States of America, **2** Mayo Clinic Arizona, Scottsdale, Arizona, United States of America, **3** Department of Pharmaceutical Sciences, Eugene Applebaum College of Pharmacy/Health Sciences, Wayne State University, Detroit, Michigan, United States of America

Abstract

We sought to evaluate the reproducibility of a liquid chromatography-tandem mass spectrometry (LC-MS/MS)-based approach to measure the stable-isotope enrichment of *in vivo*-labeled muscle ATP synthase β subunit (β-F1-ATPase), a protein most directly involved in ATP production, and whose abundance is reduced under a variety of circumstances. Muscle was obtained from a rat infused with stable-isotope-labeled leucine. The muscle was homogenized, β-F1-ATPase immunoprecipitated, and the protein was resolved using 1D-SDS PAGE. Following trypsin digestion of the isolated protein, the resultant peptide mixtures were subjected to analysis by HPLC-ESI-MS/MS, which resulted in the detection of multiple β-F1-ATPase peptides. There were three β-F1-ATPase unique peptides with a leucine residue in the amino acid sequence, and which were detected with high intensity relative to other peptides and assigned with >95% probability to β-F1-ATPase. These peptides were specifically targeted for fragmentation to access their stable-isotope enrichment based on MS/MS peak areas calculated from extracted ion chromatographs for selected labeled and unlabeled fragment ions. Results showed best linearity ($R^2 = 0.99$) in the detection of MS/MS peak areas for both labeled and unlabeled fragment ions, over a wide range of amounts of injected protein, specifically for the β-F1-ATPase$_{134-143}$ peptide. Measured stable-isotope enrichment was highly reproducible for the β-F1-ATPase$_{134-143}$ peptide (CV = 2.9%). Further, using mixtures of synthetic labeled and unlabeled peptides we determined that there is an excellent linear relationship ($R^2 = 0.99$) between measured and predicted enrichment for percent enrichments ranging between 0.009% and 8.185% for the β-F1-ATPase$_{134-143}$ peptide. The described approach provides a reliable approach to measure the stable-isotope enrichment of *in-vivo*-labeled muscle β-F1-ATPase based on the determination of the enrichment of the β-F1-ATPase$_{134-143}$ peptide.

Editor: Arto Urtti, University of Helsinki, Finland

Funding: The work conducted at the Center for Metabolic and Vascular Biology at Arizona State University and was partially supported by Arizona State University start-up funds (to CK) and by the National Institutes of Health grant R01DK47936 (to LM). The funders had no role in study design, data collection and analysis, decision to publish, or preparation of the manuscript.

Competing Interests: The authors have declared that no competing interests exist.

* E-mail: christos.katsanos@asu.edu

Introduction

There are well-established approaches to measure the protein synthesis in skeletal muscle using stable-isotope-labeled amino acids [1]. These approaches are based on the *in vivo*-metabolic labeling of proteins followed by the measurement of the stable-isotope enrichment of, typically, a mixture of proteins using gas chromatography-mass spectrometry (GC-MS). Measurement of this enrichment provides the means to determine a global/average rate of synthesis across proteins in skeletal muscle. Measurement of the enrichment of a single protein that is present in small amounts in skeletal muscle cannot, however, be practically performed using these approaches because of the low mass sensitivity of the GC-MS. Yet, it is the measurement of the enrichment at the individual protein level that it is necessary for the determination of the rate of synthesis of a specific protein.

At any time, muscle requires a continuous supply of energy in the form of adenosine triphosphate (ATP), which supports various metabolic processes within muscle and maintains muscle's function. The demand for ATP production in muscle increases several-fold as a result of physical activity. More than 90% of ATP in skeletal muscle is produced by the enzyme ATP synthase in mitochondria. The beta subunit of the ATP synthase (β-F1-ATPase) is important from a functional standpoint because it forms the catalytic site of the enzyme, and thus has a key role in determining the capacity for ATP production. In circumstances associated with reduced capacity for ATP production in muscle, such as obesity [2] and aging [3], reduced amounts of muscle β-F1-ATPase have been reported [4,5]. Reduced amounts of β-F1-ATPase have also been reported in liver and tumors in conditions associated with hypothyroidism and cancer, respectively [6,7,8].

Understanding the mechanisms behind the reduced levels of β-F1-ATPase in skeletal muscle requires knowledge of the rates of synthesis/breakdown of β-F1-ATPase. In addition to maintaining the quantity of muscle β-F1-ATPase, the rate of synthesis of β-F1-ATPase plays a key role in maintaining the quality of protein by continuously renewing the β-F1-ATPase pool in skeletal muscle. As a result of the latter process, older and/or damaged molecules of the protein are replaced by de novo synthesized β-F1-ATPase molecules. A key part in the overall process of determining the kinetics of muscle β-F1-ATPase in association with the infusion of stable-isotope amino acids is the ability to reliably measure the stable isotope enrichment of β-F1-ATPase. Recently, Bateman et al. [9] have described a liquid chromatography-tandem mass spectrometry (LC-MS/MS) approach for the measurement of amyloid-β stable-isotope enrichment from cerebrospinal fluid. However, the applicability of this approach and its reproducibility for the measurement of enrichment of a protein from skeletal muscle, such as the muscle β-F1-ATPase, has not been evaluated. Lower protein abundance and turnover rates, ultimately resulting in lower levels of labeled proteins in skeletal muscle compared to cerebrospinal fluid, as well as differences in the ionization of protein-associated peptides and their fragmentation patterns, all necessitate the need to evaluate the applicability of LC-MS/MS to specifically measure the stable-isotope enrichment of muscle β-F1-ATPase.

The overall goal of this research was, therefore, to determine the reproducibility of an LC-MS/MS method for the measurement of stable-isotope enrichment of *in vivo*-labeled muscle β-F1-ATPase. We specifically sought to determine if there is a specific β-F1-ATPase peptide that can provide reproducible measurements to be used for the determination of β-F1-ATPase enrichment. The following were evaluated for β-F1-ATPase tryptic-peptides following isolation of β-F1-ATPase from rat skeletal muscle: 1) the MS/MS signal linearity in the detection of labeled and unlabeled fragment ions of β-F1-ATPase peptides over a range of injected amounts of protein, 2) the reproducibility of the measured enrichment across β-F1-ATPase peptides, and 3) the linearity in the measured enrichment determined by using mixtures containing various ratios of labeled to unlabeled synthetic peptides for the β-F1-ATPase peptide with the most reproducible measurement of its enrichment.

Methods

Ethics statement

The study protocol and procedures were reviewed and approved by the Institutional Animal Care and Use Committee at Arizona State University (Protocol ID 07-939R).

Selection of the stable-isotope amino acid

L-[2,3,3,4,5,5,5,6,6,6-$^2H_{10}$]leucine, which is a stable-isotope of the amino acid leucine was selected to label leucine residues in the β-F1-ATPase for several reasons: 1) leucine is abundant in β-F1-ATPase and this increases the chances for detecting leucine-containing peptides that meet our criteria (discussed below); 2) the specific stable-isotope of leucine has a mass that is large enough to separate the isotope envelopes between labeled and unlabeled peptides, allowing for easier identification; 3) although not the focus of the present study, calculation of the rate of synthesis of a protein requires determination of the precursor (i.e. leucine) enrichment, which can be determined in a practical way when using the selected stable-isotope of the amino acid leucine. Specifically, in the selected d10-leucine, and as a result of the transamination reaction, the deuterium attached to the alpha carbon of the amino acid, as is also the case for the hydrogen for endogenous leucine, is rapidly exchanged for hydrogen via alpha-ketoisocaproic acid metabolism *in vivo* resulting in the production of d9-leucine inside the cell [10]. That way, both unlabeled and d9-labeled leucine are introduced intracellularly in the same way before incorporation into the protein, and under these conditions enrichment with labeled leucine that has undergone transamination measured in peripheral blood is the same as that measured in muscle [1,11], allowing for a convenient and practical determination of the precursor enrichment.

Animals and infusion protocol

Muscle for the experiments was obtained from two adult male Sprague-Dawley rats (Charles River, Wilmington, MA). Prior to the collection of muscle, d10-leucine was infused (50 mg/kg/hour; Cambridge Isotope Laboratories, Inc., Andover, MA) for seven hours intravenously in one of the rats to increase the incorporation of stable-isotope-labeled leucine into muscle β-F1-ATPase, while the other rat received a comparable rate of infusion of saline.

Muscle homogenization, β-F1-ATPase immunoprecipitation, SDS-PAGE, and in-gel digestion procedures

Approximately 100 mg of vastus lateralis muscle was homogenized in ice-cold freshly prepared buffer (1 mL/100 mg tissue; 20 mM Hepes pH 7.6, 1 mM EDTA pH 8.0, 250 mM sucrose, 5 mM NaF, 1 mM Na-pyrophosphate, 1 mM ammonium molybdate, 1 mM Na_3VO_4, 10 ug/ml aprotinin, 10 ug/ml leupeptin, 250uM PMSF) and centrifuged at 14,000 x g for 30 min at 4°C. Aliquots of supernatant were stored at -80°C. Protein concentrations in the homogenate were determined by the method of Lowry [12].

Muscle β-F1-ATPase was purified from the whole muscle homogenate by immunoprecipitation (IP) using a mouse monoclonal β-specific antibody coupled to protein G agarose beads for 1 hr at room temperature. After antibody-bead conjugation, 2 mg of muscle protein and adequate homogenization buffer was added to the mixture which was incubated overnight on rotation in 4°C. The beads were subsequently washed four times with ice-cold PBS (pH 7.4) and the proteins were denatured and eluted from the beads by incubation for 30 min at 37°C in 15 µL 2X SDS sample loading buffer [13]. Proteins were then separated by 10% SDS-polyacrylamide gel electrophoresis (PAGE) and visualized by Coomassie blue staining (Sigma Chemical Co., St. Louis, MO).

The band corresponding to the weight of β-F1-ATPase (=56 kD) was excised from the gel and cut into 1 mm cubes. Cubes from each lane were placed into microcentrifuge tubes and washed with 400 uL of deionized water. Coomassie stain was removed with 2 washes with 300 uL of 50% acetonitrile (ACN) in 40 mM NH_4HCO_3 and the gel pieces were dehydrated with 100% ACN for 15 minutes. The ACN was removed and the gel pieces were further dried in a vacuum centrifuge at 62°C for 30 min. Peptides were released from the gel by trypsin digestion (250 ng) in 30 µl of 40 mM NH_4HCO_3 and the samples maintained at 4°C for 15 minutes. 50 µL of 40 mM NH_4HCO_3 was added and the digestion was left to continue at 37°C overnight. The digestion was terminated with 10 µl of 5% formic acid (FA). Following overnight digestion, the samples were incubated at 37°C for 30 min and centrifuged for 1 min; the supernatant was transferred to a clean polypropylene tube. Another 80 µL of 0.5% FA was added to the gel pieces and the extraction procedure was repeated. Resulting peptide mixtures were purified using solid-phase extraction (C18 ZipTip; Millopore) after sample loading in 0.05% heptafluorobutyric acid:5% FA and elution with 4 uL 50% ACN:1% FA and 4 uL 80% ACN:1% FA, respectively. Eluates were combined and the sample volume was reduced to ~2 µL by vacuum centrifugation. Subsequently, 20 µl of 0.05% heptafluorobutyric acid/1% FA:2%ACN was added as loading buffer.

HPLC-ESI-MS/MS analysis

A hybrid mass spectrometer consisting of a Linear Ion Trap Mass spectrometer, LTQ, combined with a Fourier Transform Ion Cyclotron Resonance (FT-ICR) mass spectrometer (LTQ FT Ultra, Thermo Fisher; San Jose, CA) fitted with a PicoView nanospray source (New Objective, Woburn, MA) was used to perform the HPLC-ESI-MS/MS analyses. HPLC separations were accomplished with a linear gradient as described previously [14]. Labeled and unlabeled fragment ions of β-F1-ATPase peptides were quantified in the LTQ analyzer of the LTQ-FT-ICR instrument.

Initially, a full-scan spectrum was acquired followed by collision-induced dissociation mass spectra of the 10 most abundant ions in the survey scan to determine if there were β-F1-ATPase peptides with a high MS signal and a leucine residue in the sequence. No labeled-leucine peptides were detected in a data-dependent analysis, probably as a result of low abundance of stable-isotope-leucine enriched peptides in β-F1-ATPase. Since, any peptide with leucine has the potential to contain labeled-leucine, we therefore predicted that low abundance leucine-labeled peptides can be detected using a "targeted" mass spectrometry scan approach, and based on the addition of the appropriate mass to the mass of each leucine in a detected β-F1-ATPase peptide. A target list with various potential leucine-containing peptide m/z values was employed, resulting in improved efficiency in identifying stable-isotope-leucine enriched peptides.

Tandem mass spectra were extracted from Xcalibur "RAW" files and charge states were assigned using the Extract_MSN script which is a component of Xcalibur 2.0 SR2 (Thermo Fisher; San Jose, CA). The fragment mass spectra were searched against the IPI_RAT_v3.33 database (http://www.ebi.ac.uk/IPI/) using Mascot (Matrix Science, London, United Kingdom, version 2.2). The search parameters for the sample's protein peptides were as follows: 10 ppm mass tolerance for precursor ion masses and 0.5 Da for product ion masses; digestion with trypsin; a maximum of two missed tryptic cleavages; variable modifications of oxidation of methionine and phosphorylation of serine, threonine, and tyrosine, and addition of D9 or D10 labeled-leucine. Probability assessment of peptide assignments and protein identifications were made through use of Scaffold (version Scaffold-01_06_17, Proteome Software Inc., Portland, OR).

β-F1-ATPase peptides confirmation and calculation of enrichment based on b and y fragment ion analysis

Peptides released from the trypsin digestion were considered only if they met the following criteria: there is a leucine residue in the amino acid sequence; detected with high intensity relative to other peptides (top 10 peptides); assigned with >95% probability to β-F1-ATPase (assessed through Scaffold Proteome Software); there is no missed cleavage; and there is no methionine in the sequence (because of methionine oxidation). There were three β-F1-ATPase unique peptides (i.e. their amino acid sequence matches only the β-F1-ATPase protein) that met all these criteria: β-F1-ATPase$_{95-109}$ (LVLEVAQHLGESTVR; MH^{+1}, mono: 1650.9), β-F1-ATPase$_{134-143}$ (IPVGPETLGR; MH^{+1}, mono: 1038.6), and β-F1-ATPase$_{282-294}$ (VALTGLTVAEYFR; MH^{+1}, mono: 1439.8). The MS/MS peak areas of the following fragment ions from these peptides were quantified (+1 charge state): β-F1-ATPase$_{95-109}$ – b8 (890.5), b9 (1003.6), b11 (1189.7); β-F1-ATPase$_{134-143}$ – y6 (672.4), y7 (729.4) and y8 (828.5); β-F1-ATPase$_{282-294}$ – b8 (755.5), b9 (826.5), b10 (955.5), b11 (1118.6). The corresponding fragment ions MS/MS peak areas with +9 Da mass, containing d9-leucine, from the same peptides were also quantified. Extracted ion chromatographic peak areas corresponding to d9-labeled and unlabeled fragment ions for the three targeted β-F1-ATPase peptides were generated for the quantification of the amounts of d9-labeled and unlabeled fragments ions of the peptides in the β-F1-ATPase samples. Mass tolerance was set at 0.5 Da.

Signal linearity in the detection of labeled and unlabeled fragment ions

Following procedures described above, a muscle homogenate was used to create four β-F1-ATPase IP samples, each containing ~2 mg of total muscle protein. Each of the four samples was separated by SDS-PAGE and each band corresponding to the β-F1-ATPase was cut and trypsin digested. Following trypsin digestion, and after the peptide mixtures were purified using solid-phase extraction, the peptide mixtures were combined to create one sample containing an amount of β-F1-ATPase corresponding to ~8 mg of muscle protein. This sample was dried, brought to a volume of 30 uL with loading buffer and injected into the mass spectrometer at seven different volumes corresponding to 0.2, 0.3, 0.4, 0.5, 0.8, 1.1, and 2.3 mg of muscle protein. The specific peak areas for d9-leucine-labeled and unlabeled fragment ions indicated above for the β-F1-ATPase$_{95-109}$, β-F1-ATPase$_{134-143}$, and β-F1-ATPase$_{282-294}$ peptides were quantified.

Reproducibility of the measured β-F1-ATPase enrichment

Muscle homogenate corresponding to ~2 mg of total muscle protein was used to perform three separate β-F1-ATPase IP analyses. Each of the three samples containing immunoprecipitated β-F1-ATPase was loaded on a separate SDS-PAGE gel lane, and each band was treated as a separate sample. The three samples were analyzed in separate HPLC-MS/MS runs, using the same procedures in all three runs. The specific peak areas for d9-leucine-labeled and unlabeled fragment ions for the β-F1-ATPase$_{95-109}$, β-F1-ATPase$_{134-143}$, and β-F1-ATPase$_{282-294}$ peptides were quantified. The peptide enrichment was calculated by dividing the sum of the peak areas of the labeled fragment ions by the sum of the peak areas of the corresponding unlabeled fragment ions.

Linearity in the measured enrichment using synthetic peptides

To evaluate the linearity in the measured enrichment of the IPVGPETLGR peptide (i.e. β-F1-ATPase$_{134-143}$), stock solutions containing leucine-labeled and unlabeled IPVGPETLGR synthetic peptides were prepared. Given the lack of commercially available d9-leucine, d10-leucine was used to synthesize the leucine-labeled peptide. D10-leucine-labeled and unlabeled IPVGPETLGR peptides were synthesized by the Proteomics and Protein Chemistry Lab facility at Arizona State University, and dissolved in water. MALDI and LC-MS/MS mass spectrometry analyses confirmed that the synthesized peptides corresponded to unlabeled and d10-leucine-labeled IPVGPETLGR peptides (Figure S1). Stock solution containing d10-labeled peptide was serially diluted with stock solution containing the unlabeled peptide to create a series of eight samples spanning a wide range of molecular ratios of labeled to unlabeled peptide. Following solid-phase extraction, and vacuum centrifugation (procedures described above), loading buffer was added to the samples prior to their injection into the mass spectrometer. Peak areas for d10-leucine-labeled and unlabeled y6-y8 fragment ions for the IPVGPETLGR peptide were quantified. These values were used to determine the peptide enrichment, and by dividing the sum of the peak areas of the labeled fragment ions by the sum of the peak areas of the corresponding unlabeled fragment ions. The molecular content of the d10-leucine-labeled and unlabeled stock solutions were calculated based on the weighted amounts of the leucine-labeled and unlabeled IPVGPETLGR peptides added to these solutions. Given the lack of purity in the synthesized peptides used in the preparation of the stock solutions, the calculated enrichment of the IPVGPETLGR peptide with d10-leucine-containing IPVGPETLGR peptide in the resultant serial dilutions represents predicted enrichment.

Statistical analyses

Coefficient of variation (CV) values were calculated to assess the reproducibility of the enrichment measurements of the β-F1-ATPase peptides. Regression analysis was conducted to assess the linear fit between variables of interest. All the statistical analyses were performed using the Minitab® 15.1 statistical software (Minitab Inc., State College, PA).

Results

Sequence coverage of immunoprecipitated β-F1-ATPase protein by top-10 data dependent tandem mass spectrometry following these procedures was >70% (Figure 1). Figure 2a illustrates the b and y fragment ions of the representative β-F1-ATPase$_{134\text{-}143}$ peptide. Figure 2b-c depicts MS/MS scans for unlabeled and d9-leucine labeled y6-y8 fragment ions of the β-F1-ATPase$_{134\text{-}143}$ peptide, together with the corresponding extracted ion chromatographic peak areas from muscle sample collected from the rat infused with the stable-isotope-labeled leucine. No peaks were detected for fragment ions with +9 Da mass in muscle β-F1-ATPase samples from the saline infused rat for any of the three targeted β-F1-ATPase peptides.

Signal linearity in the detection of labeled and unlabeled fragment ions

For muscle β-F1-ATPase samples collected from the rat infused with the stable-isotope-labeled leucine, the MS/MS peak areas corresponding to the d9-labeled and unlabeled fragment ions of the three targeted β-F1-ATPase peptides were quantified over a wide range of amounts of protein/peptides injected into the LC-MS/MS. Regression analysis was performed between the amounts of protein analyzed, reflecting the amounts of muscle β-F1-ATPase, and each of the d9-leucine-labeled and unlabeled fragment ion peak areas of the three targeted β-F1-ATPase peptides. Calculated R^2 values from the regression analyses for the β-F1-ATPase$_{95\text{-}109}$ peptide corresponding to d9-labeled MS/MS peak areas ranged between 0.94 and 0.97, and those corresponding to unlabeled MS/MS peak areas ranged between 0.97 and 0.98. Regression analyses R^2 values for the β-F1-ATPase$_{282\text{-}294}$ peptide were between 0.62 and 0.87 for the d9-labeled MS/MS peak areas, and between 0.71 and 0.75 for the unlabeled MS/MS peak areas. R^2 values for the ATPase$_{134\text{-}143}$ peptide were 0.99 across all d9-labeled as well as unlabeled fragment ions (Figure 3). The regression analyses also indicated that for both the β-F1-ATPase$_{95\text{-}109}$ and β-F1-ATPase$_{282\text{-}294}$ peptides the y-intercept of the regression equations describing the relationship between the amounts of protein analyzed and fragment ion peak areas were significantly different from zero ($P<0.05$) for all d9-labeled and unlabeled fragment ions. However, none of the y-intercept values of the regression equations describing the corresponding relationships for the ATPase$_{134\text{-}143}$ peptide were significantly different from zero ($P>0.05$).

Reproducibility of the measured β-F1-ATPase enrichment

For each β-F1-ATPase peptide within a replicate the sum of the peak areas of its d9-leucine-labeled fragment ions was divided by the sum of the peak areas of its corresponding unlabeled fragment ions in order to determine the enrichment of the specific β-F1-ATPase peptide with d9-leucine. The average percent enrichment based on these replicates for the β-F1-ATPase$_{95\text{-}109}$ peptide was 0.564% ± 0.023% (average value ± standard deviation), with a CV value of 4.15%. These same measures corresponding to the β-F1-ATPase$_{134\text{-}143}$ peptide were 0.482% ± 0.014% and 2.6%, respectively. For the β-F1-ATPase$_{282\text{-}294}$ peptide the average percent enrichment was 0.566% ± 0.051% and the CV was 8.98%.

Linearity in the measured enrichment using synthetic peptides

Linearity in the measured enrichment was evaluated only for the β-F1-ATPase$_{134\text{-}143}$ peptide (i.e. IPVGPETLGR), because it was the only peptide that showed acceptable linearity in the signal intensity in the detection of labeled and unlabeled fragment ions (based on the results reported in the "Signal Linearity in the Detection of Labeled and Unlabeled Fragment Ions" section). The measured percent enrichment (i.e. sum of y6-y8 labeled fragment ions peak areas/sum of y6-y8 unlabeled fragment ions peak areas*100) in the LC-MS/MS-analyzed samples prepared to contain synthetic IPVGPETLGR peptides enriched with synthetic IPVGPETLGR peptides that contained labeled leucine ranged between 0.009% and 8.185%. Regression analysis indicated

1 MLSLVGRVAS ASASGALRGL NPLAALPQAH LLLRTAPAGV HPARDYAAQS SAAPKAGTAT
61 GQIVAVIGAV VDVQFDEGLP PILNALEVQG **RESRLVLEVA QHLGESTVRT IAMDGTEGLV**
120 **R**GQK**VLDSGA PIKIPVGPET LGRIMNVIGE PIDER**GPIKT K**QFAPIHAEA PEFIEMSVEQ**
180 **EIILVTGIKVV DLLAPYAKGG** KIGLFGGAGV **GKTVLIMELI NNVAKAHGGY SVFAGVGERT**
240 **REGNDLYHEM IESGVINLK**D ATSK**VALVYG QMNEPPGAR**A R**VALTGLTVA EYFRDQEGQD**
300 **VLLFIDNIFR FTQAGSEVSA LLGRIPSAVG YQPTLATDMG TMQER**ITTTK **KGSITSVQAI**
360 **YVPADDLTDP APATTFAHLD ATTVLSRAIA ELGIYPAVDP LDSTSRIMDP NIVGSEHYDV**
420 **ARGVQKILQD** YK**SLQDIIAI LGMDELSEED KLTVSRARKI QRFLSQPFQV AEVFTGHMGK**
480 LVPLKETIK**G FQQILAGDYD HLPEQAFYMV GPIEEAVAKA DKLAEEHGS**

Figure 1. Coverage map of tryptic peptides of β-F1-ATPase from rat muscle. The muscle was analyzed using the procedures described in the text. Detected peptides are depicted in bold font letters. The amino acid sequence underlined identifies the representative tryptic β-F1-ATPase$_{134\text{-}143}$ peptide (i.e. IPVGPETLGR). Percent coverage is 77%.

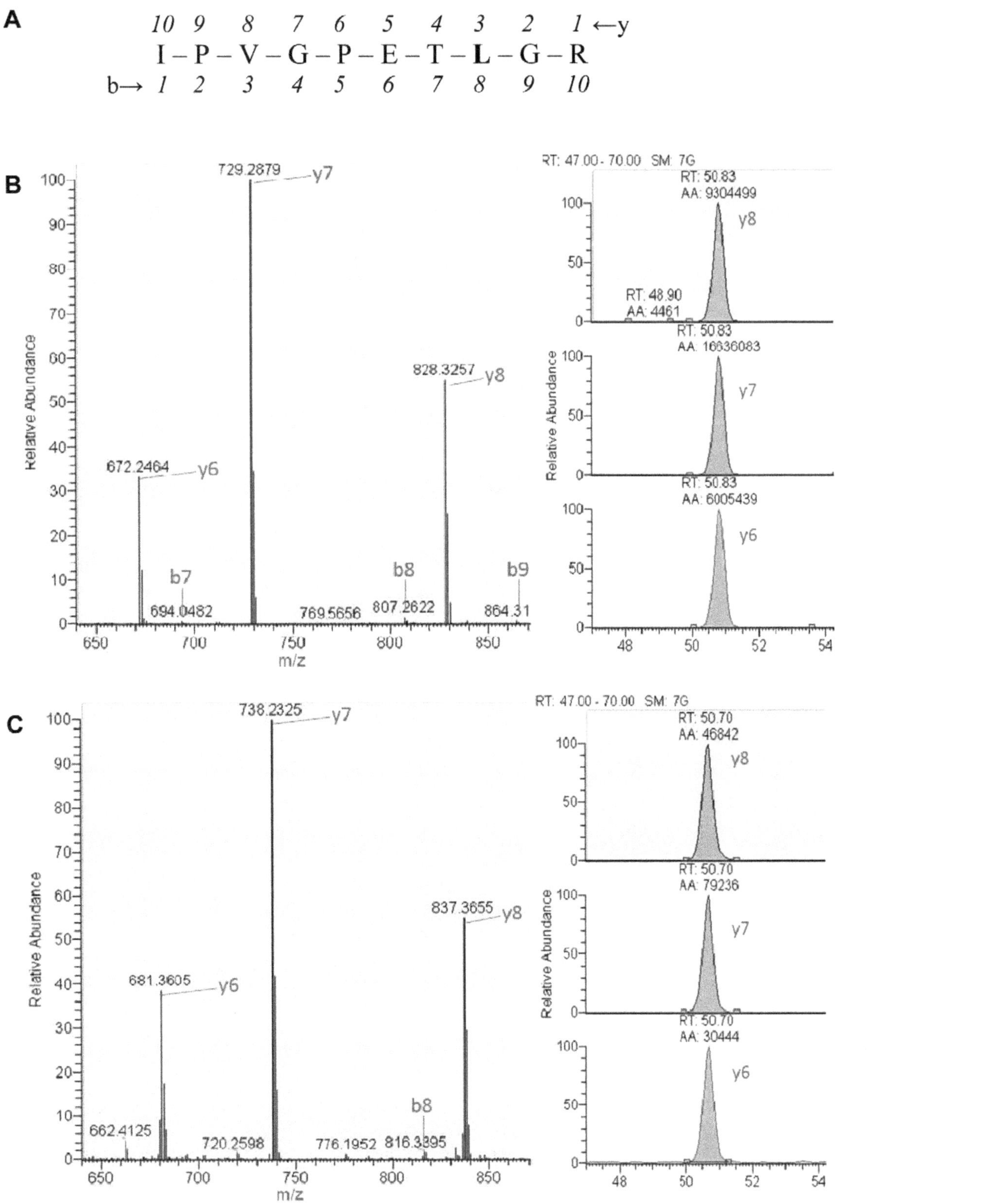

Figure 2. Representative IPVGPETLGR peptide of muscle β-F1-ATPase and LC-MS/MS spectra for IPVGPETLGR peptide enriched with d9-leucine. Tryptic peptide IPVGPETLGR, corresponding to β-F1-ATPase$_{134\text{-}143}$, showing the b and y fragment ions (A); depiction of tandem mass spectrum for fragment ions of the IPVGPETLGR peptide showing the y6-y8 unlabeled fragment ions (left) and the corresponding extracted ion chromatographic peak areas of these ions (right) used for the quantification of the enrichment of the IPVGPETLGR peptide (B); depiction of spectra for fragment ions of the IPVGPETLGR peptide showing the d9-leucine-labeled fragment ions (i.e. same fragment ions as in (B) shifted by 9 m/z) (left) and the corresponding extracted ion chromatographic peak areas of these ions (right) used for quantification of the enrichment of the IPVGPETLGR peptide (C). Fragment ions either not containing leucine (i.e. b7) or containing leucine (i.e. b8, b9) in the sequence are also shown (B). Such ions, although they contain leucine and their enrichment can theoretically be quantified (i.e. b8), were not taken into consideration because their relative low abundance (<10% of parent ion).

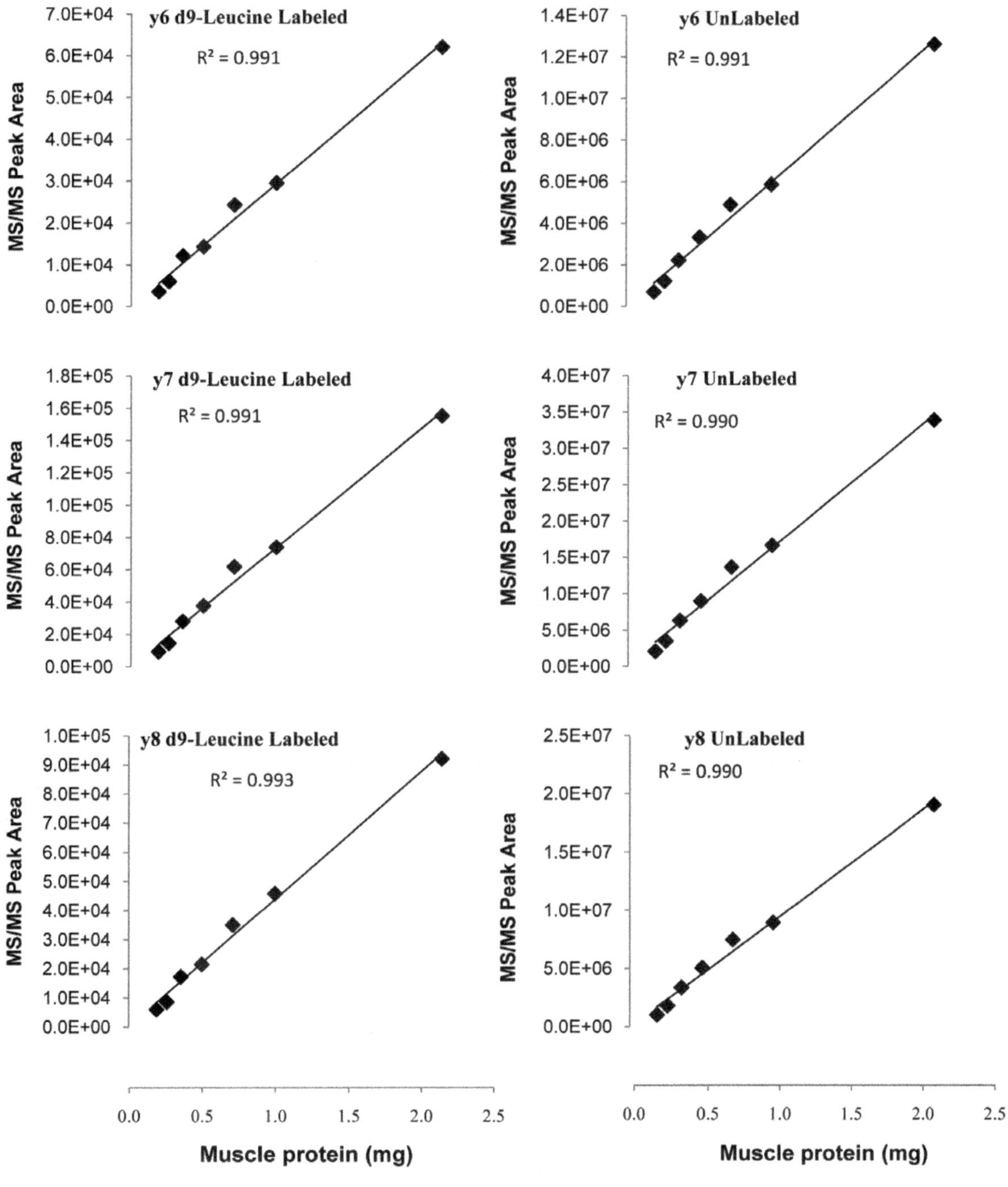

Figure 3. β-F1-ATPase$_{134-143}$ peptide fragment ions peak areas as a function of the amount of protein analyzed. D9-leucine-labeled fragment ions (left) and corresponding unlabeled fragment ions (right) were quantified in association with various amounts of muscle protein processed. Data were analyzed using regression analysis. Corresponding R^2 values are shown for each fragment ion. The y-intercept of the regression equations is not significantly different from zero (P>0.05).

excellent linear relationship between the measured and predicted IPVGPETLGR percent enrichment throughout the range of the prepared samples (Figure 4).

Discussion

The overall goal of this study was to evaluate the reproducibility of an LC-MS/MS approach to quantify the enrichment of *in vivo*-labeled muscle β-F1-ATPase based on the enrichment of its tryptic peptides. Measurement of this enrichment is a key methodological step in order to evaluate the translational regulation of β-F1-ATPase by means of stable-isotope methodology. The described approach is based on immunoprecipitation of β-F1-ATPase, separation of β-F1-ATPase by 1-D SDS-PAGE, and determination of the enrichment of fragment ions of selected β-F1-ATPase peptides using LC-MS/MS.

An important finding of the present study is that two (i.e. β-F1-ATPase$_{95-109}$ and β-F1-ATPase$_{282-294}$) of the three tryptic β-F1-ATPase peptides that met our initial criteria to be used for the determination of the reproducibility of β-F1-ATPase peptide enrichment demonstrated low, and in some cases unacceptable (i.e. $R^2 = 0.62$), linearity in the signal intensity with which amounts

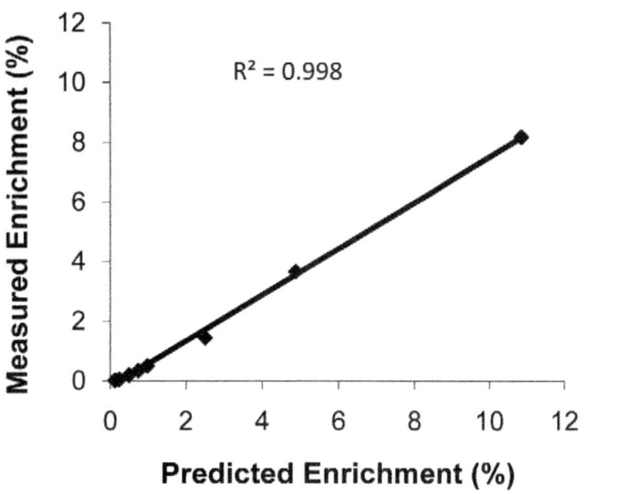

Figure 4. Relationship between measured and predicted enrichment for the synthetic peptide IPVGPETLGR. D10-leucine labeled and unlabeled IPVGPETLGR synthetic peptides were mixed at eight different molar ratios, processed, and analyzed by LC-MS/MS. The peak areas for d10-leucine-labeled and unlabeled y6-y8 fragment ions for the IPVGPETLGR peptide were quantified. Peptide percent enrichment was calculated by dividing the sum of the peak areas of the labeled fragment ions by the sum of the peak areas of the corresponding unlabeled fragment ions. Data between measured and predicted percent enrichments were analyzed using regression analysis.

of labeled and unlabeled fragment ions were detected. For these two peptides, there was proportionally less increase in MS/MS fragment ion signal intensity with increased amounts of protein/peptides injected into the LC-MS/MS. On the other hand, there was an excellent linearity (i.e. $R^2 = 0.99$) in the signal intensity with which labeled and unlabeled amounts of fragment ions were detected for the β-F1-ATPase$_{134-143}$ peptide. Also, none of the y-intercept values of the regression equations describing the MS/MS fragment ion intensities as a function of the amounts of protein/peptide injected into the LC-MS/MS was significantly different from zero for the β-F1-ATPase$_{134-143}$ peptide. The latter finding further underscores an accurate detection of generated amounts of labeled and unlabeled fragment ions for the β-F1-ATPase$_{134-143}$ peptide over a wide range of labeled and unlabeled amounts of protein analyzed. Although the specific reason(s) for the lack of linear function in the signal intensity with which labeled and unlabeled amounts of fragment ions were detected for the β-F1-ATPase$_{95-109}$ and β-F1-ATPase$_{282-294}$ peptides are not known, visual inspection of their MS/MS spectra showed poor quality spectra in association with greater noise when compared to the MS/MS spectra of the β-F1-ATPase$_{134-143}$ peptide.

The results obtained with respect to the percent enrichment of the β-F1-ATPase$_{134-143}$ peptide in the reproducibility experiments ("Reproducibility of the Measured β-F1-ATPase Enrichment" section) are similar to those when the percent enrichment (0.470% ± 0.012%) and CV (2.6%) for the β-F1-ATPase$_{134-143}$ peptide were calculated based on the data obtained across the various amounts of the protein/peptide analyzed in the MS/MS signal linearity experiment ("Signal Linearity in the Detection of Labeled and Unlabeled Fragment Ions" section). The latter findings enhance the results obtained from the reproducibility experiments for the β-F1-ATPase $_{134-143}$ peptide and provide additional evidence for an excellent reproducibility in the measurement of *in vivo*-labeled muscle β-F1-ATPase enrichment using LC-MS/MS when this measurement is based on the enrichment of the β-F1-ATPase $_{134-143}$ peptide. The presence of more than one leucine residues that can be labeled with d9-leucine in both the β-F1-ATPase$_{95-109}$ and β-F1-ATPase$_{282-294}$ peptides could be the reason for the poorer reproducibility of the enrichment measurement using those two peptides. There was also greater average percent enrichment measured using the β-F1-ATPase$_{95-109}$ and the β-F1-ATPase$_{282-294}$ peptides compared to that measured using the β-F1-ATPase $_{134-143}$ peptide that is attributed to the existence of isobaric peptides for the β-F1-ATPase$_{95-109}$ and the β-F1-ATPase$_{282-294}$. The probability of having a singly labeled leucine in any given peptide increases with the number of amino acid (i.e. leucine) residues that can incorporate labeled leucine, and the β-F1-ATPase$_{95-109}$ and the β-F1-ATPase$_{282-294}$ peptides both contain more than one leucine residues in their sequence.

Previously published approaches to measure the enrichment of proteins labeled *in vivo* with stable isotopes in animal models [15] were done in conjunction with high isotopic enrichment of the proteins (i.e. feeding the animals highly enriched isotopes for very long periods of time), none of which is practical in short-term (i.e. only hours-long) animal or human experiments. Also, contrary to the circumstance related to the determination of the enrichment of a protein with a fast turnover rate such as the amyloid-β in human cerebrospinal fluid [9], the slow turnover rate, and thus the lower abundance of labeled β-F1-ATPase molecules in muscle, constitutes a challenge to reproducibly measure the stable-isotope enrichment of muscle β-F1-ATPase, and in a way that is practical when investigating its kinetics in animals or humans. The present study builds on the previously described approach [9], and extents the methodology as well as its applicability to a muscle protein. Evaluation of the reproducibility of the measurement of the stable-isotope enrichment of peptides of a specific muscle protein using LC-MS/MS is prerequisite for the use of any peptide-based approach that is employed to measure the kinetics of such protein in association with stable-isotope infusion experiments in animals or humans. Our results show that the stable-isotope enrichment of *in-vivo*-labeled muscle β-F1-ATPase can reliably be measured based on the determination of the enrichment of the β-F1-ATPase$_{134-143}$ peptide.

Recently Jaleel et al. [16], have reported an MS/MS approach to determine the stable-isotope enrichment of muscle proteins labeled *in vivo*. This approach involves two separate steps, with the first being the identification of a protein that is separated by two-dimensional gel electrophoresis using LC-MS/MS, and the second the measurement of the enrichment of the identified protein using GC-MS/MS. Protein enrichment in the latter step is determined by measuring the stable-isotope enrichment of an amino acid following hydrolysis of the identified protein. Besides using a single MS/MS step in the analysis of a pre-selected peptide, the approach described herein is based on the measurement of the stable-isotope enrichment of a unique peptide of the muscle protein. This is important, because when determination of protein enrichment is based on the determination of the enrichment of its amino acids the accuracy of the measured enrichment depends on the purity of the isolated protein in the sample (i.e. contamination of the amino acids of the protein of interest by hydrolyzed amino acids from unrelated proteins). A major advantage of the approach described herein is, therefore, the specificity in the determination of the enrichment of the protein of interest.

It is noted that the stable-isotope of the amino acid leucine evaluated in the present study contains nine deuterium atoms, which ensures practically zero d9-leucine β-F1-ATPase enrichment at the initiation of the infusion period. Therefore, a single muscle biopsy performed at the end of the period of interest [17,18,19] in combination with frequent peripheral blood sampling can be used to accurately describe the rate of synthesis of β-F1-ATPase. Because the same peptide (IPVGPETLGR) is observed in muscle from both rats (β-F1-ATPase$_{134\text{-}143}$) and humans (β-F1-ATPase$_{134\text{-}143}$) [13], measurement of β-F1-ATPase enrichment based on the enrichment of this peptide is readily applicable to human experiments. Although we have specifically been interested in β-F1-ATPase, this approach can be applied in the determination of the enrichment of any protein from skeletal muscle that has been labeled in vivo, and after evaluating the existence of a peptide that can reliably be used to describe the enrichment of the protein of interest. With respect to the applicability of the approach described herein only the abundance of the protein of interest in the sample is a concern. In the present study, 2 mg of muscle protein (i.e. 20 mg of wet muscle) are sufficient for the determination of β-F1-ATPase enrichment. However, a larger muscle sample can increase the overall amount of labeled proteins found in smaller quantities in skeletal muscle. In this regard, and although in the present studies mitochondrial β-F1-ATPase was immunoprecipitated from whole muscle homogenate, isolation of the mitochondria can enrich the protein homogenate with mitochondrial proteins that are targeted for quantitative determination several fold [20].

In conclusion, our results show that the reproducibility of the determination of the stable-isotope enrichment of a protein using LC-MS/MS may vary based on the specific peptide chosen to describe the protein enrichment. The β-F1-ATPase$_{134\text{-}143}$ peptide provides reproducible results for the determination of the stable-isotope enrichment of *in vivo*-labeled muscle β-F1-ATPase. Although we have been interested in the enrichment of β-F1-ATPase, this approach can be extended in the study of numerous other proteins from skeletal muscle.

Supporting Information

Figure S1 MALDI and LC-MS/MS mass spectrometry analysis of synthetic peptides. Synthetic IPVGPETLGR peptides, corresponding to the tryptic peptide of the β-F1-ATPase (i.e. β-F1-ATPase), were prepared to contain unlabeled and d10-labeled-leucine. The top part shows MALDI spectra for the unlabeled synthetic peptide (A) and the labeled-leucine synthetic peptide (B), the latter shifted by 10 Da. The bottom part shows LC-MS/MS spectra for fragment ions for the unlabeled synthetic peptide (C) and the labeled-leucine synthetic peptide (D), the latter shifted by 10 Da. These mass spectrometry data provide evidence that the synthesized peptides correspond to unlabeled and d10-leucine-labeled IPVGPETLGR peptides.

Author Contributions

Conceived and designed the experiments: CK LM ZY. Performed the experiments: CK SE CR ML. Analyzed the data: CK SE ZY PL. Contributed reagents/materials/analysis tools: ZY LM. Wrote the paper: CK SE ZY PL LM CR ML.

References

1. Wolfe RR, Chinkes DL (2004) Isotope tracers in metabolic research: principles and practice of kinetic analysis. New York: Wiley-Liss.
2. Newcomer BR, Larson-Meyer DE, Hunter GR, Weinsier RL (2001) Skeletal muscle metabolism in overweight and post-overweight women: an isometric exercise study using (31)P magnetic resonance spectroscopy. Int J Obes Relat Metab Disord 25: 1309–1315.
3. Petersen KF, Befroy D, Dufour S, Dziura J, Ariyan C, et al. (2003) Mitochondrial dysfunction in the elderly: possible role in insulin resistance. Science 300: 1140–1142.
4. Hwang H, Bowen BP, Lefort N, Flynn CR, De Filippis EA, et al. (2010) Proteomics analysis of human skeletal muscle reveals novel abnormalities in obesity and type 2 diabetes. Diabetes 59: 33–42.
5. Lombardi A, Silvestri E, Cioffi F, Senese R, Lanni A, et al. (2009) Defining the transcriptomic and proteomic profiles of rat ageing skeletal muscle by the use of a cDNA array, 2D- and Blue native-PAGE approach. J Proteomics 72: 708–721.
6. Izquierdo JM, Jimenez E, Cuezva JM (1995) Hypothyroidism affects the expression of the beta-F1-ATPase gene and limits mitochondrial proliferation in rat liver at all stages of development. Eur J Biochem 232: 344–350.
7. Lin PC, Lin JK, Yang SH, Wang HS, Li AF, et al. (2008) Expression of beta-F1-ATPase and mitochondrial transcription factor A and the change in mitochondrial DNA content in colorectal cancer: clinical data analysis and evidence from an in vitro study. Int J Colorectal Dis 23: 1223–1232.
8. Willers IM, Isidoro A, Ortega AD, Fernandez PL, Cuezva JM (2010) Selective inhibition of beta-F1-ATPase mRNA translation in human tumours. Biochem J 426: 319–326.
9. Bateman RJ, Munsell LY, Chen X, Holtzman DM, Yarasheski KE (2007) Stable isotope labeling tandem mass spectrometry (SILT) to quantify protein production and clearance rates. J Am Soc Mass Spectrom 18: 997–1006.
10. Pratt JM, Petty J, Riba-Garcia I, Robertson DH, Gaskell SJ, et al. (2002) Dynamics of protein turnover, a missing dimension in proteomics. Mol Cell Proteomics 1: 579–591.
11. Chinkes D, Klein S, Zhang XJ, Wolfe RR (1996) Infusion of labeled KIC is more accurate than labeled leucine to determine human muscle protein synthesis. Am J Physiol 270: E67–71.
12. Lowry OH, Rosebrough NJ, Farr AL, Randall RJ (1951) Protein measurement with the Folin phenol reagent. J Biol Chem 193: 265–275.
13. Hojlund K, Yi Z, Lefort N, Langlais P, Bowen B, et al. (2010) Human ATP synthase beta is phosphorylated at multiple sites and shows abnormal phosphorylation at specific sites in insulin-resistant muscle. Diabetologia 53: 541–551.
14. Yi Z, Langlais P, De Filippis EA, Luo M, Flynn CR, et al. (2007) Global assessment of regulation of phosphorylation of insulin receptor substrate-1 by insulin in vivo in human muscle. Diabetes 56: 1508–1516.
15. Doherty MK, Whitehead C, McCormack H, Gaskell SJ, Beynon RJ (2005) Proteome dynamics in complex organisms: using stable isotopes to monitor individual protein turnover rates. Proteomics 5: 522–533.
16. Jaleel A, Short KR, Asmann YW, Klaus KA, Morse DM, et al. (2008) In vivo measurement of synthesis rate of individual skeletal muscle mitochondrial proteins. Am J Physiol Endocrinol Metab 295: E1255–1268.
17. Burd NA, West DW, Rerecich T, Prior T, Baker SK, et al. (2011) Validation of a single biopsy approach and bolus protein feeding to determine myofibrillar protein synthesis in stable isotope tracer studies in humans. Nutr Metab (Lond) 8: 15.
18. Boirie Y, Short KR, Ahlman B, Charlton M, Nair KS (2001) Tissue-specific regulation of mitochondrial and cytoplasmic protein synthesis rates by insulin. Diabetes 50: 2652–2658.
19. Guillet C, Delcourt I, Rance M, Giraudet C, Walrand S, et al. (2009) Changes in basal and insulin and amino acid response of whole body and skeletal muscle proteins in obese men. J Clin Endocrinol Metab 94: 3044–3050.
20. Lefort N, Yi Z, Bowen B, Glancy B, De Filippis EA, et al. (2009) Proteome profile of functional mitochondria from human skeletal muscle using one-dimensional gel electrophoresis and HPLC-ESI-MS/MS. J Proteomics 72: 1046–1060.

Permissions

All chapters in this book were first published in PLOS ONE, by The Public Library of Science; hereby published with permission under the Creative Commons Attribution License or equivalent. Every chapter published in this book has been scrutinized by our experts. Their significance has been extensively debated. The topics covered herein carry significant findings which will fuel the growth of the discipline. They may even be implemented as practical applications or may be referred to as a beginning point for another development.

The contributors of this book come from diverse backgrounds, making this book a truly international effort. This book will bring forth new frontiers with its revolutionizing research information and detailed analysis of the nascent developments around the world.

We would like to thank all the contributing authors for lending their expertise to make the book truly unique. They have played a crucial role in the development of this book. Without their invaluable contributions this book wouldn't have been possible. They have made vital efforts to compile up to date information on the varied aspects of this subject to make this book a valuable addition to the collection of many professionals and students.

This book was conceptualized with the vision of imparting up-to-date information and advanced data in this field. To ensure the same, a matchless editorial board was set up. Every individual on the board went through rigorous rounds of assessment to prove their worth. After which they invested a large part of their time researching and compiling the most relevant data for our readers.

The editorial board has been involved in producing this book since its inception. They have spent rigorous hours researching and exploring the diverse topics which have resulted in the successful publishing of this book. They have passed on their knowledge of decades through this book. To expedite this challenging task, the publisher supported the team at every step. A small team of assistant editors was also appointed to further simplify the editing procedure and attain best results for the readers.

Apart from the editorial board, the designing team has also invested a significant amount of their time in understanding the subject and creating the most relevant covers. They scrutinized every image to scout for the most suitable representation of the subject and create an appropriate cover for the book.

The publishing team has been an ardent support to the editorial, designing and production team. Their endless efforts to recruit the best for this project, has resulted in the accomplishment of this book. They are a veteran in the field of academics and their pool of knowledge is as vast as their experience in printing. Their expertise and guidance has proved useful at every step. Their uncompromising quality standards have made this book an exceptional effort. Their encouragement from time to time has been an inspiration for everyone.

The publisher and the editorial board hope that this book will prove to be a valuable piece of knowledge for researchers, students, practitioners and scholars across the globe.

List of Contributors

Lars Steinstraesser, Tobias Hirsch, Matthias Schulte, Maximilian Kueckelhaus, Frank Jacobsen and Evgenija A. Mersch
Department of Plastic Surgery, BG University Hospital Bergmannsheil, Ruhr University Bochum, Bochum, Germany

Ingo Stricker
Institut for Pathology, BG University Hospital Bergmannsheil, Ruhr University Bochum, Bochum, Germany

Nicole Afacan, Robert E. W. Hancock and Jason Kindrachuk
Centre for Microbial Diseases and Immunity Research, University of British Columbia, Vancouver, British Columbia, Canada

Havard Jenssen
Department of Science, Systems, and Models, Roskilde University, Roskilde, Denmark

Andrea Venerando, Giorgio Cozza and Lorenzo A. Pinna
Department of Biomedical Sciences, University of Padova, and CNR Institute of Neurosciences, Padova, Italy

Cinzia Franchin and Giorgio Arrigoni
Department of Biomedical Sciences, University of Padova, and CNR Institute of Neurosciences, Padova, Italy
Proteomic Center of Padova University, VIMM, and Padova University Hospital, Padova, Italy

Natasha Cant, Ateeq Al-Zahrani and Robert C. Ford
Faculty of Life Sciences, University of Manchester, Manchester, United Kingdom

Mario A. Pagano
Department of Molecular Medicine, University of Padova, Padova, Italy

Kendra Tosoni and Anil Mehta
Division of Cardiovascular and Diabetes Medicine, Ninewells Hospital and Medical School, University of Dundee, Dundee, United Kingdom

Kourosh Honarmand Ebrahimi, Peter-Leon Hagedoorn and Wilfred R. Hagen
Department of Biotechnology, Delft University of Technology, Delft, The Netherlands

John S. Tzartos
Department of Biochemistry, Hellenic Pasteur Institute, Athens, Greece
Department of Neurology, General Hospital "Red Cross", Athens, Greece

Christos Stergiou and Paraskevi Zisimopoulou
Department of Biochemistry, Hellenic Pasteur Institute, Athens, Greece

Konstantinos Kilidireas
Department of Neurology, Aeginition Hospital, School of Medicine, National and Kapodistrian University, Athens, Greece

Thomas Thomaidis
Department of Neurology, General Hospital "Red Cross", Athens, Greece

Socrates J. Tzartos
Department of Biochemistry, Hellenic Pasteur Institute, Athens, Greece
Department of Pharmacy, University of Patras, Patras, Greece

Coby M. M. Laarakkers, Erwin T. Wiegerinck, Siem Klaver, Harold Tjalsma and Dorine W. Swinkels
Laboratory of Genetic, Endocrine and Metabolic Diseases, Department of Laboratory Medicine, Radboud University Medical Centre, Nijmegen, The Netherlands, Hepcidinanalysis.com, Nijmegen, The Netherlands

Maria Kolodziejczyk, Hendrik Gille and Andreas M. Hohlbaum
Pieris AG, Freising, Germany

Leonia Bozzacco, Christine Trumpfheller, Ralph M. Steinman and Svetlana Mojsov
Laboratory of Cellular Physiology and Immunology and Chris Browne Center, The Rockefeller University, New York, New York, United States of America

Haiqiang Yu, Henry A. Zebroski III, Nawei Zhang and Haiteng Deng
Proteomics Resource Center, The Rockefeller University, New York, New York, United States of America

Jörn Dengjel and Victoria Küttner
Freiburg Institute for Advanced Studies, University of Freiburg, Freiburg, Germany
Center for Biological Systems Analysis, University of Freiburg, Freiburg, Germany

Beatrix M. Ueberheide and Brian T. Chait
Laboratory of Mass Spectrometry and Gaseous Ion Chemistry, The Rockefeller University, New York, New York, United States of America

David Fenyo
Laboratory of Computational Proteomics, Center for Health Informatics and Bioinformatics, New York University Medical Center, New York, New York, United States of America

Arton Berisha, Bernhard Spengler and Andreas Römpp
Institute of Inorganic and Analytical Chemistry, Justus Liebig University, Giessen, Germany

Krishnendu Mukherjee and Andreas Vilcinskas
Institute of Phytopathology and Applied Zoology, Justus Liebig University, Giessen, Germany

Annelize Z. B. Aragão, Marília Belloni, Fernando M. Simabuco, Mariana R. Zanetti, Sami Yokoo, Adriana F. Paes Leme, Romênia R. Domingues, Rebeca Kawahara, Bianca A. Pauletti and Anderson Gonçalves
Mass Spectrometry Laboratory, Brazilian Biosciences National Laboratory - CNPEM, Campinas, Brazil

Michelle Agostini, Edgard Graner and Ricardo D. Coletta
Piracicaba School of Dentistry, University of Campinas (UNICAMP), Piracicaba, Brazil

Jay W. Fox
School of Medicine, University of Virginia, Charlottesville, Virginia, United States of America

Changchun Chen and Shuhui Wang
School of Pharmacy, The Center for Antibody Medicine of Ministry of Education, Shanghai Jiao Tong University, Shanghai, People's Republic of China

Huajing Wang, Tiancheng Zhang, Guanghui Ji, Xin Shi, Tian Xia and Dapeng Zhang
International Joint Cancer Institute, The Second Military Medical University, Shanghai, People's Republic of China
National Engineering Research Center for Antibody Medicine, State Key Laboratory of Antibody Medicine and Targeting Therapy and Shanghai Key Laboratory of Cell Engineering, Shanghai, People's Republic of China

Xiaoyan Mao and Weijia Lu
Lanzhou Institute of Biological Products, Lanzhou, Gansu, People's Republic of China

Jianxin Dai and Yajun Guo
International Joint Cancer Institute, The Second Military Medical University, Shanghai, People's Republic of China
National Engineering Research Center for Antibody Medicine, State Key Laboratory of Antibody Medicine and Targeting Therapy and Shanghai Key Laboratory of Cell Engineering, Shanghai, People's Republic of China
PLA General Hospital Cancer Center, Beijing, People's Republic of China

Mark-Oliver Rödel and Mareike Hirschfeld
Museum für Naturkunde Berlin, Leibniz Institute for Research on Evolution and Biodiversity, Berlin, Germany
Theodor-Boveri-Institute (Biocenter of the University), Department of Animal Ecology and Tropical Biology (Zoology III), Würzburg, Germany

Christian Brede
Theodor-Boveri-Institute (Biocenter of the University), Department of Animal Ecology and Tropical Biology (Zoology III), Würzburg, Germany
Medizinische Klinik und Poliklinik II, Zentrum für Experimentelle Molekulare Medizin, Würzburg, Germany

Thomas Schmitt
Theodor-Boveri-Institute (Biocenter of the University), Department of Animal Ecology and Tropical Biology (Zoology III), Würzburg, Germany
Ecological Networks, Biology, Technische Universität Darmstadt, Darmstadt, Germany

Philippe Favreau and Reto Stöcklin
Atheris Laboratories, Bernex, Switzerland

Cora Wunder and Dietrich Mebs
Institute of Legal Medicine, Universitym of Frankfurt, Frankfurt/M., Germany

Alicia P. Higueruelo, Adrian Schreyer, G. Richard J. Bickerton and Tom L. Blundell
Department of Biochemistry, University of Cambridge, Cambridge, United Kingdom

Will R. Pitt
Department of Biochemistry, University of Cambridge, Cambridge, United Kingdom
UCB Pharma, Slough, United Kingdom

Silvana Maria Alban and Vanete Thomaz-Soccol
Department of Biotechnology and Bioprocess Engineering, Federal University of Parana, Curitiba, Parana, Brazil

Juliana Ferreira de Moura
Basic Pathology Department, Federal University of Parana, Curitiba, Parana, Brazil

Samira Bührer Sékula and Larissa Magalhães Alvarenga
Immunology Department, Tropical Pathology and Public Health Institute, Federal University of Goias, Goiania, Goias, Brazil

Marcelo Távora Mira
Health Sciences Postgraduate Program, School of Medicine, Pontifical Catholic University of Parana, Curitiba, Parana, Brazil

Carlos Chávez Olortegui
Department of Biochemistry and Immunology, Federal University of Minas Gerais, Belo Horizonte, Minas Gerais, Brazil

João Carlos Minozzo
Center for Production and Research of Immunobiological Products, Parana State Department of Health, Piraquara, Parana, Brazil

Mariya Y. Radeva, Daniela Kugelmann, Volker Spindler and Jens Waschke
Institute of Anatomy and Cell Biology, Ludwig-Maximilians-University Munich, Munich, Germany

Rizwan Aslam, Céline Marban and Marie-Hélène Metz- Boutigue
Inserm UMR-1121, Universitéde Strasbourg, Strasbourg, France

Christian Corazzol, François Jehl and Gilles Prévost
EA-7290, Virulence bactérienne précoce, Fédération de Médecine Translationnelle de Strasbourg, Institut de Bactériologie, Université de Strasbourg - CHRU Strasbourg, Strasbourg, France

François Delalande and Alain Van Dorsselaer
CNRS UMR-7178, Institut Pluridisciplinaire Hubert Curien, Université de
Strasbourg, Strasbourg, France

Youssef Haïkel
Inserm UMR-1121, Universitéde Strasbourg, Strasbourg, France
Faculté de chirurgie dentaire, Université de Strasbourg, Strasbourg, France

Corinne Taddei
Faculté de chirurgie dentaire, Université de Strasbourg, Strasbourg, France

Francis Schneider
Service de Réanimation Médicale, Hôpital de Hautepierre, Universitéde Strasbourg, Strasbourg, France

Alejandro Barriga, Elena Lorente, Carolina Johnstone, Carmen Mir and Daniel López
Centro Nacional de Microbiología, Instituto de Salud Carlos III, Majadahonda, Madrid, Spain

Margarita del Val
Centro de Biología Molecular Severo Ochoa, CSIC/ Universidad Autónoma de Madrid, Madrid, Spain

Hannah Priyadarshini Gideon, Kathryn Wood, Dominique Pepper, Tolu Oni, Ronnett Seldon and Claire Banwell
Clinical Infectious Disease Research Initiative, Institute of Infectious Diseases and Molecular Medicine, University of Cape Town, Cape Town, South Africa

Melissa Shea Hamilton and Paul R. Langford
Division of Medicine, Imperial College London, Norfolk Place, London, United Kingdom

Robert J. Wilkinson
Clinical Infectious Disease Research Initiative, Institute of Infectious Diseases and Molecular Medicine, University of Cape Town, Cape Town, South Africa
Division of Medicine, Imperial College London, Norfolk Place, London, United Kingdom
MRC National Institute for Medical Research, London, United Kingdom

Katalin A. Wilkinson
Clinical Infectious Disease Research Initiative, Institute of Infectious Diseases and Molecular Medicine, University of Cape Town, Cape Town, South Africa,
MRC National Institute for Medical Research, London, United Kingdom

Liliana Cantini, Christopher C. Attaway, Betsy Butler, Lourdes M. Andino, Melissa L. Sokolosky and Andrew Jakymiw
Depatment of Craniofacial Biology and Center for Oral Health Research, Hollings Cancer Center, Medical University of South Carolina, Charleston, South Carolina, United States of America

Michael Beer and Chun-Qiang Liu
Human Protection and Performance Division, Defence Science and Technology Organisation, Fishermans Bend, Victoria, Australia

Jose Rojas-Caraballo
Parasite and Molecular Immunology Laboratory, Tropical Disease Research Centre, Universidad de Salamanca (IBSAL-CIETUS), Salamanca, Spain
Molecular Biology and Immunology Department, Fundación Instituto de Inmunología de Colombia (FIDIC), Bogotá, Colombia

Julio López-Abán, Luis Pérez del Villar, Belén Vicente, Pedro Fernández-Soto and Antonio Muro
Parasite and Molecular Immunology Laboratory, Tropical Disease Research Centre, Universidad de Salamanca (IBSAL-CIETUS), Salamanca, Spain

Carolina Vizcaíno
Molecular Biology and Immunology Department, Fundación Instituto de Inmunología de Colombia (FIDIC), Bogotá, Colombia

Esther del Olmo
Pharmaceutical Chemistry Department, Tropical Disease Research Centre, Universidad de Salamanca (IBSAL-CIETUS), Salamanca, Spain

Manuel Alfonso Patarroyo
Molecular Biology and Immunology Department, Fundación Instituto de Inmunología de Colombia (FIDIC), Bogotá, Colombia
Basic Sciences Department, School of Medicine and Health Sciences, Universidad del Rosario, Bogotá, Colombia

James S. Testa, Vivekananda Shetty, Julie Hafner, Zacharie Nickens, Shivali Kamal, Gomathinayagam Sinnathamby and Ramila Philip
Immunotope, Inc., Doylestown, Pennsylvania, United States of America

Olivier Barré, Antoine Dufour, Ulrich Eckhard and Reinhild Kappelhoff
Centre for Blood Research, Department of Oral Biological & Medical Sciences, University of British Columbia, Vancouver, BC, Canada

François Béliveau and Richard Leduc
Department of Pharmacology, Faculty of Medicine and Health Sciences, Université de Sherbrooke, Sherbrooke, QC, Canada

Christopher M. Overall
Centre for Blood Research, Department of Oral Biological & Medical Sciences, University of British Columbia, Vancouver, BC, Canada
Department of Biochemistry and Molecular Biology, University of British
Columbia, Vancouver, BC, Canada

Sarah Everman, Paul Langlais, Lawrence J. Mandarino, Moulun Luo, Christine Roberts and Christos S. Katsanos
Center for Metabolic and Vascular Biology, School of Life Sciences, Arizona State University, Tempe, Arizona, United States of America
Mayo Clinic Arizona, Scottsdale, Arizona, United States of America

Zhengping Yi
Center for Metabolic and Vascular Biology, School of Life Sciences, Arizona State University, Tempe, Arizona, United States of America
Mayo Clinic Arizona, Scottsdale, Arizona, United States of America
Department of Pharmaceutical Sciences, Eugene Applebaum College of Pharmacy/Health Sciences, Wayne State University, Detroit, Michigan, United States of America

Index